NEW RESEARCH ON DNA REPAIR

NEW RESEARCH ON DNA REPAIR

BREEHN R. LANDSEER
EDITOR

Nova Science Publishers, Inc.
New York

For permission to use material from this book please contact us:
Telephone 631-231-7269; Fax 631-231-8175
Web Site: http://www.novapublishers.com

Library of Congress Cataloging-in-Publication Data

New research on DNA repair / Breehn R. Landseer, editor.
p. ; cm.
Includes bibliographical references and index.
ISBN 13 978-1-60021-385-4
SBN 10 1-60021-385-5
1. DNA repair. I. Landseer, Breehn R.
[DNLM: 1. DNA Repair--physiology. QU 475 N532 2006]
QH467.N4922 2006
572.8'6459--dc22 2006023966

Published by Nova Science Publishers, Inc. ✣New York

CONTENTS

PREFACE

As a major defense against environmental damage to cells DNA repair is present in all organisms including bacteria, yeast, drosophila, fish, amphibians, rodents and humans. DNA repair is involved in processes that minimize cell killling, mutations, replication errors, persistence of DNA damage and genomic instability. Abnormalities in these processes have been implicated in cancer and aging. This new book presents leading-edge research from around the world in this frontal field.

DNA repair is a complex process involving the orchestration of several signal transduction cascades initiated by the detection of damaged DNA, which can be manifest as numerous different lesions. Each of these cascades involves several unique or overlapping factors generally classified as sensors, mediators, transducers and effectors that ultimately lead to cell cycle arrest, initiation of DNA repair and/or apoptotic processes. However, many tumours can evade such checkpoints and survive normally toxic doses of therapeutic DNA damaging agents such as ionising radiation and chemotherapeutic drugs. The past few years has witnessed an advance in the development of molecular targeting strategies aimed at compromising such DNA repair processes in order to improve the therapeutic index of a given modality. Such strategies include antisense, ribozymes, siRNA and inhibitory peptides/antibodies. These can be combined with both existing and novel viral and non-viral delivery vectors, which are the key to successful gene therapy. Chapter 1 aims to give an overview of mammalian DNA repair and how specific targeting of factors within the signalling cascades can be inhibited by molecular targeting strategies to improve the biological effectiveness of a given treatment. Also, a summary of suitable delivery systems for such gene therapies will be discussed.

Lung cancer is a major cause of cancer-related death in the developed countries and the overall survival rate has still an extremely poor. Although cigarette smoking is the main cause of lung cancer, not all smokers develop lung cancer, and a fraction of lifelong nonsmokers will die from lung cancer. Genetic host factors have recently been implicated to account for some of the observed differences in lung cancer susceptibility. As discussed in chapter 2, Various DNA alterations can be caused by exposure to environmental and endogenous carcinogens. Most of these alterations, if not repaired, may result in genetic instability, mutagenesis and cell death. DNA repair mechanisms are important for maintaining DNA integrity and preventing carcinogenesis. Recent genetic association studies on lung cancer risk have focused on identifying effects of single nucleotide polymorphisms

(SNPs) in candidate genes, among which DNA repair genes are increasingly studied. Genetic variations in DNA repair genes are thought to modulate DNA repair capacity and are suggested to be related to lung cancer risk. The authors identified a sufficient number of epidemiologic studies on lung cancer to conduct meta-analyses for genetic polymorphisms in DNA repair genes. They found increased risk of lung cancer among subjects carrying excision repair cross complementing group 2 (*ERCC2/ xeroderma pigmentosum group D (XPD)*) 751Gln/Gln genotype (odds ratio (OR) = 1.30, 95% confidence interval (CI) = 1.14 - 1.49). They found a protective effect of *XPA* 23G/G genotype (OR = 0.75, 95% CI = 0.59 - 0.95). *X-ray cross-complementing group 1 (XRCC1)* 399Gln/Gln genotype was associated with an increased risk of lung cancer among Asians (OR = 1.34, 95% CI = 1.16 - 1.54). Little evidence of associations has been found between other DNA repair genes and lung cancer risk. Considering the data available, it can be conjectured that if there is any risk association between single SNP, such as the *XRCC1* Arg399Gln, the *XPA* G23A or the *ERCC2/XPD* Lys751Gln, and lung cancer, this risk increase/decrease will probably be minimal. Advances in identification of new polymorphisms and in high-throughput genotyping techniques will facilitate analysis of multiple genes in multiple DNA repair pathways. Therefore, it is likely that the defining feature of future epidemiologic studies will be the simultaneous analysis of large samples of cases and controls.

DNA repair systems include enzymes that recognize and remove DNA adducts, correct the DNA sequence and re-join strand breaks. These systems provide to repair oxidative and alkylated DNA lesions and include base-excision repair (BER), nucleotide excision repair (NER), mismatch repair and O^6-methylguanine-DNA methyltransferase. Specific mechanisms for strand breaks repair include non-homologous end-joining and homologous recombination. Current evidence suggests that DNA repair systems are very sensitive targets for several carcinogenic agents such as heavy metals (nickel, cadmium, cobalt, chromium and arsenic), PAHs, UV-B, ionizing radiations, aromatic amines, cigarette smoking and others. These agents cause diminished removal of endogenous and exogenously induced DNA damage increasing risk of tumour formation. Such compounds or their metabolites inhibit DNA repair probably by oxidative damage or direct interactions with reparative enzyme. In the last years DNA repair activity and inhibition has been widely studied by comet test, a rapid and sensitive technique that allows to evaluate direct and oxidative DNA damage and its repair on single whole cell. In particular, this technique allows the detection of single and double strand breaks, alkali-labile sites and some its applications, such as Fpg or endo III modified Comet assay, allow to evaluate oxidative DNA damage measuring oxidized purine and pyrimidine respectively. This technique does not require DNA isolation avoiding spurious DNA damage due to the sample handling. Comet assay permits to study DNA repair evaluating the persistence of either direct and oxidative DNA damage in relation to repair time, performing the test on cells immediately after induction of DNA damage and after DNA recovery time. Therefore this technique is an useful tool to study the influence of xenobiotics on the different steps involved in DNA damage induction and repair. This review in chapter 3, will present the most important applications of this technique to evaluate DNA repair alterations due to various xenobiotics particularly heavy metals such as Nickel compounds for which DNA repair inhibition was demonstrated also by a recent study. This technique and its modifications, applicable to all cellular types, represent an useful, versatile, cheap and sensitive tool to contribute to clarify the mechanisms of inhibitory processes of enzymes involved in carcinogenesis.

Investigation of dietary compounds that enhance DNA repair has been a largely underdeveloped area of research. DNA damage is regarded as a major contributing factor to cancer and aging. Unrepaired DNA damage in cells that duplicate can cause errors of replication leading to mutation and ultimately to cancer. Unrepaired DNA damages can also cause loss of gene expression, loss of cell function and cell death (apoptosis) leading ultimately to aging. Genetic defects in numerous human genes encoding DNA repair proteins substantially increase the risk of cancer or cause premature aging, thus indicating the importance of DNA repair processes in preventing cancer and premature aging. Recent studies have identified several dietary compounds that increase expression of DNA repair genes, and in some cases these compounds appear to be associated with reduced cancer incidence or increased longevity.The authors in chapter 4, have tested 19 potential chemopreventive compounds for their effect on expression of several DNA repair proteins in a colon cell line. On the basis of these results they focused on a group of compounds related to chlorogenic acid. Chlorogenic acid is found in a variety of fruits and vegetables, and is particularly high in coffee. They found that chlorogenic acid and two of its metabolites formed by the action of intestinal microbial flora, *m*-coumaric acid and 3-(*m*-hydroxyphenyl)propionic acid, significantly increased the expression of the DNA repair enzyme poly(ADP-ribose) polymerase (PARP). The repair enzyme Pms2 (post-meiotic segregation 2) was also significantly increased by *m*-coumaric acid and 3-(*m*-hydroxyphenyl)propionic acid, as well as by another metabolite of chlorogenic acid, caffeic acid. Evidence suggests that PARP activity is associated with increased longevity and cancer prevention, and that Pms2 activity is associated with cancer prevention. Thus, the increased expression of these DNA repair enzymes by chlorogenic acid-related compounds may be beneficial for promoting longevity and preventing cancer.

A significant contribution to human mutagenesis and carcinogenesis may come from DNA damage of endogenous, rather than exogenous, origin. Efficient repair mechanisms have evolved to cope with it. As discussed in chapter 5, the major pathway involved in DNA repair of endogenous damage is DNA base excision repair (BER). In addition, an important contribution is given by O^6 alkylguanine DNA alkyltranferase that repairs specifically the miscoding base O^6 alkylguanine. BER is rather efficient in human cells but a certain amount of endogenous damage inevitably escapes mending and likely contributes to human carcinogenesis. Several attempts have been carried out to enhance the efficiency of repair of endogenous DNA damage by overexpressing in mammalian cells single enzymatic activities. In some cases (e.g. formamidopyrimidine DNA glycosylase or O^6 alkylguanine DNA alkyltransferase) this approach turned out successful in improving cellular protection from endogenous and exogenous mutagens while overexpression of other enzymatic activities (e.g alkyl N-purine DNA glycosylase or DNA polymerase β) was detrimental and even produced a genome instability phenotype. Prophylaxis of endogenous DNA damage and mutation could be achieved in at-risk individuals by expression of appropriate DNA repair genes. Currently available vehicles for this peculiar type of gene therapy are surveyed.

Translesion DNA synthesis (TLS) is a fundamental biological process whereby arrested DNA replication caused by the presence of base damage that cannot be readily bypassed by replicative DNA polymerases, is relieved by an alternative DNA replication that utilizes DNA polymerases that are specialized for DNA synthesis past such lesions. Multiple such polymerases have been identified in all life forms. In particular, vertebrate cells contain no less than 10 such specialized enzymes. Chapter 6 reviews our present knowledge of these specialized DNA polymerases in general and their specific role in TLS.

The functional characterization of mammalian radiosensitivity genes has been greatly facilitated by using rodent cells derived from mutants sensitive to either ultraviolet or ionizing radiation and has led to the molecular identification of two important groups of genes called ERCC and XRCC. Numerous genetic studies have used cells from patients with hereditary disorders. However, the genetic heterogeneity of human samples remains one of the greatest impediments to the identification and characterization of human radiosensitivity genes. Over the last two years, RNA interference (RNAi) has emerged as a promising technology in biological science which has broad applications, in particular in the field of reverse genetics. RNAi is the process by which double-strand RNA (dsRNA) silences gene expression by inducing the sequence-specific degradation of the complementary mRNA. RNAi is becoming a state-of-the-art method whereby researchers can knock down the expression of a specific gene in order to untangle its biological functions. In chapter 7, the authors have developed new pEBVsiRNA plasmids to impose stable gene silencing in human cells. These plasmids, which remain as independent replicons in transfected human cells, greatly enhance RNAi by allowing efficient, specific and very long-term gene silencing, even after more than 500 days in culture. They successfully silenced several genes involved in the nucleotide excision repair pathway (e.g. *XPA*, *XPC*, *hHR23A or hHR23B),* in the non-homologous end-joining pathway (e.g. *DNA-PKcs* and *XRCC4)* and in the homologous recombination repair pathway (e.g. *Rad52).* Other genes involved in signaling and other types of recovery after irradiation (e.g. *P53*, *MRE11*, *Rad50*, *NBS1*) were also targeted. Stable knock-down clones were propagated in culture and monitored for several months. They displayed the expected DNA repair deficiency several months after transfection. At present, they are constructing more than 140 vectors to target 30 genes. They are using these cells to unravel the complex regulatory relationships between the different DNA repair pathways. For the first time, they have created a set of stable isogeneic human cells deficient in DNA repair genes without the use of viral vehicle.

Base excision repair (BER) is one of the DNA repair types, by means of which most of damages in DNA structure are removed. BER is performed through short patch or long patch pathways, the choice of which depends on the kind of the DNA damage and the level of the repair DNA polymerases in a cell. In chapter 8, the authors were interested in the elucidation of mechanisms of the repair DNA polymerases action in embryogenesis of teleost fish *Misgurnus fossilis* (loach). They identified repair DNA polymerase β (pol β) and DNA polymerase δ (pol δ) in loach eggs and embryos. The pol β activity was extremely low in eggs and embryos (7.2×10^{-4} U per an egg or 3.6 U per gram of protein) up to gastrula stage. The pol δ activity 55 times exceeded the pol β activity at the same period of development, that is why pol δ became the object of their close attention. Repair properties of pol δ were investigated with the use of the model substrates – DNA duplexes containing gaps, the length of which (1-9 nucleotides, nt) corresponded with the length of gaps forming in BER. Pol δ displayed unexpected properties upon the interaction with such duplexes: this enzyme processively filled the gaps of 5 nt and 9 nt in the absence of processivity factors (PCNA, RFC, and ATP), while it elongated primers on single stranded DNA templates in a distributive manner, as it was expected. The mechanism of the unusual processivity of pol δ was studied by electrophoretic mobility shift assay. Pol δ was shown to interact with both ends of 5 nt and 9 nt gaps, demonstrating more affinity to the latter. This interaction stabilizes the complex between pol δ and DNA duplex, allowing pol δ to reveal the processivity.

To confirm the participation of pol δ in BER *in vivo*, after the beginning of cell division, embryos were affected by superoxide radicals in the xanthine – xanthine oxidase system generating mainly single strand breaks in DNA, which are repaired by long patch pathway. In response to this influence, the activity of pol δ (but not replicative pol α) increased by 370% in embryos at the middle gastrula stage. Thus, pol δ is the main repair DNA polymerase in loach embryos and this enzyme seems to be able to carry out long patch pathway of BER by earlier unknown mechanism, which is important for extreme conditions at low levels of PCNA and PCNA-independent repair DNA polymerases in a cell.

Eukaryotic general transcription factor (TF) IIH is composed of ten proteins, seven of which are also required for nucleotide excision repair (NER) of UV radiation-induced DNA damage in human cells. Although plant counterparts of these seven human TFIIH subunits have been identified by database screening, only the roles in NER of the *Arabidopsis thaliana* XPD and p44 homologs as well as its two XPB paralogs have been investigated. In chapter 9, the authors review studies that addressed the repair and transcription functions of those proteins, as well as new findings concerning Arabidopsis homologs of the human TFIIH components p62 and TTDA. The authors and others have determined that AtXPB1, AtXPB2, AtXPD and AtGTF2H2 can function in NER of UV photoproducts, but UV treatment does not increase *AtXPB1, AtXPB2, AtXPD, AtGTF2H2, AtTFB1A* or *AtTFB5* transcript levels. Furthermore, AtGTF2H2 also can perform all essential transcription functions of yeast Ssl1 (p44 ortholog) in its absence, and AtXPD also can operate in transcription in yeast. In contrast, the newly isolated *AtTFB1A* cDNA does not complement the UV sensitivity or transcription defect conferred by a conditional yeast *tfb1* mutation, possibly because of the relatively low level of homology between Tfb1 and AtTFB1A. However, they have detected interactions between AtGTF2H2, AtXPD, AtTFB1A and AtTFB5, as well as interactions between these plant proteins and various of their yeast homologs in two-hybrid assays. Collectively, these observations suggest that AtXPB1, AtXPB2, AtXPD, AtGTF2H2, AtTFB1A and AtTFB5 may participate in NER and transcription in plants as part of TFIIH, and underpin the initial development of a structural model for its NER subassembly.

Among the large spectrum of DNA damage induced by radiation, DNA double-strand breaks (DSBs) are considered, to date, as the key-lesions responsible for the cell killing. However, all along these last 50 years, although it was always intuitive for radiobiologists, such conclusion has resulted from a long and difficult process of development of new techniques and novel concepts. In chapter 10, the authors endeavored to translate these different steps of thinking, notably by strengthening on the number of technical artifacts, biases due to yeast and rodent data, the variety of phenotypes observed in human cells and the extra-complexity encountered in human tumors. Particularly, the most recent techniques based on immunofluorescence have revolutionized the picture of the molecular events occurring early after irradiation but has also raised the crucial question of the choice of the technique assessing DSB repair and its specificity for particular steps of the repair process. While the homologous recombination repair pathway is predominant in yeasts, its importance in normal and tumor human cells appears to be less and less obvious, raising the problem of the existence of randomized and/or uncontrolled events that may produce much more errors than one could have imagined from the yeasts. DSB repair impairments are now observed in a plethora of various genetic diseases, leading to radiosensitivity, immunodeficiency and sometimes cancer-proneness, but the low frequency and the pleiotropism of such diseases make difficult the development of a *single* predictive assay. Lastly, the balance, or rather the interplay, between recombination and DNA end-joining repair pathways appears to be strongly dependent upon cell cycle, genotype and oxidative stress that can accompany the DSBs formation. Hence, although DSB repair appears, to date, undoubtedly crucial for cell survival, further investigations are still needed to offer to physicians some tools to secure clinical transfer of anti-cancer approaches by avoiding the risk of deleterious adverse effects of radiation.

DNA double-strand breaks are defined are two simultaneous nicks in opposite strands of the DNA helix, which are in sufficiently close proximity to one another that base-pairing and chromatin structure are unable to maintain the broken DNA ends juxtaposed. These lesions arise in the course of normal and pathological endogenous DNA metabolism or can result from exogenous agents such as ionizing radiation and certain chemotherapeutic drugs. They represent the most severe type of DNA damage with respect to preservation of genomic integrity and there is growing evidence suggesting that in higher eukaryotes aberrant repair of double-strand breaks is a key event in carcinogenesis. Therefore, cells have evolved complex mechanisms that include cell cycle arrest, activation and increased expression of various genes including those associated with DNA repair and in certain cases induction of the apoptotic pathway, to respond to double-strand breaks. Given the intense research effort in this field in the past few years our knowledge of the molecular details relating to the detection, signalling and repair of double-strand breaks has greatly increased. In chapter 11 they discuss recent progress in our understanding of molecular responses to DNA double-strand breaks with a particular emphasis on homologous recombination and non-homologous end-joining, the major double-strand break repair pathways. In addition, the role of changes in chromatin architecture in response to double-strand breaks is currently emerging and already numerous links between changes in chromatin – resulting from both the contribution of histone tail modifications and chromatin remodelling complexes – and repair pathways have been made. Therefore, the importance of epigenetic modifications in the double-strand break repair pathways will also be discussed.

The p53 tumor-suppressor protein plays a central role in coordinating the cellular response to DNA damage and in maintaining genomic stability. A variety of genotoxic stresses, including DNA double-strand breaks (DSBs) induced by ionizing radiation, transiently stabilize the wild-type p53 protein, cause it to accumulate in the nucleus, and activate it as a transcription factor. Depending on the extent of DNA damage and the genetic background of the cells, p53 may activate transient cell-cycle checkpoints whose role in maintaining genomic stability is traditionally embedded in the concept that this provides time for DNA-repair enzymes to remove damage from the genome before the cellular decision to either resume proliferation or undergo permanent growth arrest or apoptosis. However, there is a considerable emerging literature to suggest that p53 also directly participates in various DNA-repair pathways, including the base excision repair pathway that, among other functions, removes simple base lesions caused by ionizing radiation, as well as the RAD51-dependent homologous-recombination repair pathway and the non-homologous end-joining pathway, two distinct but complimentary mechanisms for DSB repair. In chapter 12, the authors review recent experimental data supporting a role for p53 in these various DNA-repair processes, with a focus on the repair of DNA damage caused by exposure of human cells to different doses of ionizing radiation, including doses relevant to cancer therapeutic applications. They will also consider the role of p53 in DNA repair in the context of our current knowledge of low-dose responses to ionizing radiation. Finally, they consider some of the implication of these findings to the relative responses of normal and cancerous cells to ionizing radiation.

In: New Research on DNA Repair
Editor: Breehn R. Landseer, pp. 1-37

ISBN 1-60021-385-5

Chapter I

Molecular Targeting of DNA Repair: A Means to Improve Therapeutic Index

Spencer J. Collis*

DNA Damage Response Lab, Clare Hall Laboratories, Cancer Research UK, Blanche Lane, South Mimms, EN6 3LD, UK

Abstract

DNA repair is a complex process involving the orchestration of several signal transduction cascades initiated by the detection of damaged DNA, which can be manifest as numerous different lesions. Each of these cascades involves several unique or overlapping factors generally classified as sensors, mediators, transducers and effectors that ultimately lead to cell cycle arrest, initiation of DNA repair and/or apoptotic processes. However, many tumours can evade such checkpoints and survive normally toxic doses of therapeutic DNA damaging agents such as ionising radiation and chemotherapeutic drugs. The past few years has witnessed an advance in the development of molecular targeting strategies aimed at compromising such DNA repair processes in order to improve the therapeutic index of a given modality. Such strategies include antisense, ribozymes, siRNA and inhibitory peptides/antibodies. These can be combined with both existing and novel viral and non-viral delivery vectors, which are the key to successful gene therapy. This chapter aims to give an overview of mammalian DNA repair and how specific targeting of factors within the signalling cascades can be inhibited by molecular targeting strategies to improve the biological effectiveness of a given treatment. Also, a summary of suitable delivery systems for such gene therapies will be discussed.

* Correspondence concerning this article should be addressed to Spencer J. Collis, DNA Damage Response Lab, Clare Hall Laboratories, Cancer Research UK, Blanche Lane, South Mimms, EN6 3LD, UK; E-mail: spencer.collis@cancer.org.uk. Tel: +44 (0)1707 625 773; Fax: +44(0)207 269 3801.

INTRODUCTION

The aim of this chapter is to highlight how DNA repair mechanisms can be inhibited by novel molecular targeting strategies and integrated with existing anti-cancer treatments in order to increase their effectiveness (Figure 1). Potential gene therapy (GT) approaches that provide tumour restricted delivery and/or targeting, when combined with precision clinical techniques such as three dimensional conformal radiotherapy, could provide significantly enhanced rates of cancer eradication, impeding the systemic dissemination of the disease [1]. At the clinical level, this would be an attractive strategy to help reduce normal cell toxicity without compromising tumour eradication, and has led to many studies targeting various pathways involved in radiation responses [2]. Specific molecular targeting can be achieved through several different approaches including antisense, ribozyme and siRNA's, resulting in a reduction in cellular expression of the target protein. A brief outline of the current delivery methods available for these types of approaches will be given, but an in-depth discussion of these topics is beyond the scope of this chapter and will be referenced herein.

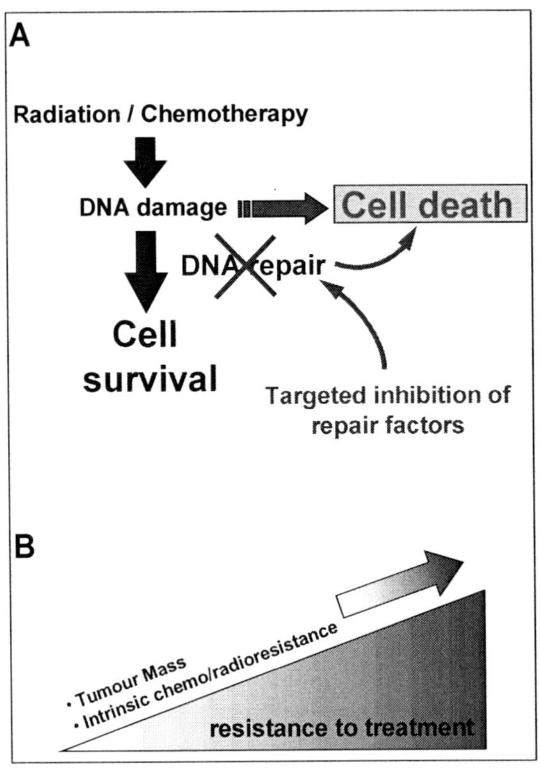

Figure 1. General concept of molecular targeting of solid tumours. A; Scheme showing how molecular targeting of DNA repair factors aims to augment the cytotoxic effects of therapeutic DNA damaging agents. B; Diagram showing how tumour resistance to such agents is both a factor of its size as well as its intrinsic sensitivity to certain agents, which is dependent on many cellular factors. Eradication of as much of the tumour mass as possible is desired in order to prevent re-development following initial treatment.

DNA DAMAGE AND REPAIR

In order to understand how molecular targeting strategies work, it is first necessary to give a brief overview of how DNA damage is sensed and efficiently dealt with. Cells incur DNA damage on a daily basis either from endogenous sources such as toxic free radicals produced during normal metabolic cellular processes or from exogenous processes arising from environmental stress. Such damage must be rapidly and, where possible, accurately repaired in order to maintain cellular viability. If the amount of damage incurred exceeds a certain threshold, then the cell is destroyed in order to prevent potential mutagenic lesions from being passed on to daughter cells and maintain genome integrity.

As such, mammalian cells have developed complex cellular pathways to aid both the detection and subsequent repair of a wide range of DNA adducts (Figure 2). Although such pathways can be involved in quite specific types of damage, overlapping functions of more than one pathway are usually required to deal with complex lesions arising from many DNA damaging agents e.g. ionising radiation, alkylating and interstrand cross-linking agents. As such, many tumourogenic phenotypes involve alterations in these DNA repair pathways allowing malignant cells to evade damage-induced death. The orchestration of these DNA repair processes must also be coordinated with specific damage checkpoints and subsequent cell cycle arrest. The inherent ability of a cell to correctly repair sites of potentially lethal DNA damage ultimately determines the cellular sensitivity to various DNA-damaging agents and it is this ability that molecular strategies can be designed to abrogate. This is the basic premise for adjuvant gene therapy where such molecular targeting strategies can be used in conjunction with existing therapies to improve their effectiveness at reducing or even eradicating tumour mass [1]. At the clinical level, this is an attractive strategy to help reduce normal cell toxicity without compromising tumour eradication, and has led to many studies targeting various pathways involved in radiation responses [2].

As highlighted in Figure 2, there are many different specialised DNA repair pathways which are tempo-spatially organised into discrete compartments within the cell [3]. Each DNA repair pathway has its own set of unique factors combining with those that possess overlapping functions [4]. The two main DNA repair pathways involved in repair of DNA double-strand breaks (DSBs), the foremost toxic lesion associated with ionising radiation damage [5], are the Non-homologous end-joining (NHEJ) and homologous recombination repair (HRR) pathways [6-8]. The NHEJ and HRR pathways respectively function primarily at the G_1/S and G_2/M cell cycle checkpoints, with different proteins involved in each pathway owing to the fact of the nature of repair taking place. As one would expect given that the originating lesion is the same, several proteins appear to play a role in both repair pathways (Figure 2B).

The base excision repair (BER) pathway employs specific glycosylases to repair base damage caused by reactive oxygen species produced from IR exposure and non-bulky adducts caused from alkylating agents. The prior processing by these specific glycosylases results in the production of a more uniform intermediate known as an apurinic/apyrimidinic (AP) site, which can then be further processed and repaired [9].

Nucleotide excision repair (NER) removes bulky adducts from DNA arising from ultra violet (UV) radiation exposure and alkylating agents. As with many other DNA repair pathways, deficient in specific factors associated with NER lead to increased cancer susceptibility disease syndromes [10]. Another role for the NER pathway is in the removal of damaged bases on the transcribed strand of DNA, a process known as transcription-coupled repair (TCR; [11]). Defects in the NER/TCR pathway have been shown to lead to the human disease Cockayne's syndrome.

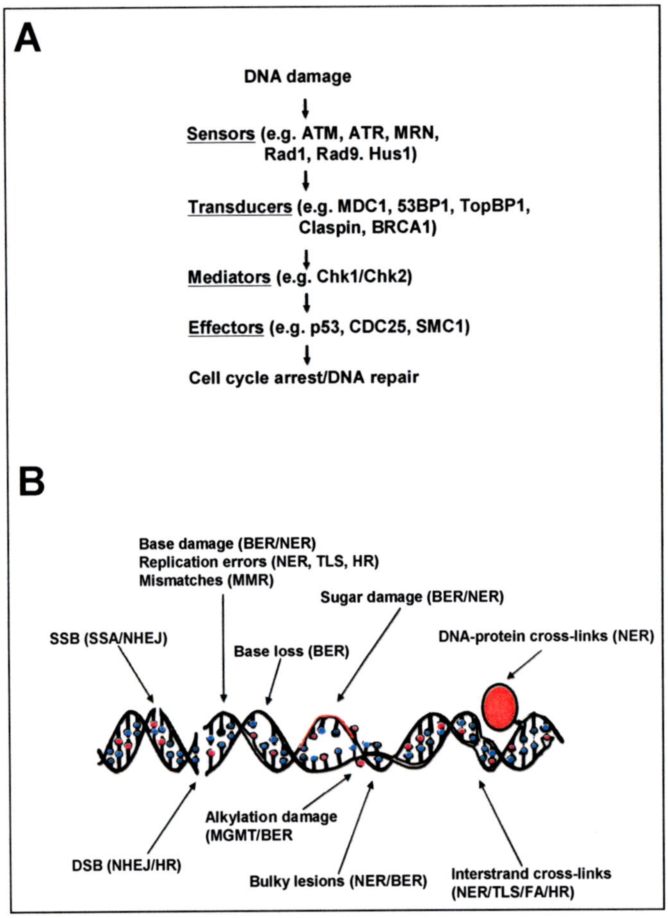

Figure 2. Cellular response to the various types of DNA damage. A; Scheme showing how detection of DNA damage is relayed through a series of signalling pathways, ultimately leading to a co-ordained arrest in the progression of the normal cell cycle with recruitment of DNA repair factors to sites of DNA damage. B; Diagram showing the many different types of damage that can occur to the DNA molecule and which pathways/factors are involved in the repair of these lesions (aadapted from a European radiation protection and training slide by Georg Gerber, European commission, 1999). BER; base excision repair, DSB; double-strand break, FA; Fanconi Anaemia, HR, homologous recombination, MGMT; *O*6-methylguanine DNA-methyltransferase MMR; mis-match reapir, NER; nucleotide excision repair, NHEJ; non-homologous end-joining, SSA; single-strand annealing, SSB; single-strand break, TLS; trans-lesion synthesis.

Perhaps one of the most complex types of lesions the cell has to deal with is the interstrand crosslink (ICL), a lesion which can be produced from exposure to anti-cancer agents such as *cis*-platin, mitomycin C and IR. ICLs are particularly difficult to repair in the context of DNA replication and as such, factors associated with both and HRR are required in order to successfully remove ICLs and efficiently restart replication [12]. Prior to restart of replication, a specific family of error-prone DNA polymerases are needed in order to bypass the processed lesion as well as a family of proteins associated with the rare recessive disorder Fanconi Anaemia an breast cancer susceptibility proteins [13,14].

Another pathway associated with DNA replication is the mis-match repair (MMR) pathway, that acts to remove base insertions, deletions and mis-matched bases arising from replication errors e.g. 'slippage' of the DNA polymerase, particularly within regions of short, highly repeated DNA sequences [15]. As with NER, defects in the MMR pathway increase susceptibility to cancer, manifest mainly as hereditary non-polyposis colon cancer (HNPCC).

As well as the NER pathway, specific base damage can also be dealt with by the direct repair of specific alkylguanine adducts by *O*6-alkylguanine DNA-alkyltransferase (ATase; a.k.a. Methylguanine methyl transferase or MGMT) [16]. Un-repaired *O*6-alkylguanine adducts can lead to base mis-pairing which then requires a functional MMR system in order to restore the correct DNA sequence, thus highlighting again that many of these repair pathways share overlapping functions in the many different pathways.

It is important to emphasise that following a exposure to a DNA damaging agent such as IR, large areas of DNA will contain more than one type of damage mentioned above [17]. As such, a complex and rapid process of repair must take place that involves many different factors from several different pathways. This perhaps goes some way to explain a functional overlap of many DNA repair proteins as pointed out above.

Perhaps not surprisingly, given the important role these pathways play in maintaining genome integrity, mutations in many of the different factors associated with them lead to a plethora of human diseases associated with an increased risk of cancer progression [18] + (see [4,7,19-21] and references therein). As highlighted above, given the roles these proteins play in DNA repair processes, such genetic disorders associated with defects in these pathways also manifest as an elevated sensitivity to various DNA damaging agents (see [18,22] and references therein for further details). Perhaps the most classic example of this characteristic is highlighted by patients who possess defects in ATM, the gene defective in the cancer pre-disposing disorder ataxia-telangiectasia and which plays a major role in detecting DNA breaks (Figure 2 and [7,8,22]). These patients and cell lines derived from them are hypersensitive to ionizing radiation compared to respective normal cells (see [23] and references therein). If one can mimic this phenotype specifically within the tumour cell microenvironment, then theoretically a much greater amount of tumour eradication could be achieved and may permit a reduction in total radiation exposure, reducing normal cell toxicity side effects. As we shall see, this simple paradigm is the basis of an increasing number of different strategies that aim to target specific DNA repair proteins.

As well as the many different and specialised DNA repair factors that maintain genome integrity in the battle against mutagenesis/tumourogenesis, there are many other factors whose role it is to preserve the integrity of the genome. The most common of these is perhaps p53 which is involved in orchestrating cell cycle arrest to allow DNA repair to proceed, and has been shown to be defective in over 50% of all cancers [24]. Furthermore, dysregulation of factors involved in DNA the damage and repair processes mentioned above may contribute to the increased resistance of tumour cells to radiation and/or certain chemotherapeutic agents. Several studies have shown that the expression and/or activity of specific DNA repair proteins in certain cancers correlates with tumour responses to such treatment regimes. These include; BRCA1/2 in breast and ovarian carcinomas [25-27], RAD51 in lung, pancreatic and breast carcinomas [28-30], components of the DNA-PK complex in lung, oesophageal, oral and cervical cancers [31-34], Fanconi Anaemia factors in breast and ovarian cancer [35,36], MMR factors in tumours resistant to alkylating agents [37] and ATM, and components of the MRN complex in breast carcinoma [38]. Together, these data strongly suggest that targeted inhibition of specific repair factors within tumour cells in an attractive approach to augment the effectiveness of radiation and certain chemotherapeutic agents.

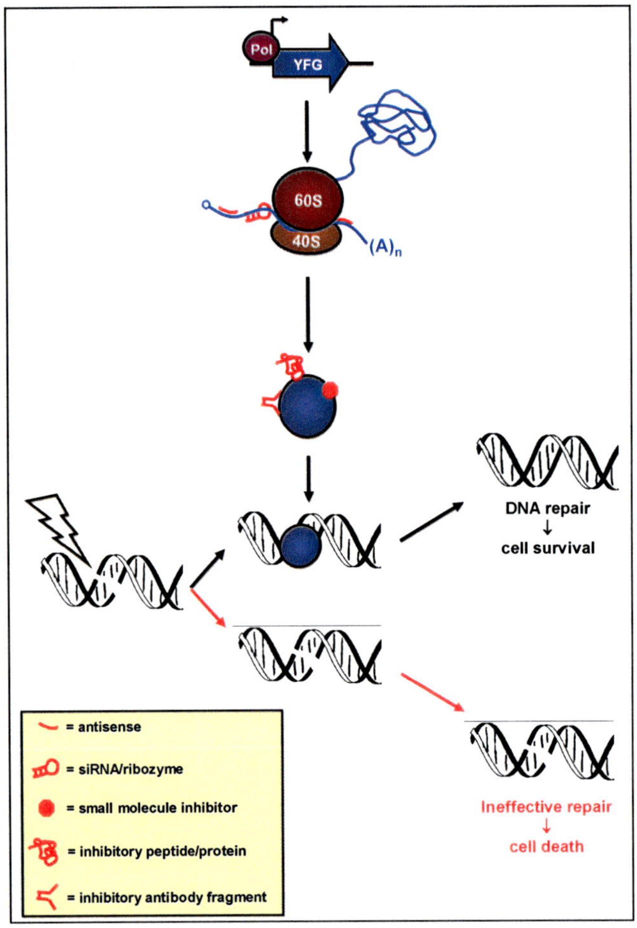

Figure 3. Inhibition of a specific DNA repair factor by molecular targeting strategies and/or small molecule inhibitors. Shown is a schematic representation of production of 'your favourite DNA repair gene' (YFG) from gene through mRNA to protein and inhibition of these processes by either molecular targeting strategies such as; antisense, siRNA, inhibitory peptide/proteins and antibody fragments or small molecule inhibitors. Antisense and siRNA species inhibit the production of protein by causing degradation of the mRNA (by different mechanisms) leading to decreased levels of YFG within the cell. Antisense (and possibly siRNA) can also sterically hinder ribosome-mediated translation. Small molecule inhibitors and inhibitory peptides/proteins and antibody fragments inhibit YFG function by binding to specific regions of the protein complex. Ultimately, such inhibition prevents the normal cellular function of the YFG, leading to inefficient DNA repair and increased cell death following treatment with therapeutic DNA damaging agents.

MOLECULAR TARGETING APPROACHES

Molecular targeting can be defined as the targeting of a specific protein or DNA/RNA encoding this protein (Figure 3). This specific targeting leads to a reduction in protein production and/or activity that ultimately causes alterations in cellular characteristics. Such attenuations could include modification of cell growth/death pathways and resistance/ sensitivity to various DNA damaging agents such as ionising radiation, UV, heat and chemotherapeutic agents. Molecular targeting can be achieved in a variety of ways including; antisense, ribozyme and small inhibitory RNA (siRNA) approaches to target the DNA/RNA as well as the exogenous delivery of inhibitory antibodies and dominant-negative forms of the protein target to 'attack' the protein and small molecule chemical inhibitors (Figure 3). The three main or preferred methods for achieving specific targeting over the last 10-15 years has been the use of antisense molecules, ribozymes and the newly discovered siRNA. The modes of action of these three techniques are highlighted in Figure 4 and are described in more detail below.

Antisense Strategies

Antisense oligonucleotides are short synthetic, chemically modified deoxyribonucleic acids designed to hybridise to complementary mRNA sequences encoding the target protein. For an antisense molecule to be successful in binding its target mRNA, the region of the mRNA targeted must be in a conformation that allows the antisense molecule accessibility to the target site (which is also true for ribozymes and siRNA; see below and Figure 4). However, mRNA structure is difficult to predict and is one of the major problems associated with using antisense, ribozymes and siRNA strategies [39]. Folding of the target and accessibility to the target site may also be affected by the many protein factors associated with mRNA, especially as most antisense, ribozyme and siRNA strategies target mRNA regions close to the translational initiation site as this is deemed in an 'open' configuration to allow the translational machinery to bind. Once an antisense molecule has bound to its target sequence, inhibition of protein expression is generally imposed as a consequence of the action of RNase H present within the cell. RNase H is an endonuclease that recognises and cleaves the RNA component of RNA-DNA duplexes [40,41], (see Figure 4). Antisense-mediated inhibition of protein expression can also occur *via* the physical blockage of mRNA processing events and/or translation [41,42]. The most important part of the design of antisense molecules is to ensure that the antisense oligonucleotide is specific for its given target mRNA sequence and thus does not lead to any undesirable 'non-antisense effects' [43]. Antisense strategies represent the most clinically advanced of the molecular targeting techniques described here. Perhaps the best example of how successful antisense strategies can be demonstrated by the targeting of the anti-apoptotic protein BCL-2. Over many years, researchers have moved from proof of principle studies in the lab to large-scale clinical studies which have demonstrated the feasibility of targeting BCL-2 with antisense molecules as an anti-cancer treatment (see [44,45] and references therein). For further reading of the use of antisense oligonucleotides for the treatment of cancer, the reader is advised to read recent review articles and references therein [45-48].

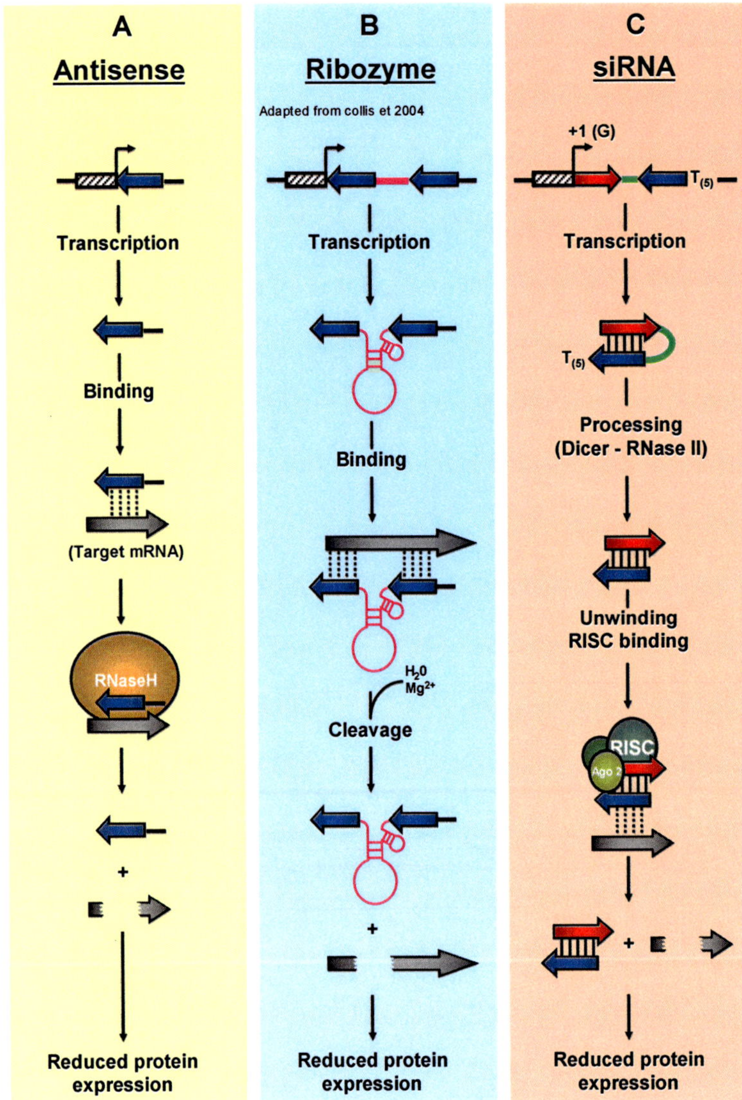

Figure 4. The three most adopted approaches of specific molecular targeting. A). Antisense strategies normally involve a plasmid encoding a full-length (inverted cDNA) or one based on partial mRNA target sequenced under the control of a constitutive or selective promoter. Upon binding to the target mRNA sequence through homology, the endonuclease RNaseH specifically cleaves the RNA component of DNA-RNA hybrid complexes, leaving the antisense intact to inhibit further substrates. Steric blockage of the translational machinery is another means by which antisense molecules can inhibit protein production. B). Ribozymes (shown is an example of the hammerhead sub-type) are composed of hybridisation arms that bind to the target sequence and a conserved catalytic domain. As with the other two approaches, these can be synthesised and administered as 'naked' species or transcribed from pol II or pol III promoters within a suitable expression vector (plasmid or viral-based). In the presence of water and magnesium ions present within the cell, the catalytic region chemically breaks the phosphodiester bond without the aid of any protein co-factors. Dissociation from the cleaved

mRNA (which is later degraded) allows the free ribozyme to bind more target mRNA species. C). siRNA using small hairpin RNAs transcribed from a pol III promoter. Inverted repeats are separated by a small loop region with an immediate termination sequence ($T_{(5)}$), producing a small hairpin structure. This is processed by the Dicer enzyme into 21-23nt double-stranded RNA species which can bind specific target mRNA when complexed with the RNA induced silencing complex (RISC). The target mRNA is cleaved, releasing the small inhibitory RNA species, which can then target further mRNAs. Figure originally appears in [235], with the kind permission of Springer Science and Business Media.

Ribozymes

Ribozymes are catalytic RNA species with site-specific cleavage activity that are found in numerous plant and animal pathogens (see [49-51] and references therein). As with antisense molecules, they bind RNA molecules e.g. mRNA, *via* complementary base sequences. Once bound, cleavage of the bound RNA occurs at a specific tri-nucleotide sequence know as the 'target site' (Figure 4). The target site has a specific sequence; for the most commonly used 'hammerhead' class of ribozyme, this is generally NUX, where N is any nucleotide and X is any nucleotide apart from G. The ribozyme then dissociates from the cleaved product, which is then degraded by cellular mechanisms. The seminal paper by Haseloff and Gerlach highlighted the potential use of hammerhead ribozymes to target any mRNA molecules as long as the transcript sequence is known [52]. Once the target mRNA is cleaved, the ribozyme molecule is then free to bind further RNA targets. Exogenous ribozymes used in potential therapies are likely to be eventually degraded within the cell, but hopefully not until they have first cleaved many specific mRNA species. As with antisense oligonucleotides, several modifications to the ribozyme RNA can be made in order to increase its cellular stability [53-55]. Although such modifications cannot be made when either antisense or ribozymes are expressed within the cell from exogenously delivered plasmid or viral-based vectors (see later). The catalytic activity exhibited by ribozymes has made them an attractive alternative to antisense oligonucleotides over the last several years, as a means of modifying the cellular levels of specific proteins *in vitro* and *in vivo* in both pre-clinical and clinical studies [49-51,56-59].

siRNA

Perhaps the most exciting and advancing field of small inhibitory RNA species is that of RNAi interference (RNAi). RNAi was first noted in *Caenorhabditis elegans* and plants as a novel mechanism to control post-transcriptional gene silencing, and has since been discovered in many eukaryotes [60-63]. As with antisense and ribozme strategies, there are two main forms of RNAi; 1) the use of 'naked' small RNAs or 2) production of self-complementary short hairpin RNAs (shRNA) by pol II, or pol III-based promoters from a suitable vector [64]. Both techniques produce short RNA species that are complementary to a specific mRNA sequence. These small inhibitory RNAs (siRNA) induce the activity of the cellular RNase II enzyme dicer, which cleaves the RNA into 21-23 double-stranded RNA species. These cleaved species are capable of binding their target mRNA and inducing the activity of the nucleotide RNA interference silencing complex (RISC). The RISC protein complex then cleaves the target mRNA/siRNA complex leading to down-regulation of the target protein (Figure 4).

Rapid progress has been made in the use of siRNA as a means of attenuating the expression of specific proteins both *in vitro* and *in vivo,* using a variety of plasmid-based and viral-based vectors [65-70]. The inherent specificity of this approach provides a powerful method of target protein down-regulation and evidence exists to demonstrate that siRNA can be incorporated into several existing viral and nonviral vector delivery platforms. Given the recent sequencing of the human genome and creation of gene sequence databases as well as companies specialising in the production of siRNAs, any protein, either already characterised or a newly discovered factor involved in DNA repair, can now be easily targeted by siRNA strategies. It is perhaps not surprising then that in the space of a only few years, siRNA has shown promise as a potentially powerful technique that holds much promise for molecular medicine [71,72].

As well as antisense, ribozyme and siRNA strategies, specific inhibition of a target protein can also be achieved by both inhibitory antibodies and dominant-negative peptide/protein moieties as well as small molecule inhibitors designs through combinatorial chemical approaches (Figure 3). Examples of such approaches will be given in a later section of this chapter.

It is important to add a cautionary note, that as with all these molecular strategies, expression of the exogenous species must be tightly regulated in order to reduce the risk of potential deleterious side effects [73,74]. Furthermore, recent concerns over gene therapy must be addressed if such molecular adjuvant strategies are to become commonplace within the clinic [75-78]. However, the types of gene therapy discussed here within the realms of molecular targeting are intended to be used as an adjuvant to pre-existing treatments and not for long-term gene correction via integration into the host cells genome, which is connected to recent adverse clinical events [75-78].

DELIVERY

For a GT protocol to be successful within the clinical setting, then efficient delivery of the transgene to the appropriate target site following either local or systemic administration is paramount [79-81]. There are an increasing number of different vectors and delivery methods available, which can be generally grouped into viral and non-viral delivery systems (Table 1). Viral delivery systems include the use of adenoviruses, adeno-associated viruses, retroviruses, herpes simplex virus, vaccinia virus, whereas non-viral delivery approaches include the use of cationic and anionic liposomes and their derivatives, protein-DNA complexes and direct injection or electrotransfer of 'naked' RNA/DNA. As one would expect, each approach comes with its own set of merits and caveats that must be respectively realised and overcome if efficient delivery of the transgene is to be achieved (Table 1). Generally, vector attributes to consider include; maximal transgene size permissible, maximal titre attainable, tendency to provoke inflammatory/immune responses, persistence of gene expression, the ability to deliver the transgene to non-dividing cells, target cell specificity and transduction efficiency (Table 1).

Table 1. Main advantages and disadvantages of current gene therapy vectors (adapted from [234]), with the kind permission of Springer Science and Business Media.

Vector	Advantages	Disadvantages
	Viral-based strategies	
Adenovirus	Broad target cell range High transduction efficiency Can infect non-dividing cells Relatively simple production and manipulation No insertional mutagenesis	Elicits host immune responses; limits transgene expression and impedes re-administration
Adeno-associated virus	Non-pathogenic, no immune response elicited Broad target cell range High transduction efficiency Can infect non-dividing cells	Small insert size (2-4 kb) Requires help adenovirus for successful transduction Integration into host genome could lead to mutagenesis
Retrovirus	Potential long-term transgene expression due to integration into host genome Not highly immunogenic	Do not infect non-dividing cells Potential for insertional mutagenesis Low transduction efficiency
Herpes simple virus	Infects non-diving cells Exists episomally, no insertional mutagenesis Potential for a large insert size	Elicits host immune response Severe symptoms associated with wild-type virus
Vaccnia virus	Large insert size Replicates *in vivo*	Elicits host immune response Limited transgene expression
	Non-viral-based strategies	
Liposome-based	Very large insert sizes No immune response so can re-administer Cost effective	Associated with very low efficiency of delivery Limited transgene expression
DNA-protein complexes	Non-toxic Cell targeting strategies	Associated with a low efficiency and short-lived expression Cellular degradation
Naked DNA/RNA	Low cost and simple production Non-toxic	Very low efficiency Cellular degradation

Viral-Based Delivery Systems

Adenovirus and Adeno-Associated Viruses

Probably the most extensively used of all available gene therapy vectors are Adenoviral (Ad) and Adeno-associated (AAV) vectors [82-86]. As outlined in Table 1, the advantages of these delivery systems are a broad host range, the ability to infect both proliferating and non-proliferating cells and the relatively high cellular expression of transgenes. Furthermore, they possess the potential for genetic manipulation to confer targeting to specific cell types [87-89] and can be engineered to deliver tissue-specific expression of the transgene [90-93].

Of the more than 50 different Ad serotypes, the most exploited have been the Ad2 and Ad5 subclass, mainly due to an extensive understanding of their biology derived from many years of extensive research. The adenoviral genome comprises of linear double-stranded DNA of approximately 36 kb which is encased in a caspid made up of several different proteins. Protruding from the caspid are fiber knobs that interact with the widely expressed cellular coxsackie adenovirus receptor (CAR). Together with cellular-expressed integrins, this interaction mediates viral internalization [94]. Virions then make their way via endosomes to the nucleus where replication of the viral genome occurs prior to packaging of progeny virions and finally cell lysis in order to release the progeny virions (if the virus is replication competent). Most pre-clinical and clinical trials involving the use of Ad have involved replication defective Ad vectors through deletions of the E1 and E3 regions (see [83,85] and references therein). Also, replication competent Ad (either by deletion of the E3 or E1b region of the viral genome) have been successfully used to target a range of tumour types in oncolytic viral therapies [90,91,93,95,96]. These strategies aim to reduce the tumour bulk from virus-mediated cell bursting, which can then increase the overall effectiveness of a cancer treatments such as IR/chemotherapy (Figure 2B).

In contrast to adenovirus, adeno-associated virus comprises of a single-stranded DNA genome of approximately 4.5 kb that is not enveloped. AAV are replication defective and therefore require a helper virus such as replication competent adenovirus for replication/lytic life cycle [83,97]. In the absence of helper virus, AAV integrates site-specifically into chromosome 19 of the human genome [98]. This feature of AAV together with their lack of immunogenicity has made them appealing for researches interested in long-term transgene expression or gene therapy protocols that involve re-administration of the given transgene (see [83] and references therein). To date, AAVs have been successfully used in gene replacement strategies for phenotypic correction of a number of different conditions including Parkinson's disease, cystic fibrosis and haemophilia [99-101]. Gene therapy strategies have also been reported using AAV to target several different cancer cell types [102-104]. Furthermore, as with Ad, AAV has been shown to specifically target p53 negative cells as a novel oncolytic approach [105].

Retroviruses and Lentiviruses

Retroviruses (RV) are comprised of a single-stranded RNA genome of 7-10 kb which randomly integrates into the host genome via its long terminal repeat regions. The viral genome has to be reverse transcribed into double-stranded DNA before it can be integrated into the host cell genome, a process which requires active host cell proliferation [106]. RVs were the first viral vector used in human gene therapy trials initiated in late 1990 for the ex vivo replacement of the adenosine deaminase (ADA) gene [107-109]. However, as eluded to in the previous section, the initial success of these and subsequent RV gene therapy trials has met with recent setbacks as several patients receiving ex vivo RV gene replacement treatment have developed Leukemia due to insertional mutagenesis caused by the mis-integration of the RV within the patients genome [76-78].

Lentivirus (LV) are a sub-family of retroviruses which contains the human immunodeficiency virus type I [110]. Modified gene therapy vectors based on HIV-1 and other LVs have been used for a range of applications over many years (see [110] and references therein). Recent work has demonstrated that modification of HIV-1 based retroviral vectors can yield promising results on targeting of tumours such as prostate cancer cells [111,112]. However, as with other RV vectors, the concerns regarding insertional mutagenesis remains and until this can be safeguarded against their routine use in future human gene therapy trials remains in doubt.

Herpes Simplex Viruses

Herpes simplex virus (HSV) is an enveloped virus with a genome comprising of 150 kb double-stranded DNA that preferentially transduces cells of neurological origin. Following entry into the cell via specific interactions between viral glycoproteins and cell surface glycosaminoglycan moieties [113], the viral genome is episomally replicated in the host cell nucleus under tight regulation of viral immediate early genes and the viral structural protein VP16 (see [114] and references therein). GT strategies involving the use of replication defective HSV vectors have increased over that last ten years (see [115,116]). However, most progress has been made in developing either replication defective or conditionally replicating HSV vectors for oncolytic approaches (reviewed in [117]). Also, specific products of HSV such as thymidine kinase (TK) an enzyme that assists in the conversion of the pro-drug ganciclovir (GCV) into the toxic product ganciclovir triphosphate, have been delivered using other viral vectors for pro-apoptotic approaches (discussed later). Finally, the HSV intercellular transport protein vp22 has been used to improve the bystander effects (the transfer of a toxic product produced within a cell to neighbouring cells) of both pro-apoptotic and gene-replacement therapies [118,119].

Vaccinia Virus

Vaccinia is a DNA virus that is naturally cytopathic but as with the other viral vectors discussed, can be engineered to a non-cytopathic form that still retains infectious activity [120,121]. Vaccinia has been commonly used to delivery antigens to tumours that elicit host cell immune responses as a means of targeting the cancer cells for immune-based destruction, given that many cancer types use a variety of mechanisms to evade normal cellular immune responses [122,123]. Also, McCart *et al* have recently reported the use of a recombinant vaccinia virus engineered to be cyotoxic to highly proliferating cells, with treatment of murine colorectal adenocarcinoma xenografts led to an approximate 15-fold decrease in tumour volume compared to control virus treated animals [124]. The main caveat with the use of vaccinia viruses in human GT protocols is the strong immune response that can be elicited and pre-existing immunity from child-hood vaccination to smallpox (certainly within the US).

Non-Viral Based Delivery Systems

Lipid-Based Approaches

Liposomes have been used in several GT clinical protocols mainly due to the low costs associated with their synthesis, large insert size permitted and their lack of immunogenicity (Table 1). Liposomes consist of DNA or RNA species surrounded by a liposomal coat that protects the DNA from degradation by cellular endonucleases [125,126]. Low efficiency of transgene delivery is considered the Achilles heal of liposome systems although specific modifications to the liposome molecule can be made including the fusion to ligands to aid cell targeting and increase transduction efficiency [125,127]. So-called 'stealth' liposomes are liposmes that have been modified by the inclusion of sialic acid residues in an attempt to reduce preferential uptake of lioposomal vectors by the endoplasmic reticulum [128]. One possible advantage of modified liposomes is their large size, which may aid their delivery more specifically to tumours, which are surrounded by 'leaky' vasculature compared to normal tissues [129,130]. Finally, recent studies have demonstrated the use of liposomes to deliver the cytokine tumour necrosis factor-alpha to enhance the effects of radiation and chemotoxic agents both in vitro and in vivo [131-133]

DNA-Protein Complexes

DNA-protein gene therapy strategies involve the use of a protein/ligand conjugated to the transgene that facilitates its delivery to a specific cell type, with minimal activation of the host immune response. However, DNA-protein complexes are associated with relatively low transduction efficiency due to their degradation following endosome/lysosome targeting. Although specific delivery of ribozymes and antisense molecules (described above) has been achieved by conjugating them to an antibody specific for the transferrin receptor [134,135]. Also, receptor-independent mediated delivery of viral vectors has been demonstrated by conjugating them to so-called 'Trojan peptides' [136]. These are short peptides derived from transduction domains of proteins such as the Drosophilae homeodomain of Antennapedia and can be conjugated to either peptides or oligonucleotides for successful delivery to a range of cell types [137].

Naked DNA/RNA

The most basic form of GT delivery strategy is to directly deliver the DNA encoding your gene of interest or RNA (ribozyme or RNAi molecule) to the desired site i.e. direct injection of a plasmid containing your transgene and suitable promoter directly into the tumour site. Although this approach is cheap and simple to perform, as one might expect, the cellular uptake of the plasmid DNA/RNA and subsequent expression of transgene occur with very low efficiency. However, this low frequency of transduction may be improved by recent advancements in in vivo DNA electroporation (for a recent review see [138] and references therein). Furthermore, DNA which is internalized into the cell is extremely susceptible to endogenous endonuclease, thus limiting the duration of expression of any transgene. On the plus side, no known immunogenic response is associated with this approach, which has fuelled continuing advancements in this field [139]. Perhaps the most successful use of plasmid DNA has been in the in vivo delivery of 'vaccine' vectors that either express lymphokines or immunogenic molecules that get internalized and expressed on the tumour cell surface [140,141]. Indeed, a recent phase I/II clinical study by Mincheff and colleagues has demonstrated that plasmid DNA vectors can be used to prime the immune response prior to boosting with a viral vector [141].

EXAMPLES OF MOLECULAR TARGETING STRATEGIES

In this section a few examples will be given of how molecular targeting approaches have been successfully used to specifically target factors associated with DNA repair, both *in vitro* and *in vivo*. This is by no means is meant to serve as a comprehensive list of successful strategies but shows proof of principle for many factors.

Targeting of DSB Repair Factors

Given that the most toxic DNA lesion is the DSB and the of both the NHEJ and HRR pathways in the repair of this lesion, factors involved in these pathways have been the focus of many molecular targeting strategies. In particular are those factors that are involved in the predominant mammalian DNA DSB repair pathway NHEJ, such as the MRN complex, ATM, DNA-PK$_{cs}$, and Ku protcins (Figurc 2). Methods employed for specific targeted inhibition include; antisense, ribozyme, siRNA and dominant-negative protein strategies. Although this review focuses on the use of such techniques to increase cellular sensitivity to therapeutic DNA damaging agents, the reader is encouraged to bear in mind that these tools can also prove extremely valuable in the continuing efforts to dissect the intricate processes of DNA DSB repair.

MRN Complex

The MRE11-NBS1-RAD50 complex plays a vital role in both early sensing of DNA damage and subsequent activation of the ATM pathway, as well as in the processing of DNA DSBs [142-145]. Mutations in NBS and MRE11 lead to the radiosensitive, cancer-prone disorders ataxia-telangiectasia like disease (A-TLD) and Nijmegen breakage syndrome (NBS) respectively [146]. As such, several groups have targeted components of the MRN complex as a means to increase the sensitivity of cancer cells to therapeutic DNA damaging agents.

Kim *et al* used a full-length RAD50 cDNA antisense-encoding plasmid to create stable cell lines with reduced RAD50 expression. They showed that these cell lines were more sensitive (~2-fold) to IR and certain alkylating agents, but not to UV, which is mainly detected by the ATR-mediated signalling pathway [147]. More recently O'Malley and colleagues used adenoviruses to exogenously deliver a dominant-negative form of NBS1 (a.k.a. p95/nibrin) to disrupt the MRN complex in head and neck cancer cell lines [148]. Viral-mediated expression of the mutant form of NBS1 led to cell cycle perturbations but only a modest increased sensitivity to IR. Finally, Xu *et al* have recently used siRNA technology to transiently reduce MRE11 expression by ~70% in human colon carcinoma cell lines and confer an increased sensitivity to IR [149].

ATM

Given it's early role in sensing DNA damage (Figure 2 and [150]) and that mutations in gene and promoter lead to the cancer predisposition and radiation sensitive syndrome A-T [151], several groups using different approaches have targeted ATM as a means to increase sensitivity to DNA damaging agents, in particular IR.

There have been four noteworthy reports over the last eight years describing the use of antisense molecules to target ATM [152-155]. Using a full-length antisense cDNA of >9 kb, Zhang *et al* demonstrated an approximate 2-4 fold-increased sensitivity of normal lymphoblastoid fibroblasts at doses up to 4Gy compared with controls. The antisense was placed in an Epstein-Barr based viral vector under the control of an inducible metallothionein promoter, which was used to make stable cell lines that could then be induced to activate transcription of the ATM antisense. Upon induction, these stable lyphoblastoid cell lines displayed many characteristics of A-T cells such as the increased IR sensitivity, however, these induced cells were not as sensitive as fibroblasts obtained from an A-T patient, which may be a consequence of residual ATM expression seen in these cells [155]. Using a smaller (1. kb) antisense fragment, Uhrhammer *et al* also made stable normal human fibroblast cell lines that were devoid of detectable ATM protein expression. Furthermore, these cells exhibited a radiosensitive phenotype on a par with that observed in an A-T fibroblast cell line [154]. A year latter, two separate groups showed that antisense ATM could be used to also confer an increased radiosensitivity to cancer cells [152,153]. Guha *et al* used a 281bp antisense ATM fragment, designed to target the PI-3 kinase domain of the protein. As with the other two studies described, stable cell lines (glioma cancer cell line U-87) were made following transfection of a plasmid containing the antisense molecule. Several stable clones showed an increased sensitivity to IR (~2-fold at 2Gy, increasing to ~5-fold at higher doses) [153]. Similar results have been obtained using a prostate cancer cell line stably expressing hammerhead ribozymes targeted to three separate regions of the ATM mRNA sequence (Collis *et al*, unpublished data). Although first three studies clearly demonstrated proof of principle concepts, the study by Fan *et al* probably lends itself best to potential clinical evaluation due to the fact that they used an adenovirus to exogenously deliver the antisense to a relatively radioresistant prostate cancer cell line [152]. They targeted either a region close to the translational start site of ATM or the PI-3 kinase domain, or a combination of the two, which had roughly equal effectiveness. Although these antisense molecules only achieved an ~50-60% reduction in ATM expression and a subsequent ~3-fold increase in sensitivity to IR, these results more represent the type of challenge such approaches face in the clinical setting.

As well as antisense and ribozyme strategies, ATM has also been successfully targeted by siRNA. Work by Collis *et al* has demonstrated the effectiveness of siRNAs targeting ATM, which yield up to ~90% reduction in ATM expression following transient transfection of plasmid DNA, leading to an approximate 3-fold increase in radiosensitivity at dose ranges of 0-6Gy [156]. Work is currently ongoing to develop adenoviral vectors expressing these siRNAs to improve their delivery to tumours *in vivo* as part of further pre-clinical testing. More recently, Mukhopadhyay *et al* used retrovirus to stably transfect p53-defective prostate cancer cells and normal fibroblast with anti-siRNAs targeting ATM and Chk1 [157]. They showed that the p53-defective cancer cells were more sensitive to the chemotherapeutic agent doxorubicin compared to normal, p53 positive cells, thus highlighting the potential use of such approaches to specifically target cancer cells. Two separate studies have shown that the exogenous delivery (via transient transfection or stable integration) of dominate-negative forms of the ATM protein, that inhibit the activity of the native protein, can also be used to confer an increased radiosensitivity to cancer cells lines [158,159]. Finally, Hickson and colleagues have recently shown that small molecule chemical inhibition of ATM renders cells sensitive to certain therapeutic DNA damaging agents [160].

In addition targeting of molecular DNA damage sensors as described above, several groups have used the same approaches to target other factors of the DSB signalling cascade (Figure 2A). These include the mediators MDC1, H2AX, 53BP1, 14-3-3, HDAC4 [161-167] as well as the transducers CHK1/2 and effector kinase CDC25A [168,169]. Although most of these studies have used such approaches to study the function of the factors in DNA damage response mechanisms, several of these studies show that molecular disruption of these factors renders cells sensitive to various DNA damaging agents and as such, could prove to be potential therapeutic targets.

DNA-PK

The DNA-PK complex has been the focus of many investigators interested in modulating cellular response to DNA damaging agents by molecular targeting strategies. DNA-PK is comprised of the heterodimer Ku70/Ku80 (note: Ku80 is also known as Ku86) and the catalytic sub-unit DNA-PKcs (see [170] and references therein). Over the last 5-6 years, at least five independent reports have been published describing the use of antisense molecules to specifically target one of the sub-units of DNA-PK. Marangoni and colleagues used a full-length inverted cDNA to target Ku80 in human fibroblast cell lines, which led to a marked decrease in Ku80 expression following both transient and stable cell line producing transfections [171]. This inhibition rendered transfected cells ~2.5 fold (transients) and ~8-fold (stables) more sensitive to IR-mediated killing than control cells/cell lines, which was likely a consequence of a reduced DNA end binding capacity and DSB repair rates observed in these cells. More recently, Belenkov *et al* used the epigenetic human glioma cell lines M059K and M059J (respectively proficient and deficient in DNA-PK expression) to demonstrate the effectiveness of transiently transfected naked antisense oligonucleotides that targeted several different regions of Ku80 mRNA [172]. The most effective antisense led to an ~10-fold increased radiosensitivity, which was shown only in the M059K cells, and not in the M059J cells, demonstrating the molecular specificity of the approach used. Furthermore, these cells were shown not to be rendered sensitive to the DNA cross-linking agent (chlorambucil), highlighting how such an approach can modulate cellular sensitivity to a particular type of DNA lesion.

The Ku70 sub-unit has also been recently targeted both *in vitro* and *in vivo* by antisense molecules in two independent studies [173,174]. Omori *et al* created stable lung carcinoma cell lines expressing an antisense molecule targeted to the first ~500bp of the Ku70 mRNA, resulting in a significant reduction in protein expression. They showed that cell lines expressing the antisense molecule were ~5-10 fold more sensitive to IR and as well as other

DNA damaging agents, such as the radiation mimetic bleomycin, compared to parental control cells or stable cell lines expressing the same antisense fragment but in the sense orientation [174]. More recently, Li and colleagues used a heat-restricted (via hsp70 promoter) replication-defective adenovirus to control the expression of a Ku70 antisense molecule exogenously delivered to cultured mouse tumour cells and xenografts [173]. *In vitro* studies demonstrated an ~80% reduction in Ku70 protein expression upon heat shock that gave rise to an ~7-10 fold increased radiosensitivity. Similar results were obtained *in vivo* with a significant growth delay observed (up to 6 days) in tumours treated with virus/heat shock and radiation compared to appropriate control groups. This study is one of many that highlight the feasibility of exogenously delivering molecular targeting gene therapies to augment the killing effects of IR-treated tumours.

As well as targeting the Ku sub-units of DNA-PK, the catalytic sub-unit DNA-PK_{cs} has also been shown to be a valid target for increasing cellular sensitively to DNA damaging agents. Recently, Sak *et al* used small (~20bp) antisense oligonucleotides targeted to three separate regions of the mRNA in order to down-regulate DNA-PK_{cs} in several human tumour cell lines [175]. Specific inhibition of DNA-PK_{cs} via treatment of high doses of naked antisense RNA targeting the AUG translational start region of the mRNA led to deficient DNA DSB repair following 30Gy IR exposure, which was comparable to pre-treatment of cells to the non-specific PI-3 kinase inhibitor Wortmannin, which is known to inhibit DNA-PK activity [176].

Recently, siRNA-based approaches have been employed to specifically target DNA-PK_{cs} [156,177,178]. Peng *et al* used consecutive transfection of normal fibroblast with naked RNA targeted to two different regions of the mRNA sequence. They demonstrated that DNA-PK_{cs} protein expression was reduced by ~80-90% at 72hrs following the second round of transfections, which caused an increased sensitivity to IR, chromosomal aberrations and mutation frequency [177]. Collis *et al* simultaneously but independently used a plasmid-based approach to deliver siRNAs targeted to 3 different regions of the DNA-PK_{cs} mRNA sequence. Similar to the data reported by Peng *et al*, the nadir of reduced protein expression (~80-90%) was shown to occur 72hrs following a single transfection of plasmid DNA, which could yield an equivalent, if not slightly greater enhanced radiosensitivity of human prostate cancer cells than that reported by Peng and colleagues [156]. As with the ATM siRNA, adenoviral vectors expressing these anti- DNA-PK_{cs} siRNAs are currently being developed a as means to improve the delivery and aid further pre-clinical testing both *in vitro* and *in vivo*. Finally, recent work by An *et al*, has shown that inhibition of DNA-PK_{cs} also impacts on other pathways, such as down-regulation of the oncogenic factor c-myc [178].

In addition to antisense and siRNA techniques, several groups have demonstrated that inhibition of DNA-PK can also be achieved by the exogenous delivery of inhibitory peptides, dominant-negative forms of the protein or inhibitory antibody fragments [179-181]. Stable Chinese hamster ovary cell lines expressing a 32 kDa portion of the C-terminal region (representing amino acids 449–732) of Ku80 were constructed by Marangoni *et al*, which were shown to yield dominant negative effects on normal DNA-PK function, presumably by abrogating Ku binding to DNA-PK_{cs}, and inhibited the heterodimer's ability to bind DNA breaks [181]. These stable cell lines were shown to be rendered sensitive to IR and unable to successfully repair DNA DSBs [181]. More recently, Kim and colleagues have synthesized a dominant-negative peptide representing amino acids 720-732 of Ku80, (encompassing the same region used by Marangoni *et al*) to inhibit DNA-PK activity following transfection into human breast cancer cell lines [179]. As with the dominant-negative Ku80 protein fragment, this peptide was shown to reduce DNA-PK_{cs} kinase activity, ultimately leading to inefficient recognition of DNA DSBs. As expected these cells also showed an increased sensitivity to IR. Li *et al* have demonstrated inhibition of DNA-PK_{cs} function by use of a single chain variable antibody fragment [180]. Unlike the small molecule inhibitors described above that target the active site of DNA-PK_{cs}, this antibody fragment bound to a regulatory region that is unique to DNA-PK_{cs} and not present in the PI-3-related kinases ATM or ATR. Although exogenous delivery of the antibody fragment did not drastically abrogate kinase activity of DNA-PK, it did lead to inhibition of NHEJ in a human skin melanoma cell line following low levels of radiation exposure [180].

It is worth mentioning that certain viruses which can be used for such GT approaches appear to have evolved to target DNA repair mechanisms as a means to increase viral infection i.e. inhibition of the DSB repair processes that can be activated by the presence of linear viral DNA within the cell. Examples of this are the herpes simplex virus type (HSV-1) and adenovirus serotype 5 (Ad5). The immediate early gene product ICP0 (a.k.a. Vmw110) of HSV-1 has been shown to bind DNA-PK_{cs} and promote its rapid ubiquitin-mediated degradation [182,183]. This inhibition leads to an increased sensitivity to IR in *in vitro*, although high levels of expression of ICP0 is toxic to cells (Collis *et al*, unpublished data). However, the oncolytic HSV-1 virus G207, which contains the ICP0 gene under control of it's native promoter, did not yield an increased sensitivity to IR in the *in vivo* prostate TRAMP model [184], suggesting further investigation into this mechanism is required. Recent data has now shown that ICP0 inhibits ATR-mediated DNA damage response during the lytic cycle of HSV infection [185].

In addition to HSV-1 ICP0, the adenoviral gene product E4orf6 (E4-34K) has been shown to bind DNA-PKcs and inhibit V(D)J recombination processes [186]. As with HSV, this may aid increased viral replication and prevent the linear double-stranded viral genome from being "repaired" by the host cell DSB repair machinery. Recent work by Mohammadi *et al* have shown that under an inducible promoter, E4-34K is capable of inhibiting DSB repair [187]. However, Collis *et al* showed that exogenous delivery and expression of E4-34K alone does not confer a radiosensitive phenotype in several human tumour cell lines [188]. Subsequent to this publication, Stracker *et al* demonstrated that E4-34K and another adenoviral protein E4orf3 bound to an additional adenoviral gene product E1B-55k to cause respective degradation and mis-localisation of components of the MRN complex [189]. It is interesting to note that the cell line used by Mohammadi *et al* for E4-34K induction was the transformed human embryonic kidney cell line 293, which constitutively expresses high levels of E1B-55K, which likely explains why E4-34K shows DNA-PK inhibitory effects in these cells [187]. However, as with expression of ICP0, co-expression of high levels of E4-34K and E1B-55K under the control of CMV promoters is toxic to cells that do not normally express these proteins [188]. Collectively, these data may help explain the synergistic effects of oncolytic adenoviruses when used in combination with IR [190]. More recent work has also shown that the Poliovirus also targets DNA-PK_{cs} for degradation in order to augment its infection [191], indicating that inhibition of DNA-PK function appears to be a key event in the evolutionary development of many different viruses, including retrovirus [192].

In addition to molecular targeting of DNA-PK_{cs} and Ku, several groups have used chemical (small molecule) inhibitors that bind the protein and inhibit its kinase activity, which is essential for its DSB repair activity [193-195]. Such work has shown the development from non-specific, general PI-3 kinase inhibitors such as Wortmannin and LY294002 [176] to highly specific and effective inhibitors derived from both natural compounds and synthesised using combinatorial chemistry (see and [170] references therein for further information). These compounds hold much promise for future modification of DNA-PK function in the clinical setting.

RAD51

The examples described above concern molecular targeting of proteins mainly involved in the NHEJ repair pathway. There is a growing amount of data showing that molecular inhibition of components of the HRR repair pathway (mainly RAD51) can also give rise to an equivalent biologically significant increased sensitivity to DNA damaging agents such as IR.

The first anti-RAD51 antisense reported was by Taki *et al* whose used short (14bp) antisense oligonucleotides to reduced RAD51 expression in murine cell lines following transient transfection of naked antisense DNA molecules [196]. They demonstrated a dose-response of antisense with an ~5-10 increased sensitivity of antisense-treated cells compared to sense and scrambled controls. Two years later, the same group extended on their initial studies and used these RAD51 antisense molecules to confer a radiosensitive phenotype to resistant glioma cell lines and *in vivo* xenografts [197]. An approximate 3-fold increased sensitivity to IR was observed in cultured cells compared to sense and scrambled control treated cells, which were more sensitive than untreated cells. They then went on to show that mice bearing glioma cancer cell xenografts had a greater mean survival when treated with RAD51 antisense molecules (17 days compared to 13 for scrambled-treated mice). In combination with 6Gy IR, the RAD51 antisense significantly increased the survival of tumour-bearing mice, with half of them still surviving at 36 days post inoculation (all mice in IR + sense/scrambled controls groups were deceased by ~day 16).

Based on this successful application of the RAD51 antisense data reported, Collis *et al* used a plasmid-based delivery system to drive the expression of hammerhead ribozymes targeting the same region within the RAD51 mRNA sequence [198]. Ribozyme-expressing prostate cancer cells had an ~3-fold increased sensitivity to IR, which was shown to correlate with the residual amounts of RAD51 protein expression following ribozyme-mediated destruction of mRNA. More recently, Ito and colleagues have used a haemagglutinating virus to deliver anti-RAD51 siRNA both *in vitro* and *in vivo*, which produced and increased sensitivity to the chemotherapeutic agent *cis*-paltin [199]. Watson and colleagues have reported the use of a dominant-negative protein approach, similar to that described above for Ku80, to inhibit RAD51 function in glioblastoma cells [200]. Glioblastoma cell lines stably expressing the mutant form of RAD51 (defective in ATP binding and hydrolysis, essential for its function) were reported to be ~3.5-fold more sensitive to IR compared to control cell lines. Finally, Russell and colleagues have recently shown that the specific c-Abl inhibitor Gleevec, which is a currently effective treatment for chronic myelogenous leukemia, also causes reduction in cellular levels of RAD51 and thus leads to subsequent sensitization to IR [201].

As well as targeting DNA DSB repair pathways, there are many examples of where other DNA repair and related pathways have been targeted by such molecular approaches. These include; MGMT [202], Fanconi Anaemia [203], histone deacetylases [204,205], as well as many other DNA repair, cell cycle checkpoint, anti-apoptotic, angiogenesis factors that are attractive targets for molecular targeting (for more information, see [2,206-208] and references therein).

Targeting of Other Factors

In addition to the examples described above where key components of the DNA repair have been targeted at the molecular level to augment radiation-mediated tumour cell killing, there are several reports of using similar approaches aimed to target proteins involved in other cellular responses to DNA damage. Such cellular responses include a whole host of signalling, anti-apoptotic pathways, DNA synthesis, cell cycle and transcriptional regulation events [209-219].

RAF

A large amount of work has been carried out over several years in the development of antisense oligonucleotides targeting the serine-threonine kinase RAF-1, which is activated upon cellular exposure to IR (recently reviewed by [210]). The observation that molecular targeting of RAF-1 could lead to an increased sensitivity to IR was first reported by Kasid and colleagues in 1989 [211]. Several years later, Gokhale et al used stabilised antisense oligonucleotides designed to target the translation start site of RAF-1 to reduce its activity in vivo [220]. The success of these and other experiments (see [210] and references therein) has aided the development and licensing of anti-RAF-1 antisense oligonucleotides (LErafAON), which have been accessed in phase I/II clinical trials [221].

Anti-Apoptotic Factors

Several groups have targeted anti-apoptotic factors as a means to increase the amount of cell death following exposure to DNA damaging agents. Antisense oligonucleotides targeting the anti-apoptotic protein Clusterin have been shown to give a modest increase in radiation sensitivity of prostate cancer cells, with a more impressive in vivo prostate tumour growth delay (>50% reduction in tumour volume) [222]. Another anti-apoptotic protein Survivin has been shown to be involved in determining the intrinsic radioresistance of several tumour types [218,223-227]. Asanuma et al showed that radioresistant PANC-1 pancreatic carcinoma cells expressing a dominant-negative mutant form of Survivin were more sensitive to IR than their respective controls [223]. More recently, Pennati et al have employed ribozyme technology to mediate down-regulation of Survivin expression in human melanoma cells [228]. Stable cell lines expressing a hammerhead ribozyme targeted to a specific Survivin mRNA sequence were ~2-fold more sensitive to IR than either normal cell or table cell lines expressing a mutant, inactive variant of the anti-Survivin ribozyme. Antisense targeting of the anti-apoptotic protein Bcl-2 is yet another way in which investigators have successfully augmented the cytotoxic effects of radiation. Such strategies are currently being assessed in clinical trials for certain haematopoietic malignancies (see [44,45] and references therein).

MnSOD

Other viable targets for molecular targeting strategies include proteins involved in the scavenging of free-radicals/reactive oxygen species produced by the interaction of radiation with cellular constituents. One such example is manganese superoxide dismutase (MnSOD), which can confer a radio-protective effect to many tissue types following exposure to IR (reviewed by [229]). Kuninaka and colleagues used antisense molecules targeting a ~200bp 5′ region of the MnSOD mRNA sequence. Stable colon cancer cell lines expressing the antisense were approximately 2-3 fold more sensitive to low doses (1-3Gy) of ionising radiation than their respective empty vector control cell lines [230]. A year later, Ueta et al showed that mouse xenografts derived from cells expressing MnSOD antisense grew on average twice as slowly as non-antisense expressing xenografts and the mice bearing these tumours survived approximately 1.5 times longer as empty vector transfected xenografts-bearing mice [231]. Moreover, they also reported that the expression of the Bcl-2 and Cox-2 was diminished in the antisense expressing xenografts, while the expression of pro-apoptotic proteins such as Bax and Bak were up regulated compared to the control xenografts. Related to these findings are the data showing that inhibition of Cox-2, a protein involved in prostaglandin synthesis, augments the killing effects of radiation of tumours and as such, there is a growing number of specific small-molecule inhibitors of Cox-2 which are being assessed both in pre-clinical and clinical trials (reviewed by [232]). One such example is that of Krysen et al, who demonstrated the effectiveness of Cox-2 antisense molecules (full length cDNA) in rendering human lung and squamous cell cancer cell lines 2-3 fold more sensitive to IR than control cell lines [233]. The antisense expressing tumours also grew significantly slower as xenografts in nude mice, although in vivo radiation sensitivity was unfortunately not studied.

CONCLUSION

The examples given above clearly demonstrate that there are multiple DNA repair proteins that can be targeted by several different approaches for the development of potent adjuvant molecular targeting GT regimes. It may prove that targeting several different components of the DNA repair pathways described could yield even greater potential therapeutic advantages over targeting a single repair factor e.g. inhibiting key components of both the NHEJ and HRR pathways.

It is evident that progress has been made in the inhibition of DNA repair pathways and/or interlinked signalling cascades as a strategy to increase tumour cell killing, but there are numerous challenges that lay ahead in the development of effective adjuvant radio/chemo-gene therapy approaches. For small molecule inhibitors, improved specificity and concomitant reduction in toxicity must be addressed if such compounds are to find their way from the bench to the bedside. For molecular targeted approaches, improvements in vector development will continue to be critical in establishing less immunogenic viruses/liposomes with the capacity to carry larger transgenes. Modification of the surface fibers of viruses will hopefully yield improved delivery and more specific targeting of the tumour cells. Finally, the increasing knowledge of human genetic information following the sequencing of the human genome may give rise to new tumour and/or cell specific promoters to augment the controlled expression of such molecular therapeutics. If these challenges can be met, then their combined use with further advancements in the therapeutic treatment of cancer such as improvements in conformal radiotherapy regimes and newly developed chemotherapeutic agents could provide significantly enhanced local control of tumours and reduction/prevention of metastatic disease.

REFERENCES

[1] Wheldon TE, Mairs RJ, Rampling RP, Barrett A. Modelling the enhancement of fractionated radiotherapy by gene transfer to sensitize tumour cells to radiation. *Radiother Oncol* 1998;48 (1):5-13.

[2] Ma BB, Bristow RG, Kim J, Siu LL. Combined-modality treatment of solid tumors using radiotherapy and molecular targeted agents. *J Clin Oncol* 2003;21 (14):2760-76.

[3] Bekker-Jensen S, Lukas C, Kitagawa R, Melander F, Kastan MB, Bartek J, Lukas J. Spatial organization of the mammalian genome surveillance machinery in response to DNA strand breaks. *J Cell Biol* 2006;173 (2):195-206.

[4] Hoeijmakers JH. Genome maintenance mechanisms for preventing cancer. *Nature* 2001;411 (6835):366-74.

[5] Iliakis G. The role of DNA double strand breaks in ionizing radiation-induced killing of eukaryotic cells. *Bioessays* 1991;13 (12):641-8.

[6] Jackson SP. Detecting, signalling and repairing DNA double-strand breaks. *Biochem Soc Trans* 2001;29 (Pt 6):655-61.

[7] Petrini JH, Stracker TH. The cellular response to DNA double-strand breaks: defining the sensors and mediators. *Trends Cell Biol* 2003;13 (9):458-62.

[8] Valerie K, Povirk LF. Regulation and mechanisms of mammalian double-strand break repair. *Oncogene* 2003;22 (37):5792-812.

[9] Sung JS, Demple B. Roles of base excision repair subpathways in correcting oxidized abasic sites in DNA. *Febs J* 2006;273 (8):1620-9.

[10] Park CJ, Choi BS. The protein shuffle. Sequential interactions among components of the human nucleotide excision repair pathway. *Febs J* 2006;273 (8):1600-8.

[11] Mellon I. Transcription-coupled repair: a complex affair. *Mutat Res* 2005;577 (1-2):155-61.

[12] McHugh PJ, Spanswick VJ, Hartley JA. Repair of DNA interstrand crosslinks: molecular mechanisms and clinical relevance. *Lancet Oncol* 2001;2 (8):483-90.

[13] D'Andrea AD. The Fanconi Anemia/BRCA signaling pathway: disruption in cisplatin-sensitive ovarian cancers. *Cell Cycle* 2003;2 (4):290-2.

[14] Levitus M, Joenje H, de Winter JP. The Fanconi anemia pathway of genomic maintenance. *Cell Oncol* 2006;28 (1-2):3-29.

[15] Kunkel TA, Erie DA. DNA mismatch repair. *Annu Rev Biochem* 2005;74:681-710.

[16] [McElhinney RS, McMurry TB, Margison GP. O6-alkylguanine-DNA alkyltransferase inactivation in cancer chemotherapy. *Mini Rev Med Chem* 2003;3 (5):471-85.

[17] Ward JF. DNA damage produced by ionizing radiation in mammalian cells: identities, mechanisms of formation, and reparability. *Prog Nucleic Acid Res Mol Biol* 1988;35:95-125.

[18] Kastan MB, Bartek J. Cell-cycle checkpoints and cancer. *Nature* 2004;432 (7015):316-23.

[19] Levitt NC, Hickson ID. Caretaker tumour suppressor genes that defend genome integrity. *Trends Mol Med* 2002;8 (4):179-86.

[20] Moses RE. DNA damage processing defects and disease. *Annu Rev Genomics Hum Genet* 2001;2:41-68.

[21] Thompson LH, Schild D. Recombinational DNA repair and human disease. *Mutat Res* 2002;509 (1-2):49-78.

[22] Shiloh Y. ATM and related protein kinases: safeguarding genome integrity. *Nat Rev Cancer* 2003;3 (3):155-68.

[23] Rotman G, Shiloh Y. ATM: from gene to function. *Hum Mol Genet* 1998;7 (10):1555-63.

[24] Soussi T, Beroud C. Assessing TP53 status in human tumours to evaluate clinical outcome. *Nat Rev Cancer* 2001;1 (3):233-40.

[25] Powell SN, Kachnic LA. Roles of BRCA1 and BRCA2 in homologous recombination, DNA replication fidelity and the cellular response to ionizing radiation. *Oncogene* 2003;22 (37):5784-91.

[26] Powell SN, Willers H, Xia F. BRCA2 keeps Rad51 in line. High-fidelity homologous recombination prevents breast and ovarian cancer? *Mol Cell* 2002;10 (6):1262-3.

[27] Yoshikawa K, Ogawa T, Baer R, Hemmi H, Honda K, Yamauchi A, Inamoto T, Ko K, Yazumi S, Motoda H, Kodama H, Noguchi S, Gazdar AF, Yamaoka Y, Takahashi R. Abnormal expression of BRCA1 and BRCA1-interactive DNA-repair proteins in breast carcinomas. *Int J Cancer* 2000;88 (1):28-36.

[28] Hansen LT, Lundin C, Spang-Thomsen M, Petersen LN, Helleday T. The role of RAD51 in etoposide (VP16) resistance in small cell lung cancer. *Int J Cancer* 2003;105 (4):472-9.

[29] Maacke H, Jost K, Opitz S, Miska S, Yuan Y, Hasselbach L, Luttges J, Kalthoff H, Sturzbecher HW. DNA repair and recombination factor Rad51 is over-expressed in human pancreatic adenocarcinoma. *Oncogene* 2000;19 (23):2791-5.

[30] Maacke H, Opitz S, Jost K, Hamdorf W, Henning W, Kruger S, Feller AC, Lopens A, Diedrich K, Schwinger E, Sturzbecher HW. Over-expression of wild-type Rad51 correlates with histological grading of invasive ductal breast cancer. *Int J Cancer* 2000;88 (6):907-13.

[31] Shintani S, Mihara M, Li C, Nakahara Y, Hino S, Nakashiro K, Hamakawa H. Up-regulation of DNA-dependent protein kinase correlates with radiation resistance in oral squamous cell carcinoma. *Cancer Sci* 2003;94 (10):894-900.

[32] Sirzen F, Nilsson A, Zhivotovsky B, Lewensohn R. DNA-dependent protein kinase content and activity in lung carcinoma cell lines: correlation with intrinsic radiosensitivity. *Eur J Cancer* 1999;35 (1):111-6.

[33] Wilson CR, Davidson SE, Margison GP, Jackson SP, Hendry JH, West CM. Expression of Ku70 correlates with survival in carcinoma of the cervix. *Br J Cancer* 2000;83 (12):1702-6.

[34] Zhao HJ, Hosoi Y, Miyachi H, Ishii K, Yoshida M, Nemoto K, Takai Y, Yamada S, Suzuki N, Ono T. DNA-dependent protein kinase activity correlates with Ku70 expression and radiation sensitivity in esophageal cancer cell lines. *Clin Cancer Res* 2000;6 (3):1073-8.

[35] Barroso E, Milne RL, Fernandez LP, Zamora P, Arias JI, Benitez J, Ribas G. FANCD2 associated with sporadic breast cancer risk. *Carcinogenesis* 2006.

[36] Taniguchi T, Tischkowitz M, Ameziane N, Hodgson SV, Mathew CG, Joenje H, Mok SC, D'Andrea AD. Disruption of the Fanconi anemia-BRCA pathway in cisplatin-sensitive ovarian tumors. *Nat Med* 2003;9 (5):568-74.

[37] Fedier A, Fink D. Mutations in DNA mismatch repair genes: implications for DNA damage signaling and drug sensitivity (review). *Int J Oncol* 2004;24 (4):1039-47.

[38] Angele S, Treilleux I, Bremond A, Taniere P, Hall J. Altered expression of DNA double-strand break detection and repair proteins in breast carcinomas. *Histopathology* 2003;43 (4):347-53.

[39] Pierce ML, Ruffner DE. Construction of a directed hammerhead ribozyme library: towards the identification of optimal target sites for antisense-mediated gene inhibition. *Nucleic Acids Res* 1998;26 (22):5093-101.

[40] Dash P, Lotan I, Knapp M, Kandel ER, Goelet P. Selective elimination of mRNAs in vivo: complementary oligodeoxynucleotides promote RNA degradation by an RNase H-like activity. *Proc Natl Acad Sci U S A* 1987;84 (22):7896-900.

[41] Walder RY, Walder JA. Role of RNase H in hybrid-arrested translation by antisense oligonucleotides. *Proc Natl Acad Sci U S A* 1988;85 (14):5011-5.

[42] Myers KJ, Dean NM. Sensible use of antisense: how to use oligonucleotides as research tools. *Trends Pharmacol Sci* 2000;21 (1):19-23.

[43] Branch AD. A good antisense molecule is hard to find. *Trends Biochem Sci* 1998;23 (2):45-50.

[44] Chanan-Khan A. Bcl-2 antisense therapy in B-cell malignancies. *Blood Rev* 2005;19 (4):213-21.

[45] Stein CA, Benimetskaya L, Mani S. Antisense strategies for oncogene inactivation. *Semin Oncol* 2005;32 (6):563-72.

[46] Gleave M, Miyake H, Zangemeister-Wittke U, Jansen B. Antisense therapy: current status in prostate cancer and other malignancies. *Cancer Metastasis Rev* 2002;21 (1):79-92.

[47] Jansen B, Zangemeister-Wittke U. Antisense therapy for cancer--the time of truth. *Lancet Oncol* 2002;3 (11):672-83.

[48] Vidal L, Blagden S, Attard G, de Bono J. Making sense of antisense. *Eur J Cancer* 2005;41 (18):2812-8.

[49] Cech TR. Ribozymes, the first 20 years. *Biochem Soc Trans* 2002;30 (Pt 6):1162-6.

[50] Sioud M. Ribozymes and siRnas: from structure to preclinical applications. *Handb Exp Pharmacol* 2006 (173):223-42.

[51] Tafech A, Bassett T, Sparanese D, Lee CH. Destroying RNA as a therapeutic approach. *Curr Med Chem* 2006;13 (8):863-81.

[52] Haseloff J, Gerlach WL. Simple RNA enzymes with new and highly specific endoribonuclease activities. *Nature* 1988;334 (6183):585-91.

[53] Heidenreich O, Pieken W, Eckstein F. Chemically modified RNA: approaches and applications. *Faseb J* 1993;7 (1):90-6.

[54] Paolella G, Sproat BS, Lamond AI. Nuclease resistant ribozymes with high catalytic activity. *Embo J* 1992;11 (5):1913-9.

[55] Usman N, Blatt LM. Nuclease-resistant synthetic ribozymes: developing a new class of therapeutics. *J Clin Invest* 2000;106 (10):1197-202.

[56] Amado RG, Mitsuyasu RT, Symonds G, Rosenblatt JD, Zack J, Sun LQ, Miller M, Ely J, Gerlach W. A phase I trial of autologous CD34+ hematopoietic progenitor cells transduced with an anti-HIV ribozyme. *Hum Gene Ther* 1999;10 (13):2255-70.

[57] Sandberg JA, Parker VP, Blanchard KS, Sweedler D, Powell JA, Kachensky A, Bellon L, Usman N, Rossing T, Borden E, Blatt LM. Pharmacokinetics and tolerability of an antiangiogenic ribozyme (ANGIOZYME) in healthy volunteers. *J Clin Pharmacol* 2000;40 (12 Pt 2):1462-9.

[58] Welch PJ, Barber JR, Wong-Staal F. Expression of ribozymes in gene transfer systems to modulate target RNA levels. *Curr Opin Biotechnol* 1998;9 (5):486-96.

[59] Wong-Staal F, Poeschla EM, Looney DJ. A controlled, Phase 1 clinical trial to evaluate the safety and effects in HIV-1 infected humans of autologous lymphocytes transduced with a ribozyme that cleaves HIV-1 RNA. *Hum Gene Ther* 1998;9 (16):2407-25.

[60] Elbashir SM, Harborth J, Lendeckel W, Yalcin A, Weber K, Tuschl T. Duplexes of 21-nucleotide RNAs mediate RNA interference in cultured mammalian cells. *Nature* 2001;411 (6836):494-8.

[61] Hammond SM, Caudy AA, Hannon GJ. Post-transcriptional gene silencing by double-stranded RNA. *Nat Rev Genet* 2001;2 (2):110-9.

[62] Hannon GJ. RNA interference. *Nature* 2002;418 (6894):244-51.

[63] Paddison PJ, Hannon GJ. siRNAs and shRNAs: skeleton keys to the human genome. *Curr Opin Mol Ther* 2003;5 (3):217-24.

[64] Banan M, Puri N. The ins and outs of RNAi in mammalian cells. *Curr Pharm Biotechnol* 2004;5 (5):441-50.

[65] Abbas-Terki T, Blanco-Bose W, Deglon N, Pralong W, Aebischer P. Lentiviral-mediated RNA interference. *Hum Gene Ther* 2002;13 (18):2197-201.

[66] Lewis DL, Hagstrom JE, Loomis AG, Wolff JA, Herweijer H. Efficient delivery of siRNA for inhibition of gene expression in postnatal mice. *Nat Genet* 2002;32 (1):107-8.

[67] McCaffrey AP, Meuse L, Pham TT, Conklin DS, Hannon GJ, Kay MA. RNA interference in adult mice. *Nature* 2002;418 (6893):38-9.

[68] Paddison PJ, Caudy AA, Bernstein E, Hannon GJ, Conklin DS. Short hairpin RNAs (shRNAs) induce sequence-specific silencing in mammalian cells. *Genes Dev* 2002;16 (8):948-58.

[69] Shen C, Buck AK, Liu X, Winkler M, Reske SN. Gene silencing by adenovirus-delivered siRNA. *FEBS Lett* 2003;539 (1-3):111-4.

[70] Zhao LJ, Jian H, Zhu H. Specific gene inhibition by adenovirus-mediated expression of small interfering RNA. *Gene* 2003;316:137-41.

[71] Ill CR, Chiou HC. Gene therapy progress and prospects: recent progress in transgene and RNAi expression cassettes. *Gene Ther* 2005;12 (10):795-802.

[72] Pai SI, Lin YY, Macaes B, Meneshian A, Hung CF, Wu TC. Prospects of RNA interference therapy for cancer. *Gene Ther* 2006;13 (6):464-77.

[73] Couzin J. Molecular biology. RNAi safety comes under scrutiny. *Science* 2006;312 (5777):1121.

[74] Grimm D, Streetz KL, Jopling CL, Storm TA, Pandey K, Davis CR, Marion P, Salazar F, Kay MA. Fatality in mice due to oversaturation of cellular microRNA/short hairpin RNA pathways. *Nature* 2006;441 (7092):537-41.

[75] Ariga T. Gene therapy for primary immunodeficiency diseases: recent progress and misgivings. *Curr Pharm Des* 2006;12 (5):549-56.

[76] Check E. A tragic setback. *Nature* 2002;420 (6912):116-8.

[77] Check E. Cancer fears cast doubts on future of gene therapy. *Nature* 2003;421 (6924):678.

[78] Kaiser J. Gene therapy. Seeking the cause of induced leukemias in X-SCID trial. *Science* 2003;299 (5606):495.

[79] Gilmore IR, Fox SP, Hollins AJ, Akhtar S. Delivery strategies for siRNA-mediated gene silencing. *Curr Drug Deliv* 2006;3 (2):147-5.

[80] Greco O, Scott SD, Marples B, Dachs GU. Cancer gene therapy: 'delivery, delivery, delivery'. *Front Biosci* 2002;7:d1516-24.

[81] Sioud M. On the delivery of small interfering RNAs into mammalian cells. *Expert Opin Drug Deliv* 2005;2 (4):639-51.

[82] Breyer B, Jiang W, Cheng H, Zhou L, Paul R, Feng T, He TC. Adenoviral vector-mediated gene transfer for human gene therapy. *Curr Gene Ther* 2001;1 (2):149-62.

[83] Lai CM, Lai YK, Rakoczy PE. Adenovirus and adeno-associated virus vectors. *DNA Cell Biol* 2002;21 (12):895-913.

[84] Vorburger SA, Hunt KK. Adenoviral gene therapy. *Oncologist* 2002;7 (1):46-59.

[85] Ghosh SS, Gopinath P, Ramesh A. Adenoviral vectors: a promising tool for gene therapy. *Appl Biochem Biotechnol* 2006;133 (1):9-29.

[86] Kaplan JM. Adenovirus-based cancer gene therapy. *Curr Gene Ther* 2005;5 (6):595-605.

[87] Barnett BG, Crews CJ, Douglas JT. Targeted adenoviral vectors. *Biochim Biophys Acta* 2002;1575 (1-3):1-14.

[88] Hemminki A, Kanerva A, Liu B, Wang M, Alvarez RD, Siegal GP, Curiel DT. Modulation of coxsackie-adenovirus receptor expression for increased adenoviral transgene expression. *Cancer Res* 2003;63 (4):847-53.

[89] Krasnykh V, Dmitriev I, Navarro JG, Belousova N, Kashentseva E, Xiang J, Douglas JT, Curiel DT. Advanced generation adenoviral vectors possess augmented gene transfer efficiency based upon coxsackie adenovirus receptor-independent cellular entry capacity. *Cancer Res* 2000;60 (24):6784-7.

[90] DeWeese TL, van der Poel H, Li S, Mikhak B, Drew R, Goemann M, Hamper U, DeJong R, Detorie N, Rodriguez R, Haulk T, DeMarzo AM, Piantadosi S, Yu DC, Chen Y, Henderson DR, Carducci MA, Nelson WG, Simons JW. A phase I trial of CV706, a replication-competent, PSA selective oncolytic adenovirus, for the treatment of locally recurrent prostate cancer following radiation therapy. *Cancer Res* 2001;61 (20):7464-72.

[91] Johnson L, Shen A, Boyle L, Kunich J, Pandey K, Lemmon M, Hermiston T, Giedlin M, McCormick F, Fattaey A. Selectively replicating adenoviruses targeting deregulated E2F activity are potent, systemic antitumor agents. *Cancer Cell* 2002;1 (4):325-37.

[92] Kubo H, Gardner TA, Wada Y, Koeneman KS, Gotoh A, Yang L, Kao C, Lim SD, Amin MB, Yang H, Black ME, Matsubara S, Nakagawa M, Gillenwater JY, Zhau HE, Chung LW. Phase I dose escalation clinical trial of adenovirus vector carrying osteocalcin promoter-driven herpes simplex virus thymidine kinase in localized and metastatic hormone-refractory prostate cancer. *Hum Gene Ther* 2003;14 (3):227-41.

[93] Li Y, Yu DC, Chen Y, Amin P, Zhang H, Nguyen N, Henderson DR. A hepatocellular carcinoma-specific adenovirus variant, CV890, eliminates distant human liver tumors in combination with doxorubicin. *Cancer Res* 2001;61 (17):6428-36.

[94] Greber UF, Willetts M, Webster P, Helenius A. Stepwise dismantling of adenovirus 2 during entry into cells. *Cell* 1993;75 (3):477-86.

[95] Bischoff JR, Kirn DH, Williams A, Heise C, Horn S, Muna M, Ng L, Nye JA, Sampson-Johannes A, Fattaey A, McCormick F. An adenovirus mutant that replicates selectively in p53-deficient human tumor cells. *Science* 1996;274 (5286):373-6.

[96] Freytag SO, Khil M, Stricker H, Peabody J, Menon M, DePeralta-Venturina M, Nafziger D, Pegg J, Paielli D, Brown S, Barton K, Lu M, Aguilar-Cordova E, Kim JH. Phase I study of replication-competent adenovirus-mediated double suicide gene therapy for the treatment of locally recurrent prostate cancer. *Cancer Res* 2002;62 (17):4968-76.

[97] Richardson WD, Westphal H. A cascade of adenovirus early functions is required for expression of adeno-associated virus. *Cell* 1981;27 (1 Pt 2):133-41.

[98] Kotin RM, Siniscalco M, Samulski RJ, Zhu XD, Hunter L, Laughlin CA, McLaughlin S, Muzyczka N, Rocchi M, Berns KI. Site-specific integration by adeno-associated virus. *Proc Natl Acad Sci U S A* 1990;87 (6):2211-5.

[99] Conrad CK, Allen SS, Afione SA, Reynolds TC, Beck SE, Fee-Maki M, Barrazza-Ortiz X, Adams R, Askin FB, Carter BJ, Guggino WB, Flotte TR. Safety of single-dose administration of an adeno-associated virus (AAV)- CFTR vector in the primate lung. *Gene Ther* 1996;3 (8):658-68.

[100] Kirik D, Georgievska B, Burger C, Winkler C, Muzyczka N, Mandel RJ, Bjorklund A. Reversal of motor impairments in parkinsonian rats by continuous intrastriatal delivery of L-dopa using rAAV-mediated gene transfer. *Proc Natl Acad Sci U S A* 2002;99 (7):4708-13.

[101] Snyder RO, Miao C, Meuse L, Tubb J, Donahue BA, Lin HF, Stafford DW, Patel S, Thompson AR, Nichols T, Read MS, Bellinger DA, Brinkhous KM, Kay MA. Correction of hemophilia B in canine and murine models using recombinant adeno-associated viral vectors. *Nat Med* 1999;5 (1):64-70.

[102] Davidoff AM, Nathwani AC, Spurbeck WW, Ng CY, Zhou J, Vanin EF. rAAV-mediated long-term liver-generated expression of an angiogenesis inhibitor can restrict renal tumor growth in mice. *Cancer Res* 2002;62 (11):3077-83.

[103] Kanazawa T, Mizukami H, Okada T, Hanazono Y, Kume A, Nishino H, Takeuchi K, Kitamura K, Ichimura K, Ozawa K. Suicide gene therapy using AAV-HSVtk/ganciclovir in combination with irradiation results in regression of human head and neck cancer xenografts in nude mice. *Gene Ther* 2003;10 (1):51-8.

[104] Kunke D, Grimm D, Denger S, Kreuzer J, Delius H, Komitowski D, Kleinschmidt JA. Preclinical study on gene therapy of cervical carcinoma using adeno- associated virus vectors. *Cancer Gene Ther* 2000;7 (5):766-77.

[105] Raj K, Ogston P, Beard P. Virus-mediated killing of cells that lack p53 activity. *Nature* 2001;412 (6850):914-7.

[106] Miller DG, Adam MA, Miller AD. Gene transfer by retrovirus vectors occurs only in cells that are actively replicating at the time of infection. *Mol Cell Biol* 1990;10 (8):4239-42.

[107] Blaese RM, Culver KW, Miller AD, Carter CS, Fleisher T, Clerici M, Shearer G, Chang L, Chiang Y, Tolstoshev P, et al. T lymphocyte-directed gene therapy for ADA-SCID: initial trial results after 4 years. *Science* 1995;270 (5235):475-80.

[108] Hock RA, Miller AD, Osborne WR. Expression of human adenosine deaminase from various strong promoters after gene transfer into human hematopoietic cell lines. *Blood* 1989;74 (2):876-81.

[109] Muul LM, Tuschong LM, Soenen SL, Jagadeesh GJ, Ramsey WJ, Long Z, Carter CS, Garabedian EK, Alleyne M, Brown M, Bernstein W, Schurman SH, Fleisher TA, Leitman SF, Dunbar CE, Blaese RM, Candotti F. Persistence and expression of the adenosine deaminase gene for 12 years and immune reaction to gene transfer components: long-term results of the first clinical gene therapy trial. *Blood* 2003;101 (7):2563-9.

[110] Quinonez R, Sutton RE. Lentiviral vectors for gene delivery into cells. *DNA Cell Biol* 2002;21 (12):937-51.

[111] Loimas S, Toppinen MR, Visakorpi T, Janne J, Wahlfors J. Human prostate carcinoma cells as targets for herpes simplex virus thymidine kinase-mediated suicide gene therapy. *Cancer Gene Ther* 2001;8 (2):137-44.

[112] Yu D, Chen D, Chiu C, Razmazma B, Chow YH, Pang S. Prostate-specific targeting using PSA promoter-based lentiviral vectors. *Cancer Gene Ther* 2001;8 (9):628-35.

[113] Rajcani J, Vojvodova A. The role of herpes simplex virus glycoproteins in the virus replication cycle. *Acta Virol* 1998;42 (2):103-18.

[114] Taylor TJ, Brockman MA, McNamee EE, Knipe DM. Herpes simplex virus. *Front Biosci* 2002;7:d752-64.

[115] Burton EA, Fink DJ, Glorioso JC. Gene delivery using herpes simplex virus vectors. *DNA Cell Biol* 2002;21 (12):915-36.

[116] Lachmann R. Herpes simplex virus-based vectors. *Int J Exp Pathol* 2004;85 (4):177-90.

[117] Burton EA, Bai Q, Goins WF, Glorioso JC. Replication-defective genomic herpes simplex vectors: design and production. *Curr Opin Biotechnol* 2002;13 (5):424-8.

[118] Marples B, Greco O, Joiner MC, Scott SD. Molecular approaches to chemo-radiotherapy. *Eur J Cancer* 2002;38 (2):231-9.
[119] Wybranietz WA, Gross CD, Phelan A, O'Hare P, Spiegel M, Graepler F, Bitzer M, Stahler P, Gregor M, Lauer UM. Enhanced suicide gene effect by adenoviral transduction of a VP22- cytosine deaminase (CD) fusion gene. *Gene Ther* 2001;8 (21):1654-64.
[120] Peplinski GR, Tsung K, Norton JA. Vaccinia virus for human gene therapy. *Surg Oncol Clin N Am* 1998;7 (3):575-88.
[121] Guo ZS, Bartlett DL. Vaccinia as a vector for gene delivery. *Expert Opin Biol Ther* 2004;4 (6):901-17.
[122] Hodge JW, Abrams S, Schlom J, Kantor JA. Induction of antitumor immunity by recombinant vaccinia viruses expressing B7-1 or B7-2 costimulatory molecules. *Cancer Res* 1994;54 (21):5552-5.
[123] Zaloudik J, Li W, Jacob L, Kieny MP, Somasundaram R, Acres B, Song H, Zhang T, Li J, Herlyn D. Inhibition of tumor growth by recombinant vaccinia virus expressing GA733/CO17-1A/EpCAM/KSA/KS1-4 antigen in mice. *Cancer Gene Ther* 2002;9 (4):382-9.
[124] McCart JA, Ward JM, Lee J, Hu Y, Alexander HR, Libutti SK, Moss B, Bartlett DL. Systemic cancer therapy with a tumor-selective vaccinia virus mutant lacking thymidine kinase and vaccinia growth factor genes. *Cancer Res* 2001;61 (24):8751-7.
[125] Smyth Templeton N. Liposomal delivery of nucleic acids in vivo. *DNA Cell Biol* 2002;21 (12):857-67.
[126] Chen WC, Huang L. Non-viral vector as vaccine carrier. *Adv Genet* 2005;54:315-37.
[127] Hope MJ, Mui B, Ansell S, Ahkong QF. Cationic lipids, phosphatidylethanolamine and the intracellular delivery of polymeric, nucleic acid-based drugs (review). *Mol Membr Biol* 1998;15 (1):1-14.
[128] Lasic DD, Martin FJ, Gabizon A, Huang SK, Papahadjopoulos D. Sterically stabilized liposomes: a hypothesis on the molecular origin of the extended circulation times. *Biochim Biophys Acta* 1991;1070 (1):187-92.
[129] Brown JM, Giaccia AJ. The unique physiology of solid tumors: opportunities (and problems) for cancer therapy. *Cancer Res* 1998;58 (7):1408-16.
[130] Wu NZ, Da D, Rudoll TL, Needham D, Whorton AR, Dewhirst MW. Increased microvascular permeability contributes to preferential accumulation of Stealth liposomes in tumor tissue. *Cancer Res* 1993;53 (16):3765-70.
[131] Kim DW, Andres ML, Kajioka EH, Dutta-Roy R, Miller GM, Seynhaeve AL, ten Hagen TL, Gridley DS. Modulation of innate immunological factors by STEALTH liposome- encapsulated tumor necrosis factor-alpha in a colon tumor xenograft model. *Anticancer Res* 2002;22 (2A):777-88.
[132] Kim DW, Andres ML, Li J, Kajioka EH, Miller GM, Seynhaeve AL, Ten Hagen TL, Gridley DS. Liposome-encapsulated tumor necrosis factor-alpha enhances the effects of radiation against human colon tumor xenografts. *J Interferon Cytokine Res* 2001;21 (11):885-97.
[133] ten Hagen TL, Seynhaeve AL, van Tiel ST, Ruiter DJ, Eggermont AM. Pegylated liposomal tumor necrosis factor-alpha results in reduced toxicity and synergistic

antitumor activity after systemic administration in combination with liposomal doxorubicin (Doxil) in soft tissue sarcoma-bearing rats. *Int J Cancer* 2002;97 (1):115-20.

[134] Hudson AJ, Normand N, Ackroyd J, Akhtar S. Cellular delivery of hammerhead ribozymes conjugated to a transferrin receptor antibody. *Int J Pharm* 1999;182 (1):49-58.

[135] Penichet ML, Kang YS, Pardridge WM, Morrison SL, Shin SU. An antibody-avidin fusion protein specific for the transferrin receptor serves as a delivery vehicle for effective brain targeting: initial applications in anti-HIV antisense drug delivery to the brain. *J Immunol* 1999;163 (8):4421-6.

[136] Gratton JP, Yu J, Griffith JW, Babbitt RW, Scotland RS, Hickey R, Giordano FJ, Sessa WC. Cell-permeable peptides improve cellular uptake and therapeutic gene delivery of replication-deficient viruses in cells and in vivo. *Nat Med* 2003;9 (3):357-63.

[137] Derossi D, Chassaing G, Prochiantz A. Trojan peptides: the penetratin system for intracellular delivery. *Trends Cell Biol* 1998;8 (2):84-7.

[138] Trezise AE. In vivo DNA electrotransfer. *DNA Cell Biol* 2002;21 (12):869-77.

[139] Herweijer H, Wolff JA. Progress and prospects: naked DNA gene transfer and therapy. *Gene Ther* 2003;10 (6):453-8.

[140] Mincheff M, Altankova I, Zoubak S, Tchakarov S, Botev C, Petrov S, Krusteva E, Kurteva G, Kurtev P, Dimitrov V, Ilieva M, Georgiev G, Lissitchkov T, Chernozemski I, Meryman HT. In vivo transfection and/or cross-priming of dendritic cells following DNA and adenoviral immunizations for immunotherapy of cancer--changes in peripheral mononuclear subsets and intracellular IL-4 and IFN-gamma lymphokine profile. *Crit Rev Oncol Hematol* 2001;39 (1-2):125-32.

[141] Mincheff M, Tchakarov S, Zoubak S, Loukinov D, Botev C, Altankova I, Georgiev G, Petrov S, Meryman HT. Naked DNA and adenoviral immunizations for immunotherapy of prostate cancer: a phase I/II clinical trial. *Eur Urol* 2000;38 (2):208-17.

[142] Carson CT, Schwartz RA, Stracker TH, Lilley CE, Lee DV, Weitzman MD. The Mre11 complex is required for ATM activation and the G(2)/M checkpoint. *Embo J* 2003;22 (24):6610-20.

[143] Lee JH, Paull TT. Direct activation of the ATM protein kinase by the Mre11/Rad50/Nbs1 complex. *Science* 2004;304 (5667):93-6.

[144] Lee JH, Paull TT. ATM activation by DNA double-strand breaks through the Mre11-Rad50-Nbs1 complex. *Science* 2005;308 (5721):551-4.

[145] Paull TT, Lee JH. The Mre11/Rad50/Nbs1 complex and its role as a DNA double-strand break sensor for ATM. *Cell Cycle* 2005;4 (6):737-40.

[146] Stracker TH, Theunissen JW, Morales M, Petrini JH. The Mre11 complex and the metabolism of chromosome breaks: the importance of communicating and holding things together. *DNA Repair (Amst)* 2004;3 (8-9):845-54.

[147] Kim YC, Koh JT, Shin BA, Ahn KY, Choi BK, Kim CG, Kim KK. An antisense construct of full-length human RAD50 cDNA confers sensitivity to ionizing radiation and alkylating agents on human cell lines. *Radiat Res* 2002;157 (1):19-25.

[148] O'Malley BW, Jr., Li D, Carney J, Rhee J, Suntharalingam M. Molecular disruption of the MRN(95) complex induces radiation sensitivity in head and neck cancer. *Laryngoscope* 2003;113 (9):1588-94.

[149] Xu M, Myerson R, Hunt C, Kumar S, Moros E, Straube B, Roti JR. Treatment of cells with Mre11 siRNA increases radiation sensitivity and reduces heat induced radiosensitization. *Int J Radiat Oncol Biol Phys* 2003;57 (2 Suppl):S144-5.

[150] Bakkenist CJ, Kastan MB. DNA damage activates ATM through intermolecular autophosphorylation and dimer dissociation. *Nature* 2003;421 (6922):499-506.

[151] Shiloh Y. The ATM-mediated DNA-damage response: taking shape. *Trends Biochem Sci* 2006.

[152] Fan Z, Chakravarty P, Alfieri A, Pandita TK, Vikram B, Guha C. Adenovirus-mediated antisense ATM gene transfer sensitizes prostate cancer cells to radiation. *Cancer Gene Ther* 2000;7 (10):1307-14.

[153] Guha C, Guha U, Tribius S, Alfieri A, Casper D, Chakravarty P, Mellado W, Pandita TK, Vikram B. Antisense ATM gene therapy: a strategy to increase the radiosensitivity of human tumors. *Gene Ther* 2000;7 (10):852-8.

[154] Uhrhammer N, Fritz E, Boyden L, Meyn MS. Human fibroblasts transfected with an ATM antisense vector respond abnormally to ionizing radiation. *Int J Mol Med* 1999;4 (1):43-7.

[155] Zhang N, Chen P, Gatei M, Scott S, Khanna KK, Lavin MF. An anti-sense construct of full-length ATM cDNA imposes a radiosensitive phenotype on normal cells. *Oncogene* 1998;17 (7):811-8.

[156] Collis SJ, Swartz MJ, Nelson WG, DeWeese TL. Enhanced radiation and chemotherapy-mediated cell killing of human cancer cells by small inhibitory RNA silencing of DNA repair factors. *Cancer Res* 2003;63 (7):1550-4.

[157] Mukhopadhyay UK, Senderowicz AM, Ferbeyre G. RNA silencing of checkpoint regulators sensitizes p53-defective prostate cancer cells to chemotherapy while sparing normal cells. *Cancer Res* 2005;65 (7):2872-81.

[158] Chen S, Paul P, Price BD. ATM's leucine-rich domain and adjacent sequences are essential for ATM to regulate the DNA damage response. *Oncogene* 2003;22 (41):6332-9.

[159] Morgan SE, Lovly C, Pandita TK, Shiloh Y, Kastan MB. Fragments of ATM which have dominant-negative or complementing activity. *Mol Cell Biol* 1997;17 (4):2020-9.

[160] Hickson I, Zhao Y, Richardson CJ, Green SJ, Martin NM, Orr AI, Reaper PM, Jackson SP, Curtin NJ, Smith GC. Identification and characterization of a novel and specific inhibitor of the ataxia-telangiectasia mutated kinase ATM. *Cancer Res* 2004;64 (24):9152-9.

[161] Goldberg M, Stucki M, Falck J, D'Amours D, Rahman D, Pappin D, Bartek J, Jackson SP. MDC1 is required for the intra-S-phase DNA damage checkpoint. *Nature* 2003;421 (6926):952-6.

[162] Kao GD, McKenna WG, Guenther MG, Muschel RJ, Lazar MA, Yen TJ. Histone deacetylase 4 interacts with 53BP1 to mediate the DNA damage response. *J Cell Biol* 2003;160 (7):1017-27.

[163] Lou Z, Chini CC, Minter-Dykhouse K, Chen J. Mediator of DNA damage checkpoint protein 1 regulates BRCA1 localization and phosphorylation in DNA damage checkpoint control. *J Biol Chem* 2003;278 (16):13599-602.

[164] Lou Z, Minter-Dykhouse K, Wu X, Chen J. MDC1 is coupled to activated CHK2 in mammalian DNA damage response pathways. *Nature* 2003;421 (6926):957-61.

[165] Qi W, Martinez JD. Reduction of 14-3-3 proteins correlates with increased sensitivity to killing of human lung cancer cells by ionizing radiation. *Radiat Res* 2003;160 (2):217-23.

[166] Stewart GS, Wang B, Bignell CR, Taylor AM, Elledge SJ. MDC1 is a mediator of the mammalian DNA damage checkpoint. *Nature* 2003;421 (6926):961-6.

[167] Taneja N, Davis M, Choy JS, Beckett MA, Singh R, Kron SJ, Weichselbaum RR. Histone H2AX phosphorylation as a predictor of radiosensitivity and target for radiotherapy. *J Biol Chem* 2003;15:15.

[168] Yu Q, Rose JH, Zhang H, Pommier Y. Antisense inhibition of Chk2/hCds1 expression attenuates DNA damage-induced S and G2 checkpoints and enhances apoptotic activity in HEK-293 cells. *FEBS Lett* 2001;505 (1):7-12.

[169] Zhao H, Watkins JL, Piwnica-Worms H. Disruption of the checkpoint kinase 1/cell division cycle 25A pathway abrogates ionizing radiation-induced S and G2 checkpoints. *Proc Natl Acad Sci U S A* 2002;99 (23):14795-800.

[170] Collis SJ, DeWeese TL, Jeggo PA, Parker AR. The life and death of DNA-PK. *Oncogene* 2005;24 (6):949-61.

[171] Marangoni E, Le Romancer M, Foray N, Muller C, Douc-Rasy S, Vaganay S, Abdulkarim B, Barrois M, Calsou P, Bernier J, Salles B, Bourhis J. Transfer of Ku86 RNA antisense decreases the radioresistance of human fibroblasts. *Cancer Gene Ther* 2000;7 (2):339-46.

[172] Belenkov AI, Paiement JP, Panasci LC, Monia BP, Chow TY. An antisense oligonucleotide targeted to human Ku86 messenger RNA sensitizes M059K malignant glioma cells to ionizing radiation, bleomycin, and etoposide but not DNA cross-linking agents. *Cancer Res* 2002;62 (20):5888-96.

[173] Li GC, He F, Shao X, Urano M, Shen L, Kim D, Borrelli M, Leibel SA, Gutin PH, Ling CC. Adenovirus-mediated heat-activated antisense Ku70 expression radiosensitizes tumor cells in vitro and in vivo. *Cancer Res* 2003;63 (12):3268-74.

[174] Omori S, Takiguchi Y, Suda A, Sugimoto T, Miyazawa H, Tanabe N, Tatsumi K, Kimura H, Pardington PE, Chen F, Chen DJ, Kuriyama T. Suppression of a DNA double-strand break repair gene, Ku70, increases radio- and chemosensitivity in a human lung carcinoma cell line. *DNA Repair (Amst)* 2002;1 (4):299-310.

[175] Sak A, Stuschke M, Wurm R, Schroeder G, Sinn B, Wolf G, Budach V. Selective inactivation of DNA-dependent protein kinase with antisense oligodeoxynucleotides: consequences for the rejoining of radiation-induced DNA double-strand breaks and radiosensitivity of human cancer cell lines. *Cancer Res* 2002;62 (22):6621-4.

[176] Stein RC. Prospects for phosphoinositide 3-kinase inhibition as a cancer treatment. *Endocr Relat Cancer* 2001;8 (3):237-48.

[177] Peng Y, Zhang Q, Nagasawa H, Okayasu R, Liber HL, Bedford JS. Silencing expression of the catalytic subunit of DNA-dependent protein kinase by small

interfering RNA sensitizes human cells for radiation-induced chromosome damage, cell killing, and mutation. *Cancer Res* 2002;62 (22):6400-4.

[178] An J, Xu QZ, Sui JL, Bai B, Zhou PK. Downregulation of c-myc protein by siRNA-mediated silencing of DNA-PKcs in HeLa cells. *Int J Cancer* 2005;117 (4):531-7.

[179] Kim CH, Park SJ, Lee SH. A targeted inhibition of DNA-dependent protein kinase sensitizes breast cancer cells following ionizing radiation. *J Pharmacol Exp Ther* 2002;303 (2):753-9.

[180] Li S, Takeda Y, Wragg S, Barrett J, Phillips A, Dynan WS. Modification of the ionizing radiation response in living cells by an scFv against the DNA-dependent protein kinase. *Nucleic Acids Res* 2003;31 (20):5848-57.

[181] Marangoni E, Foray N, O'Driscoll M, Douc-Rasy S, Bernier J, Bourhis J, Jeggo P. A Ku80 fragment with dominant negative activity imparts a radiosensitive phenotype to CHO-K1 cells. *Nucleic Acids Res* 2000;28 (23):4778-82.

[182] Lees-Miller SP, Long MC, Kilvert MA, Lam V, Rice SA, Spencer CA. Attenuation of DNA-dependent protein kinase activity and its catalytic subunit by the herpes simplex virus type 1 transactivator ICP0. *J Virol* 1996;70 (11):7471-7.

[183] Parkinson J, Lees-Miller SP, Everett RD. Herpes simplex virus type 1 immediate-early protein vmw110 induces the proteasome-dependent degradation of the catalytic subunit of DNA-dependent protein kinase. *J Virol* 1999;73 (1):650-7.

[184] Jorgensen TJ, Katz S, Wittmack EK, Varghese S, Todo T, Rabkin SD, Martuza RL. Ionizing radiation does not alter the antitumor activity of herpes simplex virus vector G207 in subcutaneous tumor models of human and murine prostate cancer. *Neoplasia* 2001;3 (5):451-6.

[185] Wilkinson DE, Weller SK. Herpes simplex virus type I disrupts the ATR-dependent DNA-damage response during lytic infection. *J Cell Sci* 2006;119 (Pt 13):2695-703.

[186] Boyer J, Rohleder K, Ketner G. Adenovirus E4 34k and E4 11k inhibit double strand break repair and are physically associated with the cellular DNA-dependent protein kinase. *Virology* 1999;263 (2):307-12.

[187] Mohammadi ES, Ketner EA, Johns DC, Ketner G. Expression of the adenovirus E4 34k oncoprotein inhibits repair of double strand breaks in the cellular genome of a 293-based inducible cell line. *Nucleic Acids Res* 2004;32 (8):2652-9.

[188] Collis SJ, Ketner GW, Hicks JL, Nelson WG, Demarzo AM, Deweese TL. Expression of the DNA-PK binding protein E4-34K fails to confer radiation sensitivity to mammalian cells. *Int J Radiat Biol* 2003;79 (1):53-60.

[189] Stracker TH, Carson CT, Weitzman MD. Adenovirus oncoproteins inactivate the Mre11-Rad50-NBS1 DNA repair complex. *Nature* 2002;418 (6895):348-52.

[190] Chen Y, DeWeese T, Dilley J, Zhang Y, Li Y, Ramesh N, Lee J, Pennathur-Das R, Radzyminski J, Wypych J, Brignetti D, Scott S, Stephens J, Karpf DB, Henderson DR, Yu DC. CV706, a prostate cancer-specific adenovirus variant, in combination with radiotherapy produces synergistic antitumor efficacy without increasing toxicity. *Cancer Res* 2001;61 (14):5453-60.

[191] Graham KL, Gustin KE, Rivera C, Kuyumcu-Martinez NM, Choe SS, Lloyd RE, Sarnow P, Utz PJ. Proteolytic cleavage of the catalytic subunit of DNA-dependent protein kinase during poliovirus infection. *J Virol* 2004;78 (12):6313-21.

[192] Skalka AM, Katz RA. Retroviral DNA integration and the DNA damage response. *Cell Death Differ* 2005;12 Suppl 1:971-8.

[193] Beamish HJ, Jessberger R, Riballo E, Priestley A, Blunt T, Kysela B, Jeggo PA. The C-terminal conserved domain of DNA-PKcs, missing in the SCID mouse, is required for kinase activity. *Nucleic Acids R*es 2000;28 (7):1506-13.

[194] Kienker LJ, Shin EK, Meek K. Both V(D)J recombination and radioresistance require DNA-PK kinase activity, though minimal levels suffice for V(D)J recombination. *Nucleic Acids Res* 2000;28 (14):2752-61.

[195] Kurimasa A, Kumano S, Boubnov NV, Story MD, Tung CS, Peterson SR, Chen DJ. Requirement for the kinase activity of human DNA-dependent protein kinase catalytic subunit in DNA strand break rejoining. *Mol Cell Biol* 1999;19 (5):3877-84.

[196] Taki T, Ohnishi T, Yamamoto A, Hiraga S, Arita N, Izumoto S, Hayakawa T, Morita T. Antisense inhibition of the RAD51 enhances radiosensitivity. *Biochem Biophys Res Commun* 1996;223 (2):434-8.

[197] Ohnishi T, Taki T, Hiraga S, Arita N, Morita T. In vitro and in vivo potentiation of radiosensitivity of malignant gliomas by antisense inhibition of the RAD51 gene. *Biochem Biophys Res Commun* 1998;245 (2):319-24.

[198] Collis SJ, Tighe A, Scott SD, Roberts SA, Hendry JH, Margison GP. Ribozyme minigene-mediated RAD51 down-regulation increases radiosensitivity of human prostate cancer cells. *Nucleic Acids Res* 2001;29 (7):1534-8.

[199] Ito M, Yamamoto S, Nimura K, Hiraoka K, Tamai K, Kaneda Y. Rad51 siRNA delivered by HVJ envelope vector enhances the anti-cancer effect of cisplatin. *J Gene Med* 2005;7 (8):1044-52.

[200] Watson G, Jensen R. Effect of hRAD51 dominant negative mutant expression in a glioblastoma model system: altered sensitivity to ionizing radiation. *Int J Radiat Oncol Biol Phys* 2003;57 (2 Suppl):S145.

[201] Russell JS, Brady K, Burgan WE, Cerra MA, Oswald KA, Camphausen K, Tofilon PJ. Gleevec-mediated inhibition of Rad51 expression and enhancement of tumor cell radiosensitivity. *Cancer Res* 2003;63 (21):7377-83.

[202] Potter PM, Harris LC, Remack JS, Edwards CC, Brent TP. Ribozyme-mediated modulation of human O6-methylguanine-DNA methyltransferase expression. *Cancer Res* 1993;53 (8):1731-4.

[203] Chirnomas D, Taniguchi T, de la Vega M, Vaidya AP, Vasserman M, Hartman AR, Kennedy R, Foster R, Mahoney J, Seiden MV, D'Andrea AD. Chemosensitization to cisplatin by inhibitors of the Fanconi anemia/BRCA pathway. *Mol Cancer Ther* 2006;5 (4):952-61.

[204] Kim MS, Blake M, Baek JH, Kohlhagen G, Pommier Y, Carrier F. Inhibition of histone deacetylase increases cytotoxicity to anticancer drugs targeting DNA. *Cancer Res* 2003;63 (21):7291-300.

[205] Zhang Y, Jung M, Dritschilo A, Jung M. Enhancement of radiation sensitivity of human squamous carcinoma cells by histone deacetylase inhibitors. *Radiat Res* 2004;161 (6):667-74.

[206] Camphausen K, Tofilon PJ. Combining radiation and molecular targeting in cancer therapy. *Cancer Biol Ther* 2004;3 (3):247-50.

[207] Chinnaiyan P, Allen GW, Harari PM. Radiation and new molecular agents, part II: targeting HDAC, HSP90, IGF-1R, PI3K, and Ras. *Semin Radiat Oncol* 2006;16 (1):59-64.

[208] Choudhury A, Cuddihy A, Bristow RG. Radiation and new molecular agents part I: targeting ATM-ATR checkpoints, DNA repair, and the proteasome. *Semin Radiat Oncol* 2006;16 (1):51-8.

[209] Brunner TB, Gupta AK, Shi Y, Hahn SM, Muschel RJ, McKenna WG, Bernhard EJ. Farnesyltransferase inhibitors as radiation sensitizers. *Int J Radiat Biol* 2003;79 (7):569-76.

[210] Kasid U, Dritschilo A. RAF antisense oligonucleotide as a tumor radiosensitizer. *Oncogene* 2003;22 (37):5876-84.

[211] Kasid U, Pfeifer A, Brennan T, Beckett M, Weichselbaum RR, Dritschilo A, Mark GE. Effect of antisense c-raf-1 on tumorigenicity and radiation sensitivity of a human squamous carcinoma. *Science* 1989;243 (4896):1354-6.

[212] Li Q, Zhu GD. Targeting serine/threonine protein kinase B/Akt and cell-cycle checkpoint kinases for treating cancer. *Curr Top Med Chem* 2002;2 (9):939-71.

[213] McKenna WG, Muschel RJ. Targeting tumor cells by enhancing radiation sensitivity. *Genes Chromosomes Cancer* 2003;38 (4):330-8.

[214] McKenna WG, Muschel RJ, Gupta AK, Hahn SM, Bernhard EJ. The RAS signal transduction pathway and its role in radiation sensitivity. *Oncogene* 2003;22 (37):5866-75.

[215] Shangary S, Johnson DE. Recent advances in the development of anticancer agents targeting cell death inhibitors in the Bcl-2 protein family. *Leukemia* 2003;17 (8):1470-81.

[216] Tenzer A, Pruschy M. Potentiation of DNA-damage-induced cytotoxicity by G2 checkpoint abrogators. *Curr Med Chem Anti-Canc Agents* 2003;3 (1):35-46.

[217] Tenzer A, Zingg D, Riesterer O, Vuong V, Bodis S, Pruschy M. Signal transduction inhibitors as radiosensitizers. *Curr Med Chem Anti-Canc Agents* 2002;2 (6):727-42.

[218] Zaffaroni N, Daidone MG. Survivin expression and resistance to anticancer treatments: perspectives for new therapeutic interventions. *Drug Resist Updat* 2002;5 (2):65-72.

[219] Zhou BB, Anderson HJ, Roberge M. Targeting DNA checkpoint kinases in cancer therapy. *Cancer Biol Ther* 2003;2 (4 Suppl 1):S16-22.

[220] Gokhale PC, McRae D, Monia BP, Bagg A, Rahman A, Dritschilo A, Kasid U. Antisense raf oligodeoxyribonucleotide is a radiosensitizer in vivo. *Antisense Nucleic Acid Drug Dev* 1999;9 (2):191-201.

[221] Moreira JN, Simoes S. Technology evaluation: LErafAON, NeoPharm. *Curr Opin Mol Ther* 2003;5 (5):547-52.

[222] Zellweger T, Chi K, Miyake H, Adomat H, Kiyama S, Skov K, Gleave ME. Enhanced radiation sensitivity in prostate cancer by inhibition of the cell survival protein clusterin. *Clin Cancer Res* 2002;8 (10):3276-84.

[223] Asanuma K, Kobayashi D, Furuya D, Tsuji N, Yagihashi A, Watanabe N. A role for survivin in radioresistance of pancreatic cancer cells. *Jpn J Cancer Res* 2002;93 (9):1057-62.

[224] Asanuma K, Moriai R, Yajima T, Yagihashi A, Yamada M, Kobayashi D, Watanabe N. Survivin as a radioresistance factor in pancreatic cancer. *Jpn J Cancer Res* 2000;91 (11):1204-9.

[225] Rodel C, Haas J, Groth A, Grabenbauer GG, Sauer R, Rodel F. Spontaneous and radiation-induced apoptosis in colorectal carcinoma cells with different intrinsic radiosensitivities: survivin as a radioresistance factor. *Int J Radiat Oncol Biol Phys* 2003;55 (5):1341-7.

[226] Sarela AI, Verbeke CS, Ramsdale J, Davies CL, Markham AF, Guillou PJ. Expression of survivin, a novel inhibitor of apoptosis and cell cycle regulatory protein, in pancreatic adenocarcinoma. *Br J Cancer* 2002;86 (6):886-92.

[227] Scott SL, Higdon R, Beckett L, Shi XB, deVere White RW, Earle JD, Gumerlock PH. BCL2 antisense reduces prostate cancer cell survival following irradiation. *Cancer Biother Radiopharm* 2002;17 (6):647-56.

[228] Pennati M, Binda M, Colella G, Folini M, Citti L, Villa R, Daidone MG, Zaffaroni N. Radiosensitization of human melanoma cells by ribozyme-mediated inhibition of survivin expression. *J Invest Dermatol* 2003;120 (4):648-54.

[229] Greenberger JS, Epperly MW, Gretton J, Jefferson M, Nie S, Bernarding M, Kagan V, Guo HL. Radioprotective gene therapy. *Curr Gene Ther* 2003;3 (3):183-95.

[230] Kuninaka S, Ichinose Y, Koja K, Toh Y. Suppression of manganese superoxide dismutase augments sensitivity to radiation, hyperthermia and doxorubicin in colon cancer cell lines by inducing apoptosis. *Br J Cancer* 2000;83 (7):928-34.

[231] Ueta E, Yoneda K, Kimura T, Tatemoto Y, Doi S, Yamamoto T, Osaki T. Mn-SOD antisense upregulates in vivo apoptosis of squamous cell carcinoma cells by anticancer drugs and gamma-rays regulating expression of the BCL-2 family proteins, COX-2 and p21. *Int J Cancer* 2001;94 (4):545-50.

[232] Choy H, Milas L. Enhancing radiotherapy with cyclooxygenase-2 enzyme inhibitors: a rational advance? *J Natl Cancer Inst* 2003;95 (19):1440-52.

[233] Krysan K, Merchant FH, Zhu L, Dohadwala M, Luo J, Lin Y, Heuze-Vourc'h N, Pold M, Seligson D, Chia D, Goodglick L, Wang H, Strieter R, Sharma S, Dubinett S. COX-2-dependent stabilization of survivin in non-small cell lung cancer. *Faseb J* 2003;3:3.

[234] Collis SJ, Khater K, DeWeese TL. Novel therapeutic strategies in prostate cancer management using gene therapy in combination with radiation therapy. *World J Urol* 2003;21 (4):275-89.

[235] Collis SJ, DeWeese TL. Enhanced radiation response through directed molecular targeting approaches. *Cancer Metastasis Rev* 2004;23 (3-4):277-92.

In: New Research on DNA Repair
Editor: Breehn R. Landseer, pp. 39-84

ISBN 1-60021-385-5

Chapter II

DNA REPAIR AND LUNG CANCER

Chikako Kiyohara[1,*], Koichi Takayama[2] and Yoichi Nakanishi[2]

[1]Department of Preventive Medicine, Graduate School of Medical Sciences, Kyushu University, 3-1-1 Maidashi, Higashi-ku, Fukuoka 812-8582, Japan

[2]Research Institute for Diseases of the Chest, Graduate School of Medical Sciences, Kyushu University, 3-1-1 Maidashi, Higashi-ku, Fukuoka 812-8582, Japan

ABSTRACT

Lung cancer is a major cause of cancer-related death in the developed countries and the overall survival rate has still an extremely poor. Although cigarette smoking is the main cause of lung cancer, not all smokers develop lung cancer, and a fraction of lifelong nonsmokers will die from lung cancer. Genetic host factors have recently been implicated to account for some of the observed differences in lung cancer susceptibility. Various DNA alterations can be caused by exposure to environmental and endogenous carcinogens. Most of these alterations, if not repaired, may result in genetic instability, mutagenesis and cell death. DNA repair mechanisms are important for maintaining DNA integrity and preventing carcinogenesis. Recent genetic association studies on lung cancer risk have focused on identifying effects of single nucleotide polymorphisms (SNPs) in candidate genes, among which DNA repair genes are increasingly studied. Genetic variations in DNA repair genes are thought to modulate DNA repair capacity and are suggested to be related to lung cancer risk. The authors identified a sufficient number of epidemiologic studies on lung cancer to conduct meta-analyses for genetic polymorphisms in DNA repair genes. They found increased risk of lung cancer among subjects carrying excision repair cross complementing group 2 (*ERCC2/ xeroderma pigmentosum group D (XPD)*) 751Gln/Gln genotype (odds ratio (OR) = 1.30, 95% confidence interval (CI) = 1.14 - 1.49). They found a protective effect of *XPA* 23G/G genotype (OR = 0.75, 95% CI = 0.59 - 0.95). *X-ray cross-complementing group 1*

* Correspondence concerning this article should be addressed to Chikako Kiyohara, Department of Preventive Medicine, Graduate School of Medical Sciences, Kyushu University, 3-1-1 Maidashi, Higashi-ku, Fukuoka 812-8582, Japan. E-mail: chikako@phealth.med.kyushu-u.ac.jp.

(XRCC1) 399Gln/Gln genotype was associated with an increased risk of lung cancer among Asians (OR = 1.34, 95% CI = 1.16 - 1.54). Little evidence of associations has been found between other DNA repair genes and lung cancer risk. Considering the data available, it can be conjectured that if there is any risk association between single SNP, such as the *XRCC1* Arg399Gln, the *XPA* G23A or the *ERCC2/XPD* Lys751Gln, and lung cancer, this risk increase/decrease will probably be minimal. Advances in identification of new polymorphisms and in high-throughput genotyping techniques will facilitate analysis of multiple genes in multiple DNA repair pathways. Therefore, it is likely that the defining feature of future epidemiologic studies will be the simultaneous analysis of large samples of cases and controls.

Keywords: Lung cancer, base excision repair, nucleotide excision repair, double-strand DNA break repair, mismatch repair, genetic polymorphism, molecular epidemiology, meta-analysis

INTRODUCTION

Sporadic cancer is a multifactorial disease that results from complex interactions between many genetic and environmental factors [1]. This means that there will not be single gene or single environmental factor that has large effects on cancer susceptibility. Environmental factors (e.g. tobacco smoke, dietary factors, infectious agents and radiation) add to carcinogenic load that humans are exposed, but exact numbers are generally less well established.

Cancer is the result of a series of DNA alternations in single cell or clone of that cell, which lead to loss of normal function, aberrant or uncontrolled cell growth and often metastases. Several of the genes, which is frequently lost or mutated, have been identified including those function is to induce cell proliferation under specific circumstances (e.g. ras and myc proto-oncogenes) and genes which are programmed to halt proliferation in damaged cells (e.g. *TP53* and *RB1* tumor suppressor genes). Other mutations in genes involved in DNA repair are also necessary. About 150 human DNA repair genes have been identified to date [2], but the real number is probably higher, since less than 50% of known and putative genes have an identified function. The association between defects in DNA repair and cancer was established by Cleaver in 1968 [3], who showed that xeroderma pigmentosum (XP) is caused by deficient nucleotide excision repair (NER). For more than a quarter of a century, it was thought that only rare syndromes, such as XP, Cockayne syndrome (CS) and ataxia telangiectasia (AT), were associated with DNA repair defects [4]. Novel, common polymorphisms in DNA repair genes are being identified [5] and these polymorphisms may play a pivotal role in sporadic carcinogenesis. A growing body of literature, including observations of inter-individual differences in measures of DNA damage, suggests that these polymorphisms may alter the functional properties of DNA repair enzymes.

The aim of this article is to review and evaluate, in a comprehensive manner, the best available published evidence on the relative contribution of genetic polymorphisms involved in DNA repair capacity to lung cancer risk in humans.

1) DNA Repair Genes

At least four pathways of DNA repair operate on specific types of damaged of DNA. Base excision repair (BER) operates on small lesions. The NER pathway repairs bulk lesions. Mismatch repair (MMR) corrects replication errors. Double-strand DNA break repair (DSBR) consists of two pathways – homologous recombination (HR) and non-homologous end-joining (NHEJ). The NHEJ repair pathway involves direct ligation of the two double strand break end while HR is a process by which double-strand DNA breaks are repaired through the alignment of homologous sequences of DNA. The following sections review the literature on DNA repair genes in more detail.

a) BER Gene

It is believed that the predominant pathway used for removal of oxidized and many of the alkylated bases is BER. The process of BER is initiated by DNA glycosylases [e.g. 8-oxoguanine DNA glycosylase 1 (OGG1), endonuclease III homolog 1, thymine glycol-DNA-glycosylase], which are often promiscuous as far as their substrate specificity is concerned. The glycosylase hydrolyzes the *N*-glycosylic bond between the oxidized base and sugar moiety thus releasing the free damaged base and giving rise to an apurinic/apyrimidinic (AP) site. AP endonuclease (APE) acts upon the AP site generating a single strand break by cleaving the phosphodiester backbone 5′ to the AP site, leaving behind a 3′-hydroxyl group and a 5′-deoxyribose phosphate group. At this point the BER pathway can proceed through two different sub-pathways: short-patch and long-patch BER. These pathways are differentiated by the enzymes involved and the number of nucleotides removed. Short-patch BER replaces a single nucleotide by polymerase β and the newly synthesized DNA sealed by DNA ligase III/X-ray cross-complementing group 1 (XRCC1) heterodimer [6] Long-patch BER inserts 2 - 13 nucleotides by concordant action of polymerase β, proliferating cell nuclear antigen, flap endonuclease 1, and ligase 1.

b) NER Gene

NER is a versatile DNA repair system that removes a wide range of DNA lesions including UV-induced lesions. There are two subpathways in NER. One is transcription-coupled DNA repair (TCR), which preferentially removes DNA damage that blocks ongoing transcription in the transcribed DNA strand of active genes and the other is global genome repair (GGR), which removes lesions throughout the genome including those from the nontranscribed strand in the active gene [7]. Three rare, autosomal recessive inherited human disorders are associated with impaired NER activity: XP, CS and trichothiodystrophy (TTD) [8]. XP has been studied most extensively. XP patients develop skin tumors at an extremely high frequency (1000 fold increase as compared to normal individuals) because of their inability to repair UV-induced DNA lesions. These clinical findings are associated with cellular defects, including hypersensitivity to killing and mutagenic effects of UV and the inability of XP cells to repair UV-induced DNA damage [9,10]. Approximately 80% of XP patients who have been classified have a defect in the NER pathway. These patients are said to have "classical" XP, in contrast to the remaining 20% of patients who are designated as XP variants (XPV) and most likely have a defect in postreplication repair. In XPV patients, DNA replication stops or is interrupted at UV-damaged sites and *de novo* DNA synthesis opposite cyclobutane pyrimidine dimers lesions is prone to errors, leading to the fixation of multiple DNA mutations and ultimately to cancer. Seven different DNA NER genes, which correct seven distinct genetic XP complementation groups [XPA, XPB (also known as excision repair cross complementing group 3, ERCC3), XPC, XPD (ERCC2), XPE, XPF (ERCC4) and XPG (ERCC5, this gene causes CS)] and XPV have been identified [8]. XPA, ERCC3/XPB, ERCC2/XPD, ERCC4/XPF and ERCC5/XPG have a defect in TCR and GGR while XPC and XPE have a defect in GGR only. ERCC6 and ERCC8 are also known as CS type B (CSB) and CSA, respectively. Approximately 20% of patients have been assigned to the CSA complementation group In a word, CS shows some overlap with certain forms of XP. In contrast to XP and TTD, the NER defect in CS is limited to the TCR pathway. As with XP, TTD involves mutations in XP genes, usually *XPD*, which encodes a component of the transcription factor TFIIH [11]. However, it has been suggested that the functions of XPD associated with TTD are distinct from those of XP. Approximately half of the patients with TTD display photosensitivity, correlated with the NER defect.

c) DSBR Gene

A DSB is a critical lesion that can promote genomic instability. Basic strategies for repair of DSBs involve either NHEJ or HRR. These two major repair pathways function to maintain the integrity of DNA after endogenous DSBs or those arising from an external agent such as ionizing radiation [12-14].

Deficiencies of the HRR proteins in humans are associated with several genetic diseases that cause progressive illnesses and premature aging. These include AT mutated (ATM) gene, an autosomal recessive disorder characterized by cerebellar ataxia, acute radiation sensitivity, immunodeficiency, and high incidence of cancer and leukemia [15,16]. The ATM protein is a member of the phosphatidylinositol-3 kinase family of proteins that respond to DNA damage by phosphorylating key substrates involved in DNA repair and/or cell cycle control. Other inherited syndromes are Nijmegen breakage syndrome (NBS), and BRCA1 and BRCA2 deficiency associated with breast and ovarian cancer [17]. BLM and WRN are members of the RecQ DNA helicase family of proteins, whose mutations lead to Werner's and Bloom's syndromes with severe premature aging and cancer predisposition [18-20].

Compared to HRR, NHEJ repairs DNA breaks without the need for homology between the segments to be rejoined. In addition to the repair of random DSB, NHEJ repairs DSB formed at specific sites of V(D)J recombination during early lymphocyte development [21]. NHEJ is mediated by the *XRCC5*, *XRCC6* and *XRCC7* genes encoding components of a DNA-dependent protein kinase (DNA-PK), consisting of the DNA-targeting Ku heterodimer (Ku70 - Ku80) and the catalytic subunit DNA-PKcs. In addition, the recently discovered

Artemis nuclease interacts with DNA-PK and, finally, the ligation step is carried out by the XRCC4-ligase IV complex [22-24].

d) MMR Gene

MMR plays critical roles in the maintenance of genomic stability in prokaryotes, simple eukaryotes and metazoan organisms such as humans [25-27]. Interest in DNA MMR and its mechanisms of action exploded in the early 1990s with the observation that germline mutations in MMR genes cause hereditary non-polyposis colorectal cancer (HNPCC) [28-31].

There are nine mammalian MMR genes (*hMLH* (human MutL homolog) *1*, *hMLH3*, *hPMS* (human postmeiotic segregation) *1 - 2* and *hMSH* (human MutS homolog) *2 - 6*) [27,32]. The loss of normal MMR is thought to promote tumorigenesis by accelerating the accumulation of mutations in oncogenes and tumor suppressor genes [26,27]. This pathway of tumorigenesis was first recognized in colorectal carcinomas, sporadic and HNPCCs. In these tumors, defective MMR results from genetic or epigenetic alterations that inactivate the MMR genes *hMLH1* or *hMSH2*, and rarely *hMSH6* or *hPMS2* [28-30]. MMR is a type of DNA repair that corrects single base mismatches and small insertion and deletion loops that occur during normal DNA replication [25]. MMR recognizes and repairs the base-base mismatches and small insertion-deletion mispairs generated during DNA synthesis. The MMR system is composed of several protein complexes, including MutS, MutL, and MutH. MSH2 heterodimerizes with MSH3 or MSH6 to form MutS complexes. These two complexes have different mispair recognition properties. MSH2-MSH3 heterodimer recognizes insertion-deletion mispairs, whereas MSH2-MSH6 heterodimer interacts with both single-base-pair mismatches and insertion-deletion mispairs. There are two MutL-related heterodimeric complexes known to be involved in MMR: MLH1-PMS2 and MLH1-MLH3. MLH1-PMS2 heterodimer interacts with both MSH2-MSH3 and MSH2-MSH6 complexes and plays a major role in MMR. MLH1-MLH3 also plays a role in repairing mispairs [33] and promoting meiotic crossing-over [34]. Microsatellite instability (MSI) has also been reported in a number of non-colonic sporadic cancers (e.g. endometrial, gastric and lung cancers) [35]. In the absence of MMR, polymerase replication errors in simple repeat sequences result in the MSI phenotype [36]. A mutator phenotype causing increased mutation rates for single-base substitution and insertion/deletion mismatches have traditionally been thought to be the main functions in MMR cancer prevention. Defects in the insertion/deletion repair function create a phenotype that is largely thought to be unique to MMR. Since this function was originally characterized on short repetitive DNA microsatellite sequences, it is often referred to as Microsatellite Instability or MSI [37-39]. MSI is divided into MSI-Low and MSI-High subtypes depending on the percentage of microsatellite markers that are mutated, but only the MSI-High subtype is clearly associated with defective MMR.

2) DNA Repair Gene and Lung Cancer Risk

Although cigarette smoking is the main cause of lung cancer, not all smokers develop lung cancer, and a fraction of lifelong nonsmokers will die from lung cancer. Genetic host factors have recently been implicated to account for some of the observed differences in susceptibility. Cigarette smoke contains several thousand chemicals that are known to chemically modify DNA [40] and lead to the formation of mutations [41]. Most of these compounds are procarcinogens that must be activated by Phase I enzymes, such as cytochrome P450s. All activated carcinogens can bind to DNA and form DNA adducts that are capable of inducing mutations and initiating carcinogenesis. The capacity to repair DNA damage induced by activated carcinogens appears to be one of host factors that may influence lung cancer risk. A critical cellular response that counteracts the carcinogenic effects of DNA damage is DNA repair. As stated earlier, there are several known pathways of DNA repair, all of which act to remove DNA lesions and prevent mutations, thereby restoring genetic integrity.

Several studies have investigated whether reduced DNA repair capacity (DRC) is associated with an increased risk of cancer [42,43]. Reduced DRC of benzo(a)pyrene-7,8-diol-9,10-epoxide (an active form of benzo(a)pyrene)-DNA adducts is associated with an increased risk of lung cancer (2.1-fold, 95% confidence interval (CI) = 1.5 - 3.0) [44]. Reduced DRC has been shown to be associated with a 5.7-fold (95% CI = 2.1 - 15.7) increased risk of developing lung cancer [45]. Likewise, reduced DRC of bleomycin-induced damage was found to be associated with an increased risk of lung cancer [46]. These studies suggested that low DRC of various DNA repair mechanisms predisposes individuals to lung cancer and prompted us to search for defined DNA repair activities that may be risk factors for lung cancer. Polymorphisms in DNA repair genes may be associated with differences in the DRC of DNA damage and may influence an individual's risk of lung cancer because the variant genotype in those polymorphisms might destroy or alter repair function.

To estimate the association between the polymorphisms in DNA repair genes and lung cancer risk, we conducted meta-analyses of identified studies. We calculated the crude odds ratio (OR) and 95% CI for each study whenever possible. The meta-analyses were performed on crude ORs, since the adjusted ORs were not comparable because of different covariates' being included in the multivariate regression models. Using individuals with the homozygous common genotype as the reference group, we calculated ORs for individuals with the heterozygous genotype and homozygous rare genotype separately whenever possible (information available in at least two studies). In some cases, we combined the heterozygous genotype with the homozygous rare genotype due to a low prevalence of the rare allele in several polymorphisms. Data were combined using both fixed effects (Mantel-Haenszel) and random effects (DerSimonian and Laird method) models [47]. Fixed effects and random effects analyses address fundamentally different research questions. The former asks what the best estimate of the true effect size is in the population studied, whereas the latter asks what the range and distribution of effect sizes in the sample of populations studied. Therefore, the calculation of the mean of the distribution of population effect sizes (random effects model) provides different information from the calculation of the mean of the distribution of sample effect sizes (fixed effects model). Random effects incorporate an estimate of the between-study variance and tend to provide wider confidence intervals, when the results of the constituent studies differ among themselves. The random effects model, compared to the fixed effects model, reduces the weight for each individual study proportion to the difference in effect size of an individual study from the pooled estimate of the effect for all other studies. The summary ORs were estimated using the random effects model, which is appropriate when heterogeneity is present [48]. Thus, the summary OR values are basically based on random effects model. We estimated the summary OR when there were data from at least three studies available. The number of studies may appear to be different in each stratum, depending on the amount of information provided in the published articles. A test of heterogeneity was performed for each summary estimate. We evaluated publication bias by Begg's and Egger's tests [49]. All statistical analyses were conducted with Stata software, version 8.2 (Stata Corporation, College Station, Texas).

a) Genetic Polymorphisms in the BER Pathway and Lung Cancer Risk

a-1) OGG1

Oxidative stress induces a cellular redox imbalance which has been found to be present in various cancer cells compared with normal cells; the redox imbalance thus may be related to oncogenic stimulation. DNA mutation is a critical step in carcinogenesis and elevated levels of oxidative DNA lesions have been noted in various tumors, strongly implicating such damage in the etiology of cancer. It appears that the DNA damage is predominantly linked with the initiation process. By far the most studied oxidative DNA lesions are 8-hydroxyguanine (8-OH-Gua) and its 2′-deoxynucleoside equivalent, 7,8-dihydro-8-oxo-2′-deoxyguanosine. It appears that, quantitatively, BER is the most important route for the removal of the majority of oxidative lesions. The glycosylase considered to have the primary responsibility for the removal of 8-OH-Gua in human cells, is the OGG1 [50]. The activity of OGG1 is complemented by another enzyme, denoted OGG2, which removes the 8-OH-Gua from the nascent strand in 8-OH-Gua:A or 8-OH-Gua:G pairs, arising from misincorporation of 8-OH-dGTP [51].

The *OGG1* maps on chromosome 3, at 3p26.2. At least 20 validated sequence variants have been described to date. Among those, a C/G sequence variant leading to an amino acid change from serine to cysteine at codon 326 (Ser326Cys, Database of SNPs (dbSNP) no. rs1052133) has been studied most frequently.

The OGG1 enzyme shows a common genetic polymorphism with a variant Ser326Cys, which in complementation assays *in vitro* appears to increase susceptibility to mutagenic properties of reactive oxygen species considerably, whereas 8-OH-dG levels and incision activity in leukocytes and some target tissues generally show no difference between the genotypes [52]. Several *in vivo* or *in vitro* studies have examined the association between *OGG1* genotypes and enzyme activity, though the results have been inconsistent, as reviewed by Weiss et al. [52]. The Ser326Cys polymorphism has not yet been convincingly shown to cause decreased OGG1 activity in humans, although a recent study found higher levels of oxidized guanine in human lymphocyte DNA from subjects with the 326Ser/Ser genotype as compared with subjects with 326Cys/Cys genotype or 326Cys/Ser genotype after *ex vivo* treatment with sodium dichromate [53].

Table 1-1. Genetic polymorphisms in the BER pathway and lung cancer risk: *OGG1* Ser326Cys polymorphism.

Author, published year [reference number]	OR (95% CI)*		Ethnicity
	Ser/Cys	Cys/Cys	
Sugimura et al., 1999 [54]	0.80 (0.52 - 1.21)	1.13 (0.63 - 2.02)	Asian
Wirkman et al., 2000 [58]	0.66 (0.37 - 1.17)	2.20 (0.41 - 11.8)	Caucasian
Ito et al., 2002 [55]	1.02 (0.63 - 1.67)	0.85 (0.46 - 1.56)	Asian
Sunaga et al., 2002 [56]	1.49 (0.91 - 2.43)	0.98 (0.54 - 1.77)	Asian
Le Marchand et al., 2002 [60]	0.90 (0.65 - 1.26)	1.76 (1.15 - 2.71)	Admixed population
Lan et al., 2004 [57]	1.96 (1.10 - 3.48)	1.84 (0.83 - 4.06)	Asian
Park et al., 2004 [61]	1.89 (1.27 - 2.80)	4.10 (1.65 - 10.2)	Mostly composed of Caucasians
Hung et al., 2005 [59]	0.90 (0.79 - 1.03)	1.15 (0.84 - 1.57)	Mostly composed of Caucasians
Hunng et al., 2005 [62]	1.09 (0.86 - 1.40)	1.37 (1.02 - 1.53)	Meta-analysis (Based on the above 8 individual populations)
Vogel et al., 2004 [65]	1.09 (0.75 - 1.60)	0.78 (0.35 - 1.72)	Caucasian
Zienoldddiny et al., 2006 [66]	0.91 (0.64 - 1.29)	0.63 (0.40 - 0.97)	Caucasian
Matullo et al., 2006 [67]	1.26 (0.83 - 1.91)	0.82 (0.21 - 2.33)	Caucasian
Summary**			
All	1.08 (0.90 - 1.29)	1.17 (0.88 - 1.56)	Based on the above 11 individual populations
Caucasian (mostly composed of Caucasians)	1.06 (0.83 - 1.36)	1.14 (0.69 - 1.88)	Based on the above 6 individual populations
Asian	1.21 (0.82 - 1.78)	1.09 (0.79 - 1.49)	Based on the above 4 individual populations

*Crude odds ratio and 95% confidence interval.

**Based on random effects model.

The first study among Japanese indicated the Ser allele frequency was similar between controls (0.59) and lung cancer cases (0.59) and OR of the 326Cys/Cys genotype was 1.13 (95% CI = 0.63 - 2.02) [54]. Other Asian studies also found no association between the OGG1 Ser326Cys polymorphism and lung cancer risk [55-57]. Two white (mostly composed of whites) studies reported that the Cys allele was not significantly associated with an increased risk of lung cancer [58,59]. Compared with the Ser/Ser genotype, the Cys/Cys genotype was associated with an increased risk of lung cancer (OR = 1.76, 95% CI = 1.15 - 2.71) among a mixed population [60]. Likewise, significantly increased risk for lung cancer was observed for the Cys/Cys genotype (OR = 4.10, 95% CI = 1.65 - 10.2) among mostly composed of Caucasians [61]. The individual ORs from each study for *OGG1* Ser326Cys are shown in Table 1-1. In a narrative review, the Ser326Cys polymorphism has inconsistently been associated with risk of lung cancer as recently reviewed with ORs above unity for subjects with two mutant alleles in five studies, of which two were statistically significant, and two studies like the present showing an OR below unity [52]. Combining data from all eight studies on lung cancer, the summary ORs were 1.09 (95 % CI = 0.86 - 1.40) for the

Ser/Cys genotype and 1.37 (95% CI = 1.02 - 1.82) for the Cys/Cys genotype [62]. After removing one outlying study (OR = 4.10, 95% CI 1.65 - 10.2) [61], an increased risk of lung cancer for the Cys/Cys genotype (OR = 1.24; 95%CI =1.01 - 1.53) based on 3,253 cases and 3,371 controls was still observed [62]. There was an increased risk of lung cancer among subjects with the *OGG1* 326Cys/Cys genotype, which is consistent with experimental evidence that this isoform exhibits decreased the BER activity [63,64].

Three studies were not included in the meta-analysis by Hung et al. [62]. No significant effect of the Ser 326Cys polymorphism on lung cancer risk was found in Caucasians (adjusted OR = 0.65, 95% CI = 0.30 - 1.41; Cys/Cys genotype vs. Ser/Ser genotype) [65]. A protective effect of the Cys/Cys genotype was also seen among two Caucasian populations [66,67]. As shown in Table 1-1, combining data from all 11 populations above-mentioned, the summary ORs were 1.08 (95% CI = 0.90 - 1.29) for Ser/Cys carriers and 1.17 (95% CI = 0.88 - 1.56) for Cys/Cys carriers. The risk has been no longer significant by introducing the most recent studies into our meta-analysis. The *OGG1* Ser326Cys polymorphism may not be associated with lung cancer risk.

a-2) XRCC1

XRCC1 protein is an important component of the BER pathway, which fixes base damage and DNA single strand breaks caused by ionizing radiation and alkylating agents. The XRCC1 protein has no known catalytic activity but serves to orchestrate BER via its role as a central scaffolding protein physically associated with DNA ligase III at its COOH terminus, DNA polymerase ß at its NH_2 terminus, APE, polynucleotide kinase, and poly(ADP-ribose) polymerase, and via its function in recognizing and binding to single strand breaks [68-71].

The human *XRCC1* gene has 17 exons, spans approximately 31.9 kb and is located on chromosome 19q13.2. This gene was found to restore DRC in Chinese hamster ovary mutant EM9 cells [72]. A G to A transition of the *XRCC1* gene at codon 399 (dbSNP no. rs25487) is located at the COOH-terminal side of the poly(ADP-ribose) polymerase-interacting domain within a relatively nonconserved region between conserved residues of the BRCT domain [69] and leads to amino acid substitution of Arg to Gln [73]. Lunn et al. [74] measured higher levels of aflatoxin B1 adducts in the *XRCC1* Arg399Gln polymorphism and suggested that this might result in a deficient DRC. Although this polymorphism results in amino acid substitutions, there is no direct evidence on its functional consequences. Two other *XRCC1* polymorphisms, Arg194Trp (db SNP no. rs1799782) and Arg280His (db SNP no. rs25489), have been also determined and the functional effect of these polymorphisms is also unclear, even though some studies have revealed that amino acid changes at the evolutionary conserved regions can alter its function [75]. Lamerdin et al. [76] reported that the *XRCC1* Arg194Trp polymorphism occurred at a conserved residue in humans, indicating the functional significance of this evolutionary conservation.

Table 1-2. Genetic polymorphisms in the BER pathway and lung cancer risk: *XRCC1* Arg399Gln polymorphism.

Author (published year)	OR (95% CI)*		Ethnicity
	Arg/Gln	Gln/Gln	
Ratnasinghe et al., 2001 [78]	1.00 (0.60 - 1.60)	1.40 (0.50 - 1.70)	Asian
David-Beabes et al, 2001 [79]	1.03 (0.66 - 1.60)	0.52 (0.14 - 1.97)	African-American
David-Beabes et al, 2001 [79]	0.75 (0.52 - 1.08)	0.63 (0.34 - 1.14)	Caucasian
Divine et al., 2001 [80]	0.76 (0.47 - 1.22)	1.64 (0.80 - 3.36)	Caucasian
Chen et al., 2002 [81]	1.02 (0.57 - 1.80)	0.67 (0.20 - 2.26)	Asian
Park et al., 2002 [82]	1.27 (0.81 - 2.04)	2.30 (0.87 - 6.09)	Asian
Misra et al., 2003 [83]	1.10 (0.78 - 1.54)	0.84 (0.45 - 1.58)	Caucasian
Zhou et al., 2005 [88]	1.00 (0.80 - 1.20)	1.30 (1.00 - 1.70)	Caucasian
Ito et al., 2004 [84]	1.01 (0.70 - 1.45)	1.39 (0.70 - 2.76)	Asian
Popanda et al., 2004 [85]	0.89 (0.67 - 1.17)	0.87 (0.58 - 1.29)	Caucasian
Harms et al., 2004 [86]	0.73 (0.44 - 1.25)	1.07 (0.39 - 2.96)	Caucasian
Zhang et al., 2005 [87]	0.95 (0.79 - 1.14)	1.14 (0.84 - 1.55)	Asian
Hung et al., 2005 [59]	1.12 (0.98 - 1.28)	1.01 (0.83 - 1.23)	Mostly composed of Caucasians
Hung et al., 2005 [62]	1.01 (0.93 - 1.09)	1.07 (0.93 - 1.23)	Meta-analysis (Based on the above 13 individual populations)
Vogel et al., 2004 [65]	0.79 (0.54 - 1.17)	0.81 (0.46 - 1.41)	Caucasian
Schneider et al., 2005 [89]	0.94 (0.72 - 1.23)	0.83 (0.54 - 1.26)	Caucasian
Shen et al., 2005 [90]	0.59 (0.33 - 1.05)	0.75 (0.13 - 4.23)	Asian
Zienolddiny et al., 2006 [66]	1.08 (0.78 - 1.49)	0.67 (0.39 - 1.14)	Caucasian
Matullo et al., 2006 [67]	1.14 (0.75 - 1.73)	0.52 (0.19 - 1.19)	Caucasian
Summary**			
All	0.99 (0.93- 1.06)	1.02 (0.88 - 1.19)	Based on the above 18 individual populations
Caucasian (mostly composed of Caucasians)	0.99 (0.90 - 1.08)	0.94 (0.80 - 1.11)	Based on the above 11 individual populations
Asian	0.97(0.84- 1.11)	1.34 (1.16 - 1.54)	Based on the above 6 individual populations

*Crude odds ratio and 95% confidence interval.

**Based on random effects model.

Three polymorphisms in *XRCC1* (Arg194Trp (exon 7), Arg280His (exon 10) and Arg399Gln (exon 11) have been frequently examined in the studies on cancer epidemiology. The *XRCC1* Arg399Gln polymorphism was the most common sequence variant among the three *XRCC1* polymorphisms. No association between the *XRCC1* Arg399Gln polymorphism and lung cancer risk was found (distribution of genotypes was not given) [77]. As shown in Table 1-2, 11 studies have reported that the variant genotype of *XRCC1* Arg399Gln polymorphism was not associated with risk of lung cancer [59,78-87]. In the largest Caucasian study, the adjusted OR of *XRCC1* Arg399Gln polymorphism was 1.3 (95% CI = 1.0 - 1.8; Gln/Gln vs. Arg/Arg) [88]. Stratified analyses revealed that the ORs decreased as pack-years increased. For nonsmokers, the adjusted OR was 2.4 (95% CI = 1.2 - 5.0), whereas for heavy smokers ($\geq$55 pack-years), the OR decreased to 0.5 (95% CI = 0.3 - 1.0) [88]. In 2005 Hung et al. conducted the meta-analysis of 11 studies in13 different ethnic populations mentioned above [62]. They did not observe an association between the

Arg399Gln polymorphism and t lung cancer risk (OR = 1.07, 95% CI = 0.93 -1.23; Gln/Gln genotype vs. Arg/Arg genotype). There is one study previously published, which were not included in the meta-analysis by Hung et al. [62]. No associations was found between the *XRCC1* Arg399Gln polymorphism (adjusted OR = 0.94, CI = 0.55 - 1.66; Gln/Gln genotype vs. Arg/Arg genotype) and risk of lung cancer among Danes [65]. Five studies have been published after the meta-analysis by Hung et al. [62]. Lung cancer patients who carried at least one minor allele of *XRCC1* Arg399Gln (OR = 0.99, CI = 0.73 - 1.34) did not show any elevated risk. When analyzed by histology, no individual subtype of lung cancer was significantly associated with the polymorphism [89]. Individuals with the *XRCC1* 399Gln allele had a decreased risk (OR = 0.60, 95% CI = 0.35 - 1.02) of lung cancer compared with those with the Arg/Arg genotype [90]. Likewise, no association was observed between the Arg399Gln polymorphism and lung cancer risk (adjusted OR = 0.67, 95% CI = 0.40 - 1.10; Gln/Gln genotype vs. Arg/Arg genotype) [66]. Although the Arg399Gln polymorphism was not associated with lung cancer risk (adjusted OR = 0.50, 95% CI = 0.20 - 1.21; Gln/Gln genotype vs. Arg/Arg genotype), the recessive model in the stepwise multivariate analysis revealed a protective effect of the 399Gln/Gln genotype in lung cancer (adjusted OR = 0.22, 95% CI = 0.05 - 0.98) [67]. The summary OR for the 339Gln/Gln genotype among 18 different ethnic populations was 1.02 (95% CI = 0.88 - 1.19). The summary OR for the 339Gln/Gln genotype among Caucasians was 0.94 (95% CI = 0.80 - 1.11) while the summary OR for the 339Gln/Gln genotype among Asians was 1.34 (95% CI = 1.16 - 1.54). The *XRCC1* Arg399Gln polymorphism has been associated with risk of breast cancer among African Americans, but not among Caucasians [91,92], indicating that *XRCC1* Arg399Gln may be linked to another biologically effective mutation.

As for *XRCC1* Arg194Trp polymorphism (Table 1-3), Ratnasinghe et al. observed a nonsignificant decreased risk of lung cancer associated with the194Trp/Trp genotype (OR = 0.70, 95% CI = 0.30 - 1.60) among Chinese tin miners [78]. Subjects with the 194Arg/Trp genotype were at statistically significantly decreased risk of lung cancer relative to subjects with the 194Arg/Arg genotype among African-Americans (OR = 0.40, 95% CI =0.19 - 0.83) [79]. They observed no gene dosage effect, but they had too few the Trp/Trp genotype to reliably estimate the association in this category. No material association was observed among Caucasians (OR = 1.05, 95% CI =0.62 - 1.78; 194Arg/Gln genotype vs. 194Arg/Arg genotype), however [79]. The 194Trp/Trp genotype was associated with a borderline increased risk of lung cancer (adjusted OR = 3.06, 95% CI = 0.94 - 9.92) [81]. The *XRCC1* Arg194Trp polymorphism was not associated with the risk of lung cancer in the whole study population (adjusted OR = 1.16, 95% CI = 0.46 - 2.92; Trp/Trp genotype vs. Arg/Arg genotype) [59]. In 2005 Hung et al. conducted the meta-analysis of 4 studies in different ethnic populations mentioned above [62]. They did not observe an association between the Arg194Trp polymorphism and the risk of lung cancer (OR = 1.05, 95% CI = 0.58 - 1.91; Trp/Trp genotype vs. Arg/Arg genotype). Four studies have not included in the meta-analysis by Hung et al. [62]. Lung cancer patients who carried at least one 194Trp allele did not show any elevated risk (OR = 1.03, 95% CI = 0.66 - 1.61) [89]. No association was also observed between the Arg194Trp polymorphism and lung cancer risk (OR = 0.87, 95% CI = 0.51 - 1.49; Arg/Trp and Trp/Trp genotypes combined vs. Arg/Arg genotype) [66]. Similarly, the Arg194Trp polymorphism was not associated with lung cancer risk (OR = 0.94, CI = 0.46 -

1.90; Arg/Trp and Trp/Trp genotypes combined vs. Arg/Arg genotype) [67]. Our meta-analysis based on 8 studies in 9 different ethnic populations shows no significant association between the Arg194Trp polymorphisms and lung cancer risk. The summary ORs for the Arg/Trp genotype and Trp/Trp genotype were 0.89 (95% CI = 0.78 - 1.03) and 1.19 (95% CI = 0.76 - 1.86), respectively.

Table 1-3. Genetic polymorphisms in the BER pathway and lung cancer risk: *XRCC1* Arg194Trp polymorphism.

Author (published year)	OR (95% CI)*		Ethnicity
	Arg/Trp	Trp/Trp	
Ratnasinghe et al., 2001 [78]	0.70 (0.40 - 1.20)	0.70 (0.30 - 1.60)	Asian
David-Beabes et al, 2001 [79]	0.40 (0.19 - 0.83)	1.44 (0.20 - 10.37)	African-American
David-Beabes et al, 2001 [79]	1.05 (0.62 - 1.78)	—	Caucasian
Chen et al., 2002 [81]	1.31 (0.73 - 2.32)	2.61 (0.85 - 8.04)	Asian
Hung et al., 2005 [59]	0.86 (0.72 - 1.03)	0.81 (0.35 - 1.88)	Mostly composed of Caucasians
Hung et al., 2005 [62]	0.85 (0.64 - 1.12)	1.05 (0.58 - 1.91)	Meta-analysis (Based on the above 5 individual populations)
Schneider et al., 2005 [89]	0.99 (0.67 - 1.46)	1.86 (0.31 - 12.8)	Caucasian
Schen et al., 2005 [90]	1.01 (0.56 - 1.83)	1.48 (0.51 - 4.45)	Asian
Zienolddiny et al., 2006 [66]	0.88 (0.50 - 1.55)	0.60 (0.01 - 11.5)	Caucasian
Matullo et al., 2006 [67]	1.10 (0.59 - 1.94)	9.70 (0.69 - 134.6)	Caucasian
Summary**			
All	0.89 (0.78 - 1.03)	1.19 (0.76 - 1.86)	Based on the above 9 individual populations
Caucasian (mostly composed of Caucasians)	0.90 (0.78 - 1.05)	1.24 (0.50 - 3.11)	Based on the above 5 individual populations
Asian	0.96 (0.67 - 1.39)	1.29 (0.60 - 2.81)	Based on the above 3 individual populations

*Crude odds ratio and 95% confidence interval.

**Based on random effects model.

As for the *XRCC1* Arg280His polymorphism (Table 1-4), the Arg/His and His/His genotypes combined was associated with an increased risk of lung cancer (adjusted OR = 1.8, 95% CI = 1.0 - 3.4) [78]. No direct association between lung cancer risk and the *XRCC1* Arg280His polymorphism (OR = 1.17, 95% CI = 0.73 - 1.87; Arg/His and His/His genotypes combined vs. Arg/Arg genotype) [83]. The *XRCC1* Arg280His polymorphism was not associated with the risk of lung cancer (adjusted OR = 1.25, 95% CI = 0.38 - 4.08) [59]. Hung et al. conducted a meta-analysis of 3 published studies mentioned above [62]. They did not observe an association between the Arg280His polymorphism and the risk of lung cancer (OR = 1.10, 95% CI = 0.84 - 1.43; Arg/His and His/His genotypes combined vs. Arg/Arg genotype). Four studied have published after the meta-analysis by Hung et al. [62]. No association was found between the *XRCC1* Arg280His polymorphism (adjusted OR = 1.06, CI = 0.60 - 1.87; Arg/His and His/His genotypes combined vs. Arg/Arg genotype) and risk of lung cancer among Danes [65]. Lung cancer patients who carried at least one 280His allele did not show any elevated risk (OR=0.95, 95% CI = 0.57 - 1.60) [89]. Likewise, the

280Arg/His and 280His/His genotypes combined was not associated with an increased risk of lung cancer (adjusted OR = 1.20, 95% CI = 0.66 -2.19) [90]. No association was also observed between the Arg280His polymorphism and lung cancer risk (adjusted OR = 1.54, 95% CI = 0.88 - 2.68; Arg/His and His/His genotypes combined vs. Arg/Arg genotype) [66]. The summary OR for the Arg/His vs. the Arg/Arg genotype among 7 studies was 1.03 (95% CI = 0.88 - 1.20). The summary OR for the Arg/His and His/His genotypes combined vs. the Arg/Arg genotype was 1.06 (95% CI = 0.91 - 1.23). Among Asians, the summary OR for the Arg/His and His/His genotypes combined vs. the Arg/Arg genotype was associated with a borderline increased risk of lung cancer (OR = 1.45, 95% CI = 0.94 - 2.22).

Table 1-4. Genetic polymorphisms in the BER pathway and lung cancer risk: *XRCC1* Arg280His polymorphism.

Author (published year)	OR (95% CI)*		Ethnicity
	Arg/His	Arg/His or His/His	
Ratnasinghe et al., 2001 [78]	1.40 (0.70 - 2.60)	1.60 (0.90 - 2.90)	Asian
Mirsa et al., 2003 [83]	1.12 (0.70 - 1.80)	1.17 (0.73 - 1.87)	Caucasian
Hung et al., 2005 [59]	0.95 (0.77 - 1.18)	0.95 (0.77 - 1.17)	Mostly composed of Caucasians
Hung et al., 2005 [62]	—	1.10 (0.84 - 1.43)	Meta-analysis (Based on the above 3 individual populations)
Vogel et al., 2004 [65]	0.98 (0.53 - 1.79)	1.01 (0.56 - 1.85)	Caucasian
Schneider et al., 2005 [89]	0.93 (0.59 - 1.44)	0.97 (0.63 - 1.53)	Caucasian
Shen et al., 2005 [90]	1.14 (0.60 - 2.18)	1.29 (0.69 - 2.41)	Asian
Zienoldddiny et al., 2006 [66]	1.53 (0.85 - 2.78)	1.45 (0.82 - 2.56)	Caucasian
Summary**			
All	1.03 (0.88 - 1.20)	1.06 (0.91 - 1.23)	Based on the above 7 individual populations
Caucasian (mostly composed of Caucasians)	1.00 (0.85 - 1.18)	1.02 (0.86 - 1.19)	Based on the above 5 individual populations
Asian	1.26 (0.80 - 2.00)	1.45 (0.94 - 2.22)	Based on the above 2 individual populations

*Crude odds ratio and 95% confidence interval.

**Based on random effects model.

Several epidemiologic studies have examined *XRCC1* polymorphisms and lung cancer risk, but only limited data have been available for the *XRCC1* polymorphisms. Our result for the *XRCC1* Arg399Gln polymorphism did not replicate the result of Hung et al. [62], who reported that the399Gln/Gln genotype was not associated with an increased risk of lung cancer. It is not clear whether individuals with the *XRCC1* 399Gln allele have a decreased risk of lung cancer or not as of this moment. Results of previous studies that examined the association between the *XRCC1* Arg399Gln polymorphism and lung cancer risk were inconsistent, possibly owing to the large random error in several small studies. This discrepancy might be due, in part, to differences in the prevalence of smokers. Further investigations of the combined effects of polymorphisms within these DNA repair genes, smoking and other risk factors may help to clarify the influence of genetic variation in the carcinogenic process.

a-3) APE1

The AP sites are common mutagenic and cytotoxic DNA lesions that arise from the loss of normal bases, either through spontaneous processes, or excision of damaged bases by DNA glycosylases during the DNA repair process [93,94]. The AP sites are repaired by a series of reactions involving the multiple repair proteins of the BER pathway [93,95]. APE1, the major AP endonuclease of human cells, plays a central role in BER by hydrolyzing the phosphodiester backbone immediately 5' to the AP site [96,97]. This incision generates a normal 3'-hydroxyl group and an abasic deoxyribose-5-phosphate, which is processed by enzymes in the subsequent steps of the BER pathway [98]. APE1 also acts as a 3'-phosphodiesterase to initiate repair of single strand breaks [96,99,100], damage that may result from direct attack and fragmentation of deoxyribose residues in DNA by free radicals [96,101-103].

The *APE1* gene consists of 5 exons and 4 introns spanning 2.21 kb. This gene is located on chromosome 14q11.2 - q12. By hydrolyzing 3'-blocking fragments from oxidized DNA, APE1 produces normal 3'-hydroxyl nucleotide termini that are necessary for DNA repair synthesis and ligation at single- or double-strand breaks [96,104]. A common *APE1* polymorphism, a T to G transversion (Asp148Glu, dbSNP no. rs3136820), was shown to have 94% of the endonuclease activity relative to the Asp allele, and 21% less binding activity [97]. The *APE1* Asp148Glu polymorphism is the only known common nonsynonymous *APE1* coding region variant. In a small study on X-ray exposure to lymphocytes and DNA repair gene polymorphisms on chromosome aberrations, samples from individuals with the 148Asp/Glu or 148Glu/Glu genotype showed higher levels of damage with respect to all of the studied measures: aberrant cells, chromatid breaks, chromatid exchanges, deletions and dicentrics [105]. Prolonged cell cycle delay was also associated with the number of *APE1* variant alleles [106]. The *APE1* Asp148Glu polymorphism was not associated with lung cancer risk among male smokers in Finland (adjusted OR = 0.92, 95% CI = 0.56 - 1.51; Glu/Glu genotype vs. Asp/Asp genotype) [83]. The crude ORs for *APE1* Asp/Glu and Glu/Glu as compared with Asp/Asp genotype were not statistically significant and the risks were virtually unchanged after adjustment for age, sex, smoking status and pack-years of smoking (adjusted OR = 0.99, 95% CI = 0.66 - 1.49 for the Asp/Glu genotype and adjusted OR = 1.29, 95% CI = 0.75 - 2.21 for the Glu/Glu genotype) in Japanese [84]. An interaction between *APE1* genotype and smoking was seen in the Japanese study such that the genotypic effect was apparent primarily among light smokers (those with 40 pack-years of consumption or less), as compared with never smokers and heavy smokers. For individuals with the 148Glu/Gle genotype, a protective effect was seen in Germans (adjusted OR = 0.77, 95% CI = 0.51 - 1.16) [85]. Hung et al. reported that the summary OR for *APE1* Asp148Glu based on the above 3 studies was 0.94 (95% CI = 0.77 - 1.14) [62]. To date, 6 studies, including the studies mentioned above, have examined the association between *APE1* Asp148Glu polymorphism and lung cancer risk. The 148Glu/Glu genotype had a borderline significant 2-fold risk of lung cancer compared with the 148Asp/Asp genotype in Chinese (adjusted OR = 2.13, 95% CI = 0.96 - 4.74) [90]. The minor genotypes of Asp148Glu (OR = 0.78, 95% CI = 0.53 - 1.14) and Gln51His (OR = 0.86, 95% CI = 0.33 - 2.22) polymorphisms were not associated with lung cancer risk while the minor genotype of Ile64Val polymorphism was significantly associated with a decreased risk (OR = 0.10, 95% CI = 0.01

- 0.81) in Norwegians [66]. Finally, the *APE1* Asp148Glu polymorphism was not also associated with the risk of lung cancer in European Prospective Investigation into Cancer and Nutrition study (adjusted OR = 0.83, 95% CI = 0.44 - 1.54; Asp/Glu and Glu/Glu genotypes combined vs. Asp/Asp genotype) [67]. In our meta-analysis, the summary ORs for the Asp/Glu genotype and Asp/Glu and Glu/Glu genotypes combined were 1.01 (95% CI = 0.81 - 1.20) and 0.97 (95% CI = 0.83 - 1.14), respectively (Table 1-5). No clear association between *APE1* Asp148Glu polymorphism and lung cancer risk was observed. Ethnic-deference in the role of *APE1* Asp148Glu polymorphism was not observed.

Given the inconclusive evidence, the interaction between *APE1* Asp148Glu polymorphism and smoking status in the Japanese study, the dearth of studies on Asp148Glu polymorphism and lung cancer risk, and effect modification of the association between other genes and risk of lung by smoking [107,108], more evidence regarding the association between *APE1* polymorphism and lung cancer risk is warranted.

Table 1-5. Genetic polymorphisms in the BER pathway and lung cancer risk: *APE1* Asp148Glu polymorphism.

Author (published year)	OR (95% CI)*		Ethnicity
	Asp/Glu	Asp/Glu or Glu/Glu	
Misra et al., 2003 [83]	1.06 (0.69 - 1.63)	1.05 (0.70 - 1.58)	Caucasian
Ito et al., 2004 [84]	0.95 (0.65 - 1.40)	1.03 (0.71 - 1.48)	Asian
Popanda et al., 2004 [85]	0.88 (0.65 - 1.20)	0.84 (0.62 - 1.12)	Caucasian
Hung et al., 2005 [62]	—	0.94 (0.77 - 1.14)	Meta-analysis (Based on the above 3 individual populations)
Shen et al., 2005 [90]	1.23 (0.65 - 2.35)	1.41 (0.77 - 2.61)	Asian
Zienolddiny et al., 2006 [66]	1.31 (0.84 - 2.06)	0.95 (0.68 - 1.34)	Caucasian
Matullo et al., 2006 [67]	1.00 (0.62 - 1.62)	0.99 (0.64 -1.56)	Caucasian
Summary**			
All	1.01 (0.85 - 1.20)	0.97 (0.83 - 1.14)	Based on the above 6 individual populations
Caucasian	1.01 (0.83 - 1.23)	0.93 (0.78 - 1.12)	Based on the above 4 individual populations
Asian	1.02 (0.73 - 1.41)	1.12 (0.82 - 1.53)	Based on the above 2 individual populations

*Crude odds ratio and 95% confidence interval.

**Based on random effects model.

b) Genetic Polymorphisms in the NER Pathway and Lung Cancer Risk

b-1) XPA

The heterotrimeric replication protein A (RPA) is required for NER and has been suggested to play a role in the damage recognition process [109-111]. The XPA protein is also required for NER and is involved in the DNA damage recognition process [112-114]. Both RPA and XPA preferentially bind damaged DNA, and because RPA and XPA directly interact in the absence of DNA, the RPA-XPA complex has been implicated as a key component in the earliest stage of damage recognition [114-119]. There is also evidence that the XPC-hHR23B protein complex may initiate recognition of DNA damage for the global genomic repair pathway of NER [120]. Recent evidence also implicates the damaged DNA binding protein heterodimer in damage recognition because the complex binds damaged DNA with high affinity [121] and can dramatically increase the repair rate of certain DNA adducts, including cyclobutane pyrimidine dimers in conjunction with XPA and RPA [122].

Table 2-1. Genetic polymorphisms in the NER pathway and lung cancer risk: *XPA* G23A polymorphism.

Author (published year)	OR (95% CI)*		Ethnicity
	G/A	G/G	
Park et al., 2002 [126]	1.00 (0.62 - 1.62)	0.62 (0.35 - 1.10)	Asian
Wu et al., 2003 [125]	0.65 (0.48 - 0.87)	0.74 (0.55 - 1.01)	Caucasian
Wu et al., 2003 [125]	0.31 (0.09 - 1.00)	0.40 (0.13 - 1.25)	Mexican-American
Wu et al., 2003 [125]	0.54 (0.16 - 1.68)	0.49 (0.15 - 1.49)	African-American
Popanda et al., 2004 [85]	0.77 (0.48 - 121)	0.82 (0.52 - 1.30)	Caucasian
Vogel et al., 2005 [127]	0.78 (0.41 - 1.49)	0.57 (0.30 - 1.06)	Caucasian
Zienolddiny et al., 2006 [66]	0.87 (0.48 - 1.57)	1.41 (0.79 - 2.52)	Caucasian
Summary**			
All	0.73 (0.61 - 0.89)	0.75 (0.59 - 0.95)	Based on the above 7 individual populations
Caucasian	0.72 (0.58 - 0.89)	0.82 (0.61 - 1.11)	Based on the above 4 individual populations

*Crude odds ratio and 95% confidence interval.

**Based on random effects model.

The *XPA* maps on chromosome 9, at 9q22.3. In the *XPA* gene, a polymorphic site was identified that was in the 5' untranslated region of the gene and consisted of a G-to-A substitution in the fourth nucleotide before the ATG start codon (dbSNP rs 1800975) [123]. The polymorphism, termed *XPA* G23A polymorphism at position 23 in the transcript, four nucleotides upstream of the start codon), is in the Kozak sequence near the start codon and thus may affect the XPA protein levels in cells [124]. A functional association between *XPA* G23A polymorphism and DRC has been reported [125], which showed significantly higher repair proficiency in healthy subjects with at least one 23G allele. A significant protective effect of the *XPA* G23A polymorphism on lung cancer risk was also reported in Americans (adjusted OR = 0.69, 95% CI = 0.53 - 0.90; A/G and G/G genotypes combined vs. A/A genotype) and Mexican-Americans (adjusted OR = 0.32, 95% CI = 0.12 - 0.83; A/G and G/G genotypes combined vs. A/A genotype) [125]. Likewise, a protective effect was seen among Koreans (adjusted OR = 0.58, 95% CI = 0.32 - 1.06; G/G genotype vs. A/A genotype) [126]. The risk estimates for the 23A/A genotype were 1.53 in all cases (95% CI = 0.94- 2.50), 1.62 in adenocarcinoma patients (95% CI = 0.91 - 2.88), and 1.67 among small cell carcinoma cases (95% CI = 0.86 - 3.26) [85]. Homozygous carriers of the 23A allele of *XPA* G23A polymorphism were at 1.72-fold (CI = 0.95 - 3.13) higher risk of lung cancer than homozygous carriers of the 23G allele, although the association was not statistically significant among a Danish population [127]. The 23G/A and 23A/A genotypes combined of *XPA* G23A polymorphism was associated with a significant reduction of the risk of lung cancer (OR = 0.63, 95% CI = 0.44 - 0.89) in a Norwegian population [66]. In our meta-analysis, the summary ORs for the 23G/A genotype and 23G/G genotype among 5 studies in 7 different ethnic populations were0.73 (95% CI = 0.61 - 0.89) and 0.75 (95% CI = 0.59 - 0.95), respectively (Table 2-1). Among Caucasian studies, the summary ORs for the 23G/A genotype and the 23A/A genotype were 0.72 (95% CI = 0.58 - 0.89) and 0.82 (95% CI = 0.82 - 1.11), respectively. The *XPA* G23A polymorphism may be associated with risk of lung cancer.

b-2) ERCC1

The *ERCC1* coding region is 1.1 kb long and comprises 10 exons [128,129]. This gene is located on 19q13.2 - q13.3. Shen et al. [73] have identified polymorphisms of three of exons of the *ERCC1* gene, all of which resulted in silent mutations. No amino acid substitutions were observed among the *ERCC1* polymorphisms [130]. The functional effects of the silent polymorphisms in *ERCC1* have not been fully elucidated; however, some of variant alleles of the polymorphisms in DNA repair genes may be associated with reduced DRC, increased cancer susceptibility [131] and with response to treatment or survival time after cancer diagnosis [132]. The studies have focused on polymorphisms of 3′ UTR (C8092A, dbSNP no. rs3212986) and codon 118 (Asn118Asn, T19007C, dbSNP no. rs11615) in *ERCC1*.

For the T19007 C (Asn118Asn) polymorphism, although the 19007T/T genotype generates the less commonly associated triplet codon sequence encoding the amino acid and has been termed the "variant" by convention, the 19007T/T genotype indeed has higher frequencies in a number of studies [133-136]. Hence, the 19007T/T genotype is used as reference in this paper. The 8092C/C genotype of the C8092A polymorphism is used as reference on the same score.

Lack of association between the T19007C polymorphism and lung cancer risk in a Dane population (adjusted OR = 0.85, 95% CI = 0.49 - 1.46; C/C genotype vs. T/T genotype) [137]. No association was found between the T19007C polymorphism and lung cancer risk in a large study (OR = 1.02, 95% CI = 0.82 - 1.27; C/C genotype vs. T/T genotype) [138]. The C/C genotype of T19007C polymorphism was not associated with an increased risk of lung cancer in a Chinese population (adjusted OR = 0.71, 95% CI = 0.39 - 1.32) [139]. There was no difference in risk estimates according to histological type of lung cancer [139]. The C/C genotype of the T19007C polymorphism was associated with a significantly decreased risk of lung cancer (adjusted OR = 0.32, 95% CI = 0.19 - 0.55) [66]. It was found no substantial relationship between the T19007C polymorphism and lung cancer risk (adjusted OR = 0.71, 95% CI = 0.36 - 1.40; C/C genotype vs. T/T genotype) [67]. As shown in Table 2-2, the summary ORs for the 19007T/C genotype and the19007C/C genotype were 0.82 (95% CI = 0.62 - 1.08) and 0.72 (95% CI = 0.46- 1.11), respectively. Even if the analysis was restricted to Caucasian studies, the ORs did not materially change. No clear association between *ERCC1* T19007C polymorphism and lung cancer risk was observed.

As for the C8092A polymorphism, no association was found between the polymorphism and lung cancer risk (OR = 1.06; 95% CI = 0.78 - 1.44; A/A genotype vs. C/C genotype) [138]. There was not a positive association between the C8092A polymorphism and lung cancer risk (adjusted OR = 0.71, 95% CI = 0.35 - 1.4; A/A genotype vs. C/C genotype) [66]). The C8092A and T19007C polymorphisms have been reported to be in linkage disequilibrium [138]. Taken together, the results suggested that both *ERCC1* polymorphisms may not be associated with lung cancer risk. However, the results need to be confirmed in other independent studies and further studies are needed to detect the function of these polymorphisms.

Table 2-2. Genetic polymorphisms in the NER pathway and lung cancer risk: *ERCC1* T19007C and C8092A polymorphisms.

Author (published year)	OR (95% CI)*		Ethnicity
T19007C polymorphism	T/C	C/C	
Vogel et al., 2004 [137]	0.99 (0.67 - 1.47)	0.86 (0.49 - 1.50)	Caucasian
Zhou et al., 2005 [138]	1.00 (0.86 - 1.17)	1.02 (0.82 - 1.27)	Caucasian
Yin et al., 2006 [139]	0.39 (0.08 - 1.53)	0.49 (0.11 - 1.84)	Asian
Zienolddiny et al., 2006 [66]	0.50 (0.31 - 0.79)	0.35 (0.20 - 0.61)	Caucasian
Matullo et al., 2006 [67]	0.82 (0.52 - 1.27)	0.85 (0.46 - 1.51)	Caucasian
Summary**			
All	0.82 (0.62 - 1.08)	0.72 (0.46 - 1.11)	Based on the above 5 individual populations
Caucasian	0.84 (0.63 - 1.11)	0.74 (0.46 - 1.17)	Based on the above 4 individual populations
C8092A polymorphism	C/A	A/A	
Zhou et al., 2005 [138]	1.00 (0.86 - 1.16)	1.06 (0.78 - 1.44)	Caucasian
Zienolddiny et al., 2006 [66]	1.07 (0.77 - 1.47)	1.48 (0.71 - 3.11)	Caucasian
Summary**			
All	1.01 (0.88 - 1.16)	1.11 (0.84 - 1.48)	Based on the above 2 individual populations

*Crude odds ratio and 95% confidence interval.

**Based on random effects model.

b-3) ERCC2/XPD

The ERCC2/XPD protein takes part in the NER pathway, which recognizes and repairs a wide range of structurally unrelated lesions such as bulky adducts and thymidine dimers. ERCC2/XPD works as an ATP-dependent (5'→3') helicase joined to the basal TFIIH complex to separate double helix [140]. The ERCC2/XPD protein is necessary for normal transcription initiation and NER [141]. *ERCC2/XPD* maps on chromosome 19, at 19q13.3 and covers 21.14 kb. Mutations in the *ERCC2* gene can diminish the activity of TFIIH complexes giving rise to repair defects, transcription defects, and abnormal responses to apoptosis [142].

A number of polymorphisms in the *ERCC2/XPD* gene have been reported. Whereas the codons 199, 201 and 575 polymorphisms are rare, those in codons 156, 312, 711 and 751 are common. Two *ERCC2/XPD* polymorphisms, Asp312Asn (db SNP no. rs1799793) and Lys751Gln (db SNP no. rs13181), have mainly been investigated in relation to phenotypic endpoints relevant to lung carcinogenesis. Concerning the Asp312Asn polymorphism, most of the reported data indicate a higher level of DNA adducts in subjects with Asn allele, which is interpreted as a lower DRC for the 312Asn allele than the 312Asp allele. This is also true for the *ERCC2/XPD* Lys751Gln polymorphism. The 751Gln allele is associated with a higher DNA adduct level or lower DRC.

The Asp/Asp genotype of *ERCC2/XPD* Asp312Asn polymorphism was found to have almost twice the risk of lung cancer when the Asp/Asn and Asn/Asn genotypes combined served as reference (OR = 1.86, 95% CI =1.02 - 3.40) [77]. No association with this polymorphism was seen in an American study [143]. No relationship between *ERCC2/XPD* Asp312Asn polymorphism and lung cancer risk was also seen in a smaller Swedish study of lung cancer [144]. A large American lung-cancer study reported an elevated risk (adjusted OR = 1.5, 95% CI = 1.1 - 2.0; Asn/Asn genotype vs. Asp/Asp genotype) [145]. Subjects homozygous for the 312Asn/Asn genotype had an increased risk of lung cancer (adjusted OR = 10.33, 95% CI = 1.29 - 82.50) compared with subjects homozygous for the 312Asp/Asp genotype. Stratification analysis revealed that the increased risk was mainly confined to lung squamous cell carcinoma, with the ORs being 20.50 (95% CI = 2.25 - 179.05) for the 312Asn/Asn genotype [146]. No association between lung cancer risk and the Asp312Asn polymorphism was found among Finnish smoking men [83]. Two meta-analyses have been published in 2004 [147] and 2005 [148], respectively. Both of them are based on the same published data from 6 individual case-control studies. The first meta-analysis showed that individuals with the 312Asn/Asn genotype had a 27% (95% CI = 1.04 - 1.56) increased risk of lung cancer compared with individuals with the 312 Asp/Asp genotype. The results supported the hypothesis that individuals with the 312 Asn/Asn genotype are at higher risk of developing lung cancer [147]. The second meta-analysis was somewhat different from the first one because unadjusted ORs were summarized in the first one. Summary OR associated with the 312Asn/Asn genotype was 1.18 (95% CI = 0.84 - 1.67). No significant association between the *ERCC2/XPD* Asp312Asn polymorphism and lung cancer was found in the second meta-analysis [148]. Anyway, these meta-analyses indicate that the excess lung cancer risk from the 312Asn/Asn genotype may be about 23%.

Six studies have been reported after the two meta-analyses. Subjects with at least one 312Asn variant allele had an increased risk of squamous cell carcinoma of the lung compared with those with the 312Asp/Asp genotype (adjusted OR =1.80, 95% CI = 1.10 - 2.96) [149]. Individuals with 312Asn/Asn genotype did not show an increased risk for lung cancer (adjusted OR = 1.05, 95% CI = 0.68 - 161) [85]. Homozygous carriers of the 312Asn allele of the *ERCC2/XPD* Asp312Asn polymorphism was at 1.30-fold (95% CI = 0.80 - 2.11) higher risk of cancer than homozygous carriers of the 312Asp allele [137]. The Asp312Asn polymorphism was not associated with lung cancer risk (adjusted OR = 0.62, 95% CI = 0.25 - 1.53; Asp/Asn genotype vs. Asp/Asp genotype) [150]. No association was observed between the Asp312Asn genotypes and lung cancer risk (OR = 1.09, 95% CI 0.68 - 1.73; Asn/Asn genotype vs. Asp/Asp genotype) [66]. Similarly, the Asp312Asn polymorphism was not associated with lung cancer risk among Caucasians (OR = 0.94, 95% CI = 0.48 - 1.85) [67]. The summary ORs associated with the *ERCC2/XPD* Asp312Asn polymorphism are shown in Table 2-3. No significant association between lung cancer and the heterozygous 312Asp/Asn genotype was found for all of the studies combined or by ethnicity. The summary OR of 1.14 (95% CI = 0.95 - 1.37) was found for subjects with the homozygous Asn/Asn genotype.

Table 2-3. Genetic polymorphisms in the NER pathway and lung cancer risk: *ERCC2* Asp312Asn polymorphism.

Author (published year)	OR (95% CI)*		Ethnicity
	Asp/Asn	Asn/Asn	
Butkiewicz et al., 2001 [77]	0.49 (0.24 - 0.98)	0.71 (0.29 - 1.74)	Caucasian
Spitz et al., 2001 [143]	0.92 (0.62 - 1.36)	1.54 (0.78 - 3.05)	Admixed population
Hou et al., 2002 [144]	1.27 (0.78 - 2.05)	0.88 (0.43 - 1.84)	Caucasian
Zhou et al., 2002 [145]	0.98 (0.82 - 1.17)	1.41 (1.06 - 1.86)	Caucasian
Laing et al., 2003 [146]	0.98 (0.76 - 1.28)	11.2 (1.45 - 87.2)	Asian
Misra et al., 2003 [83]	0.76 (0.53 - 1.07)	0.94 (0.56 - 1.59)	Caucasian
Hu et al., 2004 [147]	—	1.27 (1.04 - 1.56)	Meta-analysis (Based on the above 6 individual populations)
Benhamou and Sarasin, 2005 [148]	0.92 (0.78 - 1.08)	1.18 (0.84 - 1.67)	Meta-analysis (Based on the above 6 individual populations)
Xing et al., 2002 [149]	1.30 (0.82 - 2.05)	—	Asian
Popanda et al., 2004 [85]	1.14 (0.77 - 1.68)	1.03 (0.70 - 1.51)	Caucasian
Vogel et al., 2004 [137]	1.27 (0.86 - 1.89)	1.09 (0.63 - 1.86)	Caucasian
Shen et al., 2005 [150]	0.58 (0.21 - 1.52)	—	Asian
Zienoldding et al., 2006 [66]	0.85 (0.58 - 1.25)	1.11 (0.68 - 1.81)	Caucasian
Matullo et al., 2006 [67]	0.81 (0.52 - 1.26)	0.95 (0.51 - 1.71)	Caucasian
Summary**			
All	0.97 (0.86 - 1.09)	1.14 (0.95 - 1.37)	Based on the above 12 individual populations
Caucasian	1.12 (0.95 - 1.32)	1.12 (0.95 - 1.32)	Based on the above 8 individual populations
Asian	1.02 (0.78 - 1.34)	—	Based on the above 3 individual populations

*Crude odds ratio and 95% confidence interval.

**Based on random effects model.

As for Lys751Gln polymorphism, David-Beades et al. reported that an appreciable association between 751Gln/Gln genotype and lung cancer risk in neither Caucasians nor African-Americans [151]. They combined the data for Caucasians and African Americans because they observed no evidence of heterogeneity by ethnic group (P > 0.30), and found no association [151]. No association with this polymorphism was seen in an American study (adjusted OR = 1.36, 95% CI = 0.84 - 2.20; 751Gln/Gln genotype vs. 751Lys/Lys genotype) [143]. A deceased risk of lung cancer was seen among Chinese carriers of the 751Gln allele [81]. No associations with the Lys751Gln polymorphism were seen in two Caucasian studies [144,145]. The 751Gln/Gln genotype was associated with an increased risk for lung cancer compared with the 751Lys/Lys genotype (adjusted OR = 2.71, 95% CI = 1.01 - 7.24). Stratification analysis revealed that the increased risk was mainly confined to lung squamous cell carcinoma, with the OR being 4.24 (95% CI = 1.34 - 13.38) for the 751Gln/Gln genotype [146]. No association between lung cancer risk and the Lys751Gln polymorphism was found among Finns [83]. The meta-analysis by Hu et al. showed that the751Gln/Gln genotype had a 21% (95% CI = 1.02 - 1.43) increased risk of lung cancer compared with individuals with the 751Lys/Lys genotype [147]. The meta-analysis by Benhamou and Sarasin reported that summary OR for the 751Gln/Gln genotype was 1.18 (95% CI = 0.95 - 1.47) [147]. No significant association between the *ERCC2/XPD* Lys751Gln polymorphism and lung cancer was found in the two meta-analyses [148]. These meta-analyses indicate that the excess lung cancer risk from the 751Gln/Gln genotype may be about 20%. Seven studies (one study has been reported before the two meta-analyses) are appended to Table 2-4. Compared with subjects having the 751Lys/Lys genotype, subjects carrying at least one variant 751Gln allele were at a borderline increased risk of squamous cell carcinoma of the lung (adjusted OR = 1.52; 95% CI = 0.94 - 2.46) [149]. Individuals with the 751GlnGln genotype showed an increased risk for lung cancer (adjusted OR = 1.59, 95% CI = 0.95 - 2.67) [85]. The adjusted OR for 751Gln/Gln genotype was 0.95 (95% CI = 0.26 - 3.45) [86]. The 751Gln/Gln genotype was also at 2.01-fold (95% CI = 1.20 - 3.35) higher risk of lung cancer than the 751Lys/Lys genotype [137]. However, individuals with the Gln allele had a 61% (95% CI = 14 - 83) reduction of lung cancer risk among a Chinese population [150]. The 751Gln/Gln genotype was associated with a significantly increased risk of lung cancer (adjusted OR = 1.60, 95% CI = 1.10 - 2.30) [66]. The Lys751Gln polymorphism was not associated with lung cancer risk among Caucasians (adjusted OR = 1.23, 95% CI = 0.63 - 2.42; Gln/Gln genotype vs. Lys/Lys genotype) [67]. The summary ORs for the 751Gln/Gln genotype and 751Lys/Gln genotype were 1.07 (95% CI = 0.98 - 1.17) and 1.30 (95% CI = 1.14 - 1.49), respectively. The effect of the 751Gln/Gln genotype was stronger in Caucasian (OR = 2.25, 95% CI = 0.97 - 5.23) than in Asian (OR = 1.18, 95% CI = 0.35 - 3.93) but this may be only due to difference in sample sizes.

Several studies have investigated the possible association of *ERCC2/XPD* Asp312Asn and Lys751Gln polymorphisms with lung cancer, with inconsistent results. The Lys751Gln polymorphism has been more studies than the Asp312Asn because the frequency of the 751Gln allele is more prevalent than the 312Asn allele. The inconsistent associations in previous studies observed for the *ERCC2/XPD* polymorphisms could be due to differences in study populations, the small sample sizes of earlier studies and possible environmental interactions.

b-4) ERCC4/XPF

ERCC4/XPF is an essential in the NER pathway, which is responsible for removing UV-C photoproducts and bulky adducts from DNA. Among the NER enzymes, the ERCC4/XPF and ERCC1 are also uniquely involved in removing DNA interstrand cross-linking damage. ERCC4/XPF-ERCC1 complex, which makes incisions at the 5′ end of DNA loops [152-154], may contribute to the repair of large trinucleotide repeat containing loops that are generated due to replication slippage and are too long to be repaired by the postreplicative DNA mismatch repair system [155]. Polymorphisms in enzymes involved in large loop repair could be responsible for the observed variation in stability of similar-sized trinucleotide repeat disease alleles among different individuals [156]. The *ERCC4/XPF* gene is evolutionarily conserved. Extensive homology exists between human ERCC4/XPF, *Drosophila* Mei-9, *Saccharomyces cerevisiae* RAD1, and *S. pombe* Rad16 [152,157], all of which have similar functions in NER.

Table 2-4. Genetic polymorphisms in the NER pathway and lung cancer risk: *ERCC2* Lys751Gln polymorphism.

Author (published year)	OR (95% CI)*		Ethnicity
	Lys/Gln	Gln/Gln	
David-Beades et al., 2001 [151]	1.14 (0.77 - 1.71)	1.72 (1.00 - 2.94)	Caucasian
David-Beades et al., 2001 [151]	1.14 (0.73 - 1.78)	1.39 (0.54 - 3.55)	African-American
Spitz et al., 2001 [143]	1.07 (0.78 - 1.46)	1.36 (0.84 - 2.20)	Admixed population
Chen et al., 2002 [81]	0.79 (0.17 - 1.11)	0.44 (0.17 - 1.11)	Asian
Hou et al., 2002 [144]	1.22 (0.75 - 2.00)	1.11 (0.58 - 2.13)	Caucasian
Zhou et al., 2002 [145]	1.01 (0.84 - 1.21)	1.17 (0.90 - 1.51)	Caucasian
Laing et al., 2003 [146]	0.93 (0.73 - 1.18)	2.36 (0.90 - 6.17)	Asian
Misra et al., 2003 [83]	0.87 (0.60 - 1.26)	1.06 (0.64 - 1.76)	Caucasian
Hu et al., 2004 [147]	—	1.21 (1.02 - 1.43)	Meta-analysis (Based on the above 8 individual populations)
Benhamou and Sarasin, 2005 [148]	1.00 (0.90 - 1.11)	1.18 (0.95 - 1.47)	Meta-analysis (Based on the above 8 individual populations)
Xing et al., 2002 [149]	1.33 (0.86 - 2.06)	1.91 (0.37 - 12.4)	Asian
Popanda et al., 2004 [85]	1.14 (0.86 - 1.52)	1.37 (0.93 - 2.02)	Caucasian
Harms et al., 2004 [86]	1.34 (0.79 - 2.49)	1.07 (0.34 - 3.38)	Caucasian
Vogel et al., 2004 [137]	1.57 (1.05 - 2.34)	1.73 (1.01 - 2.96)	Caucasian
Shen et al., 2005 [150]	0.44 (0.18 - 1.03)	—	Asian
Zienolddiny et al., 2006 [66]	1.20 (0.84 - 1.73)	1.56 (1.06 - 2.31)	Caucasian
Matullo et al., 2006 [67]	1.23 (0.78 - 1.96)	1.17 (0.63 - 2.11)	Caucasian
Summary**			
All	1.07 (0.98 - 1.17)	1.30 (1.14 - 1.49)	Based on the above 15 individual populations
Caucasian	1.11 (1.00 - 1.24)	2.25 (0.97 - 5.23)	Based on the above 9 individual populations
Asian	0.92 (0.66 - 1.27)	1.18 (0.35 - 3.93)	Based on the above 4 individual populations

*Crude odds ratio and 95% confidence interval.

**Based on random effects model.

The *ERCC4/XPF* gene contains 11 exons, spans 28.2 kb and is located on chromosome 16p13.2 - p13.13. Several polymorphisms exist in the coding region of *ERCC4/XPF*, a few of which have been associated with cancer risks. Genetic instability of simple repeated sequences might also be influenced by the *ERCC4/XPF* polymorphisms. The *ERCC4/XPF* G1244A polymorphism is a G-to-A change in exon 8 (Arg415Gln, dbSNP no. rs1800067) that results in a change from arginine to glutamine. The *ERCC4/XPF* polymorphism in exon 8 has been reported to be associated with an increased risk for developing breast cancer [158]. The T2505C polymorphism is a T-to-C change in exon 11 (Ser835Ser, dbSNP no. rs1799801) that results in no amino acid change (Serine is conserved) [159]. Functionally significant SNPs in the *ERCC4/XPF* gene may also contribute to individual differences in the fine details of DNA repair. Lack of association between the G1244A (Arg415Gln) polymorphism and lung cancer risk (adjusted OR = 1.11, 95% CI = 0.59 - 2.07; Arg/Gln genotype vs. Arg/Arg genotype) [66]. The 2505C/C genotype of the T2505C polymorphism was nonsignificantly associated with an increase risk of lung cancer (adjusted OR = 1.71, 95% CI = 0.52 - 5.58) [150].

b-5) ERCC5/XPG

ERCC5/XPG is responsible for a 1186 amino acid structure-specific endonuclease activity that is essential for the 2 incision steps in NER [160-162]. The ERCC5/XPG nuclease has been suggested to act on the single-stranded region created as a result of the combined action of the XPB helicase and the ERCC2/XPD helicase at the DNA damage site. In human cells, the ERCC5/XPG catalyses incision at approximately 5 nucleotides 3' to the site of damage but is also involved non-enzymatically in the subsequent 5' incision [161,162]. It is also involved in the stabilization of a pre-incision complex on the damaged DNA [163].

The *ERCC5/XPG* gene contains 17 exons, spans 32 kb and is located on chromosome 13q32.3 -q33.1. Several polymorphisms in the coding sequence of the *EECC5/XPG* gene have been identified. The association between lung cancer and two common polymorphisms, T335C (His46His, dbSNP no. rs1047768) and G3507C (Asp1104His, dbSNP no. rs17655), have been investigated. The functional effects of these two SNPs are still unknown. However, it is likely that the SNPs in the coding DNA sequences may result in a subtle structural alteration of the ERCC5/XPG activity and modulation of lung cancer susceptibility.

The 1104Asp/Asp genotype of the Asp1104His polymorphism was associated with a significantly decreased risk of lung cancer in a Korean population (adjusted OR = 0.60, 95% CI = 0.38 - 0.95) [164]. Similarly, the 1104Asp/Asp genotype was inversely associated with lung cancer (adjusted OR = 0.65, 95% CI = 0.39 - 1.1) in an admixed population (mostly composed of whites) [165]. In a Chinese population, the 1104Asp/Asp genotype was not associated with lung cancer risk (adjusted OR = 1.03, 95% CI = 0.49 - 2.13) [150]. As for T335C polymorphism, the 335C/C genotype was associated with a slightly increased risk of lung cancer (adjusted OR = 1.48, 95% = 0.87 - 2.52) [150]. The 335C/C genotype was also associated with a significantly increased risk of lung cancer (adjusted OR = 1.79, 95% CI = 1.19 - 2.63) in Caucasians [66].

c) Genetic Polymorphisms in the DSBR Pathway and Lung Cancer Risk

c-1) XRCC3

XRCC3 participates in homologous recombination repair of DNA double strand breaks and cross-links. It is a member of an emerging family of Rad-51-related proteins that may take part in homologous recombination to maintain chromosome stability and to repair DNA damage [166]. The XRCC3 was shown to interact directly with HsRad51 and, as with Rad55 and Rad57 in yeast, may cooperate with HsRad51 during recombination repair [167]. The *XRCC3*-deficient cells were found to be unable to form Rad51 foci after radiation damage and demonstrated genetic instability and increased sensitivity to DNA-damaging agents [168].

The *XRCC3* is located on chromosome 14q32.3 and covers 17.87 kb. The Thr241Met (C18067T, dbSNP no. rs861539) substitution in *XRCC3* is due to a C to T transition at exon 7 and is a nonconservative change but does not reside in the ATP-binding domains, which are the only functional domains that have been identified in the protein at this time [73]. The Thr241Met polymorphism is the most frequent polymorphism in the *XRCC3*, which may affect the coding enzyme's function and/or its interaction with other proteins involved in the DNA repair. The *XRCC3* Met241 variant allele was significantly associated with higher DNA adduct levels [169] and homology-directed repair activity [170].

Table 3-1. Genetic polymorphisms in the DSBR pathway and lung cancer risk: *XRCC3* Thr241Met polymorphism.

Author (published year)	OR (95% CI)*		Ethnicity
	Thr/Met	Met/Met	
David-Beades et al., 2001 [151]	0.86 (0.58 - 1.27)	0.82 (0.45 - 1.43)	Caucasian
David-Beades et al., 2001 [151]	0.93 (0.59 - 1.46)	1.36 (0.47 - 3.88)	African-American
Misra et al., 2003 [83]	0.86 (0.61 - 1.22)	1.17 (0.62 - 2.22)	Caucasian
Popanda et al., 2004 [85]	0.87 (0.65- 1.16)	1.20 (0.82 - 1.75)	Caucasian
Harms et al., 2004 [86]	0.76 (0.43 - 1.32)	1.33 (0.52 - 3.39)	Caucasian
Jacobson et al., 2004 [172]	1.29 (0.88 - 1.91)	1.05 (0.61 - 1.82)	Caucasian
Zienolddiny et al., 2006 [66]	0.82 (0.54 - 1.22)	0.67 (0.32 - 1.40)	Caucasian
Matullo et al., 2006 [67]	0. 90 (0.58 - 1.39)	0.83 (0.43 - 1.56)	Caucasian
Summary**			
All	0.91 (0.79 - 1.04)	1.03 (0.83 - 1.27)	Based on the above 8 individual populations
Caucasian	0.90 (0.78 - 1.04)	1.02 (0.82 - 1.26)	Based on the above 7 individual populations

*Crude odds ratio and 95% confidence interval.

**Based on random effects model.

The distribution of genotypes in the Thr241Met polymorphism did not differ between lung cancer cases and controls (three genotype distributions ware not provided) [77]. David-Beabes et al. also found no significant association between the Thr241Met polymorphism and lung carcinoma in African Americans, whites, or both ethnic groups combined [151]. Similarly, a negative study has been published on the association between the *XRCC3* Thr241Met polymorphism and lung cancer risk (adjusted OR = 1.12, 95% CI = 0.59 - 2.12; 241Met/Met genotype vs. 241Thr/Thr genotype) [83]. Individuals who had the 241Thr/Met or 241Met/Met genotypes did not show a statistically significantly greater risk of developing lung carcinoma compared with individuals who had the wild type 241Thr/Thr genotype (OR = 1.25, 95% CI = 0.72 - 2.15, three genotype distributions ware not provided) [171]. Individuals homozygous for the 241Met showed a slightly higher risk for lung cancer overall (adjusted OR = 1.29, 95% CI = 0.85 - 1.98). The OR was increased for individuals with 241Met/Met genotype among patients with adenocarcinoma (adjusted OR = 1.65, 95% CI = 0.99 - 2.75) [85]. There was no significant association between the 241Met/Met genotype and increased risk of lung cancer (adjusted OR = 1.25, 95% CI = 0.47 - 3.32) [86]. Heterozygous (adjusted OR = 1.54, 95% CI=1.05 – 2.26) and homozygous (adjusted OR = 1.46, 95% CI=0.87 – 2.47) carriers of the 241Met allele were both at relatively high risk of lung cancer [172]. The adjusted OR for the 241Met/Met genotype was OR = 0.61 (95% CI = 0.30 - 1.24) in non-small cell lung cancer patients [66]. A significant association between the Thr241Met polymorphism and lung cancer was not observed (adjusted OR = 1.35, 95% CI = 0.63 - 2.93;

Met/Met genotype vs. Thr/Thr genotype) [67]. None of the ORs of the 241Met/Met genotype shown in Table 3-1 were statistically significant.

c-2) Other DSBR Genes

Although the role of the XRCC2 in DNA repair is unclear, the XRCC2, a family of proteins related to human RAD51, may be essential components of the DSB repair pathway. The *XRCC2* maps on chromosome 7, at 7q36.1 and covers 31.58 kb. A Arg188His (dbSNP no. rs3218536) polymorphism has been reported to be associated with breast cancer risk [173]. The 188Arg/His and 188His/His genotypes combined was associated with significantly increased risk of non-small cell lung cancer (adjusted OR = 2.82, 95% CI = 1.91 - 4.18) [66]. Carriage of the rare 188His allele was associated with an increased risk of non-small cell lung cancer

Fanconi anemia (FA) comprises 8 distinct complementation groups [174]. Four of the FA genes, including the *FANCA*, *FANCG*, *FANCC*, and *FANCF* genes, have been cloned [175-179]. The *FANCG* gene was found to be identical to human *XRCC9*, a gene proposed to be involved in cell cycle regulation or post-replication repair [180]. This gene maps on chromosome 9, at 9p13 and covers 6.20 kb. The variant allele of the *XRCC9* Thr297Ile polymorphism (dbSNP no. rs2237857) may play a protective role and were associated with a significantly decreased risk of developing non-small cell lung cancer (OR = 0.11, 95% CI = 0.01 - 0.91; Thr/Ile genotype vs. Thr/Thr genotype) [66].

The Nijmegen breakage syndrome 1 (NBS1), also known as p95, is part of the MRE11/RAD50 complex and is involved in DNA double strand break repair and damage-induced checkpoint activation [181,182]. The *NBS1* is located on chromosome 8q21 and covers 51.37 kb. Individuals homozygous for the 6285A allele of G6285A (Leu34Leu, dbSNP no. rs1063045) or the 185Glu of the Gln185Glu polymorphism (G11122C, dbSNP no. rs1805794) were found to have an increased risk of lung cancer (OR = 2.15, 95% CI = 0.91 - 5.10 and OR = 2.53, 95% CI = 1.05 - 6.08, respectively) in a Chinese population [183]. No association was observed between lung cancer risk and the *NBS1* Glu185Gln polymorphism (OR= 0.74, 95% CI = 0.44 - 1.25; 185Gln/Gln genotype vs. 185Glu/Glu) in a Caucasian population [66]. It has been suggested that heterozygous carriers of the founder mutation of the *NBS1* (657del5 allele, 5 nucleotides in exon 6), which results in protein truncation, may be at increased risk of cancer. However, the association between this common polymorphism and lung cancer risk has not yet been examined.

d) Genetic Polymorphisms in the MMR Pathway and Lung Cancer Risk

Genetic and epigenetic inactivation of the MMR genes has been implicated in the etiology of hereditary nonpolyposis colorectal cancer syndrome and also in a wide variety of sporadic tumors such as colorectal, ovarian, and endometrial cancers [184,185]. However, the pathogenic role of the MMR genes in the environment-induced cancers such as lung cancer has not been well defined [35,186]. In addition to their established role in the repair of postreplicative DNA errors, the MMR proteins are also involved in a variety of other vital cellular processes such as the induction of apoptosis in response to exogenous DNA damage

[187,188] and the transcription-coupled nucleotide excision repair of bulky DNA adducts [189,190]. Therefore, a subtle defect in the DNA repair capacity that is caused by functional polymorphisms in the MMR genes that are neither necessary nor sufficient for the development of lung cancer could place some individuals at an increased risk of lung cancer. To our knowledge few data are available about the biological and clinical significance of the MMR gene in non-small lung cancer patients. Wang et al suggested that *hMLH1* is the major altered the MMR gene involved in tumorigenesis of non-small lung cancer, and that promoter methylation is the predominant mechanism in hMLH1 and hMSH2 deregulation [191]. hMLH1 plays a central role in MMR through its interaction with hMSH2 and hMSH6 [192,193]. Several polymorphisms have been identified in the *hMLH1* gene [194-197]. While the functional effects of these polymorphisms are not known, it is possible that some of these variants could have an effect on DRC, thereby modulating the susceptibility to lung cancer.

The *hMLH1* maps on chromosome 3, at 3p21.3 and covers 57.51 kb. Absent, which can result in the transcriptional silencing of the target gene, or low *hMLH1* expression as a consequence of promoter CpG methylation has been shown in colorectal cancer [198,199]. A highly significant correlation was found between *hMLH1* gene methylation and negative protein and mRNA expression [191]. The -93A/A genotype of G-93A polymorphism (dbSNP no. unidentified, G to A transition at position -93 in the promoter region) was associated with a significantly increased risk for overall lung cancer when the G/A and G/G genotypes combined was used as the reference group (adjusted OR = 1.57, 95% CI = 1.14 - 2.18). When stratified by histological type, the -93A/A genotype was associated with a significantly increased risk for squamous cell carcinoma compared with either the -93G/G genotype (adjusted OR = 2.02, 95% CI = 1.15 - 3.55) or the combined genotype (adjusted OR = 1.83, 95% CI = 1.24 - 2.71). However, there was no significant association between the G-93A polymorphism and the risk of adenocarcinoma or small cell carcinoma. These results suggest that the *hMLH1* -93G>A polymorphism could be used as a marker of genetic susceptibility to squamous cell carcinoma of the lung among a Korean population [200], although the functional effects of the polymorphism are not fully elucidated.

The *hMSH2* maps on chromosome 2, at 2p22-p21 and covers 159.34 kb. The *hMSH2* gene encodes the human homologue of the bacterial MutS protein, which is responsible for recognizing DNA mismatches [191,201]. To date, more than 10 polymorphisms in the *hMSH2* gene have been reported [202]. The functional effects of these polymorphisms have not been known. No significant association was found between four polymorphisms [-118T>C (T to C substitution 118 bases 5' of the ATG start codon), IVS1+9G>C (G to C substitution at nucleotide +9 of intron 1), IVS10+12A>G (A to G substitution at nucleotide +12 of intron 10), and IVS12–6T>C (T to C substitution at nucleotide -6 of intron 12)] and the risk of overall lung cancer among Koreans. However, when stratified by histological type, those individuals with at least one the G allele of the IVS10+12A>G polymorphism were at a significantly decreased risk of adenocarcinoma (adjusted OR = 0.59, 95% CI = 0.40 - 0.88), and those individuals with at least one the C allele of the IVS12-6T>C polymorphism were at a significantly increased risk of adenocarcinoma (adjusted OR = 1.52, 95% CI = 1.02 - 2.27). The A/G or G/G genotype of IVS10+12 A>G and the T/T genotype of IVS12-6 T>C polymorphisms combined was associated with a significantly lower risk of adenocarcinoma compared with the A/A genotype of IVS10+12 A>G and the T/C or C/C genotype of IVS12-6T>C (adjusted OR = 0.47, 95% CI = 0.28 - 0.78) [202].

Additional studies with larger sample sizes are required to confirm these findings. Future studies of other polymorphisms in the MMR genes and their association with DNA repair capacity phenotypes are also needed to understand the role of the MMR genes in determining the risk of lung cancer.

DISCUSSION

Epidemiological studies of common polymorphisms in DNA repair genes, if large and unbiased, can provide insight into the *in vivo* relationships between DNA repair genes and lung cancer risk. Such studies may identify empirical associations indicating that a polymorphism in a gene of interest has an impact on lung cancer, independent of metabolic regulatory mechanisms and other genetic and environmental variability. Findings from epidemiological studies can complement *in vitro* analyses of the various polymorphisms, genes, and pathways. In addition, epidemiological studies of common polymorphisms can lead to increased understanding of the public health dimension of DNA-repair variation.

We conducted a systematic literature review to evaluate the associations between sequence variants in DNA repair genes and lung cancer risk. Firstly, we found increased risk of lung cancer among subjects carrying the *ERCC2/XPD* 751Gln/Gln genotype (OR = 1.30, 95% CI = 1.14 - 1.49). The Gln allele of the *ERCC2/XPD* Lys751Gln polymorphism is associated with a higher DNA adduct level or lower DNA repair efficiency, except in research by Duell et al.[203], who found no correlation between the *ERCC2/XPD Lys751Gln* polymorphism and the level of polyphenol-DNA adducts in human blood samples. Matullo et al. [204] demonstrated a higher level of DNA adducts, measured by ^{32}P-postlabeling, in lymphocytes from nonsmokers with the *ERCC2/XPD* 751Gln/Gln genotype. Similarly, Palli et al. [205] reported a higher level of DNA adducts in workers with at least one Gln allele who were exposed to traffic pollution in comparison with workers with the two common alleles. An increased number of aromatic DNA adducts has been found by Hou et al. [144] in peripheral blood lymphocytes from subjects with *ERCC2/XPD* 312Asn and *ERCC2/XPD* 751Gln alleles. The 312Asn/Asn and 751Gln/Gln genotypes combined showed a higher level of DNA lesions than did other genotypes. Secondly, we found a protective effect of the *XPA* 23G/G genotype (OR = 0.75, 95% CI = 0.59 - 0.95) on lung cancer risk. The *XPA* G23A polymorphism itself may alter the transcription and/or translation. Because this polymorphism is located in the vicinity of the translation initiation codon, it may alter translation efficiency. The proximal near-by nucleotides to the AUG initiation codon is important for initiation of translation because the 40S ribosomal subunit binds initially at the 5'-end of mRNA [206]. The consensus sequence around the start codon is GCCRCCAUGG, which is known as the Kozak consensus sequence [207]. The R at position -3 and the G just downstream of the start codon are especially important, and the lack of these bases causes read-through of the start codon [208]. However, there has been no precise explanation of the mechanism by which the recognition of the start codon is aided by a purine at position -3 [207,209], which is the core nucleotide of the Kozak consensus. The polymorphism *XPA*

G23A is a G/A transversion 4 nucleotides upstream of the start codon of *XPA* possibly improving the Kozak sequence [126]. The sequences (CCAGAGAUGG) around the predicted initiator methionine codon of the *XPA* gene agree with the Kozak's consensus sequence at positions -3 and +4 [210]. Although both the A and polymorphic variant G nucleotides at the -4 position of the *XPA* gene does not correspond to the original consensus Kozak sequence containing nucleotide C at position -4, it is possible that a nucleotide substitution of A to G at position -4 preceding the AUG codon may affect ribosomal binding and thus alter the efficiency of the XPA protein synthesis. To investigate whether the transition from A to G changes the translation efficiency, an *in vitro* transcription/translation analysis [211] and a primer extension assay of initiation complex [212] will be necessary in the future. Besides, a functional association between the *XPA* G23A polymorphism and DRC was reported [125], which showed significantly higher repair proficiency in healthy subjects with at least one G allele. An alternative explanation could be that the protective *XPA* 23A allele is in linkage disequilibrium with an allele from an adjacent gene, which is the true susceptible gene. The *ERCC2/XPD* and *XPA* are genes in the NER pathway. Several DNA repair pathways are involved in the maintenance of genetic stability. The most versatile and important one is the NER pathway, which detects and removes bulky DNA adducts, including those induced by cigarette smoking [213]. Lastly, we found that the *XRCC1* 399Gln/Gln genotype was associated with an increased risk of lung cancer among Asians (OR = 1.34, 95% CI = 1.16 - 1.54). The Gln allele of the *XRCC1* 3Arg399Gln polymorphism is associated with increased levels of DNA damage that may be due to reduced DRC, as reflected in a higher level of DNA adducts [74,169], glycophorin A variants [74,205] and bleomycin sensitivity [214] as well as chromatid exchange frequencies [215]. However, there are several conflicting reports on the association between this polymorphism and lung cancer risk among various populations. Although the reasons for the inconsistencies in the studies are not clear, possible explanations are: 1) low frequency of the "at risk" genotype, which reduces the statistical power and 2) small size of the studies. Ethnic differences in roles of the polymorphism may be caused by gene-gene interactions, different linkages to the polymorphisms determining lung cancer risk and different lifestyles.

The most important problems facing lung cancer research are identifying high-risk individuals and implementing clinical surveillance, prevention practices, and follow-up care. Repair pathways play an important role in lung cancer risk, and genetic variations may contribute to decreased DRC and lung cancer susceptibility. Although the increased/decreased risk associated with individual DNA repair SNPs may be small compared to that conferred by high-penetrance cancer genes, their public health implication may be large, because of their high frequency in the general population. It is essential that epidemiological investigations of DNA repair polymorphisms are adequately designed. Unfortunately a fairly good number of studies are limited by their sample size and, consequently, low power to detect effects that may truly exist. Also given the borderline significance of some associations and multiple comparisons carried out, there is a possibility that one or more findings are false-positives [216]. Large and combined analyses such as those by Healey et al. [217] and Spurdle et al. [218] are preferred to minimize the likelihood of both false-positive and false-negative results. In addition, controls should be chosen in such a way that, if they were cases, they would be included in the case group; when controls are matched to cases, it is essential to account for matching in the analysis. When appropriate, confounding should be controlled for with particular consideration for race and ethnic group. An additional major concern is the grouping of genotypes for calculation of ORs; without functional data to dictate genotype groupings, it seems prudent to present two ORs per polymorphism (one for heterozygotes vs. common-allele homozygotes and one for rare-allele homozygotes vs. common-allele homozygotes) so that dominant, codominant, or recessive patterns may be elucidated.

Continued advances in SNP maps and in high-throughput genotyping methods will facilitate the analysis of multiple polymorphisms within genes and analysis of multiple genes within pathways. The effects of polymorphisms are best represented by their haplotypes. Data from multiple polymorphisms within a gene can be combined to create haplotypes, the set of multiple alleles on a single chromosome. None of the studies reviewed here reported haplotype associations, although several studies analyzed multiple polymorphisms within a gene, sometimes with inconsistent results. The analysis of haplotypes can increase power to detect disease associations because of higher heterozygosity and tighter linkage disequilibrium with disease-causing mutations [219-221]. In addition, analysis of haplotypes offers the advantage of not assuming that any of the genotyped polymorphisms is functional; rather, it allows for the possibility of an ungenotyped functional variant to be in linkage disequilibrium with the genotyped polymorphisms [222]. An analysis of data from multiple genes within the same DNA-repair pathway (particularly those known to form complexes) can provide more comprehensive insight into the studied associations. Such an analysis may shed light on the complexities of the many pathways involved with DNA repair and lung cancer development, and providing hypotheses for future functional studies. Because of concerns over inflated type I error rates in pathway-wide or genome-wide association studies, methods of statistical analysis seeking to obviate this problem are under development [223]. The ability to include haplotype information and data from multiple genes, and to model their interactions, will provide more powerful and more comprehensive assessments of the DNA repair pathways.

Research into the role of DNA repair polymorphisms in lung cancer is still in its early stages. As suggested by IARC and Todd, priorities for studies on molecular epidemiology should include large sample size, an independent replication followed by an initial study, biological plausibility and physiologically, meaningful data supporting the functional role of the polymorphism in question [224-226]. The initial studies showed substantial variations in risk of developing lung cancer in individuals with specific genotypes. Even so, etiology of lung cancer cannot be explained by allelic variability at a single locus [227,228]. Advances in identification of new variants and in high-throughput genotyping techniques will facilitate analysis of multiple polymorphisms within the genes along the same pathway [131]. Therefore, it is likely that definitive studies in the future will require analysis of large samples of cases and controls [229,230]. Additional epidemiological analyses of these and other DNA repair-polymorphisms will provide essential information about the *in vivo* relationships between the DNA-repair mechanisms and carcinogenesis and can complement *in vitro* analysis. Large, well-designed epidemiological studies are needed to help further illuminate the complex landscape of DNA repair and lung cancer risk.

This review, which is limited by the bias against publication of null findings, highlights the complexities inherent in epidemiological research and, particularly, in molecular epidemiological research. There is evidence that some polymorphisms in DNA repair genes play a role in carcinogenesis, notably the *ERCC2/XPD* Lys751Gln *XPA* G23A and the *XRCC1* Arg399Gln. The variant allele of each of the three polymorphisms was associated with about a 30% decrease/increase in lung cancer risk. Although the summary risk for developing lung cancer in individuals at each genotype may not be large, lung cancer is such a common malignancy that even a small increase in risk can translate to a large number of excess lung cancer cases. Therefore, polymorphisms, even those not strongly associated with lung cancer, should be considered as a potentially important public health issue. In addition, a susceptibility factor in one population may not be a factor in another. There are differences in the prevalence of DNA repair polymorphisms across populations. In a population where the prevalence of an "at-risk" genotype in a given polymorphism is very low, the "at-risk" allele or "at-risk" genotype may be too infrequent to assess its associated risk. At a population level, the attributable risk must be small simply because it is an infrequent allele. Finally, the major burden of lung cancer in the population probably results from complex interaction between many genetic and environmental factors over time. Most environmental carcinogens first require metabolic activation by Phase I enzymes to their ultimate forms and then bind to DNA, forming aromatic-DNA adducts that are thought to be an early step in tumorigenesis. On the other hand, these activated forms are detoxified by Phase II enzymes. Thus, genetically determined susceptibility to lung cancer may depend on the metabolic balance among Phase I enzyme, Phase II enzyme and DNA repair enzyme [231,232]. Further investigations of the combined effects of polymorphisms between DNA repair genes and drug-metabolizing genes may help to clarify the influence of genetic variation in the carcinogenic process. Consortia and international collaborative studies, which may a way to maximize study efficacy and overcome the limitations of individual studies, are needed to help further illuminate the complex landscape of lung cancer risk and genetic variations.

Acknowledgements

This study was funded in part by a Grant-in-Aid for Scientific Research (B) (17390175) from the Ministry of Education, Science, Sports and Culture, Japan.

References

[1] Pharoah PD, Dunning A., Ponder BA, Easton DF. Association studies for finding cancer-susceptibility genetic variants. *Nat Rev Cancer* 2004; 4: 850-60.

[2] Wood RD, Mitchell M, Lindahl T. Human DNA repair genes, 2005. *Mutat Res* 2005; 577: 275-83.

[3] Cleaver JE. Defective repair replication of DNA in xeroderma pigmentosum. *Nature* 1968; 218: 652-6.

[4] de Boer J, Hoeijmakers JH. Nucleotide excision repair and human syndromes. *Carcinogenesis* 2000; 21: 453-60.

[5] Miller MC 3rd, Mohrenweiser HW, Bell DA. Genetic variability in susceptibility and response to toxicants. *Toxicol Lett* 2001; 120: 269-80.

[6] Tomkinson AE. Mackey, ZB. Structure and function of mammalian DNA ligases. *Mutat Res* 1988; 407: 1-9.

[7] Hanawalt PC. Subpathways of nucleotide excision repair and their regulation. *Oncogene* 2002; 21: 8949-56.

[8] Bootsma D, Kraemer KH, Cleaver JE, Hoeijmakers JH. *Nucleotide excision repair syndromes: xeroderma pigmentosum, Cockayne syndrome, and trichothiodystrophy.* In Vogelstein B, Kinzler KW. (eds) The Genetic Basis of Human Cancer. McGraw-Hill, New York, NY, 1998; 245-274.

[9] Kraemer KH. Sunlight and skin cancer: another link revealed. *Proc Natl Acad Sci* (USA) 1997; 94: 11-4.

[10] van Steeg H, Kraemer KH. Xeroderma pigmentosum and the role of UV-induced DNA damage in skin cancer. *Mol Med Today* 1999; 5: 86-94.

[11] Lehmann AR. DNA repair-deficient diseases, xeroderma pigmentosum, Cockayne syndrome and trichothiodystrophy. *Biochimie* 2003; 85:1101-11.

[12] Hoeijmakers JH. Genome maintenance mechanisms for preventing cancer. *Nature* 2001; 411: 366-74.

[13] Thompson LH, Schild D. Homologous recombinational repair of DNA ensures mammalian chromosome stability. *Mutat Res* 2001; 477: 131-53.

[14] Haber JE. Partners and pathwaysrepairing a double-strand break. *Trends Genet* 2000; 16: 259-64.

[15] Khanna KK. Cancer risk and the ATM gene: a continuing debate. *J Natl Cancer Inst* 2000; 92: 795-802.

[16] Boultwood J. Ataxia telangiectasia gene mutations in leukaemia and lymphoma. *J Clin Pathol* 2001; 54: 512-6.

[17] Thompson LH, Schild D. Recombinational DNA repair and human disease. *Mutat Res* 2002; 509: 49-78.

[18] Bachrati CZ, Hickson ID. RecQ helicases: suppressors of tumorigenesis and premature aging. *Biochem J* 2003; 374: 577-606.

[19] Mohaghegh P, Hickson ID. Premature aging in RecQ helicase-deficient human syndromes. *Int J Biochem Cell Biol* 2002; 34: 1496-501.

[20] Cheok CF, Bachrati CZ, Chan KL, Ralf C, Wu L, Hickson ID. Roles of the Bloom's syndrome helicase in the maintenance of genome stability. *Biochem Soc Trans* 2005; 33:1456-9.

[21] Bassing CH, Swat W, Alt FW. The mechanism andregulation of chromosomal V(D)J recombination. *Cell* 2002; 109: S45–S55.

[22] Lieber MR, Ma Y, Pannicke U, Schwarz K. Mechanism and regulation of human non-homologous DNA end-joining. *Nat Rev Mol Cell Biol* 2003; 4: 712-20.

[23] Lieber MR, Ma Y, Pannicke U, Schwarz K. The mechanism of vertebrate nonhomologous DNA end joining and its role in V(D)J recombination. *DNA repair* 2004; 3: 817-826.

[24] Burma S, Chen DJ. Role of DNA–PK in the cellular response to DNA double-strand breaks. *DNA repair 2004*; 3: 909-918.
[25] Eisen JA. A phylogenomic study of the MutS family of proteins. *Nucleic Acids Res* 1998; 26: 4291-300.
[26] Kolodner RD, Marsischky GT. Eukaryotic DNA mismatch repair. *Cur Opin Genet Dev* 1999; 9: 89-96.
[27] Muller A, Fishel R. Mismatch repair and the hereditary non-polyposis colorectal cancer syndrome (HNPCC). *Cancer Invest* 2002; 20: 102-9.
[28] Fishel R, Lescoe MK, Rao MR, Copeland NG, Jenkins NA, Garber J, Kane M, Kolodner R. The human mutator gene homolog MSH2 and its association with hereditary nonpolyposis colon cancer. *Cell* 1993; 75: 1027-38.
[29] Leach FS, Nicolaides NC, Papadopoulos N, Liu B, Jen J, Parsons R, Peltomaki P, Sistonen P, Aaltonen LA, Nystrom-Lahti M, Guan XY, Zhang J, Meltzer PS, Yu JW, Kao FT, Chen DJ, Cerosaletti KM, Fournier REK, Todd S, Lewis T, Leach RJ, Naylor SL, Weissenbach J, Mecklin JP, Jarvinen H, Petersen GM, Hamilton SR, Green J, Jass J, Watson P, Lynch HT, Trent JM, Delachapelle A, Kinzler KW, Vogelstein B. Mutations of a mutS homolog in hereditary nonpolyposis colorectal cancer. *Cell* 1993; 75: 1215-25.
[30] Bronner CE, Baker SM, Morrison PT, Warren G., Smith LG, Lescoe MK, Kane M, Earabino C, Lipford J, Lindblom A, Tannergard P, Bollag RJ, Godwin AR, Ward DC, Nordenskjold M, Fishel R, Kolodner R, Liskay RM. Mutation in the DNA mismatch repair gene homologue hMLH1 is associated with hereditary non-polyposis colon cancer. *Nature* 1994; 368: 258-61.
[31] Papadopoulos N, Nicolaides NC, Wei YF, Ruben SM, Carter KC, Rosen CA, HaseltineWA, Fleischmann RD, Fraser CM, Adams MD, Venter JC, Hamilton SR, Petersen GM, Watson P, Lynch HT, Peltomaki P, Mechlin JP, Delachapelle A, Komzler KW, Vpgelstein B. Mutation of a mutL homolog in hereditary colon cancer. *Science*1994; 263: 1625-9.
[32] Kunkel TA, Erie DA. DNA mismatch repair. *Annu Rev Biochem* 2005; 74: 681-710.
[33] Flores-Rozas H, Kolodner RD. The Saccharomyces cerevisiae MLH3 gene functions in MSH3-dependent suppression of frameshift mutations. *Proc Natl Acad Sci* (USA) 1998; 95: 12404-9.
[34] Wang TF, Kleckner N, Hunter N. Functional specificity of MutL homologs in yeast: evidence for three Mlh1-based heterocomplexes with distinct roles during meiosis in recombination and mismatch correction. *Proc Natl Acad Sci* (USA) 1999; 96: 13914-9.
[35] Boland CR, Thibodeau SN, Hamilton SR, Sidransky D, Eshleman JR, Burt RW, Meltzer SJ, Rodriguez-Bigas MA, Fodde R, Ranzani GN, Srivastava S. A National Cancer Institute Workshop on microsatellite instability for cancer detection and familial predisposition: development of international criteria for the determination of microsatellite instability in colorectal cancer. *Cancer Res* 1998; 58: 5248-57.
[36] Bocker T, Ruschoff J, Fishel R. Molecular diagnostics of cancer predisposition: hereditary non-polyposis colorectal carcinoma and mismatch repair defects. *Biochim Biophys Acta* 1999; 1423: O1–O10.

[37] Thibodeau SN, Bren G, Schaid D. Microsatellite instability in cancer of the proximal colon. *Science* 1993; 260: 816-9.

[38] Shibata D, Peinado MA, Ionov Y, Malkhosyan S, Perucho M. Genomic instability in repeated sequences is an early somatic event in colorectal tumorigenesis that persists after transformation. *Nature Genet* 1994; 6: 2273-10.

[39] Dietmaier W, Wallinger S, Bocker T, Kullmann F, Fishel R, Ruschoff J. Diagnostic microsatellite instability: definition and correlation with mismatch repair protein expression. *Cancer Res* 1997; 57: 4749-56.

[40] [Hecht SS. Tobacco smoke carcinogens and lung cancer. *J Natl Cancer Inst* 1999; 91: 1194-210.

[41] Livneh Z. DNA damage control by novel DNA polymerases: translesion replication and mutagenesis. *J Biol Chem* 2001; 276: 25639-42.

[42] Berwick M, Vineis P. Markers of DNA repair and susceptibility to cancer in humans: an epidemiologic review. *J Natl Cancer Inst* 2000; 92: 874-97.

[43] Lippman SM, Spitz MR. Lung cancer chemoprevention: an integrated approach. *J Clin Oncol* 2001; 19: S74-82.

[44] Wei Q, Cheng L, Amos CI, Wang LE, Guo Z, Hong WK, et al. Repair of tobacco carcinogen-induced DNA adducts and lung cancer risk: a molecular epidemiologic study. *J Natl Cancer Inst* 2000; 92: 1764-72.

[45] Wei Q, Cheng L, Hong WK, Spitz MR. Reduced DNA repair capacity in lung cancer patients. *Cancer Res* 1996; 56: 4103-7.

[46] Rajaee-Behbahani N, Schmezer P, Risch A, Rittgen W, Kayser KW, Dienemann H, et al. Altered DNA repair capacity and bleomycin sensitivity as risk markers for non-small cell lung cancer. *Int J Cancer* 2001; 95: 86-91.

[47] Cochran WG. The combination of estimates from different experiments. *Biometrics* 1954; 10: 101-129.

[48] DerSimonian R, Laird N. Meta-analysis in clinical trials. *Control Clin Trials.* 1986; 7: 177-188.

[49] Egger M, Davey SG, Schneider M, Minder C. Bias in meta-analysis detected by a simple, graphical test. *BMJ* 1997; 315: 629-34.

[50] Rosenquist A, Zharkov DO, Grollman AP. Cloning and characterization of a mammalian 8-oxoguanine DNA glycosylase. *Proc Natl Acad Sci* (USA) 1997; 94: 7429-34.

[51] Nishioka K, Ohtsubo T, Oda H, Fujiwara T, Kang D, Sugimachi K, Nakabeppu Y. Expression and differential intracellular localization of two major forms of human 8-oxoguanine DNA glycosylase encoded by alternatively spliced OGG1 mRNAs. *Mol Biol Cell* 1999; 10: 1637-52.

[52] Weiss JM, Goode EL, Ladiges WC, Ulrich CM. Polymorphic variation in OGG1 and risk of cancer: A review of the functional and epidemiologic literature. *Mol Carcinog* 2005; 42: 127-41.

[53] Lee AJ, Hodges NJ, Chipman JK. Interindividual variability in response to sodium dichromate-induced oxidative DNA damage: role of the Ser326Cys polymorphism in the DNA-repair protein of 8-oxo-7,8-dihydro-2'-deoxyguanosine DNA glycosylase 1. *Cancer Epidemiol Biomarkers Prev* 2005; 14: 497-505.

[54] Sugimura H, Kohno T, Wakai K, Nagura K, Genka K, Igarashi H, Morris BJ, Baba S, Ohno Y, Gao C, Li Z, Wang J, Takezaki T, Tajima K, Varga T, Sawaguchi T, Lum JK, Martinson JJ, Tsugane S, Iwamasa T, Shinmura K, Yokota J. R hOGG1 Ser326Cys polymorphism and lung cancer susceptibility. *Cancer Epidemiol Biomarkers Prev* 1999; 8: 669-74.

[55] Ito H, Hamajima N, Takezaki T, Matsuo K, Tajima K, Hatooka S, Mitsudomi T, Suyama M, Sato S, Ueda R. A limited association of OGG1 Ser326Cys polymorphism for adenocarcinoma of the lung. *J Epidemiol* 2002; 12: 258-65.

[56] Sunaga N, Kohno T, Yanagitani N, Sugimura H, Kunitoh H, Tamura T, Takei Y, Tsuchiya S, Saito R, Yokota J. Contribution of the NQO1 and GSTT1 polymorphisms to lung adenocarcinoma susceptibility. *Cancer Epidemiol Biomarkers Prev* 2002; 11: 730-8.

[57] Lan Q, Mumford JL, Shen M, Demarini DM, Bonner MR, He X, Yeager M, Welch R, Chanock S, Tian L, Chapman RS, Zheng T, Keohavong P, Caporaso N, Rothman N. Oxidative damage-related genes AKR1C3 and OGG1 modulate risks for lung cancer due to exposure to PAH-rich coal combustion emissions. *Carcinogenesis* 2004; 25: 2177-81.

[58] Wikman H, Risch A, Klimek F, Schmezer P, Spiegelhalder B, Dienemann H, Kayser K, Schulz V, Drings P, Bartsch H. hOGG1 polymorphism and loss of heterozygosity (LOH): significance for lung cancer susceptibility in a caucasian population. *Int J Cancer* 2000; 88: 932-7.

[59] Hung RJ, Brennan P, Canzian F, Szeszenia-Dabrowska N, Zaridze D, Lissowska J, Rudnai P, Fabianova E, Mates D, Foretova L, Janout V, Bencko V, Chabrier A, Borel S, Hall J, Boffetta P. Large-scale investigation of base excision repair genetic polymorphisms and lung cancer risk in a multicenter study. *J Natl Cancer Inst* 2005; 97: 567-76.

[60] Le Marchand L, Donlon T, Lum-Jones A, Seifried A, Wilkens LR. Association of the hOGG1 Ser326Cys polymorphism with lung cancer risk. *Cancer Epidemiol Biomarkers Prev* 2002; 11: 409-12.

[61] Park J, Chen L, Tockman MS, Elahi A, Lazarus P. The human 8-oxoguanine DNA N-glycosylase 1 (hOGG1) DNA repair enzyme and its association with lung cancer risk. *Pharmacogenetics* 2004; 14: 103-9.

[62] Hung RJ, Hall J, Brennan P, Boffetta P. Genetic polymorphisms in the base excision repair pathway and cancer risk: a HuGE review. *Am J Epidemiol* 2005; 162: 925-42.

[63] Kohno T, Shinmura K, Tosaka M, Tani M, Kim SR, Sugimura H, Nohmi T, Kasai H, Yokota J. Genetic polymorphisms and alternative splicing of the hOGG1 gene, that is involved in the repair of 8-hydroxyguanine in damaged DNA. *Oncogene* 1998; 16: 3219-25.

[64] Dhenaut A, Boiteux S, Radicella JP. Characterization of the hOGG1 promoter and its expression during the cell cycle. *Mutat Res* 2000; 461: 109-18.

[65] Vogel U, Nexo BA, Wallin H, Overvad K, Tjonneland A, Raaschou-Nielsen O. No association between base excision repair gene polymorphisms and risk of lung cancer. *Biochem Genet* 2004; 42: 453-60.

[66] Zienolddiny S, Campa D, Lind H, Ryberg D, Skaug V, Stangeland L, Phillips DH, Canzian F, Haugen A. Polymorphisms of DNA repair genes and risk of non-small cell lung cancer. *Carcinogenesis* 2006; 27: 560-7.

[67] Matullo G, Dunning AM, Guarrera S, Baynes C, Polidoro S, Garte S, Autrup H, Malaveille C, Peluso M, Airoldi L, Veglia F, Gormally E, Hoek G, Krzyzanowski M, Overvad K, Raaschou-Nielsen O, Clavel-Chapelon F, Linseisen J, Boeing H, Trichopoulou A, Palli D, Krogh V, Tumino R, Panico S, Bueno-De-Mesquita HB, Peeters PH, Lund E, Pera G, Martinez C, Dorronsoro M, Barricarte A, Tormo MJ, Quiros JR, Day NE, Key TJ, Saracci R, Kaaks R, Riboli E, Vineis P. DNA repair polymorphisms and cancer risk in non-smokers in a cohort study. *Carcinogenesis* 2006; 27: 997-1007.

[68] Caldecott KW, Aoufouchi S, Johnson P, Shall S. XRCC1 polypeptide interacts with DNA polymerase ß and possibly poly(ADP-ribose) polymerase, and DNA ligase III is a novel molecular 'nick-sensor' in vitro. *Nucleic Acids Res* 1996; 24: 4387-94.

[69] Masson M, Niedergang C, Schreiber V, Muller S, Menissier-de Murcia J, de Murcia G. XRCC1 is specifically associated with poly(ADP-ribose) polymerase and negatively regulates its activity following DNA damage. *Mol Cell Biol* 1998; 18: 3563-71.

[70] Whitehouse CJ, Taylor RM, Thistlethwaite A, Zhang H, Karimi-Busheri F, Lasko DD, Weinfeld M, Caldecott KW. XRCC1 stimulates human polynucleotide kinase activity at damaged DNA termini and accelerates DNA single-strand break repair. *Cell* 2001; 104: 107-17.

[71] Vidal AE, Boiteux S, Hickson ID, Radicella JP. XRCC1 coordinates the initial and late stages of DNA abasic site repair through protein-protein interactions. *EMBO J* 2001; 20: 6530-9.

[72] Thompson LH, Brookman KW. Jones J, Allen SA, Carrano AV. Molecular cloning of the human XRCC1 gene, which corrects defective DNA strand break repair and sister chromatid exchange. *Mol Cell Biol* 1990; 10: 6160-71.

[73] Shen MR, Jones IM, Mohrenweiser H. Nonconservative amino acid substitution variants exist at polymorphic frequency in DNA repair genes in healthy humans. *Cancer Res* 1998; 58: 604-8.

[74] Lunn RM, Langlois G., Hsieh LL, Thompson CL, Bell DA. XRCC1 polymorphisms: effects on aflatoxin B1-DNA adducts and glycophorin A variant frequency. *Cancer Res* 1999; 59: 2557-61.

[75] Savas S, Kim DY, Ahmad MF, Shariff M, Ozcelik H. Identifying functional genetic variants in DNA repair pathway using protein conservation analysis. *Cancer Epidemiol Biomarkers Prev* 2004; 13: 801-7.

[76] Lamerdin JE, Montgomery MA, Stilwagen SA, Scheidecker LK, Tebbs RS, Brookman KW, Thompson LH, Carrano AV. Genomic sequence comparison of the human and mouse XRCC1 DNA repair gene regions. *Genomics* 1995; 25: 547-54.

[77] Butkiewicz D, Rusin M, Enewold L, Shields PG, Chorazy M, Harris CC. Genetic polymorphisms in DNA repair genes and risk of lung cancer. *Carcinogenesis* 2001; 22: 593-7.

[78] Ratnasinghe D, Yao SX, Tangrea JA, Qiao YL, Andersen MR, Barrett MJ, Giffen CA, Erozan Y, Tockman MS, Taylor PR. Polymorphisms of the DNA repair gene XRCC1 and lung cancer risk. *Cancer Epidemiol Biomarkers Prev* 2001; 10: 119-23.

[79] David-Beabes GL, London SJ. Genetic polymorphism of XRCC1 and lung cancer risk among African-Americans and Caucasians. *Lung Cancer* 2001; 34: 333-9.

[80] Divine KK, Gilliland FD, Crowell RE, Stidley CA, Bocklage TJ, Cook DL, Belinsky SA. The XRCC1 399 glutamine allele is a risk factor for adenocarcinoma of the lung. *Mutat Res* 2001; 461: 273-8.

[81] Chen S, Tang D, Xue K, Xu L, Ma G, Hsu Y, Cho SS. DNA repair gene XRCC1 and XPD polymorphisms and risk of lung cancer in a Chinese population. *Carcinogenesis* 2002; 23:1321-5.

[82] Park JY, Lee SY, Jeon HS, Bae NC, Chae SC, Joo S, Kim CH, Park JH, Kam S, Kim IS, Jung TH. Polymorphism of the DNA repair gene XRCC1 and risk of primary lung cancer. *Cancer Epidemiol Biomarkers Prev* 2002; 11: 23-7.

[83] Misra RR, Ratnasinghe D, Tangrea JA, Virtamo J, Andersen MR, Barrett M, Taylor PR, Albanes D. Polymorphisms in the DNA repair genes XPD, XRCC1, XRCC3, and APE/ref-1, and the risk of lung cancer among male smokers in Finland. *Cancer Lett* 2003; 191: 171-8.

[84] Ito H, Matsuo K, Hamajima N, Mitsudomi T, Sugiura T, Saito T, Yasue T, Lee KM, Kang D, Yoo KY, Sato S, Ueda R, Tajima K. Gene-environment interactions between the smoking habit and polymorphisms in the DNA repair genes, APE1 Asp148Glu and XRCC1 Arg399Gln, in Japanese lung cancer risk. *Carcinogenesis* 2004; 25: 1395-401.

[85] Popanda O, Schattenberg T, Phong CT, Butkiewicz D, Risch A, Edler L, Kayser K, Dienemann H, Schulz V, Drings P, Bartsch H, Schmezer P. Specific combinations of DNA repair gene variants and increased risk for non-small cell lung cancer. *Carcinogenesis* 2004; 25: 2433-41.

[86] Harms C, Salama SA, Sierra-Torres CH, Cajas-Salazar N, Au WW. Polymorphisms in DNA repair genes, chromosome aberrations, and lung cancer. *Environ Mol Mutagen* 2004; 44: 74-82.

[87] Zhang X, Miao X, Liang G, Hao B, Wang Y, Tan W, Li Y, Guo Y, He F, Wei Q, Lin D. Polymorphisms in DNA base excision repair genes ADPRT and XRCC1 and risk of lung cancer. *Cancer Res* 2005; 65: 722-6.

[88] Zhou W, Liu G, Miller DP, Thurston SW, Xu LL, Wain JC, Lynch TJ, Su L, Christiani DC. Polymorphisms in the DNA repair genes XRCC1 and ERCC2, smoking, and lung cancer risk. *Cancer Epidemiol Biomarkers Prev* 2003; 12: 359-65.

[89] Schneider J, Classen V, Bernges U, Philipp M. XRCC1 polymorphism and lung cancer risk in relation to tobacco smoking. *Int J Mol Med* 2005; 16: 709-16.

[90] Shen M, Berndt SI, Rothman N, Mumford JL, He X, Yeager M, Welch R, Chanock S, Keohavong P, Donahue M, Zheng T, Caporaso N, Lan Q. Polymorphisms in the DNA base excision repair genes APEX1 and XRCC1 and lung cancer risk in Xuan Wei, China. *Anticancer Res* 2005; 25: 537-42.

[91] Duell EJ, Millikan RC, Pittman GS, Winkel S, Lunn RM, Tse CK, Eaton A, Mohrenweiser HW, Newman B, Bell DA. Polymorphisms in the DNA repair gene XRCC1 and breast cancer. *Cancer Epidemiol Biomarkers Prev* 2001; 10: 217-22.

[92] Nexo BA, Vogel U, Olsen A, Ketelsen T, Bukowy Z, Thomsen BL, Wallin H, Overvad K, Tjonneland A. A specific haplotype of single nucleotide polymorphisms on chromosome 19q13.2-3 encompassing the gene RAI is indicative of post-menopausal breast cancer before age 55. *Carcinogenesis* 2003; 24: 899-904.

[93] Lindahl T. Recognition and processing of damaged DNA. *J Cell Sci* 1995; 19: S73-7.

[94] Bennett RA, Wilson DM 3rd, Wong D, Demple B. Interaction of human apurinic endonuclease and DNA polymerase beta in the base excision repair pathway. *Proc Natl Acad Sci* (USA) 1997; 94: 7166-9.

[95] Mitra S, Hazra TK, Roy R, Ikeda S, Biswas T, Lock J, Boldogh I, Izumi T. Complexities of DNA base excision repair in mammalian cells. *Mol Cells* 1997; 7: 305-12.

[96] Izumi T, Hazra TK, Boldogh I, Tomkinson AE, Park MS, Ikeda S, Mitra S. Requirement for human AP endonuclease 1 for repair of 3-blocking damage at DNA single-strand breaks induced by reactive oxygen species. *Carcinogenesis* 2000; 21: 1329-34.

[97] Hadi MZ, Coleman MA, Fidelis K, Mohrenweiser HW, Wilson DM 3rd. Functional characterization of Ape1 variants identified in the human population. *Nucleic Acids Res* 2000; 28: 3871-9.

[98] Wilson DM 3rd, Sofinowski TM, McNeill DR. Repair mechanisms for oxidative DNA damage. *Front Biosci* 2003; 8: d963-81.

[99] Suh D, Wilson DM 3rd, Povirk LF. 3-phosphodiesterase activity of human apurinic/apyrimidinic endonuclease at DNA double-strand break ends. *Nucleic Acids Res* 1997; 25: 2495-500.

[100] Ramana CV, Boldogh I, Izumi T, Mitra S. Activation of apurinic/apyrimidinic endonuclease in human cells by reactive oxygen species and its correlation with their adaptive response to genotoxicity of free radicals. *Proc Natl Acad Sci* (USA) 1998; 95: 5061-6.

[101] Wilson DM 3rd, Takeshita M, Grollman AP, Demple B. Incision activity of human apurinic endonuclease (Ape) at abasic site analogs in DNA. *J Biol Chem* 1995; 270: 16002-7.

[102] Breen AP, Murphy JA. Reactions of oxyl radicals with DNA. *J Free Radic Biol Med* 1995; 18: 1033-77.

[103] Sancar A. DNA repair in humans. *Annu Rev Genet* 1995; 29: 69-105.

[104] Kelley MR, Cheng L, Foster R, Tritt R, Jiang J, Broshears J, Koch M. Elevated and altered expression of the multifunctional DNA base excision repair and redox enzyme Ape1/ref-1 in prostate cancer. *Clin Cancer Res* 2001; 7: 824-30.

[105] Au WW, Salama SA, Sierra-Torres CH. Functional characterization of polymorphisms in DNA repair genes using cytogenetic challenge assays. *Environ Health Perspect* 2003; 111: 1843-50.

[106] Hu JJ, Smith TR, Miller MS, Mohrenweiser HW, Golden A, Case LD. Amino acid substitution variants of APE1 and XRCC1 genes associated with ionizing radiation sensitivity. *Carcinogenesis* 2001; 22: 917-22.

[107] Vineis P. Molecular epidemiology: low-dose carcinogens and genetic susceptibility. *Int J Cancer* 1997; 71: 1-3.

[108] Nakachi K, Imai K, Hayashi S, Watanabe J, Kawajiri K. Genetic susceptibility to squamous cell carcinoma of the lung in relation to cigarette smoking dose. *Cancer Res* 1991; 51: 5177-80.
[109] Coverley D, Kenny MK, Lane DP, Wood RD. A role for the human single-stranded DNA binding protein HSSB/RPA in an early stage of nucleotide excision repair. *Nucleic Acids Res* 1992; 20: 3873-80.
[110] Sancar A. Excision repair in mammalian cells. *J Biol Chem* 1995; 270: 15915-8.
[111] He Z, Henricksen LA, Wold MS, Ingles CJ. RPA involvement in the damage-recognition and incision steps of nucleotide excision repair. *Nature* 1995; 374: 566-9.
[112] Aboussekhra A, Biggerstaff M, Shivji MK, Vilpo JA, Moncollin V, Podust VN, Protic M, Hubscher U, Egly JM, Wood RD. Mammalian DNA nucleotide excision repair reconstituted with purified protein components. *Cell* 1995; 80: 859-68
[113] Mu D, Park CH, Matsunaga T, Hsu DS, Reardon JT, Sancar A. Reconstitution of human DNA repair excision nuclease in a highly defined system. *J Biol Chem* 1995; 270: 2415-2418
[114] Jones CJ, Wood RD. Preferential binding of the xeroderma pigmentosum group A complementing protein to damaged DNA. *Biochemistry* 1993; 32: 12096-104.
[115] Burns JL, Guzder SN, Sung P, Prakash S, Prakash L. An affinity of human replication protein A for ultraviolet-damaged DNA. *J Biol Chem* 1996; 271: 11607-10.
[116] Patrick SM, Turchi JJ. Human replication protein A preferentially binds cisplatin-damaged duplex DNA in vitro. *Biochemistry* 1998; 37: 8808-15.
[117] Matsuda T, Saijo M, Kuraoka I, Kobayashi T, Nakatsu Y, Nagai A, Enjoji T, Masutani C, Sugasawa K, Hanaoka F, Yasui A, Tanaka K. DNA repair protein XPA binds replication protein A (RPA). *J Biol Chem* 1995; 270: 4152-7.
[118] Li L, Lu X, Peterson CA, Legerski RJ. An interaction between the DNA repair factor XPA and replication protein A appears essential for nucleotide excision repair. *Mol Cell Biol* 1995; 15: 5396-402.
[119] Wakasugi M, Sancar A. Order of assembly of human DNA repair excision nuclease. *J Biol Chem* 274; 1999: 18759-68.
[120] Sugasawa K, Ng JM, Masutani C, Iwai S, van der Spek PJ, Eker AP, Hanaoka F, Bootsma D, Hoeijmakers JH. Xeroderma pigmentosum group C protein complex is the initiator of global genome nucleotide excision repair. *Mol Cell* 1998; 2: 223-32.
[121] Reardon JT, Nichols AF, Keeney S, Smith CA, Taylor JS, Linn S, Sancar A. Comparative analysis of binding of human damaged DNA-binding protein (XPE) and Escherichia coli damage recognition protein (UvrA) to the major ultraviolet photoproducts: T[c,s]T, T[t,s]T, T[6-4]T, and T[Dewar]T. *J Biol Chem* 1993; 268: 21301-8.
[122] Wakasugi M, Shimizu M, Morioka H, Linn S, Nikaido O, Matsunaga T. Damaged DNA-binding protein DDB stimulates the excision of cyclobutane pyrimidine dimers in vitro in concert with XPA and replication protein A. *J Biol Chem* 2001; 276: 15434-40.
[123] Butkiewicz D, Rusin M, Harris CC, Chorazy M. Identification of four single nucleotide polymorphisms in DNA repair genes: XPA and XPB (ERCC3) in Polish population. *Hum Mutat* 2000; 15: 577-8.

[124] Kozak M. Interpreting cDNA sequences: some insights from studies on translation. *Mamm Genome* 1996; 7: 563-74.

[125] Wu X, Zhao H, Wei Q, et al. XPA polymorphism associated with reduced lung cancer risk and a modulating effect on nucleotide excision repair capacity. *Carcinogenesis* 2003; 24: 505-9.

[126] Park JY, Park SH, Choi JE, Lee SY, Jeon HS, Cha SI, Kim CH, Park JH, Kam S, Park RW, Kim IS, Jung TH. Polymorphisms of the DNA repair gene xeroderma pigmentosum group A and risk of primary lung cancer. *Cancer Epidemiol Biomarkers Prev* 2002; 11: 993-7.

[127] Vogel U, Overvad K, Wallin H, Tjonneland A, Nexo BA, Raaschou-Nielsen O. Combinations of polymorphisms in XPD, XPC and XPA in relation to risk of lung cancer. *Cancer Lett* 2005; 222: 67-74.

[128] van Duin M, Koken MH, van den Tol J, ten Dijke P, Odijk H, Westerveld A, Bootsma D, Hoeijmakers JH. Genomic characterization of the human DNA excision repair gene ERCC-1. *Nucleic Acids Res* 1987; 15: 9195-213.

[129] Hoeijmakers HJ. Characterization of genes and proteins involved in excision repair of human cells. *J Cell Sci* 1987; 6: S111-25.

[130] Yu JJ, Mu C, Lee KB, Okamoto A, Reed EL, Bostick-Bruton F, Mitchell KC, Reed E. A nucleotide polymorphism in ERCC1 in human ovarian cancer cell lines and tumor tissues. *Mutat Res* 1997; 382: 13-20.

[131] Goode EL, Ulrich CM, Potter JD. Polymorphisms in DNA repair genes and associations with cancer risk. *Cancer Epidemiol Biomarkers Prev* 2002; 11: 1513-30.

[132] Ryu JS, Hong YC, Han HS, Lee JE, Kim S, Park YM, Kim YC, Hwang TS. Association between polymorphisms of ERCC1 and XPD and survival in non-small-cell lung cancer patients treated with cisplatin combination chemotherapy. *Lung Cancer* 2004; 44: 311-6.

[133] Yu JJ, Lee KB, Mu C, Li Q, Abernathy TV, Bostick-Bruton F, Reed E. Comparison of two human ovarian carcinoma cell lines (A2780/CP70 and MCAS) that are equally resistant to platinum, but differ at codon 118 of the ERCC1 gene. *Int J Oncol* 2000; 16: 555-60.

[134] Park DJ, Stoehlmacher J, Zhang W, Tsao-Wei D, Groshen S, Zahedy S, Gil J, Mallik N, Lenz H-J. ERCC1 polymorphism is associated with differential ERCC1 gene expression. *Proc AACR* 2002; 93: 1591.

[135] Park DJ, Zhang W, Stoehlmacher J, Tsao-Wei D, Groshen S, Gil J, Yun J, Sones E, Mallik N, Lenz HJ. ERCC1 gene polymorphism as a predictor for clinical outcome in advanced colorectal cancer patients treated with platinum-based chemotherapy. *Clin Adv Hematol Oncol* 2003; 1: 162-6.

[136] Winsey SL, Haldar NA, Marsh HP, Bunce M, Marshall SE, Harris AL, Wojnarowska F, Welsh KI. A variant within the DNA repair gene XRCC3 is associated with the development of melanoma skin cancer. *Cancer Res* 2000; 60: 5612-6.

[137] Vogel U, Laros I, Jacobsen NR, Thomsen BL, Bak H, Olsen A, Bukowy Z, Wallin H, Overvad K, Tjonneland A, Nexo BA, Raaschou-Nielsen O. Two regions in chromosome 19q13.2-3 are associated with risk of lung cancer. *Mutat Res* 2004; 546: 65-74.

[138] Zhou W, Liu G, Park S, Wang Z, Wain JC, Lynch TJ, Su L, Christiani DC. Gene-smoking interaction associations for the ERCC1 polymorphisms in the risk of lung cancer. *Cancer Epidemiol Biomarkers Prev* 2005; 14: 491-6.

[139] Yin J, Vogel U, Guo L, Ma Y, Wang H. Lack of association between DNA repair gene ERCC1 polymorphism and risk of lung cancer in a Chinese population. *Cancer Genet Cytogenet* 2006; 164: 66-70.

[140] Egly JM. The 14th Datta Lecture. TFIIH: from transcription to clinic. *FEBS Lett* 2001; 498: 124-8.

[141] Schaeffer L, Moncollin V, Roy R, Staub A, Mezzina M, Sarasin A, Weeda G, Hoeijmakers JH, Egly JM. The ERCC2/DNA repair protein is associated with the class II BTF2/TFIIH transcription factor. *EMBO J* 1994; 13: 2388-92.

[142] Coin F, Bergmann E, Tremeau-Bravard A, Egly JM. Mutations in XPB and XPD helicases found in xeroderma pigmentosum patients impair the transcription function of TFIIH. *EMBO J* 1999; 18: 1357-66.

[143] Spitz MR, Wu X, Wang Y, Wang LE, Shete S, Amos CI, Guo Z, Lei L, Mohrenweiser H, Wei Q. Modulation of nucleotide excision repair capacity by XPD polymorphisms in lung cancer patients. *Cancer Res* 2001; 61: 1354-7.

[144] Hou SM, Falt S, Angelini S, Yang K, Nyberg F, Lambert B, Hemminki K. The XPD variant alleles are associated with increased aromatic DNA adduct level and lung cancer risk. *Carcinogenesis* 2002; 23: 599-603.

[145] Zhou W, Liu G, Miller DP, Thurston SW, Xu LL, Wain JC, Lynch TJ, Su L, Christiani DC. Gene-environment interaction for the ERCC2 polymorphisms and cumulative cigarette smoking exposure in lung cancer. *Cancer Res* 2002; 62: 1377-81.

[146] Liang G, Xing D, Miao X, Tan W, Yu C, Lu W, Lin D. Sequence variations in the DNA repair gene XPD and risk of lung cancer in a Chinese population. *Int J Cancer* 2003; 105: 669-73.

[147] Hu Z, Wei Q, Wang X, Shen H. DNA repair gene XPD polymorphism and lung cancer risk: a meta-analysis. *Lung Cancer* 2004; 46: 1-10.

[148] Benhamou S, Sarasin A. ERCC2 /XPD gene polymorphisms and lung cancer: a HuGE review. *Am J Epidemiol* 2005; 161: 1-14.

[149] Xing D, Tan W, Wei Q, Lin D. Polymorphisms of the DNA repair gene XPD and risk of lung cancer in a Chinese population. *Lung Cancer* 2002; 38: 123-9.

[150] Shen M, Berndt SI, Rothman N, Demarini DM, Mumford JL, He X, Bonner MR, Tian L, Yeager M, Welch R, Chanock S, Zheng T, Caporaso N, Lan Q. Polymorphisms in the DNA nucleotide excision repair genes and lung cancer risk in Xuan Wei, China. *Int J Cancer* 2005; 116: 768-73.

[151] David-Beabes GL, Lunn RM, London SJ. No association between the XPD (Lys751Gln) polymorphism or the XRCC3 (Thr241Met) polymorphism and lung cancer risk. *Cancer Epidemiol Biomarkers Prev* 2001; 10: 911-2.

[152] Sijbers AM, de Laat WL, Ariza RR, Biggerstaff M, Wei YF, Moggs JG, Carter KC, Shell BK, Evans E, de Jong MC, Rademakers S, de Rooij J, Jaspers NG, Hoeijmakers JH, Wood RD. Xeroderma pigmentosum group F caused by a defect in a structure-specific DNA repair endonuclease. *Cell* 1996; 86: 811-22.

[153] de Laat WL, Appeldoorn E, Jaspers NG, Hoeijmakers JH. DNA structural elements required for ERCC1-XPF endonuclease activity. *J Biol Chem* 1998; 273: 7835-42.

[154] Matsunaga T, Park CH, Bessho T, Mu D, Sancar A. Replication protein A confers structure-specific endonuclease activities to the XPF–ERCC1 and XPG subunits of human DNA repair excision nuclease. *J Biol Chem* 1996; 271: 11047-50.

[155] Modrich P, Lahue R. Mismatch repair in replication fidelity, genetic recombination, and cancer biology. *Annu Rev Biochem* 1996; 65: 101-33.

[156] Leeflang EP, Tavare S, Marjoram P, Neal CO, Srinidhi J, MacFarlane H, MacDonald ME, Gusella JF, de Young M, Wexler NS, Arnheim N. Analysis of germline mutation spectra at the Huntington's disease locus supports a mitotic mutation mechanism. *Hum Mol Genet* 1999; 8: 173-83.

[157] Brookman KW, Lamerdin JE, Thelen MP, Hwang M, Reardon JT, Sancar A, Zhou ZQ, Walter CA, Parris CN, Thompson LH. ERCC4 (XPF) encodes a human nucleotide excision repair protein with eukaryotic recombination homologs. *Mol Cell Biol* 1996; 16: 6553-62.

[158] Smith TR, Levine EA, Perrier ND, Miller MS, Freimanis RI, Lohman K, Case LD, Xu J, Mohrenweiser HW, Hu JJ. DNA-repair genetic polymorphisms and breast cancer risk. *Cancer Epidemiol Biomarkers Prev* 2003; 12: 1200-4.

[159] Fan F, Liu C, Tavare S, Arnheim N. Polymorphisms in the human DNA repair gene XPF. *Mutat Res* 1999; 406: 115-20.

[160] MacInnes MA, Dickson JA, Hernandez RR, Learmonth D, Lin GY, Mudgett JS, Park MS, Schauer S, Reynolds RJ, Strniste GF. Human ERCC5 cDNA-cosmid complementation for excision repair and bipartite amino acid domains conserved with RAD proteins of Saccharomyces cerevisiae and Schizosaccharomyces pombe. *Mol Cell Biol* 1993; 13: 6393-402.

[161] Wakasugi M, Reardon JT, Sancar A. The non-catalytic function of XPG protein during dual incision in human nucleotide excision repair. *J Biol Chem* 1997; 272: 16030-4.

[162] Matsunaga T, Mu D, Park CH, Reardon JT, Sancar A. Human DNA repair excision nuclease: analysis of the roles of the subunits involved in dual incisions by using anti-XPG and anti-ERCC1 antibodies. *J Biol Chem* 1995; 270: 20862-9.

[163] Constantinou A, Gunz D, Evans E, Lalle P, Bates PA, Wood RD, Clarkson SG. Conserved residues of human XPG protein important for nuclease activity and function in nucleotide excision repair. *J Biol Chem* 1999; 274: 5637-48.

[164] Jeon HS, Kim KM, Park SH, Lee SY, Choi JE, Lee GY, Kam S, Park RW, Kim IS, Kim CH, Jung TH, Park JY. Relationship between XPG codon 1104 polymorphism and risk of primary lung cancer. *Carcinogenesis* 2003; 24: 1677-81.

[165] Cui Y, Morgenstern H, Greenland S, Tashkin DP, Mao J, Cao W, Cozen W, Mack TM, Zhang ZF. Polymorphism of Xeroderma Pigmentosum group G and the risk of lung cancer and squamous cell carcinomas of the oropharynx, larynx and esophagus. *Int J Cancer* 2006; 118: 714-20.

[166] Brenneman MA, Weiss AE, Nickoloff JA, Chen DJ. XRCC3 is required for efficient repair of chromosome breaks by homologous recombination. *Mutat Res* 2000; 459: 89-97.

[167] Pierce AJ, Johnson RD, Thompson LH, Jasin M. XRCC3 promotes homology-directed repair of DNA damage in mammalian cells. *Genes Dev* 1999; 13: 2633-8.

[168] Griffin CS. Aneuploidy, centrosome activity and chromosome instability in cells deficient in homologous recombination repair. *Mutat Res* 2002; 504: 149-55.

[169] Matullo G, Palli D, Peluso M, Guarrera S, Carturan S, Celentano E, Krogh V, Munnia A, Tumino R, Polidoro S, Piazza A, Vineis P. XRCC1, XRCC3, XPD gene polymorphisms, smoking and (32)P-DNA adducts in a sample of healthy subjects. *Carcinogenesis* 2001; 22: 1437-45.

[170] Araujo FD, Pierce AJ, Stark JM, Jasin M. Variant XRCC3 implicated in cancer is functional in homology-directed repair of double-strand breaks. *Oncogene* 2002; 21: 4176-80.

[171] Wang Y, Liang D, Spitz MR, Zhang K, Dong Q, Amos CI, Wu X. XRCC3 genetic polymorphism, smoking, and lung carcinoma risk in minority populations. *Cancer* 2003; 98: 1701-6.

[172] Jacobsen NR, Raaschou-Nielsen O, Nexo B, Wallin H, Overvad K, Tjonneland A, Vogel U. XRCC3 polymorphisms and risk of lung cancer. *Cancer Lett* 2004; 213: 67-72.

[173] Rafii S, O'Regan P, Xinarianos G, Azmy I, Stephenson T, Reed M, Meuth M, Thacker J, Cox A. A potential role for the XRCC2 R188H polymorphic site in DNA-damage repair and breast cancer. *Hum Mol Genet* 2002; 11: 1 433-8.

[174] Joenje H, Oostra AB, Wijker M, di Summa FM, van Berkel CG, Rooimans MA, Ebell W, van Weel M, Pronk JC, Buchwald M, Arwert F. Evidence for at least eight Fanconi anemia genes. *Am J Hum Genet* 1997; 61: 940-4.

[175] de Winter JP, Waisfisz Q, Rooimans MA, van Berkel CG, Bosnoyan-Collins L, Alon N, Carreau M, Bender O, Demuth I, Schindler D, Pronk JC, Arwert F, Hoehn H, Digweed M, Buchwald M, Joenje H. The Fanconi anaemia group G gene is identical with human XRCC9. *Nat Genet* 1998; 20: 281-83.

[176] Strathdee CA, Gavish H, Shannon WR, Buchwald M. Cloning of cDNAs for Fanconi anaemia by functional complementation. *Nature* 1992; 356: 763-7.

[177] Consortium TFABC. Positional cloning of the Fanconi anaemia group A gene. *Nat Genet* 1996; 14: 324-8.

[178] Lo Ten Foe JR, Rooimans MA, Bosnoyan-Collins L, Alon N, Wijker M, Parker L, Lightfoot J, Carreau M, Callen DF, Savoia A, Cheng NC, van Berkel CG, Strunk MH, Gille JJ, Pals G, Kruyt FA, Pronk JC, Arwert F, Buchwald M, Joenje H. Expression cloning of a cDNA for the major Fanconi anaemia gene, FAA. *Nat Genet* 1996; 14: 320-3.

[179] de Winter JP, Rooimans MA, van Der Weel L, van Berkel CG, Alon N, Bosnoyan-Collins L, de Groot J, Zhi Y, Waisfisz Q, Pronk JC, Arwert F, Mathew CG, Scheper RJ, Hoatlin ME, Buchwald M, Joenje H. The Fanconi anaemia gene FANCF encodes a novel protein with homology to ROM. *Nat Genet* 2000; 24: 15-6.

[180] Liu N, Lamerdin JE, Tucker JD, Zhou ZQ, Walter CA, Albala JS, Busch DB, Thompson LH. The human XRCC9 gene corrects chromosomal instability and mutagen sensitivities in CHO UV40 cells. *Proc Natl Acad Sci* (USA) 1997; 94: 9232-7.

[181] Carney JP, Maser RS, Olivares H, Davis EM, Le Beau M, Yates JR 3rd, Hays L, Morgan WF, Petrini JH. The hMre11/hRad50 protein complex and Nijmegen breakage syndrome: linkage of double-strand break repair to the cellular DNA damage response. *Cell* 1998; 93: 477-86.

[182] Tauchi H, Matsuura S, Kobayashi J, Sakamoto S, Komatsu K. Nijmegen breakage syndrome gene, NBS1, and molecular links to factors for genome stability. *Oncogene* 2002; 21: 8967-80.

[183] Lan Q, Shen M, Berndt SI, Bonner MR, He X, Yeager M, Welch R, Keohavong P, Donahue M, Hainaut P, Chanock S. Smoky coal exposure, NBS1 polymorphisms, p53 protein accumulation, and lung cancer risk in Xuan Wei, China. *Lung Cancer* 2005; 49: 317-23.

[184] Duval A, Hamelin R. Mutations at coding repeat sequences in mismatch repair deficient human cancers: toward a new concept of target genes for instability. *Cancer Res* 2002; 62: 2447-54.

[185] Simpkins SB, Bocker T, Swisher EM, et al. MLH1 promoter methylation and gene silencing is the primary cause of microsatellite instability in sporadic cancers. *Hum Mutat Genet* 1999; 8: 661-6.

[186] Park JY, Jeon H-S, Park SH, et al. Microsatellite alteration in histologically normal lung tissue of patients with non-small cell lung cancer. *Lung Cancer* 2000; 30: 83-9.

[187] Wu J, Gu L, Wang H, et al. Mismatch repair processing of carcinogen-DNA adducts triggers apoptosis. *Mol Cell Biol* 1999; 19: 8292-301.

[188] Hickman MJ, Samson LD. Role of DNA mismatch repair and p53 in signaling induction of apoptosis by alkylating agents. *Proc Natl Acad Sci* (USA) 1999; 96: 10764-9.

[189] Mellon I, Rajpal DK, Koi M, Boland CR, Champe GN. Transcription-coupled repair deficiency and mutations in human mismatch repair genes. *Science* 1996; 272: 557-60.

[190] Bertrand P, Tishkoff DX, Filosi N, Dasgupta R, Kolodner RD. Physical interaction between components of DNA mismatch repair and nucleotide excision repair. *Proc Natl Acad Sci* (USA) 1998; 95: 14278-83.

[191] Wang YC, Lu YP, Tseng RC, Lin RK, Chang JW, Chen JT, Shih CM, Chen CY. Inactivation of hMLH1 and hMSH2 by promoter methylation in primary non-small cell lung tumors and matched sputum samples. *J Clin Invest* 2003; 111: 887-95.

[192] Kolodner RD. Mismatch repair: mechanisms and relationship to cancer susceptibility. *Trends Biochem Sci* 1995; 20: 397-401.

[193] Hsieh P. Molecular mechanisms of DNA mismatch repair. *Mutat Res* 2001; 486: 71-87.

[194] Buersteddle JM, Alday P, Torhorst J, Weber W, Muller H, Scott R. Detection of new mutations in 6 out of Swiss HNPCC families by genomic sequencing of the MSH2 and MLH1 genes. *J Med Genet* 1995; 32: 909-12.

[195] Liu B, Nicolaides NC, Markowitz S, Willson JKV, Parson RE, Jen J, Papadopoulos N, Peltromaki P, de la Chapelle A, Hamilton SR, Kinzler KW, Vogelstein B. Mismatch repair gene defects in sporadic colorectal cancers with microsatellite instability. *Nat Genet* 1995; 9: 48-55.

[196] Wehner M, Buschhausen L, Lamberti C, Kruse R, Caspari R, Propping P, Friedl W. Hereditary nonpolyposis colorectal cancer (HNPCC): eight novel germline mutations in hMSH2 or hMLH1 genes. *Hum Mutat* 1997; 10: 241-4.

[197] Ito E, Yanagisawa Y, Iwahashi Y, Suzuki Y, Nagasaki H, Akiyama Y, Sugano S, Yuasa Y, Kazuo M. A core promoter and a frequent single-nucleotide polymorphism of the mismatch repair gene hMLH1. *Biochem Biophys Res Commun* 1999; 256: 488-94.

[198] Kane MF, Loda M, Gaida GM, Lipman J, Mishra R, Goldman H, Jessup JM, Kolodner R. Methylation of the hMLH1 promoter correlates with lack of expression of hMLH1 in sporadic colon tumors and mismatch repair-defective human tumor cell lines. *Cancer Res* 1997; 57: 808-11.

[199] Deng G, Peng E, Gum J, Terdiman J, Sleisenger M, Kim YS. R Methylation of hMLH1 promoter correlates with the gene silencing with a region-specific manner in colorectal cancer. *Br J Cancer* 2002; 86: 574-9.

[200] Park SH, Lee GY, Jeon HS, Lee SJ, Kim KM, Jang SS, Kim CH, Lee WK, Kam S, Park RW, Kim IS, Jung TH, Park JY. -93G-->A polymorphism of hMLH1 and risk of primary lung cancer. *Int J Cancer* 2004; 112: 678-82.

[201] Bellacosa A. Functional interactions and signaling properties of mammalian DNA mismatch repair proteins. *Cell Death Differ* 2004; 25: 1821-7.

[202] Jung CY, Choi JE, Park JM, Chae MH, Kang HG, Kim KM, Lee SJ, Lee WK, Kam S, Cha SI, Kim CH, Han SB, Jung TH, Jeon SH, Park JY. Polymorphisms in the hMSH2 gene and the risk of primary lung cancer. *Cancer Epidemiol Biomarkers Prev* 2006; 15: 762-8.

[203] Duell EJ, Wiencke JK, Cheng TJ, Varkonyi A, Zuo ZF, Ashok TD, Mark EJ, Wain JC, Christiani DC, Kelsey KT. Polymorphisms in the DNA repair genes XRCC1 and ERCC2 and biomarkers of DNA damage in human blood mononuclear cells. *Carcinogenesis* 2000; 21: 965-71.

[204] Matullo G, Peluso M, Polidoro S, Guarrera S, Munnia A, Krogh V, Masala G, Berrino F, Panico S, Tumino R, Vineis P, Palli D. Combination of DNA repair gene single nucleotide polymorphisms and increased levels of DNA adducts in a population-based study. *Cancer Epidemiol Biomarkers Prev* 2003; 12: 674-7.

[205] Palli D, Russo A, Masala G, Saieva C, Guarrera S, Carturan S, Munnia A, Matullo G, Peluso M. DNA adduct levels and DNA repair polymorphisms in traffic-exposed workers and a general population sample. *Int J Cancer* 2001; 94: 121-7.

[206] Kozak M. Role of ATP in binding and migration of 40S ribosomal subunits. *Cell* 1990; 22: 459-67.

[207] Kozak M. Initiation of translation in prokaryotes and eukaryotes. *Gene* 1999; 234: 187-208.

[208] Kozak M. Adherence to the first-AUG rule when a second AUG codon follows closely upon the first. *Proc Natl Acad Sci* (USA) 1995; 92: 2662-6.

[209] Kozak M. Pushing the limits of the scanning mechanism for initiation of translation. *Gene* 2002; 299: 1-34.

[210] Tanaka K, Miura N, Satokata I, Miyamoto I, Yoshida MC, Satoh Y, Kondo S, Yasui A, Okayama H, Okada Y. Analysis of a human DNA excision repair gene involved in

group A xeroderma pigmentosum and containing a zinc-finger domain. *Nature* 1990; 348: 73-6.

[211] Kanaji T, Okamura T, Osaki K, Kuroiwa M, Shimoda K, Hamasaki N, Niho Y. A common genetic polymorphism (46 C to T substitution) in the 5'-untranslated region of the coagulation factor XII gene is associated with low translation efficiency and decrease in plasma factor XII level. *Blood* 1998; 91: 2010-4.

[212] Kozak M. Primer extension analysis of eukaryotic ribosome-mRNA complexes. *Nucleic Acids Res* 1998; 26: 4853-9.

[213] Sarasin A. An overview of the mechanisms of mutagenesis and carcinogenesis. *Mutat Res* 2003; 544: 99-106.

[214] Wang Y, Spitz MR, Zhu Y, Dong Q, Shete S, Wu X. From genotype to phenotype: correlating XRCC1 polymorphisms with mutagen sensitivity. *DNA Rep* 2003; 2: 901-8.

[215] Abdel-Rahman SZ, El-Zein RA. The 399Gln polymorphism in the DNA repair gene XRCC1 modulates the genotoxic response induced in human lymphocytes by the tobacco-specific nitrosamine NNK. *Cancer Lett* 2000; 159: 63-71.

[216] Wacholder S, Chanock S, Garcia-Closas M, El Ghormli L, Rothman N. Assessing the probability that a positive report is false: an approach for molecular epidemiology studies. *J Natl Cancer Inst* 2004; 96: 434-42.

[217] Healey CS, Dunning AM, Teare MD, Chase D, Parker L, Burn J, Chang-Claude J, Mannermaa A, Kataja V, Huntsman DG, Pharoah PD, Luben RN, Easton DF, Ponder BA. A common variant in BRCA2 is associated with both breast cancer risk and prenatal viability. *Nat Genet* 2000; 26: 362-4.

[218] Spurdle AB, Hopper JL, Chen X, Dite GS, Cui J, McCredie MR, Giles GG, Ellis-Steinborner S, Venter DJ, Newman B, Southey MC, Chenevix-Trench G. The BRCA2 372 HH genotype is associated with risk of breast cancer in Australian women under age 60 years. *Cancer Epidemiol Biomarkers Prev* 2002; 11: 413-6.

[219] Stephens JC, Schneider JA, Tanguay DA, Choi J, Acharya T, Stanley SE, Jiang R, Messer CJ, Chew A, Han JH, Duan J, Carr JL, Lee MS, Koshy B, Kumar AM, Zhang G, Newell WR, Windemuth A, Xu C, Kalbfleisch TS, Shaner SL, Arnold K, Schulz V, Drysdale CM, Nandabalan K, Judson RS, Ruano G, Vovis GF. Haplotype variation and linkage disequilibrium in 313 human genes. *Science* 2001; 293: 489-93.

[220] Judson R, Stephens JC, Windemuth A. The predictive power of haplotypes in clinical response. *Pharmacogenomics* 2000; 1: 15-26.

[221] Fallin D, Cohen A, Essioux L, Chumakov I, Blumenfeld M, Cohen D, Schork NJ. Genetic analysis of case/control data using estimated haplotype frequencies: application to APOE locus variation and Alzheimer's disease. *Genome Res* 2001; 11: 143-51.

[222] Khoury M, Beaty TH, Cohen BH. *Fundamentals of Genetic Epidemiology. Monographs in Epidemiology and Biostatistics*, New York, NY Oxford University Press 1993; 144-145.

[223] Hoh J, Wille A, Ott J. Trimming, weighting, and grouping SNPs in human case-control association studies. *Genome Res* 2001; 11: 2115-9.

[224] Toniolo P, Boffetta P, Shuker D, Rothman N, Hulka B, Pearce N. Application of biomarkers in cancer epidemiology (IARC Scientific Publication no. 142), 1997.

[225] Vineis P, Maltas M, Lang M, D'Errico A, Caporaso N, Cuzick J, Boffeta P. Metabolic polymorphisms and susceptibility to cancer (*IARC Scientific Publication* no. 148), 1999.

[226] Todd JA. Interpretation of results from genetic studies of multifactorial diseases. *Lancet* 1999; 354: S115-6.

[227] Kiyohara C, Yoshimasu K, Takayama K, Nakanishi Y. NQO1, MPO, and the risk of lung cancer: a HuGE review. *Genet Med* 2005; 7: 463-78.

[228] Kiyohara C, Yoshimasu K, Takayama K, Nakanishi Y. EPHX1 polymorphisms and the risk of lung cancer: a HuGE review. *Epidemiology* 2006; 17: 89-99.

[229] Caporaso NE. Why have we failed to find the low penetrance genetic constituents of common cancers? *Cancer Epidemiol Biomarkers Prev* 2002; 11: 1544-9.

[230] Brennan P. Gene-environment interaction and aetiology of cancer: what does it mean and how can we measure it? *Carcinogenesis* 2002; 23: 381-7.

[231] Kiyohara C, Otsu A, Shirakawa T, Fukuda S, Hopkin JM. Genetic polymorphisms and lung cancer susceptibility: a review. *Lung Cancer* 2002; 37: 241-56.

[232] Kiyohara C, Yoshimasu K, Shirakawa T, Hopkin JM. Genetic polymorphisms and environmental risk of lung cancer: a review. *Rev Environ Health* 2004; 19: 15-38.

In: New Research on DNA Repair
Editor: Breehn R. Landseer, pp. 85-98

ISBN 1-60021-385-5

Chapter III

COMET ASSAY AS A TOOL TO EVALUATE DNA REPAIR ACTIVITY AND INHIBITION

Delia Cavallo* and Cinzia Lucia Ursini
Department of Occupational Medicine, ISPESL National Institute for Occupational Safety and Prevention, Via Fontana Candida, 1 - 00040 Monteporzio Catone, Rome, Italy

ABSTRACT

DNA repair systems include enzymes that recognize and remove DNA adducts, correct the DNA sequence and re-join strand breaks. These systems provide to repair oxidative and alkylated DNA lesions and include base-excision repair (BER), nucleotide excision repair (NER), mismatch repair and O^6-methylguanine-DNA methyltransferase. Specific mechanisms for strand breaks repair include non-homologous end-joining and homologous recombination. Current evidence suggests that DNA repair systems are very sensitive target for several carcinogenic agents such as heavy metals (nickel, cadmium, cobalt, chromium and arsenic), PAHs, UV-B, ionizing radiations, aromatic amines, cigarette smoking and others. These agents cause diminished removal of endogenous and exogenously induced DNA damage increasing risk of tumour formation. Such compounds or their metabolites inhibit DNA repair probably by oxidative damage or direct interactions with reparative enzyme. In the last years DNA repair activity and inhibition has been widely studied by comet test, a rapid and sensitive technique that allows to evaluate direct and oxidative DNA damage and its repair on single whole cell. In particular, this technique allows the detection of single and double strand breaks, alkali-labile sites and some its applications, such as Fpg or endo III modified Comet assay, allow to evaluate oxidative DNA damage measuring oxidized purine and pyrimidine respectively. This technique does not require DNA isolation avoiding spurious DNA damage due to the sample handling. Comet assay permits to study DNA repair evaluating the persistence of either direct and oxidative DNA damage in relation to

* Correspondence concerning this article should be addressed to: Delia Cavallo, Department of Occupational Medicine, ISPESL-National Institute for Occupational Safety and Prevention, Via Fontana Candida 1 - 00040 Monteporzio Catone, Rome, Italy; E-mail: delia.cavallo@ispesl.it

repair time, performing the test on cells immediately after induction of DNA damage and after DNA recovery time. Therefore this technique is an useful tool to study the influence of xenobiotics on the different steps involved in DNA damage induction and repair. This review will present the most important applications of this technique to evaluate DNA repair alterations due to various xenobiotics particularly heavy metals such as Nickel compounds for which DNA repair inhibition was demonstrated also by an our recent study. This technique and its modifications, applicable to all cellular types, represent an useful, versatile, cheap and sensitive tool to contribute to clarify the mechanisms of inhibitory processes of enzymes involved in carcinogenesis.

Keywords: DNA repair alterations, comet assay, cancer

INTRODUCTION

The cellular DNA is subject to several modifications due to normal cell functions and to external agents that induce either direct DNA damage by formation of purine dimers, cross-links, strand breaks and adducts, or oxidative DNA damage induced by generation of reactive oxygen species (ROS). Cells are constantly exposed to oxidant from physiological processes and pathophysiological conditions such as inflammation, ischemia/reperfusion, foreign compounds metabolism and radiation (Ames et al 1995). In particular the oxidative DNA damage can be induced by endogenous processes such as reactive oxygen species generated continuously by oxygen metabolism, but also by environmental factors that can induce an increase of DNA oxidation. The defence against oxidants consists of antioxidant enzymes, radical scavengers and chain breakers. Failure of the system of enzymatic endogenous and nutritional antioxidants may lead to mutagenic oxidative DNA damage as well as deregulation of cell cycle control, resulting in carcinogenesis (Poulsen et al 2005). Oxidative stress-induced mutations are also suggested to play a major role in several chronic diseases including neurodegenerative disorders and cardiovascular diseases (Cooke et al 2003, Marnett 2000, Hussain et al 2003).

Many xenobiotics such as ionizing radiation, UV-light, alkylating agents, polycyclic aromatic hydrocarbons (PAH) and heterocyclic aromatic amines, have been identified to compromise genetic stability by inducing different type of DNA damage either directly or after metabolic conversion to reactive intermediates or by ROS induction. To reduce adverse consequences of DNA damage a complex network of different repair systems has evolved to maintain genomic integrity (Christmann et al 2003) these act in association with cell cycle check points that are activated by DNA damage and that arrest cell cycle progression to allow time for DNA repair to occur (Nyberg et al 2002).

Replication errors are repaired by mismatch repair pathway and double strand breaks by homologous or illegitimate recombination processes, while the excision repair systems (Nucleotide excision repair (NER) and base excision repair (BER)) eliminate DNA base damage and helix distortions. Several xenobiotics such as carcinogenic metal compounds (nickel, cadmium, cobalt, chromium and arsenic), PAHs, UV-B, ionizing radiations, aromatic amines, cigarette smoking are able to interfere with DNA repair systems inhibiting or reducing their activity then increasing the adverse consequences of DNA lesions.

In the last years DNA repair activity and inhibition has been widely studied by comet test, a rapid and sensitive technique that allows to evaluate direct and oxidative DNA damage and its repair on single whole cells. In particular this technique allows the detection of single and double strand break, alkali-labile sites and some of its applications, such as Fpg and endo III modified Comet assay, allow to evaluate oxidative DNA damage measuring oxidized purine and pyrimidine respectively. This technique is also used to study DNA repair evaluating the persistence of either direct and oxidative DNA damage in relation to repair time, performing the test on cells immediately after induction of DNA damage and after DNA recovery time. This technique also permits to study the eventual DNA repair inhibition activity of different xenobiotic agents evaluating the recovery of DNA damage induced by an exposure to known genotoxic agents in presence or absence of the studied xenobiotics.

This review will present the most important applications of comet test to evaluate DNA repair alterations due to various xenobiotics and special emphasis will be given to heavy metals for which DNA repair inhibition has been widely demonstrated in several studies.

DNA Repair Systems

Systems of response to DNA damage are known as DNA repair, repair enzymes that recognize and remove DNA adducts, correct the DNA sequence, and rejoin strand breaks. More than 130 genes have been identified from the human genome that contribute to DNA repair (Wood et al 2001). The disruption of these pathways is associated to increased cancer risk (colon cancer, skin cancer, lymphoma, lymphoma/leukaemia, leukaemia, breast/ovarian cancer, retinoblastoma, and others) (Digweed 2003). The main DNA repair system is the excision repair system that consists of nucleotide excision repair (NER) and base excision repair (BER). BER represents the predominant pathway used to remove oxidized and many of the alkylated bases. In the first step of BER the damaged bases are recognized and hydrolized by specific DNA N-glycosylase, resulting in the formation of apurinic/ apirimidinic (AP) sites. These sites are removed by specific apurinic endonucleases, the DNA strand breaks formed are then filled with newly synthesized DNA and the ends are ligated in the final step (Friedberg et al 1995). NER is able to repair more complex types of DNA damage such as pyrimidine dimers. In the first step of this pathway the Xeroderma pigmentosum (XP)C-associated enzyme complex recognizes the pyrimidine dimer, a nuclease nicks the DNA on both sides of the lesion and the oligonucleotide is removed by DNA helicase then the gap is repaired by DNA polymerase and DNA ligase using the complimentary strand as a template (Kao et al. 2005). Patients suffering from the DNA repair disorder Xeroderma pigmentosum (XP) have defects in certain genes coding for NER, they are extremely sensitive to sunlight and have about 1000 time higher risk for cancer development (Filipič et al 2006).

Other DNA repair pathways principally acting on direct-DNA damage (single and double strand breaks, alkyl adducts, mismatches) include: O6-alkylguanine DNA alkyltransferase (AGT) that repairs alkyl adducts such as O6-methyl guanine methyltransferase (MGMT); repair of cross-link by DNA inter-strand cross-link repair; repair of double strand breaks by homologous recombination (HR), non-homologous end joining (NHEJ) and single strand annealing (SSA) and DNA mismatch repair (MMR) that repairs mismatches and insertion/deletion loops (Kao et al. 2005).

There is a great interest on the influence of polymorphism of genes involved in base excision repair (OGG1, XRCC1), nucleotide excision repair (ERCC1, XPC, XPD, XPF) and double strand break repair (BRCA2, and XRCC3) on cancer risk. An association between the OGG1 S326C polymorphism and increased risk of lung, prostate and oesophageal cancer was found, while the XRCC1 R194W polymorphism seems associated with reduced risk of breast, lung, stomach and head and neck cancer and BRCA2 N372H polymorphism seems associated with breast cancer (Goode et al 2002). Current evidence suggests that DNA repair systems are very sensitive target for several carcinogenic agents such as heavy metals, PAHs, UV-B, ionizing radiations, aromatic amines, cigarette smoking and others. These agents cause diminished removal of endogenous and exogenously induced DNA damage increasing risk of tumour formation. Such agents or their metabolites inhibit DNA repair probably by oxidative damage or direct interaction with reparative enzymes and act as co-carcinogens, enhancing the mutagenic and carcinogenic activity of genotoxic carcinogens.

Assays of DNA Repair

Assays currently used to evaluate DNA repair capacity include principally: tests based on DNA damage induced with chemicals or physical agents (the mutagen sensitivity assay, the G2-radiation assay, induced micronuclei, comet assay); indirect tests of DNA repair (unscheduled DNA synthesis); tests that measure repair kinetics (the host cell reactivation assay); measurements of genetic variations associated with DNA repair; combination of more than one category of assay (Berwick and Vineis 2000). The most used assays are based on the in vitro evaluation of presence of DNA damage, measured as unrepaired single- or double-strand breaks or the rate of incorporation of a radioisotope, after a recovery time following the induction of DNA damage by pulse of physical or chemical agent (γ-rays, UV radiation, benzo[a]pyrene diol epoxide [BPDE], hydrogen peroxide, and methyl methane sulphonate (MMS)). Among these tests the comet assay or single cell gel electrophoresis (SCGE) that represents a rapid, cheap and sensitive technique, consents to study the DNA repair activity and inhibition at individual cell level either in vitro and in vivo on different cellular types. Moreover another advantage of this technique in repair studies is that it does not require DNA isolation avoiding spurious DNA damage due to sample handling.

Comet Assay

The comet assay or single cell gel electrophoresis (SCGE) consents to demonstrate the damaging effects of different compounds or physical treatment on DNA at individual cell level. This technique allows the detection of moderate but still appreciable damage even at low exposures, and gives a good preliminary idea of the possible early genotoxic and oxidative effects (Collins et al. 1993, 1995). In this methodology a cellular suspension is embedded in agarose on a microscope slide and lysed by detergents and high salt concentration, in order to liberate the DNA, so the slides are treated at alkaline conditions so that the DNA unwinds from sites of strand breakage. During the electrophoresis at the same alkali condition, DNA fragments induced by genotoxic agents migrate to the anode of the chamber. Subsequently the slides are washed with neutralising solution and stained with fluorescent DNA binding dyes.

Cells with damaged DNA show increased migration of DNA fragments from the nucleus generating a comet shape. The migration length of the tail and parameters as percentage of DNA in the tail and tail moment (the product of tail length and DNA tail fluorescence intensity) which are proportional to DNA damage, are commonly evaluated by a specific software. Under alkaline conditions, pH 13, comet test detects single-strand breaks (SSB), double strand breaks (DSB) and alkali-labile sites (Singh et al. 1988). This assay, modified with use of proteinase K, can be also used to evaluate DNA cross-link lesions (Merk et al. 2000) that are related to no increased or decreased DNA migration.

An application of this technique is the enzyme (endo III or Fpg) modified Comet assay to evaluate oxidative DNA damage. Endo III converts oxidized pyrimidines into strand breaks and Fpg glycosylase recognises and specifically cuts the oxidized bases principally 8-oxoguanine from DNA producing apurinic site converted in breaks by the associated AP-endonuclease activity, which can be detected by comet assay and give a specific and accurate measure of oxidative DNA damage (Evans et al. 1995).

This technique allows to study DNA repair evaluating the persistence of either direct and oxidative DNA damage in relation to repair time, performing the test on cells immediately after induction of DNA damage and after DNA recovery time. Therefore it is a useful tool to study the influence of xenobiotics on the different steps involved in DNA damage induction and repair (Collins 2004, Speit and Hartmann 2006, Kim et al. 2002). Comet assay and its modifications is applicable to all cell types and represents a useful, versatile, cheap and sensitive tool to contribute to clarify the mechanisms of inhibitory processes of enzymes, involved in carcinogenesis.

Comet Assay to Evaluate DNA Repair Alterations

The use of comet assay in carcinogenesis studies is gradually increased, in particular it has been employed to evaluate the effects of xenobiotics in relation to hypothesized or known mechanisms of action of different genotoxic substances such as direct and oxidative DNA damage and inhibition of DNA repair (Cavallo et al. 2003, 2005, Karlsson et al. 2005, Gutierrez-Castillo et al. 2006). Moreover in the last years this assay has become an important tool in human biomonitoring studies principally involving exposed workers (Andreoli et al. 1997; Garaj-Vrhovac and Zeljeric 2000, Speit et al. 2003, Cavallo et al 2002, 2006a, 2006b, Ursini et al. 2006).

This assay allows to study the influence of several xenobiotics such as metals on the different steps involved in induction of DNA damage and DNA repair.

In particular the influence on DNA repair systems can be studied in vitro, evaluating the recovery of DNA damage induced by exposure to certain DNA damaging agents (pulse with physical or chemical agents) in absence or presence of the studied xenobiotics or assessing the persistence of xenobiotics-induced DNA damage after recovery time in absence of exposure (Fig.1A). In biomonitoring studies DNA repair alterations can be studied evaluating, on lymphocytes from exposed populations treated in vitro with known DNA damaging agents, the recovery efficiency or evaluating the persistence of DNA damage after in vitro recovery time on lymphocytes from exposed populations in respect to controls (Fig. 1B).

A ***In vitro studies***

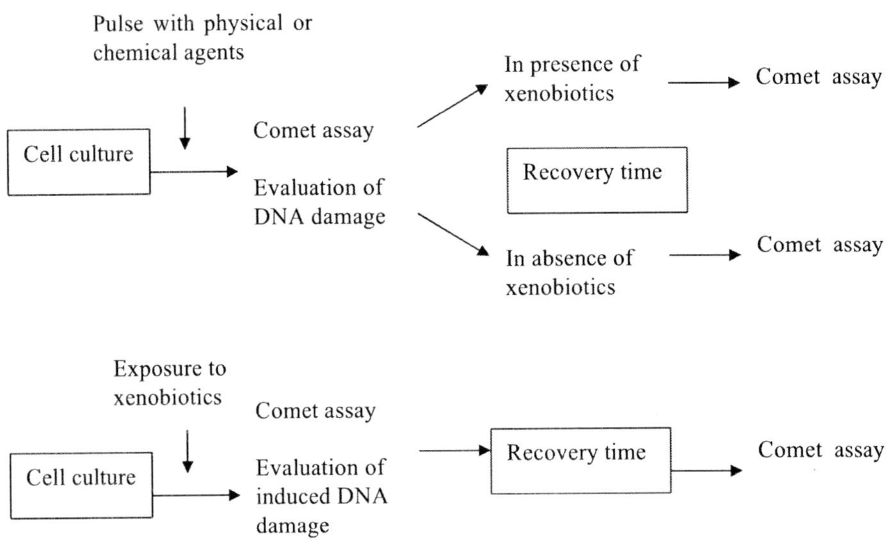

B ***Biomonitoring studies***

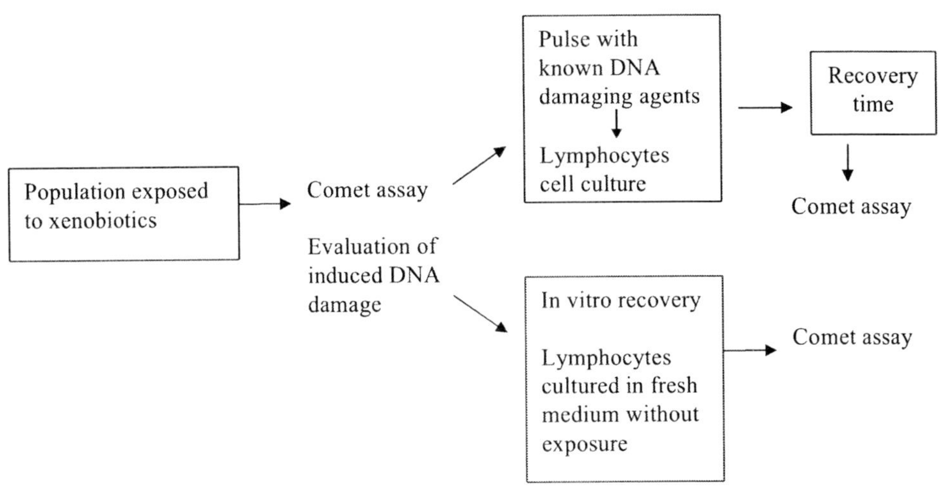

Figure 1. Experimental approaches to study by comet assay DNA repair alterations induced by xenobiotics. (A) In vitro studies (B) Biomonitoring studies

An example of the first approach used in in vitro studies is the evidence of an inhibition by arsenic and cadmium ($NaAsO_2$ and $CdSO_4$) of benzo[a]pyrene and methyl metanesulphonated-induced DNA repair process demonstrated by comet assay in human blood cells and SV40 transformed fibroblasts (Hartmann and Speit 1996).

For metal compounds an interference on DNA repair systems was showed in several studies indicating that metal compounds exert comutagenic effects which may be explained by the inhibition of different DNA repair systems (Hartwig et al 2002). During the last years there has been accumulating evidences that diverse DNA repair systems including BER and NER are inhibited at low, non cytotoxic concentrations of Ni(II), Co(II), Cd(II) and As(III).

In Table 1 we report the main in vitro studies evaluating, by comet assay, DNA repair alterations induced by the most commonly handled metal compounds.

In particular a study on TK6 human lymphoblastoid cells exposed to As(III) showed an impairment of nucleotide excision repair pathway in UV treated cells (Danaee et al. 2004).

As previously referred an inhibition by arsenic and cadmium of DNA repair process on benzo[a]pyrene and methyl metanesulphonate-induced DNA damage has been demonstrated in human blood cells and SV40 transformed fibroblasts (Hartmann and Speit 1996). The inhibition of DNA repair by Cd and As compounds has been also showed in human lung fibroblasts (MRC-5 cell line) exposed to $CdCl_2$, $CdSO_4$, $NaAsO_2$ and $C_2H_7AsO_2$ that evidenced an increase of DNA damage due to direct induction of DNA lesions or to inhibition of excision repair mechanisms (Mouròn et al. 2001).

The effect of cadmium on DNA repair systems was studied in $CdCl_2$ adapted alveolar epithelial cells that showed less efficiency at repairing oxidative DNA damage compared with non-adapted cells (Potts et al. 2001) and a decrease of DNA repair was also observed in Mytilus edulis gill cells exposed to Cd and treated with H_2O_2 (Pruski and Dixon 2002). The influence of cadmium on repair kinetics of DNA damage induced by UV-Radiation, MMS and N-methyl-N-nitrosourea (MNU) was examined on Chinese hamster ovary (CHO) cells by Fatur et al. (2003). In this study the authors found that in cells exposed to UV-irradiation alone, the formation of DNA strand break was rapid followed by a fast rejoining phase during the first 60 min after irradiation. While in irradiated cells pre-exposed to CdCl2, the formation of DNA strand breaks was slower and, after 60 min of recovery, there was no decline of the level of DNA strand breaks. Therefore the cadmium prevented the initial step of NER, the DNA damage recognition and/or incision/excision steps as well as later step of NER. A decrease of DNA repair capability induced by Cd Cl_2 was also found on rat lymphocytes and V79 chinese hamster lung cells treated with hydrogen peroxide (Jianhua et al. 2006).

A recent study has investigated, by comet assay, the level of DNA damage in human peripheral blood lymphocytes treated with Al and the impact of Al on the repair of DNA damage induced by ionizing radiation. The results of this study demonstrate that Al induces DNA damage in a dose-dependent manner. Moreover irradiated cells pre-treated with Al have an increased level of the initial yield of DNA damage, but also a decreased capacity of repair (Lankoff et al 2006).

The influence of Ni on DNA repair systems was evaluated in an our previous study performed on Jurkat cells exposed to low doses of $NiSO_4$ in which we found that Ni compound is able to induce oxidative DNA damage and inhibit DNA damage repair after treatment with H_2O_2 (Cavallo et al 2003). An inhibition of various steps of DNA excision repair after treatment with UV radiation and MNNG- was showed in human lymphocytes exposed to Nickel chloride (Wozniak and Blasiak 2004).

Table 1. In vitro studies on DNA repair alterations due to exposure to metals evaluated by comet assay.

Exposure	Observed effects	References
Exposure of TK6 human lymphoblastoid cells to As(III)	Impairment of nucleotide excision repair pathway in UV treated cells	Danaee et al. 2004
Exposure of human blood cells and SV40 transformed fibroblasts cells to arsenic and cadmium	Inhibition of benzo[a]pyrene and MMS-induced DNA repair process	Hartmann and Speit 1996
Exposure of human lung fibroblast cells to $CdCl_2$, $CdSO_4$ $NaAsO_2$ and $C_2H_7AsO_2$	Induction of DNA damage due to inhibition of excision repair mechanisms	Mouròn et al. 2001
$CdCl_2$ adapted alveolar epithelial cells	Inhibition of oxidative DNA repair	Potts et al. 2001
Exposure of Mytilus edulis gill cells to Cd	Decrease of DNA repair in cells treated with H_2O_2	Pruski and Dixon 2002
Pre-exposure to $CdCl_2$ of Chinese hamster ovary cells	Inhibition of DNA damage (by UV, MMS- and N-methyl-N-nitrosourea) repair.	Fatur et al. 2003
Exposure of rat lymphocytes and V79 chinese hamster lung cells to $CdCl_2$	Decrease of DNA repair capability in cells treted with H_2O_2	Jianhua et al. 2006
Exposure of human lymphocytes to Al Cl_3	Induction of DNA damage in a dose-dependent manner and decrease of DNA repair capacity	Lankoff et al. 2006
Exposure of Jurkat cells to $NiSO_4$	Induction of oxidative reactive species and inhibition of DNA repair capacity	Cavallo et al. 2003
Exposure of human lymphocytes to Nickel chloride	Impairment of the repair of UV- and MNNG- damaged DNA	Wozniak and Blasiak 2004
Exposure of human lymphocytes to Cobalt metal, cobalt with tungsteno carbide and cobalt chloride	Induction of DNA damage in a dose and time-dependent way. Inhibition of the repair of methylmetane-sulphonated-induced DNA damage by Co metal	De Boeck et al. 1998
Exposure of human lymphocytes to Cr(III) and Cr(VI)	Induction of oxidative DNA damage induced by hexavalent chromium and direct DNA damage by trivalent chromium. Inhibition of DNA repair by Cr(III)	Blasiak and Kowalik 2000
Exposure of human gastric mucosa cells and lymphocytes to potassium dichromate ($K_2Cr_2O_7$)	Induction of DNA damage and inhibition of DNA repair	Trzeciak et al. 2000

Table 2. Biomonitoring studies on DNA repair alterations by xenobiotics evaluated by comet assay.

Exposure	Observed effects	References
Workers of electric battery factories exposed to lead compounds	Interference with DNA repair after in vitro exposure of lymphocytes to X-rays	Restrepo et al. 2000
Workers of battery plant exposed to lead or cadmium	Induction of DNA damage due to both the exposures and reduction of DNA repair efficiency by Cd exposure	Palus et al. 2003
Workers exposed to welding fumes containing Al, Cd, Cr, Co, Pb, Mn, Ni, Zn	Induction of DNA strand breaks between the beginning and the end of the week. The DNA damage found in this study might be a direct effect or the secondary effect due to DNA repair inhibition	Iarmarcovai et al. 2005
Case-control study of workers exposed to mercury vapour and matched controls	Lymphocytes from workers had significant repair deficiency compared to the controls. The occupational exposure to mercury vapour cause base-excision repair deficiency.	Cebulska-Wasilewska et al. 2005a
Workers employed in pesticide production	Induction of DNA damage and decrease of DNA repair efficiency	Garaj-Vrhovac and Zeljezic 2000
Agricultural workers exposed to a complex mixture of pesticides	No induction of DNA damage no interference with DNA repair process after H_2O_2 induced DNA damage	Piperakis et al. 2003
City policemen exposed to PAHs	Impairment of cellular DNA repair process on lymphocytes treated in vitro with X-rays	Cebulska-Wasilewska et al. 2005b

A study on human lymphocytes exposed to Cobalt metal, cobalt with tungsten carbide and cobalt chloride showed a dose and time dependent induction of DNA damage. Moreover DNA repair capacity after treatment with methyl methanesulphonate (MMS) and further incubation with Co metal was investigated and the inhibition of the repair of MMS-induced DNA damage by Co metal was found (De Boeck et al 1998).

The interference of Cr compounds with DNA repair was studied by Blasiak and Kowalik (2000) on human lymphocytes exposed to $CrCl_3$ and $K_2Cr_2O_7$. The study shows an induction of oxidative DNA damage by hexavalent chromium and direct DNA damage by trivalent chromium with inhibition of DNA repair by Cr(III) (Blasiak and Kowalik 2000). While an induction of DNA damage and inhibition of DNA repair by Cr (VI) was found by Trzeciak et al. (2000) on human gastric mucosa cells and lymphocytes exposed to potassium dichromate ($K_2Cr_2O_7$).

Actually several biomonitoring studies using comet assay to evaluate on exposed populations the influence of exposure to xenobiotics on DNA repair efficiency are available (Table 2).

Among these, a study on workers of electric battery factories exposed to lead compounds reports genotoxic effects and an interference of lead exposure with DNA repair after in vitro exposure of lymphocytes to X-rays (Restrepo et al. 2000). In another study on workers of a battery plant exposed to lead or cadmium an induction of genotoxic effects was observed for both the exposures and an inhibition of DNA repair due to exposure to Cd was found (Palus et al. 2003).

An induction of DNA strand breaks on welders exposed to eight metals (Al, Cd, Cr, Co, Pb, Mn, Ni, Zn) present in welding fume was showed, by comet assay, during the work week. The DNA damage found in this study might be a direct effect or the secondary effect due to DNA repair inhibition (Iarmacovai et al 2005).

A case-control study of workers exposed to mercury vapour was conducted to investigate whether occupational exposure to low concentrations of mercury vapour cause genotoxicity and can be associated with defective DNA repair capacity. The comet assay analysis showed a lack of DNA strand breaks induction. To evaluate the DNA repair efficiency, lymphocytes from exposed workers and from controls were treated with UV-rays and X-rays then the irradiated cells were analyzed by comet assay for the presence of DNA strand breaks immediately and after 2h incubation in culture medium with and without phytohemagglutinin (recovery time). They found that after incubation the cells from exposed workers showed a significant reduction in the repair of X-ray- but not UV-light-induced DNA damage in respect to controls. The study suggests that the occupational exposure to mercury cause significant deficiency in DNA base-excision repair (Cebulska-Wasilewska et al 2005a).

In a study performed on workers employed in pesticide production an induction of DNA damage and a decrease of DNA repair efficiency was found in exposed workers in respect to controls by Garaj-Vrhovac and Zeljezic (2000). While agricultural workers, exposed to a complex mixtures of pesticides at low doses, did not show significant DNA damage or DNA repair alterations after in vitro DNA damage induction of lymphocytes by H_2O_2 treatment (Piperakis et al. 2003).

In another biomonitoring study on city policemen exposed to PAHs the comet assay evidenced an impairment of cellular DNA repair process after in vitro X-rays treatment of lymphocytes due to PAH exposure (Cebulska-Wasilewska et al. 2005b).

CONCLUSION

The role of DNA repair inhibition systems in the carcinogenesis of metal compounds has been largely demonstrated. Comet assay widely contributed to clarify the mechanisms of metal carcinogenesis as evidenced by the numerous recent studies available in the field. Metal compounds seem to act prevalently as co-carcinogen even if a direct carcinogenic effect was however showed. The comet assay and its modifications, allowing the evaluation of direct and oxidative DNA damage and the study of DNA repair activity, in a sensitive and reliable way, represents an useful tool to clarify the mechanisms involved in the carcinogenesis process induced by different agents. Combining this technique with the evaluation of specific DNA repair enzyme activity during recovery process from induced DNA damage in presence of specific xenobiotics, the specific inhibition system involved in the induced carcinogenesis process could be clarified.

The comet assay and its modifications, applicable to all cellular types, represent an useful, versatile, cheap and sensitive tool to contribute to clarify the mechanisms of inhibitory processes of enzymes involved in carcinogenesis. The recent wide application and the increasing use of comet assay to study the DNA damage induction and DNA repair mechanisms either in biomonitoring or experimental studies demonstrates the suitability of this tool to investigate carcinogenesis and DNA repair processes with possible important basic and clinical implications.

The elucidation of mechanisms of the inhibitory processes of repair enzymes involved in carcinogenesis could be useful to develop new anticancer approach that could act at molecular level on repair enzymatic systems impeding their inhibition and restoring and increasing their activity, this is of particular interest for susceptible and predisposed subjects. Moreover the identification of key pathways involved in DNA repair of DNA damage induced by cancer therapy treatment, representing important factors of therapeutic resistance, could bring to the identification of new drugs that target and inhibit specific DNA repair protein making anticancer agents more effective (Madhusudan and Middleton 2005).

REFERENCES

Andreoli C, Leopardi P, Crebelli R. Detection of DNA damage in human lymphocytes by alkaline single cell gel electrophoresis after exposure to benzene or benzene metabolites. *Mutat Res*, 1997, 377/1, 95-104.

Ames BN, Gold LS, Willet WC. The causes and prevention of cancer. *Proc Natl Acad Sci USA*, 1995, 92, 5258-5265.

Blasiak and Kowalik. A comparison of the in vitro genotoxicity of tri- and hexavalent chromium. *Mutat Res*, 2000, 469/1, 135-145.

Berwick M and Vineis P. Marker of DNA repair and susceptibility to cancer in humans: an epidemiologic review. *Journal of the National Cancer Inst* 2000, 92/11, 874-889.

Cavallo D, Iavicoli I, Setini A, Marinaccio A, Perniconi B, Carelli G, Iavicoli S. Evaluation of genotoxic risk and oxidative DNA damage in workers exposed to antimony trioxide. *Environ Mol Mutagen*, 2002, 40/3,184-189.

Cavallo D, Ursini CL, Setini A, Chianese C, Piegari P, Perniconi B, Iavicoli S. Evaluation of oxidative damage and inhibition of DNA repair in an *in vitro* study of nickel exposure. *Toxicol in vitro*, 2003, 17/5-6, 603-607.

Cavallo D, Ursini CL, Iavicoli S. Role of DNA repair systems inhibition in carcinogenicity of metal compounds. *Curr Enz Inhibition*, 2005, 1, 257-264.

Cavallo D, Ursini CL, Bavazzano P, Cassinelli C, Frattini A, Perniconi B, Di Francesco A, Ciervo A, Rondinone B, Iavicoli S. Sister Chromatid Exchange and oxidative DNA damage in paving workers exposed to PAHs. *Ann Occup Hyg*, 2006a, 50/3, 211-218.

Cavallo D, Ursini CL, Carelli G, Iavicoli I, Ciervo A, Perniconi B, Rondinone B, Gismondi M, Iavicoli S. Occupational exposure in airport personnel: characterization and evaluation of genotoxic and oxidative effects. *Toxicology*, 2006b, 223 /1-2, 26-35.

Cebulska-Wasilewska A, Panek A, Żabiński Z, Moszczyński P, Au WW. Occupational exposure to mercury vapour on genotoxicity and DNA repair. *Mutat Res* 2005a, 586, 102-114.

Cebulska-Wasilewska A, Wiecheć A, Panek A, Binková B, Šrám RJ, Farmer PB. Influence of environmental exposure to PAHs on the susceptibility of lymphocytes to DNA-damage induction and on their repair capacity. *Mutat Res* 2005b, 588, 73-81.

Christmann M, Tomicic MT, Roos WP, Kaina B. Mechanism of human DNA repair: an update. *Toxicology*, 2003, 193, 3-34.

Collins AR, Duthie SJ, Dobson VL. Direct enzymatic detection of endogenous base damage in human lymphocyte DNA. *Carcinogenesis* 1993, 14, 1733-1735.

Collins AR, Ai-Guo M, Duthie SJ. The kinetics of repair of oxidative DNA damage (strand breaks and oxidised pyrimidines) in human cells. *Mutat Res*. 1995, 336/1, 69-77.

Collins AR. The comet assay for DNA damage and repair: principles, applications, and limitations. *Mol Biotechnol*, 2004, 26/3, 249-261.

Cooke MS, Evans MD, Dizdaroglu M, Lunec J. Oxidative DNA damage: mechanisms, mutation and disease. *Fed Am Soc Exp Biol*, 2003, 17, 1195-1214.

Danaee H, Nelson HH, Liber H, Little JB, Kelsey KT. Low dose exposure to sodium arsenite synergistically interacts with UV radiation to induce mutations and alter DNA repair in human cells. *Mutagenesis*, 2004, 19/2, 143-148.

De Boeck M, Lison D, Kirsh-Volders M. Evaluation of the in vitro direct and indirect genotoxic effects of cobalt compounds using the alkaline comet assay. *Carcinogenesis* 1998, 19, 2021-2029.

Digweed M. Response to environmental carcinogens in DNA-repair deficient disorders. *Toxicology* 2003, 193,111-124.

Evans MD, Podmore ID, Daly GJ, Perret D, Lunec J, Herbert KE. Immunochemical detection of reactive oxygen species DNA damage. *Biochem Soc Trans*. 1995, 23/3, 482S.

Fatur T, Lah TT, Filipič M. Cadmium inhibits repair of UV-, methyl methanesulfonate- and N-methyl-N-nitrosourea-induced DNA damage on Chinese hamster ovary cells. *Mutat Res*, 2003, 529, 109-116.

Friedberg EC, Walker GC, Siede W. DNA repair and mutagenesis. *ASM Press*, 1995.

Filipič M, Fatur T and Vudrag M. Molecular mechanisms of cadmium induced mutagenicity. *Hum Exp Toxicol*, 2006, 25, 67-77.

Garaj-Vrhovac V and Zeljezic D. Evaluation of DNA damage in workers occupationally exposed to pesticides using single-cell gel electrophoresis (SCGE) assay Pesticide genotoxicity revealed by comet assay. *Mutat Res*, 2000, 469/2, 279-285.

Goode EL, Ulrich CM, Potter JD. Polymorphisms in DNA repair genes and associations with cancer risk. *Cancer Epidemiol Biomarkers Prev*, 2002, 11, 1513-1530.

Gutierrez-Castillo ME, Roubicek DA, Cebrian-Garcia ME, De Vizcaya-Ruiz A, Sordo-Cedeno M, Ostrosky-Wegman P. Effect of chemical composition on the induction of DNA damage by urban airborne particulate matter. *Environ Mol Mutagen*, 2006, 47/3, 199-211.

Hartmann A and Speit G. Effect of arsenic and cadmium on the persistence of mutagen-induced DNA lesions in human cells. *Environ Mol Mutagen*, 1996, 27, 98-104.

Hartwig A, Assmuss M, Ehleben I, Herzer U, Kostelac D, Pelzer A, Schwerdtle T, Bürkle A. Interference by toxic metals ion with DNA repair processes and cell cycle control: molecular mechanisms. *Environ Health Perspect*, 2002, 110, 797-799.

Hussain SP, Hofseth LJ, Harris CC. Radical causes of cancer. *Nat Rev Cancer*, 2003, 3, 276-285.

Iarmarcovai G, Sari-Minodier I, Chaspoul F, Botta C, De Méo M, Orsiére T, Bergé-Lefranc JL, Gallice P, Botta A. Risk assessment of welders using analysis of eight metals by ICP-MS in blood and urine and DNA damage evaluation by the comet and micronucleus assay: influence of XRCC1 and XRCC3 polymorphisms. *Mutagenesis,* 2005, 20/6, 425-432.

Jianhua Z, Lian X, Shuanlai Z, Juan D, Shuanxi Y. DNA lesion and Hprt mutant frequency in rat lymphocytes and V79 Chinese hamster cells exposed to cadmium. *J Occup Health*, 2006, 48/2, 93-99.

Kao J, Rosenstein BS, Peters S, Milano MT, Kron SJ. Cellular response to DNA damage. *Ann N Y Acad Sci* 2005, 1066, 243-258.

Karlsson HL, Nilsson L, Möller L. Subway particles are more genotoxic than street particles and induce oxidative stress in cultures human lung cells. *Chem Res Toxicol*, 2005, 18, 19-23.

Kim BS, Park JJ, Edler L, Von Fournier D, Haase W, Sautter-Bihl ML, Gotzes F, Thielmann HW. New measure of DNA repair in the single-cell gel electrophoresis (comet) assay. *Environ Mol Mutagen*, 2002, 40/1, 50-56.

Lankoff A, Banasik A, Duma A, Ochniak E, Lisowska H, Kuszewski T, Góźdź S, Wojcik A. A comet assay study reveals that aluminium induces DNA damage and inhibits the repair of radiation-induced lesions in human peripheral blood lymphocytes. *Toxicology Letters*, 2006, 161, 27-36.

Madhusudan S and Middleton MR. The emerging role of DNA repair proteins as predictive, prognostic and therapeutic targets in cancer. *Cancer Treat Rev*, 2005, 31. 603-617.

Marnett LJ. Oxyradicals and DNA damage. *Carcinogenesis,* 2000, 21, 361-370.

Merk O, Reiser K, Speit G. Analysis of chromate-induced DNA-protein crosslinks with the comet assay. *Mutat Res* 2000, 471/1-2, 71-80.

Mouròn SA, Golijow CD, Dulout FN. DNA damage by cadmium and arsenic salts assessed by the single cell gel electrophoresis. *Mutat Res*, 2001, 498, 47-55.

Nyberg KA, Michelson RJ, Putnam, CW, Weinert TA. Toward maintaining the genome: DNA damage and replication checkpoints. *Annual Rev Genet* 2002, 36, 617-656.

Palus J, Rydzynski K, Dziubaltowska E, Wyszynska K, Natarajan AT, Nilsson R. Genotoxic effects of occupational exposure to lead and cadmium. *Mutat Res*, 2003, 540, 19-28.

Piperakis SM, Petrakou E, Tsilimigaki S, Sagnou M, Monogiudis E, Haniotakis G, Karkaseli H, Sarikaki E. Biomonitoring with the comet assay of greek greenhouse workers exposed to pesticides. *Environ Mol Mutagen*, 2003, 41/2, 104-110.

Potts RJ, Bespalov IA, Wallace SS, Melamede RJ, Hart BA. Inhibition of oxidative DNA repair in cadmium-adopted alveolar epithelial cells and the potential involvement of metallothionein. *Toxicology*, 2001, 161/1-2, 25-38.

Poulsen, HE. Oxidative DNA modifications. *Exp Toxicol Pathology*, 2005, 57, 161-169.

Pruski AM and Dixon DR. Effects of cadmium on nuclear integrity and DNA repair efficiency in the gill cells of Mytilus edulis L. *Aquat Toxicol*, 2002, 57/3, 127-137.

Restrepo HG, Sicard D, Torres MM. DNA damage and repair in cells of lead exposed people. *Am J Ind Med*, 2000, 38/3, 330-334.

Speit G, Witton-Davies T, Heepchantree W, Trenz K, Hoffmann H. Investigations on the effect of cigarette smoking in the comet assay. *Mutat Res*, 2003, 542/1-2, 33-42.

Speit G and Hartmann A. The comet assay: a sensitive genotoxicity test for the detection of DNA damage and repair. *Methods Mol Biol*, 2006, 314, 275-286.

Singh NP, McCoy MT, Tice RR, Schneider EL. A simple technique for quantitation of low levels of DNA damage in individual cells. *Exp Cell Res*. 1988, 175/1, 184-191.

Trzeciak A, Kowalik J, Malecka-Panas E, Drzewoski J, Wojewódzka M, Iwaneńko T, Blasiak J. Genotoxicity of chromium in human gastric mucosa cells and peripheral blood lymphocytes evaluated by the single cell gel electrophoresis (comet assay). *Med Sci Monit*, 2000, 6/1, 24-29.

Ursini CL, Cavallo D, Colombi A, Giglio M, Marinaccio A, Iavicoli S. Evaluation of early DNA damage in healthcare workers handling antineoplastic drugs. *Intern Arch Occup Environ Health*, 2006 June 8, epub haed of print.

Wood RD, Mitchell M, Sgouros J, Lindahl T. Human DNA repair genes. *Science* 2001, 291, 1284-1289.

Wozniak K and Blasiak J. Nickel impairs the repair of UV- and MNNG-damaged DNA. *Cell Mol Biol Lett*, 2004, 9/1, 83-94.

In: New Research on DNA Repair
Editor: Breehn R. Landseer, pp. 99-113

ISBN 1-60021-385-5

Chapter IV

DIETARY COMPOUNDS THAT ENHANCE DNA REPAIR AND THEIR RELEVANCE TO CANCER AND AGING

Harris Bernstein[1,3,*], ***Cheray Crowley-Skillicorn***[1], ***Carol Bernstein***[1,4], ***Claire M. Payne***[1,3,4], ***Katerina Dvorak***[1,3,4] ***and Harinder Garewal***[2,3,4]

Departments of [1]Cell Biology and Anatomy, [2]Internal Medicine, [3]Arizona Cancer Center, College of Medicine, University of Arizona, Tucson, Arizona 85724, USA

[4]Section of Hematology/Oncology, Southern Arizona Veterans Affairs Health Care System, Tucson, Arizona 85723, USA

ABSTRACT

Investigation of dietary compounds that enhance DNA repair has been a largely underdeveloped area of research. DNA damage is regarded as a major contributing factor to cancer and aging. Unrepaired DNA damage in cells that duplicate can cause errors of replication leading to mutation and ultimately to cancer. Unrepaired DNA damages can also cause loss of gene expression, loss of cell function and cell death (apoptosis) leading ultimately to aging. Genetic defects in numerous human genes encoding DNA repair proteins substantially increase the risk of cancer or cause premature aging, thus indicating the importance of DNA repair processes in preventing cancer and premature aging. Recent studies have identified several dietary compounds that increase expression of DNA repair genes, and in some cases these compounds appear to be associated with reduced cancer incidence or increased longevity.

We have tested 19 potential chemopreventive compounds for their effect on expression of several DNA repair proteins in a colon cell line. On the basis of these results we focused on a group of compounds related to chlorogenic acid. Chlorogenic acid is found in a variety of fruits and vegetables, and is particularly high in coffee. We found that

* Correspondence concerning this article should be addressed to Carol Bernstein, Dept. of Cell Biology and Anatomy, College of Medicine, University of Arizona, Tucson, Arizona 85724 bernstein3@earthlink.net

chlorogenic acid and two of its metabolites formed by the action of intestinal microbial flora, *m*-coumaric acid and 3-(*m*-hydroxyphenyl)propionic acid, significantly increased the expression of the DNA repair enzyme poly(ADP-ribose) polymerase (PARP). The repair enzyme Pms2 (post-meiotic segregation 2) was also significantly increased by *m*-coumaric acid and 3-(*m*-hydroxyphenyl)propionic acid, as well as by another metabolite of chlorogenic acid, caffeic acid. Evidence suggests that PARP activity is associated with increased longevity and cancer prevention, and that Pms2 activity is associated with cancer prevention. Thus, the increased expression of these DNA repair enzymes by chlorogenic acid-related compounds may be beneficial for promoting longevity and preventing cancer.

Keywords: DNA repair, cancer, aging, DNA damage, phytochemicals, PARP, Pms2, chlorogenic acid, caffeic acid, m-coumaric acid, 3-(m-hydroxyphenyl)propionic acid

INTRODUCTION

DNA Repair Processes are Important for Preventing Cancer and Premature Aging

DNA repair is of substantial importance in cancer prevention. One kind of evidence in support of this concept is that inherited defects in genes whose products are employed in DNA repair cause cancer predisposition in humans. At least 26 such genes have been identified, and include such specific examples as BRCA1, ATM, BLM (homologous recombinational repair), XPB and XPD (nucleotide excision repair) and MSH2, MSH6, MLH1, PMS2 (mismatch repair) (reviewed by Bernstein et al., 2002). Epidemiological studies have also consistently indicated that deficiencies in DNA repair capacity cause increased susceptibility to cancer (reviewed by Berwick and Vineis, 2000).

DNA repair is also of substantial importance in preventing premature aging. Again, one type of evidence for this concept is that inherited defects in genes whose products are employed in DNA repair can cause premature aging. Examples of such genes are the *Werner syndrome (WS)* gene which is employed in homologous recombinational repair and non-homologous end joining and the *Cockayne syndrome B (CSB)* gene employed in transcription-coupled nucleotide excision repair and base excision repair (Kyng and Bohr, 2005). Poly(ADP-ribose) polymerase (PARP) is a DNA repair protein employed in base excision repair. PARP is of particular interest, since there is a substantial body of evidence linking PARP activity to mammalian longevity (reviewed by Burkle et al., 2005), as well as to protection against cancer (*e.g.* Rajaee-Behbahani et al., 2002; Wu et al., 1998).

Phytochemicals that Enhance DNA Repair

Naringenin, a citrus flavonoid, found in grapefruit and oranges, stimulates DNA base excision repair following oxidative damage in human prostate cancer cells (Gao et al., 2006). Two DNA repair enzymes, 8-oxoguanine-DNA glycosylase 1 and DNA polymerase beta are increased at the mRNA level by exposure of the cells to naringenin. The flavonoid myricetin, found in berries, vegetables, teas, wine and herbs, also stimulates repair of oxidatively damaged DNA and increases the expression of DNA polymerase beta at the mRNA level (Morel et al., 2001). Curcumin is a component of turmeric, a spice from the plant *Curcuma longa.* Turmeric is a standard ingredient in curry and mustard. Curcumin accelerates DNA repair efficiency in damaged cells (Chakraborty et al, 2004), and increases expression of several DNA repair genes at the mRNA level (Van Erk et al., 2004). Numerous studies have shown a substantial protective effect of curcumin against colon carcinogenesis in rats and mice (reviewed in Van Erk et al., 2004). Indole-3-carbinol (I3C) is found in cruciferous vegetables such as cabbage and cauliflower and the isoflavone genistein is derived from soy. Both I3C and genistein induce the expression of breast cancer susceptibility genes BRCA1 and BRCA2 (Fan et al., 2006). The BRCA1 and BRCA2 genes encode proteins involved in DNA recombinational repair. Induction of the BRCA genes occurs at low doses of I3C and genistein, suggesting potential relevance to cancer prevention (Fan et al., 2006).

On the basis of these findings we speculated that other dietary-related agents might also be found to enhance expression of DNA repair enzymes. Thus we tested 19 potential chemopreventive compounds, including phytochemicals, antioxidants and selective inhibitors, for their effects on expression of particular DNA repair enzymes.

MATERIALS AND METHODS

HCT-116SA, a long passage apoptosis-sensitive colon cancer cell line (Crowley-Weber et al., 2002) was maintained in DMEM (Gibco, Grand Island, NY) supplemented with 10% heat inactivated fetal calf serum (Omega Scientific, Tarzana, CA), 1% MEM non-essential amino acids, 100μg/mL streptomycin, 100U/ml penicillin, and 3.44 mg/mL L-glutamine (Gibco, Grand Island, NY). Cells were cultured at 37°C in 5% CO_2.

Chlorogenic acid, caffeic acid, *m*-coumaric acid, aspirin, catalase, curcumin, DL-α-lipoic acid, resveratrol, seleno-L-methionine, γ-tocopherol, ubiquinone-10 and 1400W were purchased from Sigma-Aldrich (St. Louis, MO). Bilirubin, caffeic acid phenethyl ester, (-)-epigallocatechin-gallate, glutathione monoethyl ester, and N-acetyl-L-cysteine were purchased from Calbiochem (La Jolla, CA), 3-(*m*-hydroxyphenyl)propionic acid was purchased from Pfaltz & Bauer (Waterbury, CT) and KT5823 was purchased from Biomol International (Plymouth Meeting, PA). All of these chemicals had been previously tested for 24 hours over a range of concentrations for induction of apoptosis in the apoptosis sensitive cell line HCT116SA (data not shown). The highest concentration of the tested agents that gave no detectable apoptosis was then used for the assays of expression of the DNA repair enzymes. For the experiments shown in Figures 2 and 3, treatment with the chemical agents was for 24 hours at the following concentrations: chlorogenic acid (200 μM), caffeic acid (100 μM), *m*-coumaric acid (100 μM) and 3-(*m*-hydroxyphenyl)propionic acid (50 μM).

Following treatment, the HCT-116SA cells were lysed using Lysis Buffer (50 mM NaCl, 0.5% NP-40, 3 mM phenylmethylsulfonyl fluoride (PMSF), 3 μg/ml leupeptin and 3 μg/ml aprotinin). Protein aliquots were loaded (30 μg/lane) on a 4-20% Tris HCl gel (Biorad, Hercules, CA) for size fractionation by electrophoresis. The proteins were blotted into

Immobilon-P PVDF transfer membrane (Millipore, Bedford, MA). Separate membranes were immunostained with anti-Pms2 monoclonal antibody (3 μg/ml, BD Biosciences, San Diego, CA), anti-PARP monoclonal antibody (1:500, Trevigen, Gaithersburg, MD), anti-APE1 monoclonal antibody (1:1000 Trevigen, Gaithersburg, MD) and anti-mtOGG1 polyclonal antibody (1:500, Abcam Inc., Cambridge, MA). The membranes were then incubated with goat anti-mouse Ig (1:20,000 – 1:40,000) antibodies conjugated to horseradish peroxidase. The antibody complexes were detected using Super Signal West Pico chemiluminescence reagents (Pierce, Rockford, IL). The protein band intensities were measured using automated densitometry (Kodak Digital Science Image Station 440). The protein band intensities derived from cells incubated in supplemented media were then compared to corresponding band intensities derived from cells incubated in control media.

RESULTS

Chlorogenic Acid and Related Compounds Increase Expression of DNA Repair Enzymes

Initially, we ran duplicate tests of nineteen potential chemopreventive agents (Table 1) on cell line HCT116SA for their effect on the level of three DNA repair proteins, PARP, redox effector factor-1/apurinic/apyrimidinic endonuclease 1 (Ref-1/APE1), and mitochondrial 8-oxoguanineglycosylase-1 (mtOGG1). Two inhibitors of the nitric oxide pathway, 1400W (a selective NOS2 inhibitor), and KT-5823 (a cGMP-dependent protein kinase inhibitor) were compared to the phytochemicals since many of the phytochemicals effectively scavange reactive nitrogen species in addition to scavenging reactive oxygen species.

We found that a group of three related compounds (chlorogenic acid, *m*-coumaric acid and 3-(*m*-hydroxyphenyl)propionic acid) stood out from among the 19 compounds tested in that they had the strongest positive effects, as indicated by Western blots, on expression of PARP, one of the DNA repair proteins initially tested. The structures of these compounds are shown in Figure 1.

The four chlorogenic acid-related compounds shown in Figure 1 were then further evaluated against PARP, as well as against another DNA repair enzyme, Post Meiotic Segregation-2 (Pms2). Pms2 is employed in DNA mismatch repair. An example of the result with PARP is shown in Figure 2. PARP expression, in the experiment shown, was increased by treatment of the cells with all four compounds compared to the untreated control.

In the experiment shown in Figure 3, Pms2 expression was increased by treatment with the four compounds tested compared to the untreated control.

Table 2 shows the mean fold-increase in expression of PARP and Pms2 due to treatment with the four compounds compared to the untreated control in multiple experiments. In four experiments, a significant mean fold-increase in expression of PARP occurred following treatment with chlorogenic acid, *m*-coumaric and 3-(*m*-hydroxyphenyl)propionic acid. Similarly, in three experiments, a significant mean fold-increase in expression of Pms2 occurred following treatment with *m*-coumaric acid, 3-(*m*-hydroxyphenyl)propionic acid and caffeic acid. It is worth noting that 3-(*m*-hydroxyphenyl)propionic acid caused significant overexpression of both PARP and Pms2, even though this compound was administered at 50μM, a concentration that was at least two-fold lower than that of the other compounds (see Materials and Methods).

Table 1. Potential chemopreventive agents tested in HCT116SA cells for their effect on expression of the DNA repair proteins, PARP, Ref-1 and mtOGG1.

Agent	Description
Chlorogenic acid	Scavenges ROS[a] and RNS[b] reduces DNA damage and inhibits chemically induced carcinogenesis in rodents (see text)
Caffeic acid	Scavenges ROS, reduces DNA damage and inhibits chemically induced carcinogenesis in rodents (see text)
m-coumaric acid	Metabolite of chlorogenic acid and caffeic acid by action of gut microflora (see Figure 1)
3-(*m*-hydroxyphenyl) propionic acid	Metabolite of chlorogenic acid and caffeic acid by action of gut microflora (see Figure 1)
Aspirin	Blocks production of prostaglandins by inhibiting cyclooxygenase
Bilirubin	Physiological antioxidant; A major end-product of hemoglobin decomposition
Caffeic acid phenethyl ester (CAPE)	NF-κB/ODC[c]/LOX[d] inhibitor; found in honeybee hives
Catalase	Catalyzes decomposition of hydrogen peroxide; natural antioxidant enzyme
Curcumin	Natural phenolic compound; antioxidant; antitumor agent
DL-α-lipoic acid	Antioxidant; coenzyme; modulates NF-κB
(-)-Epigallocatechin-gallate (EGCG)	One of the main polyphenolic constituents of green tea; antioxidant; antitumor agent
Glutathione monoethyl ester	Cell permeable derivative of glutathione; antioxidant
KT5823	Inhibitor of cGMP-dependent protein kinase
N-acetyl-L-cysteine (NAC)	Antioxidant; increases the cellular pool of free radical scavengers
Resveratrol	Phenolic antioxidant found in grapes, red wine and peanuts
Seleno-L-methionine	Regulates p53 function through a Ref-1 dependent redox mechanism; prevents oxidative DNA damage
γ-tocopherol	A major form of vitamin E; membrane antioxidant; found in many plants and nuts
Ubiquinone-10 (Coenzyme Q_{10})	Endogenous cellular antioxidant; essential component of the electron transfer chain
1400W[e]	Specific NOS2[f] inhibitor

[a]ROS = reactive oxygen species; [b]RNS = reactive nitrogen species; [c]ODC = ornithine decarboxylase; [d]LOX = lipoxygenase; [e]1400W = (N-(3-aminomethyl)benzyl)acetamidine; [f]NOS2 = inducible nitric oxide synthase.

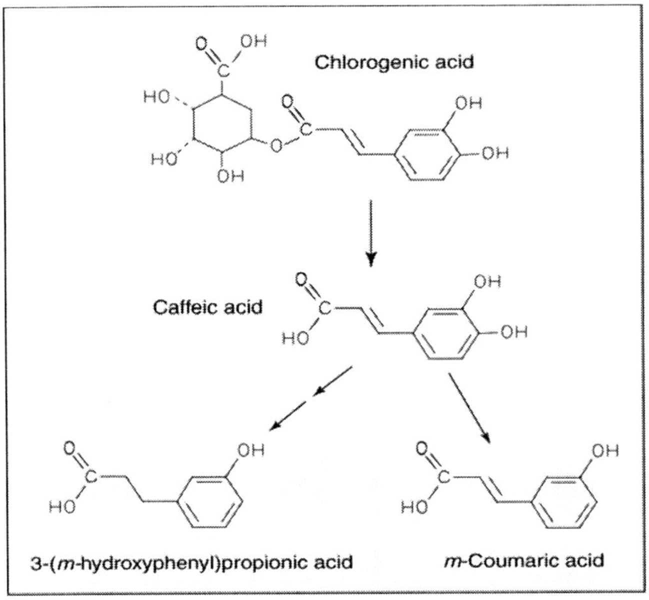

Figure 1. Chlorogenic acid, the ester of caffeic acid with quinic acid, is one of the most abundant polyphenols in the human diet with coffee, fruits and vegetables as its major sources. *m*-coumaric acid and 3-(*m*-hydroxyphenyl)propionic acid are the main metabolites of chlorogenic and caffeic acid by action of microbial flora in the intestine (Gonthier et al., 2003).

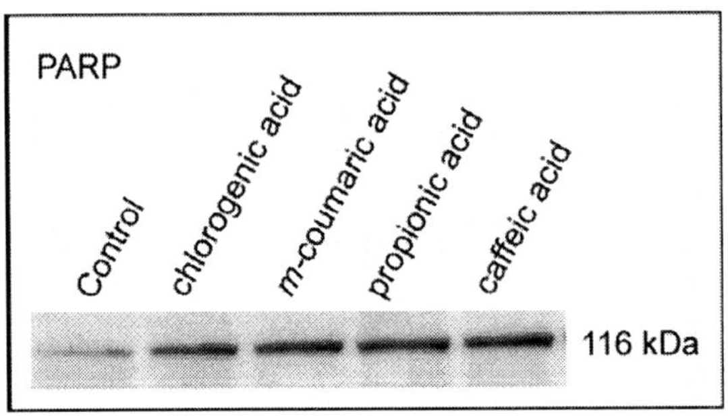

Figure 2. Western gel showing expression of PARP after treatment with chlorogenic acid, *m*-coumaric acid, 3-(*m*-hydroxyphenyl)propionic acid and caffeic acid.

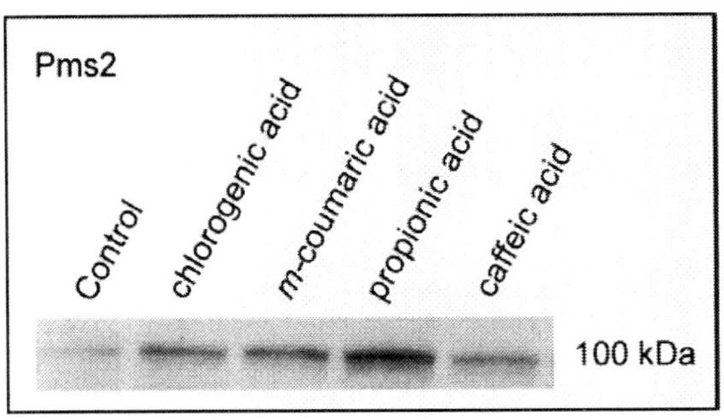

Figure 3. Western gel showing expression of Pms2 after treatment with chlorogenic acid, *m*-coumaric acid and 3-(*m*-hydroxyphenyl) propionic acid and caffeic acid.

Table 2. Mean fold-increase ± S.E. in expression of PARP or Pms2 after treatment.

DNA repair enzyme	Number of experi-ments	Phytochemicals			
		Chlorogenic acid	Caffeic acid	*m*-coumaric acid	3-(*m*-hydroxy phenyl) propionic acid
PARP	4	*1.25 ± 0.06	1.13 ± 0.14	*1.42 ± 0.05	*1.40 ± 0.01
Pms2	3	1.32 ± 0.10	*1.44 ± 0.05	*1.34 ± 0.06	*1.61 ± 0.16

*indicates $p \leq 0.05$ [using the paired t-test (one-tailed distribution)], i. e. a significant increase in expression of PARP or Pms2 in the treated cells compared to the untreated control cells.

CONCLUSION

The Role of PARP in Promoting Longevity and Protecting Against Cancer

PARP catalyzes poly(ADP-ribosyl)ation, which is the posttranslational modification of proteins with poly(ADP-ribose). DNA strand breaks induce catalytic activation of two PARP isoforms, namely PARP-1 and –2, and this activity is involved in DNA base excision repair and other repair pathways (Burkle et al., 2005). Upon activation, PARP-1 and –2 are thought to regulate the accessibility of the damaged DNA site to other repair enzymes. Evidence suggests that local chromatin relaxation is mediated by poly(ADP-ribosyl)ation of histones, or by non-covalent interaction of histones with auto-modified PARP itself (Burkle et al., 2005). PARP-1 is the prototype enzyme of the PARP family, and PARP-1 activity accounts for about 90% of the total cellular poly(ADP-ribose) (PAR) formation. Because of our finding that PARP expression is significantly increased by chlorogenic acid, *m*-coumaric acid and 3-(*m*-hydroxyphenyl)propionic acid, it is of interest to examine the role of PARP in aging and cancer.

A body of correlative data suggests a link between PARP activity and mammalian longevity (Burkle et al., 2005). The lifespan of 13 mammalian species correlates with cellular PAR formation (largely due to PARP-1 activity) (Grube and Burkle, 1992). PARP activity is also increased in human centenarians (Muiras et al., 1998).

There is a strong interaction between PARP-1 and the wild-type form of the protein mutated in Werner syndrome (WS) (von Kobbe et al., 2003), a premature-aging disorder. The wild-type WS protein is necessary for poly(ADP-ribosyl)ation mediated DNA repair (von Kobbe et al., 2003). Also PARP-1's DNA repair functions depend on the wild-type form of the protein mutated in another premature-aging disorder, Cockayne syndrome (Flohr et al., 2003). Cockayne syndrome B protein (CSB) is involved in base excision repair of oxidative DNA damages, and in the PARP-1 poly(ADP-ribosyl)ation response to such damages (Thorslund et al., 2005). These findings suggest that the DNA repair process catalyzed by PARP has a beneficial influence on aging.

PARP-1 knockout mice have increased susceptibility to nitrosamine-induced carcinogenicity (Tsutsumi et al., 2001) indicating that PARP-1 protects against carcinogenesis. In addition, individuals with reduced capacity of synthesizing PAR are at increased risk for laryngeal cancer (Rajaee-Behbahani et al., 2002). Also there is a statistically significant association between the PARP genotype and lung cancer among Mexican-Americans (Wu et al., 1998).

Another line of evidence suggesting a protective role of PARP against cancer comes from studies of niacin (vitamin B_3). The term niacin refers to nicotinic acid and nicotinamide, which are both used by the body to form NAD^+. PARP utilizes NAD^+ as a substrate for carrying out poly(ADP-ribosyl)ation of proteins leading to DNA repair. Nicotinamide protects cells against induction of apoptosis, presumably by enhancing PARP-mediated DNA repair of DNA damages (Crowley et al., 2000) and/or replenishing the NAD^+ levels in mitochondria. A large case-control study in northern Italy and Switzerland found that increased consumption of niacin is associated with a 40% decrease in oral and pharyngeal cancer (Negri et al., 2000), and a similar decrease was found in squamous-cell carcinoma of the esophagus (Franceschi et al., 2000). This suggests that niacin may protect against these forms of cancer through its role as a substrate for PARP in enhancing DNA repair.

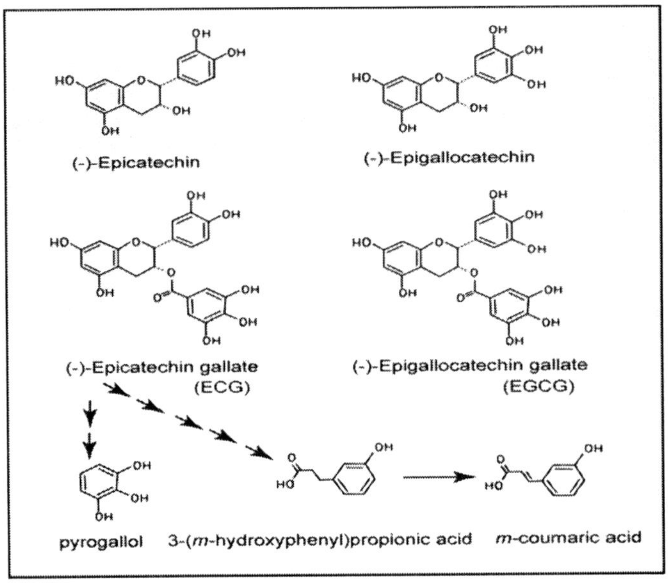

Figure 4. The structures of ECG and EGCG are indicated, as well as the conversion of ECG to 3-(*m*-hydroxyphenyl)propionic acid and *m*-coumaric acid.

(-)-Epigallocatechin Gallate (EGCG)

Other evidence for a protective role of PARP against cancer comes from studies of the consumption of tea which contains the polyphenol (-)-epigallocatechin gallate (EGCG) as a major component (see Figure 4). EGCG induces formation of poly(ADP-ribose) (PAR), the product of PARP activity, in human peripheral lymphocytes (Bertram et al., 2003). At low concentrations of EGCG, near the physiological peak values found after tea ingestion, PAR formation is pronounced, indicating increased PARP activity, while induced DNA damage is negligible. This suggests that, aside from the induction of PARP activity by DNA breaks,

EGCG might increase expression of PARP-1 by a more direct mechanism, possibly by induction of the repair mechanisms related to poly(ADP- ribosyl)ation (Bertram et al., 2003). At higher concentrations, EGCG induces both PAR and DNA damage. In animal models, green or black tea preparations inhibit tumorigenesis at different organ sites including skin, lung, oral cavity, esophagus, forestomach, stomach, small intestine, colon, pancreas and mammary gland (Yang et al, 2002). Some human epidemiological studies have also suggested a protective effect of tea consumption against cancer (e.g. Imai et al., 1997). However, in general, the results of epidemiological studies in humans have been inconsistent (Yang et al., 2002; Zaveri, 2006). We detected no significant effect of EGCG itself in our initial experiment, which may be due to the lack of bacterial conversion of EGCG to more active compounds in our in vitro cultures devoid of bacteria.

(-)-Epicatechin Gallate (ECG) Metabolites

(-)-Epicatechin gallate (ECG) is another major tea polyphenol related to EGCG. It is worthy of note that two of the five major metabolites of ECG (see Figure 4) excreted in rat urine are 3-(*m*-hydroxyphenyl)propionic acid and *m*-coumaric acid (Kohri et al., 2003). These are the two compounds that, in the present study, caused the greatest increase in expression of PARP. Thus, the beneficial effects of tea may be due, at least in part, to increased DNA repair induced by these two compounds.

Role of Pms2 in Protecting Against Cancer

Since expression of Pms2 was significantly increased by caffeic acid, *m*-coumaric acid, and 3-(*m*-hydroxyphenyl) propionic acid, it is of interest to note the role of Pms2 in cancer. Pms2 is a component of the DNA mismatch repair complex responsible both for post-replication correction of DNA single base mismatches and insertion-deletion loops as well as for early steps in apoptosis. Heterozygous germline mutations in Pms2 cause hereditary nonpolyposis colorectal carcinoma (Lynch syndrome) (Hendriks et al., 2006). Homozygous Pms2 mutations give rise to childhood cancers (De Vos et al., 2006). We have found that in the non-neoplastic flat colonic mucosa of patients with sporadic colon cancer (which is much more common than the inherited forms of colon cancer) Pms2 expression is often reduced (Bernstein et al., 2006). These findings suggest that reduced Pms2 expression in the colonic mucosa may be an early step in progression to colon cancer, and further suggest that enhancement of expression of Pms2 by chlorogenic acid-related compounds in plant foods may protect against colon cancer.

Agents Found to Increase Expression of DNA Repair Enzymes

We found that three dietary-related compounds, chlorogenic acid, *m*-coumaric acid and 3-(*m*-hydroxyphenyl)propionic acid significantly increased the expression of PARP, and that

Pms2 expression was significantly increased by caffeic acid, *m*-coumaric acid, and 3-(*m*-hydroxyphenyl)propionic acid. Chlorogenic acid and caffeic acid are antioxidants that scavenge ROS (Rice-Evans et al., 1996; Foley et al., 1999) and RNS (Kono et al., 1997). Although it has been reported that *m*-coumaric acid is a weak scavenger of hydrogen peroxide (a non-free radical oxidant) (Mansouri et al., 2005), we found no reports of robust free radical scavenging activity for either *m*-coumaric acid or 3-(*m*-hydroxyphenyl)propionic acid. This is consistent with the observations that the monophenolic structure is generally a poorer antioxidant than ortho-diphenolic structures (Konishi and Kobayashi, 2004).

As shown in Figure 1, chlorogenic acid is metabolized to caffeic acid, and then to *m*-coumaric acid and 3-(*m*-hydroxyphenyl)propionic acid by the action of intestinal microbial flora (Gonthier et al., 2003). As a result of this metabolism, the latter two compounds, particularly 3-(*m*-hydroxyphenyl)propionic acid, are more bioavailable upon ingestion of chlorogenic acid than chlorogenic acid itself (Gonthier et al., 2003). The experiments performed here involved direct exposure of the cultured cells to the four compounds in the absence of bacteria, so there was no opportunity for microbial conversion of chlorogenic acid or the further conversion of caffeic acid. Thus, the effects of *m*-coumaric acid and 3-(*m*-hydroxyphenyl)propionic acid on expression of the DNA repair proteins in our cell culture system are more likely to reflect the consequence of feeding humans, or other animals, chlorogenic acid or caffeic acid, than the effects of the non-metabolized compounds themselves.

Chlorogenic Acid, Caffeic Acid and Cancer

Chlorogenic acid and/or caffeic acid are each found at particularly high concentration in coffee and are also found in a variety of fruits (e.g. strawberry, rasberry, apple) and vegetables (e.g. potato, carrot, tomato) (Mattila and Kumpulainen, 2002). Chlorogenic acid and caffeic acid are reported to protect against DNA damage. In a rat carcinogenesis model, feeding chlorogenic acid inhibited 4-NQO-induced formation of 8-hydroxydeoxyguanosine, an indicator of oxidative DNA damage (Kasai et al., 2000). Caffeic acid reduced oxidative DNA damage as measured by the comet assay in a colon cancer cell line (Schaefer et al., 2006).

Chlorogenic acid and caffeic acid are also reported to protect against cancer in animal models. Chlorogenic acid, when administered i.p. to mice, inhibited lung carcinogenesis induced by benzo[a]pyrene (Lesca, 1983). When added to the diet of hamsters, chlorogenic acid inhibited methylazoxymethanol induced carcinogenesis of the large intestine (Mori et al., 1986). Chlorogenic acid and caffeic acid, when applied topically, inhibited skin tumor promotion in mouse skin by 12-*O*-tetradecanoylphorbol-13-acetate (TPA) (Huang et al., 1988). This inhibition was dose-dependent. In another mouse model, caffeic acid inhibited the promotional phase of 7,12-dimethylbenz[a]anthracene–induced skin tumorigenesis when topically applied, where phorbol-12-myristate-13-acetate (PMA) or mezerein were used as promoters (Kaul and Khanduja, 1998). When added to the diet, chlorogenic and caffeic acid inhibited chemically-induced carcinogenesis in the rat tongue (Tanaka et al., 1993). Furthermore, caffeic acid administered orally to nude mice suppressed the growth of HepG2 tumor xenografts *in vivo* and significantly reduced liver metastasis (Chung et al., 2004). For humans, in two large prospective cohort studies, there was a lower risk of rectal cancer associated with consumption of decaffeinated coffee (Michels et al., 2005). These apparent beneficial effects of chlorogenic acid, caffeic acid and coffee consumption against DNA damage and cancer may be due, at least in part, to the increased expression of DNA repair enzymes.

Phenolic Acids in Cocoa and Reduced Mortality

Cocoa is a rich source of polyphenol-derived phenolic acids (Rios et al., 2003). Recently it was shown, in elderly men, that cocoa intake is inversely associated with blood pressure levels and 15-year cardiovascular and all-cause mortality (Buijsse et al., 2006). Compared with the lowest tertile of cocoa intake, the adjusted relative risk for men in the highest tertile was 0.50 for cardiovascular mortality and 0.53 for all-cause mortality, a remarkably protective effect. To identify the bioavailable low molecular weight compounds formed by the microflora of the intestine from higher molecular weight polyphenols, the phenolic acids excreted in the urine of human subjects was determined (Rios et al., 2003). Chocolate intake increased the urinary excretion of six phenolic compounds including 3-(*m*-hydroxyphenyl)propionic acid, which we found to increase expression of DNA repair proteins. Thus, the effect of chocolate in reducing mortality may be due, in part, to the conversion of polyphenols found in chocolate to bioavailable compounds which are distributed to various tissues where they increase DNA repair, and then are finally excreted in the urine.

Phytochemicals as Chemopreventive Agents

Epidemiological evidence strongly indicates that consumption of fruits and vegetables considerably lowers the risk of cancer (e.g. Hertog et al., 1996). This has led to the idea that it is specific chemicals in these foods that are the effective preventive agents. In humans, successful chemopreventive strategies for reducing DNA damage would be expected to ameliorate cancer and aging. Such strategies would require chemopreventive agents that are non-toxic, since, if proven effective, they would likely be used by large numbers of healthy people for long periods. Non-toxicity is of greater concern for chemopreventive agents than for chemotherapeutic agents, where the intent is to treat severe disease such as malignant tumors. Thus, for discovery of potential chemopreventive agents, it is reasonable to give primary consideration to naturally occurring compounds in plant foods, since many of these have proven to be non-toxic to humans.

Preventing and Repairing DNA Damage in Cancer and Aging

In principle, there are two ways of reducing DNA damage in cells; (1) preventing its occurrence, and (2) repairing the damage after it has occurred. There has been considerable research effort on the role of foods in preventing the occurrence of DNA damage in the first place. There is now a large body of literature on antioxidant phytochemicals that prevent oxidative DNA damage. However, intervention trials using antioxidants with cancer as the endpoint have generally proved disappointing (see review by Collins, 2005). Thus far, phytochemicals that specifically enhance DNA repair, as distinct from preventing damage, have received relatively little attention. In this chapter, we reviewed the limited literature on phytochemicals that increase the expression of DNA repair enzymes, and presented some new data suggesting that compounds related to the polyphenol chlorogenic acid (found in a variety of plant foods) enhance DNA repair. Specifically, these compounds enhance expression of PARP, which may protect against cancer and promote longevity, and expression of Pms2, which may protect against cancer.

Acknowledgments

This work was supported in part by NIH Institutional Core Grant #CA23074, NIH PPG #CA72008, Arizona Disease Control Research Commission Grants #10016 and #6002, Arizona Biomedical Research Commission Grant #0012, VAH Merit Review Grant 2HG, NCI SPORE Grant 1 P50CA95060-01, NIH 1R21CA111513-01A1 and Biomedical Diagnostics & Research, Inc., Tucson, AZ 85719.

References

Bernstein, C; Bernstein, H; Payne, CM; Garewal, H. DNA repair/pro-apoptotic dual-role proteins in five major DNA repair pathways: fail-safe protection against carcinogenesis. *Mutat Res*, 2002, 511, 145-178.

Bernstein, H; Prasad, A; Holubec, H; Bernstein, C; Payne, CM; Ramsey, L; Dvorakova, K; Wilson, M; Warneke, JA; Garewal, H. Reduced Pms2 expression in non-neoplastic flat mucosa from patients with colon cancer correlates with reduced apoptosis competence. *Appl Immunohistochem Mol Morphol,* 2006, 14,166-172.

Bertram, B; Bollow, U; Rajaee-Behbahani, N; Burkle, A; Schmezer, P. Induction of poly(ADP-ribosyl)ation and DNA damage in human peripheral lymphocytes after treatment with (-)-epigallocatechin-gallate. *Mutat Res*, 2003, 534, 77-84.

Berwick, M; Vineis, P. Markers of DNA repair and susceptibility to cancer in humans: an epidemiologic review. *J Natl Cancer Instit*, 2000, 92, 874-897.

Buijsse, B; Feskens, EJM; Kok, FJ; Kromhout, D. Cocoa intake, blood pressure, and cardiovascular mortality. The Zutphen elderly study. *Arch Intern Med*, 2006, 166, 411-417.

Burkle, A; Diefenbach, J; Brabeck, C; Beneke, S. Ageing and PARP. *Pharmacological Research*, 2005, 52, 93-99.

Chakraborty, S; Roy, M; Bhattacharya, RK. Prevention and repair of DNA damage by selected phytochemicals as measured by single cell gel electrophoresis. *J Environ Pathol Toxicol Oncol*, 2004, 23(3), 215-226.

Chung, TW; Moon, SK; Chang, YC; Ko, JH; Lee, YC; Cho, G; Kim, SH; Kim, JG; Kim, CH. Novel and therapeutic effect of caffeic acid and caffeic acid phenyl ester on hepatocarcinoma cells: complete regression of hepatoma growth and metastasis by dual mechanism. *FASEB J*, 2004, 18, 1670-1681.

Collins, AR. Antioxidant intervention as a route to cancer prevention. *European Journal of Cancer*, 2005, 41, 1923-1930.

Crowley, CL; Payne, CM; Bernstein, H; Bernstein, C; Roe, D. The NAD^+ precursors, nicotinic acid and nicotinamide protect cells against apoptosis induced by a multiple stress inducer, deoxycholate. *Cell Death and Differentiation*, 2000, 7, 314-326.

Crowley-Weber, CL; Payne, CM; Gleason-Guzman, M; Watts, GS; Futscher, B; Waltmire, CN; Crowley, C; Dvorakova, K; Bernstein, C; Craven, M; Garewal, H; Bernstein H. Development and molecular characterization of HCT-116 cell lines resistant to the tumor promoter and multiple stress-inducer, deoxycholate. *Carcinogenesis*, 2002, 23(12), 2063-2080.

De Vos, M; Hayward, BE; Charlton, R; Taylor, GR; Glaser, AW; Picton, S; Cole, TR; Maher, ER; McKeown, CME; Mann, JR; Yates, JR; Baralle, D; Rankin, J; Bonthron, DT; Sheridan, E. PMS2 mutations in childhood cancer. *J Natl Cancer Instit*, 2006, 98, 358-361.

Fan, S; Meng, Q; Auborn, K; Carter, T; Rosen, EM. BRCA1 and BRCA2 as molecular targets for phytochemicals indole-3-carbinol and genistein in breast and prostate cancer cells. *British J Cancer*, 2006, 94, 407-426.

Flohr, C; Burkle, A; Radicella, JP; Epe, B. Poly(ADP-ribosyl)ation accelerates DNA repair in a pathway dependent on cockayne syndrome B protein. *Nucleic Acids Research*, 2003, 31(18), 5332-5337.

Foley, S; Navaratnam, S; McGarvey, DJ; Land, EJ; Truscott, TG; Rice-Evans, CA. Singlet oxygen quenching and the redox properties of hydroxycinnamic acids. *Free Radical Biology & Medicine*, 1999, 26(9/10), 1202-1208.

Franceschi, S; Bidoli, E; Negri, E; Zambon, P; Talamini, R; Ruol, A; Parpinel, M; Levi, F; Simonato, L; LaVecchia, C. Role of macronutrients, vitamins and minerals in the aetiology of squamous-cell carcinoma of the oesophagus. *Int J Cancer*, 2000, 86, 626-631.

Gao, K; Henning, SM; Niu, Y; Youssefian, AA; Seeram, NP; Xu, A; Heber, D. The citrus flavonoid naringenin stimulates DNA repair in prostate cancer cells. *J Nutr Biochem*, 2006, 17, 89-95.

Gonthier, MP; Verny, MA; Besson, C; Remesy, C; Scalbert, A. Chlorogenic acid bioavailability largely depends on its metabolism by the gut microflora in rats. *J Nutr*, 2003, 133, 1853-1859.

Grube, K; Burkle, A. Poly(ADP-ribose) polymerase activity in mononuclear leukocytes of 13 mammalian species correlates with species-specific life span. *Proc Natl Acad Sci USA*, 1992, 89, 11759-11763.

Hendriks, YMC; Jagmohan-Changur, S; van der Klift, HM; Morreau, H; Puijenbroek, MJ; Tops, C; van Os, T; Wagner, A; Ausems, MGFM; Gomez, E; Breuning, MH; Brocker-Vriends, AHJT; Vasen, HFA; Wijnen, JT. Heterozygous mutations in Pms2 cause hereditary nonpolyposis colorectal carcinoma (Lynch Syndrome). *Gastroenterology*, 2006, 130, 312-322.

Hertog, MGL; Bueno-de-Mesquita, HB: Fehily, AM; Sweetnam, PM; Elwood, PC; Kromhout, D. Fruit and vegetable consumption and cancer mortality in the Caerphilly study. *Cancer Epidemiol Biomarkers Prev*, 1996, 5, 673-677.

Huang, MT; Smart, RC; Wong, CQ; Conney, AH. Inhibitory effect of curcumin, chlorogenic acid, caffeic acid, and ferulic acid on tumor promotion in mouse skin by 12-*O*-tetradecanoylphorbol-acetate. *Cancer Res*, 1988, 48, 5941-5946.

Imai, K; Suga, K; Nakachi, K. Cancer-preventive effects of drinking green tea among a Japanese population. *Preventive Medicine*, 1997, 26, 769-775.

Kasai, H; Fukada, S; Yamaizumi, Z; Sugie, A; Mori, H. Action of chlorogenic acid in vegetables and fruits as an inhibitor of 8-hydroxydeoxyguanosine formation in vitro and in a rat carcinogenesis model. *Food and Chemical Toxicology*, 2000, 38, 467-471.

Kaul, A; Khanduja, KL. Polyphenols inhibit promotional phase of tumorigenesis: Relevance to superoxide radicals. *Nutrition and Cancer*, 1998, 32(2), 81-85.

Kohri, T; Suzuki, M; Nanjo, F. Identification of metabolites of (-)-epicatechin gallate and their metabolic fate in the rat. *J Agric Food Chem*, 2003, 51, 5561-5566.

Konishi, Y; Kobayashi, S. Transepithelial transport of chlorogenic acid, caffeic acid, and their colonic metabolites in intestinal Caco-2 monolayers. *J Agric Food Chem*, 2004, 52, 2518-2526.

Kono, Y; Kobayashi, K; Tagawa, S; Adachi, K; Ueda, A; Sawa, Y; Shibata, H. Antioxidant activity of polyphenolics in diets: Rate constants of reactions of chlorogenic acid and caffeic acid with reactive species of oxygen and nitrogen. *Biochimica et Biophysica Acta*, 1997, 1335, 335-342.

Kyng, KJ; Bohr VA. Gene expression and DNA repair in progeroid syndromes and human aging. *Ageing Research Reviews*, 4, 2005, 579-602.

Lesca, P. Protective effects of ellagic acid and other plant phenols on benzo[a]pyrene-induced neoplasia in mice. *Carcinogenesis*, 1983, 4(12), 1651-1653.

Mansouri, A; Makris, DP; Kefalas, P. Determination of hydrogen peroxide scavenging activity of cinnamic and benzoic acids employing a highly sensitive peroxyoxalate chemiluminescence-based assay: Structure-activity relationships. *Journal of Pharmaceutical and Biomedical Analysis*, 2005, 39, 22-26.

Mattila, P; Kumpulainen, J. Determination of free and total phenolic acids in plant-derived foods by HPLC with diode-array detection. *J Agric Food Chem*, 2002, 50, 3660-3667.

Michels, KB; Willett, WC; Fuchs, CS; Giovannucci, E. Coffee, tea, and caffeine consumption and incidence of colon and rectal cancer. *J Natl Cancer Instit*, 2005, 97(4), 282-292.

Morel, I; Abalea, V; Cillard, P; Cillard, J. Repair of oxidized DNA by the flavonoid myricetin. *Methods in Enzymology*, 2001, 335, 308-316.

Mori, H; Tanaka, T; Shima, H; Kuniyasu, T; Takahashi, M. Inhibitory effect of chlorogenic acid on methylazoxymethanol acetate-induced carcinogenesis in large intestine and liver of hamsters. *Cancer Letters*, 1986, 30, 49-54.

Muiras, ML; Muller, M; Schachter, F; Burkle, A. Increased poly(ADP-ribose) polymerase activity in lymphoblastoid cell lines from centenarians. *J Mol Med*, 1998, 76(5), 346-354.

Negri, E; Franceschi, S; Bosetti, C; Levi, F; Conti, E; Parpinel, M; LaVecchia, C. Selected micronutrients and oral and pharyngeal cancer. *Int J Cancer*, 2000, 86, 122-127.

Rajaee-Behbahani, N; Schmezer, P; Ramroth, H; Burkle, A; Bartsch, H; Dietz, A; Becher, H. Reduced poly(ADP-ribosyl)ation in lymphocytes of laryngeal cancer patients; Results of a case-control study. *Int J Cancer*, 2002, 98, 780-784.

Rice-Evans, CA; Miller, NJ; Paganga G. Structure-antioxidant activity relationships of flavonoids and phenolic acids. *Free Rad Biol Med*, 1996, 20(7), 933-956.

Rios, LY; Gonthier, MP; Remesy, C; Mila, I; Lapierre, C; Lazarus, SA; Williamson, G; Scalbert, A. Chocolate intake increases urinary excretion of polyphenol-derived phenolic acids in healthy human subjects. *Am J Clin Nutr*, 2003, 77, 912-918.

Schaefer, S; Baum, M; Eisenbrand, G; Dietrich, H; Will, F; Janzowski, C. Polyphenolic apple juice extracts and their major constituents reduce oxidative damage in human colon cell lines. *Mol Nutr Food Res*, 2006, 50, 24-33.

Tanaka, T; Kojima, T; Kawamori, T; Wang, A; Suzui, M; Okamoto, K; Mori, H. Inhibition of 4-nitroquinoline-1-oxide-induced rat tongue carcinogenesis by the naturally occurring plant phenolics caffeic, ellagic, chlorogenic and ferulic acids. *Carcinogenesis*, 1993, 14, 1321-1325.

Thorslund, T; von Kobbe, C; Harrigan, JA; Indig, FE; Christiansen, M; Stevnsner, T; Bohr, VA. Cooperation of the Cockayne syndrome B protein and poly(ADP-ribose) polymerase 1 in the response to oxidative stress. *Molec Cell Biol*, 2005, 25(17), 7625-7636.

Tsutsumi, M; Masutani, M; Nozaki, T; Kusuoka, O; Tsujiuchi, T; Nakagama, H; Suzuki, H; Konishi, Y; Sugimura, T. Increased susceptibility of poly(ADP-ribose) polymerase-1 knockout mice to nitrosamine carcinogenicity. *Carcinogenesis*, 2001, 22(1), 1-3.

Van Erk, MJ; Teuling, E; Staal, YCM; Huybers, S; van Bladeren, PJ; Aarts, JMMJG; van Ommen, B. Time- and dose-dependent effects of curcumin on gene expression in human colon cancer cells. *Journal of Carcinogenesis*, 2004, 3, 8.

Von Kobbe, C; Harrigan, JA; May, A; Opresko, PL; Dawut, L; Cheng, WH; Bohr, VA. Central role for the Werner syndrome protein/poly(ADP-ribose) polymerase 1 complex in the poly(ADP-ribosyl)ation pathway after DNA damage. *Molec Cell Biol*, 2003, 23(23), 8601-8613.

Wu, X; Hsu, TC; Cao, S; Lee, JJ; Amos, CI; Spitz, MR. Deletion in poly(ADP-ribose)polymerase pseudogene and lung cancer risk. *Carcinogenesis*, 1998, 19(1), 93-98.

Yang, CS; Maliakal, P; Meng, X. Inhibition of carcinogenesis by tea. *Annu Rev Pharmacol Toxicol*, 2002, 42, 25-54.

Zaveri, NT. Green tea and its polyphenolic catechins: Medicinal uses in cancer and non cancer applications. *Life Sciences*, 2006, 78, 2073-2080.

In: New Research on DNA Repair
Editor: Breehn R. Landseer, pp. 115-166

ISBN 1-60021-385-5

Chapter V

GENE PROPHYLAXIS OF ENDOGENOUS DNA DAMAGE

*Guido Frosina**
Department of Translational Oncology, Experimental Oncology "B" laboratory, Istituto Nazionale Ricerca Cancro, Largo Rosanna Benzi n. 10, 16132 Genova, Italy

ABSTRACT

A significant contribution to human mutagenesis and carcinogenesis may come from DNA damage of endogenous, rather than exogenous, origin. Efficient repair mechanisms have evolved to cope with it. The major pathway involved in DNA repair of endogenous damage is DNA base excision repair (BER). In addition, an important contribution is given by O^6 alkylguanine DNA alkyltranferase, that repairs specifically the miscoding base O^6 alkylguanine. BER is rather efficient in human cells but a certain amount of endogenous damage inevitably escapes mending and likely contributes to human carcinogenesis. Several attempts have been carried out to enhance the efficiency of repair of endogenous DNA damage by overexpressing in mammalian cells single enzymatic activities. In some cases (e.g. formamidopyrimidine DNA glycosylase or O^6 alkylguanine DNA alkyltransferase) this approach turned out successful in improving cellular protection from endogenous and exogenous mutagens while overexpression of other enzymatic activities (e.g alkyl N-purine DNA glycosylase or DNA polymerase β) was detrimental and even produced a genome instability phenotype. Prophylaxis of endogenous DNA damage and mutation could be achieved in at-risk individuals by expression of appropriate DNA repair genes. Currently available vehicles for this peculiar type of gene therapy are surveyed.

* Correspondence concerning this article should be addressed to Guido Frosina, Department of Translational Oncology, Experimental Oncology "B" laboratory, Istituto Nazionale Ricerca Cancro, Largo Rosanna Benzi n. 10, 16132 Genova, Italy. Tel.:+39.010.5737543. FAX:+39.010.5737237. E-mail: guido.frosina@istge.it.

Keywords: DNA repair; DNA base excision repair; endogenous DNA damage; oxidative DNA damage; 8-oxoguanine; glycosylase; mutation; cancer; gene therapy; O^6 alkylguanine DNA alkyltransferase; oxidation; methylation; abasic sites; rate-limiting step; transfection; retrovirus; lentivirus; adenovirus; vector.

INTRODUCTION

A significant portion of carcinogenic events may be due to genetic damage of endogenous, rather than environmental, origin. Most endogenous damage originates from products of oxygen dismetabolism [socalled Reactive Oxygen Species (ROS)], from the intrinsic lability of chemical bonds in DNA, from endogenous methylating agents or from DNA replication errors. Several mechanisms have evolved in humans to repair damage to DNA. We know since several years that, should the repair of damage induced by sunlight be defective [as in the case of the hereditary disorder Xeroderma Pigmentosum (XP)], cancer of the skin is boosted. This paradigmatic disorder may indicate what DNA repair can do to protect humans from cancer. Endogenous damage, that is abundant under physiological conditions, is mainly mended by two processes called "base excision repair (BER)" and "mismatch repair (MMR)". In addition, O^6 alkylguanine (O^6 alkG), a miscoding base that can form endogenously, is repaired by a specific repair protein called "O^6 alkylguanine DNA alkyltransferase (ATase)". Reviews on various aspects of endogenous damage [1-4], DNA BER [5-13], DNA MMR [14-16] and ATase [17-18] are available to the reader. The aim of this chapter is to survey and analyze the potential to increase the protective role of DNA repair of endogenous damage against spontaneous mutagenesis and carcinogenesis. In those cases in which data are scant or unavailable, the survey will extend to protection from induced mutagenesis and carcinogenesis. Currently available vehicles for expression of DNA repair proteins in vivo will be lastly surveyed.

ENDOGENOUS DAMAGE

Endogenous damage affects the primary, rather than the secondary structure of the double helix. Four general classes of modifications can be envisaged.

Oxidation

It is due mainly to byproducts of oxygen metabolism and is the subject of current intensive investigation, as some oxidized bases have been shown to be miscoding and formed in huge amounts under physiological conditions [e.g. 8-oxo-7,8-dihydroguanine (8-oxoG), thymine glycol and similar oxidized bases]. Perhaps the most abundant and dangerous of these lesions is 8-oxoG, that frequently mispairs with adenine (A) [2,3]. It has been calculated that approximately 1000-7,500 8-oxoG lesions may form daily in a mammalian cell (~ 5% of all oxidized lesions [3,19]). A recent thorough interlaboratory assessment [20] indicated that the median concentration of 8-oxodG in lymphocyte DNA, as determined by HPLC, was 4.24 per 10^6 guanines (2400 /cell). The median concentration of FPG-sensitive sites, measured with the comet assay, was 0.34 per 10^6 guanines (200/cell). The median values for 8-oxodGuo in HeLa cells were 2.78 per 10^6 guanines [1600/cell (by HPLC)] and 0.50 per 10^6 guanines [300/cell (by enzymic methods)]. The discrepancy between chromatographic and FPG-based approaches could reflect overestimation by HPLC (if spurious oxidation was still not completely controlled) or underestimation by the enzymic method.

A current estimate of formation (that may suffer of the same variations observed with 8-oxoG) predicts around 500 oxidized pyrimidines formed per cell per day [19]. These lesions may decompose further in fragments of bases and are essentially non-coding [21,22]. The most deeply investigated oxidized pyrimidine is thymine glycol. Thymine glycols have been shown to be essentially cytotoxic lesions in vivo [23-25], but they can also induce mutations [26].

In addition to oxidized bases, ROS may also generate strand interruptions. The extent to which single strand interruptions in DNA contribute to the spontaneous mutation rate is unknown. The most acute response to single strand breaks is exerted through activation of poly(ADP) ribose polymerase (PARP [27]), a function whose biological significance is still elusive under many respects. The 3' side of strand breaks generated by oxygen radicals or bifunctional glycosylases (Fig.1) is characterized by obstructive fragments of the sugar-phosphate backbone. These fragments must be excised by 3' repair diesterases to leave a 3' OH terminus that permits the subsequent activity of DNA polymerases [5].

Peroxidation of polyunsaturated lipids generates substances that possess DNA damaging potential. These include lipid hydroperoxides and various species with unpaired electrons such as the alkoxyl and peroxyl radicals [28]. Lipid peroxidation also produces a range of carbonyl-containing compounds such as malondialdehyde that give rise to exocyclic DNA adducts [29]. In intact biological systems a portion of endogenously produced lipid peroxidation intermediates effectively escape detoxification and damage DNA. The significance of this phenomenon with respect to spontaneous mutagenesis and carcinogenesis is undetermined [28].

Methylation

Bases can be methylated endogenously by some small molecules such as S-adenosylmethionine (SAM). Quantitatively, 7-methylguanine (7 meG) is the most important lesion (4000 residues generated per day). Although this methylated base doesn't show marked miscoding or cytotoxic properties [1], it is yet particularly abundant and persistent, its steady state level being evaluated in ~ 3000 lesions per cell [19]. On the contrary, 3 methyladenine (3-meA) that is formed endogenously at a rate of a few hundreds per day, is a block for DNA replication and must be efficiently repaired. Small amounts of the miscoding lesion O^6methylguanine (O^6 meG) are also formed endogenously, maybe as a consequence of formation of endogenous nitrosamines and may significantly contribute to spontaneous mutagenesis [30,31].

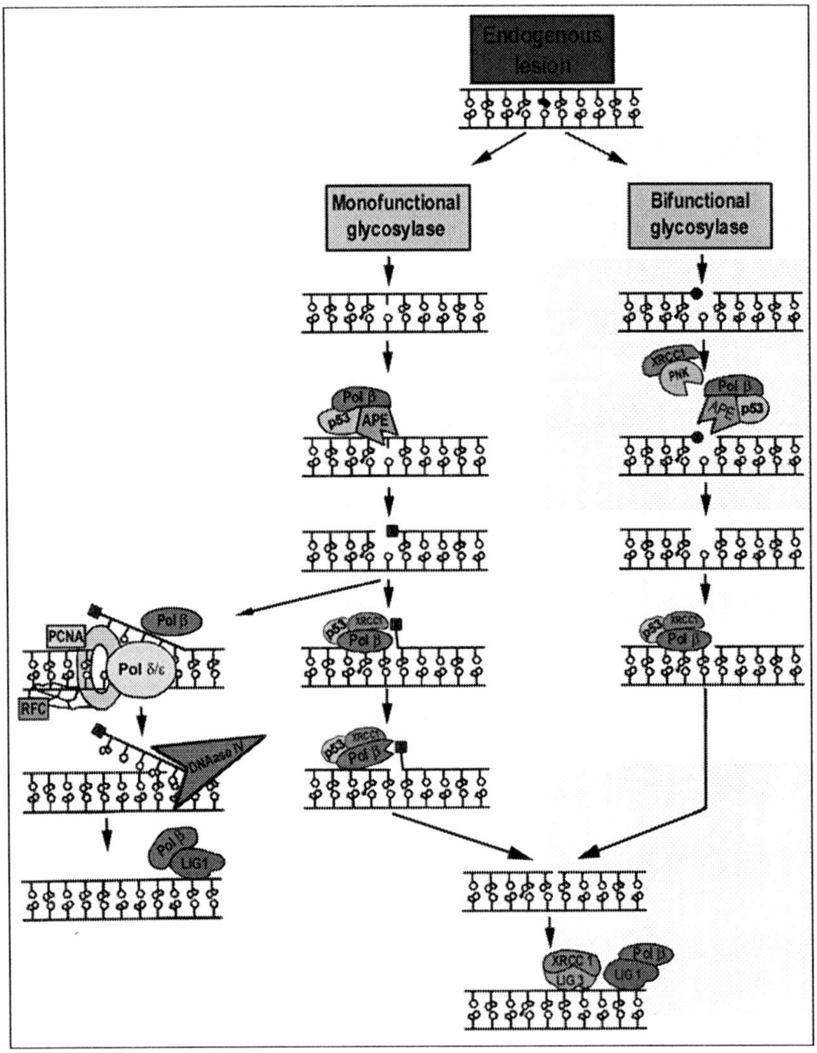

Figure 1. Outline of the DNA BER pathways in mammals [modified from [13] with permission. Altered bases are removed by either monofunctional or bifunctional DNA glycosylases. Monofunctional glycosylases (left-hand pathway) only remove the base leaving a natural AP site. The latter is incised in 5' by the major APE/Ref-1 hydrolytic AP endonuclease that leaves a 5'-terminal dRP residue and a 3' OH priming terminus. DNA pol β inserts in most cases one nucleotide and the dRP group is removed by its N-terminal dRpase activity. p53 interacts with both APE/Ref-1 and pol β and stabilizes the latter on DNA. The complex DNA ligase III/XRCC1 or DNA ligase I seals the interruption. A number of repair patches is longer (2-10 nucleotides) and require participation of proliferating cell nuclear antigen (PCNA - bottom left-hand pathway). Both pol β and the PCNA-dependent polymerases δ/ε (pol δ/ε) are involved in the long patch pathway. The damaged 2-10 nucleotides long-DNA fragment is displaced during polymerization and removed by DNAase IV (FEN1). DNA ligase I is the main sealing activity in the long-patch pathway. In the case of bifunctional glycosylases (right-hand pathway), the AP site generated by the glycosylase activity is further incised in 3' by an associated AP lyase. The 3'-blocking fragment left by AP lyases is trimmed in human cells by the 3' phosphodiesterase activity associated to APE/Ref-1 or by the complex PNK-XRCC1. Synthesis is performed via pol β only and ligation proceeds as for the short-patch pathway initiated by monofunctional glycosylases.

Hydrolysis

The base sugar bonds in DNA are relatively labile and several thousands of bases are lost every day in human cells under physiological conditions. Purines are lost more easily than pyrimidines (5% of the rate of purins [1]). Base loss sites likely represent the most frequent damage in human cells. It has been estimated that 2000-10000 AP sites are generated daily in a mammalian cell [1] and some estimates indicate that their number may even be one order of magnitude higher (50,000 to 200,000 AP sites/cell/day) [32]. Brain is the most affected organ followed by colon and heart [32].

The exocyclic amino groups of cytosine (C) and A may also be frequently lost under physiological conditions, thus generating the miscoding bases uracil (U) and hypoxanthine. Similarly, deamination of 5-methylcytosine (5-meC) generates thymine (T) opposite guanine (G). The base most frequently undergoing deamination is 5-meC followed by unmethylated C [1,19]. The resulting Ts and Us pair with dAMP during the subsequent round of DNA replication giving rise ultimately to C to T transitions, a kind of mutations frequently detected in inherited human diseases [33]. Ts resulting from deamination of 5-meC are repaired by a glycosylase (T:G DNA glycosylase) that is less efficient than that acting on U [U DNA glycosylase – (UDG)] [34]. For their marked proneness to deamination and slow repair, 5-meCs are hotspots for spontaneous point mutation in human cells [35].

Mismatches

They represent a frequent type of endogenous error but will not be dealt with in this chapter. The formation and repair of mismatches as well as the clinical consequences of defects in MMR have been the subject of several reviews [14-16,36]. Most of mammalian genes coding for mismatch repair proteins have been cloned and some overexpression studies have been performed [37,38].

MOLECULAR MECHANISMS OF DNA REPAIR OF ENDOGENOUS DAMAGE

The bypass and proofreading properties of replicative DNA polymerases likely contribute to reduce the cytotoxic and mutagenic effects of endogenous DNA lesions [39]. However, it is clear that endogenous damage cannot be tolerated in mammals if left unrepaired [40,41]. A major repair mechanism for endogenous damage is BER (Fig. 1, Table I). ATase helps by repairing the highly mutagenic lesions O^6 meG and O^4 methyl thymine (O^4 meT). Some oxidized bases may also be processed via nucleotide excision repair (NER) [6,42], but overexpression studies with NER proteins will not be included here, as the contribution of this system to repair of endogenous damage is limited.

Base Excision Repair

Base Removal and AP Site Incision

The first step of BER consists in recognition and removal of the altered base (Fig. 1). This is accomplished by DNA glycosylases. With few exceptions (e.g. T:G DNA glycosylase [43]), these enzymes have rather low molecular weight, ranging from 20 to 40 kDA.

Table I. Proteins involved in DNA repair of endogenous damage*.

BER	
Repair protein	Function
UDG	Glycosylase for U, 5-hydroxyuracil, isodialuric acid.
ANPG	Glycosylase for N-alkyllpurines, hypoxanthine, 1,N^6 ethenoadenine
T:G DNA glycosylase	Glycosylase for T or U opposite G, ethenocytosine
hOGG1	Glycosylase for 8-oxoG opposite C, FaPy residues. Associated AP lyase.
hNTH1	Glycosylase for thymine glycol, cytosine glycol, dihydrouracil, FaPy residues. Associated AP lyase
MYH	Glycosylase for A opposite 8-oxoG.
APE/HAP1/Ref-1	Hydrolytic AP endonuclease; low-efficiency repair of 3' blocking fragments
Pol β	DNA polymerase involved in both BER pathways. Removal of dRP.
Pol δ or ε	DNA polymerase involved in the PCNA –dependent pathway.
DNAase IV (FEN1)	5'-flap endonuclease. Removal of displaced region during PCNA– dependent BER.
PCNA	Accessory factor for Pol δ/ε and DNAase IV (FEN1)
XRCC1	Partner of Pol β, DNA ligase III and PARP. Possible scaffold function.
LIG1	DNA ligase involved in both BER pathways
LIG3	DNA ligase mainly involved in the single nucleotide insertion BER pathway
Drosophila ribosomal protein S3	Removal of dRP, incision of 8-oxoG
PARP	Poly(ADP-ribose) polymerase. Interacts with XRCC1
Damage reversal	
Repair protein	Function
ATase	Repair of endogenous O^6meG and O^4meT

* For simplification, proteins involved in MMR and NER are not included. From ref. [13] with permission

Some DNA glycosylases have rather narrow substrate specificity. For instance, UDG in addition to its main substrate uracil (U), can only remove similar lesions such as 5-hydroxyuracil [44] and isodialuric acid [45]. Other glycosylases have broader substrate specificity and can initiate BER of a wide range of bases. E.g. 3-alkyl N-purine glycosylase (ANPG) [46], in addition to N–methylpurines, can also remove the deamination product hypoxanthine [47] and 1,N^6 ethenoadenine, an adduct generated by lipid peroxidation intermediates [48].

DNA glycosylases can be classified as monofunctional or bifunctional, according to their association to an AP lyase activity (Fig. 1). Monofunctional glycosylases only remove the base leaving a natural AP site. In the case of bifunctional DNA glycosylases, the AP site generated by the glycosylase activity is further incised in 3' via β-elimination which produces an obstructive 3'-end (e.g. trans-4-hydroxy-2-pentenal-5-phosphate) (reviewed in [49]). Typical monofunctional glycosylases are the above mentioned UDG and ANPG whilst human 8-oxoG DNA glycosylase (hOGG1) and the human homologue of *E.coli* endonuclease III (hNTH1 – [50]) are representative bifunctional glycosylases. The activity of the bifunctional glycosylase hNTH1 has been shown to be enhanced by the xeroderma pigmentosum G protein, an endonuclease involved in NER [51], thus indicating that BER and NER probably interact at one or more levels.

In human cells, natural AP sites left by monofunctional glycosylases are incised in 5' by a major hydrolytic AP endonuclease variously termed APE or HAP1 or Ref-1 [8] that leaves a 5'-terminal 2-deoxyribose-5-phosphate (dRP) residue and a 3' OH priming terminus. This protein is further endowed with an important redox activity regulating binding of transcription factors to DNA [8]. APE/HAP1 is inducible by oxidative stress [52,53] and is finely tuned by phosphorilation [54].

Incision Processing and DNA Synthesis

Either the 3'-blocking fragment left by AP lyases or the dRP residue resulting from hydrolytic incision must be removed to permit completion of the repair process. Some 3' trimming events are performed in human cells by the 3' phosphodiesterase activity associated to APE/HAP1 [5,55] or by the human polynucleotide kinase (PNK) protein that is endowed with a phosphatase activity capable of restoring conventional 3'-OH termini to DNA single-strand breaks [56]. PNK is stimulated by XRCC1. Alternatively, removal of 3' obstructive lesions may occur by spontaneous δ-elimination followed by removal of the resulting 3'-*O*-phosphate via a 3' phosphatase activity [5,58]. The dRP group is mainly removed by the 5'-deoxyribophosphodiesterase (dRpase) activity associated to DNA polymerase β (pol β) [57,58] but other proteins have also been found to possess a DNA dRpase activity [59]. Removal of the 5' dRP group by pol β follows DNA synthesis [61]. It has been proposed that this step is rate-determining during *in vitro* repair of AP sites by the short patch base excision repair pathway [61]. The observation that the majority of endogenous AP sites in normal tissues of rats and human liver is found cleaved 5' to the AP site suggests that removal of dRP is the rate-limiting step of BER of AP sites also *in vivo* [32]. Interestingly, the *Drosophila* ribosomal protein S3, is endowed with the capacity to incise both AP sites and 8-

oxoG lesions as well as to remove a dRP group from DNA substrates containing 5' incised AP sites [59]. A reported additional 3' repair activity [59] has not been confirmed [60]. Removal of dRP by S3 protein occurs via β-elimination [59]. The S3 ribosomal protein thus functions in several steps of the DNA BER pathway in *Drosophila*.

The synthesis step involves mainly replacement of a single nucleotide [62] but a number of repair patches is longer (2-10 nucleotides) and require participation of proliferating cell nuclear antigen (PCNA) (Fig. 1) [63,64,216]. The short patch pathway is mainly performed by DNA pol β while both pol β and the PCNA-dependent polymerases δ/ε (pol δ/ε) are involved in the long patch pathway [9,65,66,216]. It has been proposed that the choice of the pathway is determined by the type of glycosylase removing the lesion [67]. According to this model, monofunctional glycosylases would be followed by both pathways because during the slow processing of 5' dRP by the dRPase activity of pol β, long patch repair events could occur in competition with the predominant single nucleotide insertions (Fig. 1). On the contrary, the formation of a 3' blocked terminus by the AP lyase activity associated to a bifunctional glycosylase would be preferentially followed by single nucleotide replacement reactions. The 2-10 nucleotides long DNA fragment that is replaced during PCNA-dependent synthesis is not chewed down but rather displaced during polymerization and cut away by DNAase IV (FEN1), a mammalian endonuclease previously identified for its role in DNA replication [68]. DNAase IV is stimulated by PCNA [68].

DNA Ligation

Once the short or long patch synthesis has been completed, the repair patches must be sealed. It is likely that any of the four DNA ligases so far characterized in mammalian cells may to a varying extent fulfil this function. DNA ligase I [69] and DNA ligase III [70,71] are probably the main activities involved in ligation of short patches. DNA ligase III is complexed to the X-ray cross complementing I (XRCC1) gene product, a protein that has no known catalytic function but for which an important structural role in assembling a possible "repairosome" has been proposed [70]. Indeed, specific interactions between all the major players of BER (DNA pol β and APE/HAP1 [72]; XRCC1 and DNA pol β [70]; XRCC1 and DNA ligase III [73]; XRCC1 and PARP [74,75] pol β and DNA ligase I [69]) have been demonstrated. The complex DNA ligase III/XRCC1 does not take part in ligation of long patches [71]. Rather, DNA ligase I is the main activity fulfilling this task [216].

ATase

A single protein capable to remove the methyl group of O^6 meG thus reconstituting normal G has been isolated in various mammalian cells and tissues (reviewed in [17,18]). This "suicide" enzyme uses one of its cysteine residues as a methyl group acceptor and is inactivated in the course of the reaction [76]. It is unable to catalyze the removal of alkyl groups from alkylphosphotriesters in DNA [77] but can remove the methyl group from the miscoding lesion O^4-meT, although with lower efficiency as compared to O^6 meG [78].

NER

NER has evolved to repair damage inflicted by UV light (reviewed in [217]). This complex repair system is particularly efficient on bulky DNA lesions that distort the double helix. However, it can also act as a backup system for endogenous damage [6,42]. When NER operates in nontranscribed regions, the lesion is probably recognized by a complex of two proteins, XPC and hHR23B. A bubble is then created around by a second complex containing the transcription/repair factor TFIIH (that in turn includes two DNA helicase activities, XPB and XPD), the XPA protein and the single-strand DNA binding heterotrimer RPA. The bubble is a suitable substrate for incision of the damaged strand by the ERCC1-XPF and XPG nucleases that cut 20-24 nucleotides 5' and 4-8 nucleotides 3' to the lesion respectively, with concomitant release of a 24- to 32- residue oligonucleotide. The resulting gap is filled in by a PCNA-dependent DNA polymerase (δ or ε) and sealed by a DNA ligase.

NER occurring in transcribed regions is 5-10 times faster than in non transcribed regions, because of active recruiting of repair proteins to the damaged site by several factors including the Cockayne syndrome A and B gene products. In this case the XPC protein does not take part to the process.

Defective Repair of Endogenous DNA Lesions and Enhanced Mutagenesis

Defective Repair of Oxidized Bases

The E.coli formamidopyrimidine DNA glycosylase (FPG) protein specifically removes imidazole ring-opened forms of purines such as 2,6-diamino-4-hydroxy-5-N-methylformamidopyrimidine (FaPy) and 8-oxoG residues (reviewed in [79]). The *FPG* gene was isolated and described as *MUT M* during a screen for mutator genes conferring elevated transversion frequencies in *E.coli* [80]. Inactivation of FPG results in an increase of the steady-state level of 8-oxoG of approximately 6-fold [81] while the spontaneous mutation frequency is elevated 10-14 fold [80,82]. Two other proteins limit spontaneous mutagenesis by 8-oxoG in *E.coli* cells: the MUT Y and the MUT T proteins. MUT Y is a DNA glycosylase removing As that have mispaired with 8-oxoG and 8-oxoadenine [2]. Hence, it acts as a safeguard system in case FPG/MUTM were not able to remove the oxidized base before passage of the replication fork. Spontaneous mutagenesis is increased 20-100 fold in *MUTY* cells [83]. MUTT is a nucleotide triphosphatase that hydrolyses 8-oxodGTP, an endogenous nucleotide triphosphate that mispairs with A. *MUTT* strains are characterized by a high frequency of AT-CG transversions [84]. Double mutants for any two of these proteins display elevated frequencies of mutation (25 to 75-fold higher) as compared to single-locus mutants thus indicating the cooperative functions of these proteins in control of spontaneous mutagenesis [82,85]. This sinergy is more marked than in other double repair mutants (e.g. the *NTH NEI* mutant – see below) thus indicating the important role of 8-oxoG in spontaneous mutagenesis. Overexpression of MUTM suppresses the mutator phenotype of

MUT Y, probably by eliminating most 8-oxoG residues before passage of the replicative polymerase [82]. In the budding yeast *Saccharomyces cerevisiae*, inactivation of the MUTM homologue yOGG1 leads to a mutator phenotype with a marked increase in spontaneously occurring GC – TA transversions [88]. The spontaneous mutation frequency could be returned to almost normal levels by introduction of a normal human *OGG1* gene [89].

The *E.coli* endonuclease III (NTH) protein is mainly involved in removing oxidized or ring-fragmented pyrimidines. *NTH* mutants display only a modest increase in spontaneous mutation rate (~ 3-fold), thus suggesting the presence of other activities with similar substrate specificity (e.g. endonuclease VIII, the product of the *NEI* gene). The double mutant *NTH NEI* shows indeed a 20-fold increase in spontaneous mutagenesis with respect to wild type cells [86]. However, Saito et al, [87] working with a sytem detecting specifically G:C to A:T transitions, found no increase of this kind of mutation in the double mutant. This may indicate that the spectrum of lesions repaired by NTH and NEI does not include miscoding bases inducing G:C to A:T transitions. Similar results have been found in the *Saccharomyces cerevisiae* mutants *NTG1* and *NTG2*, both defective in homologues of *E.coli* NTH protein [90,91]. Targeted gene disruption of either *NTG1* or *NTG2* produced mutants with greatly increased spontaneous mutation frequency relative to the wild type, and the mutation response was further increased in the double mutant. Both NTG1 and NTG2 remove thymine glycol residues and FaPy residues, while 5-hydroxycytosine is removed efficiently only by NTG2. A functional homolog of *Escherichia coli* endonuclease III (hNTH1) has been cloned in human cells [50]. The isolated open reading frame encodes a protein of 34.3 kDa that shares extensive sequence similarity with *E.coli* NTH protein and exhibits DNA glycosylase activity on double-stranded polydeoxyribonucleotides containing urea and thymine glycol residues, as well as an AP lyase activity. The sequence and functional homologies of NTH-like proteins across the phylogenetic spectrum indicates that the role of these enzymes for genome maintenance is important.

Defective Repair of Methylated Bases

ANPG knockout cells are viable and completely deficient in repair of 3-maA, 1,N6-ethenoadenine and hypoxanthine [101,104]. ANPG knockout embryonic stem cells display increased sensitivity to chromosome damage and cell killing induced by alkylating agents and a range of other DNA damaging agents thus reflecting the broad substrate specificity of the enzyme [101,102]. Unlike other BER knockouts so far developed (e.g. for APE/HAP1 and pol β), ANPG-deficient mice are viable and show no evident development abnormalities [9]. In these mice, an increased mutation frequency at the *hprt* locus of splenic T lymphocytes could be observed after exposure to methylmethanesulfonate (MMS) [103].

E.coli mutants defective in ADA and OGT DNA methyltransferases display a spontaneous rate of base substitutions elevated about four fold with respect to wild type [92–94]. The cloned *ADA* and *OGT* genes each restored wild-type spontaneous mutation rates to the methyltransferase-deficient strains [93]).

Likewise *E.coli* mutants, *Saccharomyces cerevisiae* mutants deficient in ATase display increased spontaneous mutation rates [95,96]. Spontaneous mutation is returned to wild type levels by heterologous expression of a human *ATase* gene [96].

ATase knockout mice show an increase of nitrosamine-induced tumor frequency [98] and are more sensitive to the toxic effect of MNU [99,214]. Spontaneous carcinogenesis is also increased [98-100]. Taken together the data indicate that ATase may significantly contribute to protect eukaryotic cells from spontaneous mutagenesis.

Defective Repair of U

The importance of UDG in counteracting spontaneous mutagenesis is well established in different organisms. In *E..coli*, mutations in the *UNG* gene encoding UDG, cause a 2.4 – 30 fold increase of spontaneous mutation frequency [105-106]. Similar results can be observed in other bacteria and in baker's yeast. UDG–deficient *Streptococcus pneumoniae* displaies a ten fold increase in C to T transitions [107]. *Saccharomyces cerevisiae UNG1* mutants display a 4-40 fold increase in spontaneous mutation frequency at four nuclear loci or at the plasmid locus SUP4-o [108,109].

Defective Repair of AP Sites

Bacterial cells partially deficient in AP sites incision capacity (*NTH*) display enhanced mutation rates [110] while attempts to construct an *E.coli* completely unable to incise DNA at AP sites have been unsuccessful, thus suggesting that such a cell would not be viable [186]. Likewise, the *S.cerevisiae* main AP endonuclease APN1 exerts an important role in control of spontaneous mutation. *APN1* mutant cells have a rate of spontaneous mutation 6- to 12-fold higher than that measured in wild type cells, thus indicating the steady erosion of DNA integrity occurring under physiological conditions [111]. Surprisingly, those increased spontaneous mutation rates are markedly reduced in yeast by a second mutation in the *MAG1* gene encoding 3meA DNA glycosylase [95,112] thus indicating that AP sites generated by enzymatic removal of spontaneously formed alkylated purines may exert an important mutagenic role (see below section OVEREXPRESSION OF ENZYMES INVOLVED IN REPAIR OF ENDOGENOUS DAMAGE). APN1 – deficient yeast mutants can also be complemented by the NTH protein of *E.coli* [113].

Mammalian cells made deficient in APE/HAP1 activity by expression of an *APE/HAP1* antisense construct become sensitive to a wide range of DNA damaging agents, including MMS, hydrogen peroxide, menadione and paraquat (but not ultraviolet light) and to hypoxic stress [114,115]. No data have been reported concerning spontaneous mutation.

APE/HAP1 knockout mice die at a very early gestation phase, following blastocyst formation (embryonic day 5.5 [40]) and no mutant cell lines have been reported, thus indicating the important role of this protein in cell viability and embryonic development [40,213].

Defects in other BER Functions

Knock out *DNA pol β* mice are embryonic lethal at midgestation (embryonic day 10.5) [41,116] but, unlike the *APE/HAP1* knockout, mouse embryonic fibroblasts do divide indefinitely [41]. *Pol β* knockout cells are hypersensitive to cytotoxicity and clastogenicity of alkylating agents, thus indicating that the repair function of pol β cannot be substituted efficiently by other polymerases [41]. Further, *pol β* deficient cells show increased sensitivity to crosslinking agents [99], thus indicating that this polymerase is involved in repair of a wide range of lesions.

Mutants in the *XRCC1* gene (EM9, EMC-11) are hypersensitive to MMS and ethylmethanesulfonate (EMS) [117-119]. *XRCC1* mutants have very low levels of DNA ligase III [71,119] and sensitivity to alkylating agents may in part be explained with this ligation defect. No catalytic functions of the XRCC1 protein has been found to date, but, as mentioned above, the role of this protein in assembling a repair complex may be important. *XRCC1* knockout mice are early embryonic lethal (embryonic day ~6.5) thus indicating the essential function exerted by this protein in embryonic development [9,120]. Without exposure to DNA-damaging agents, mutant embryos show increased levels of unrepaired DNA strand breaks in the egg cylinder compared with normal embryos. *XRCC1* mutant embryos that carry an additional homozygous inactivating mutation for p53, undergo developmental arrest after only slightly further development, thus indicating a p53-independent mechanism of embryonic lethality. *XRCC1 -/-* cell lines derived from mutant embryos are hypersensitive to DNA damaging agents [9,120].

Mammalian cells with knockout PARP activity show increased sensitivity to alkylating agents [121]. Knockout mice are viable but display reduced litter size and adult weight and a number of characteristics indicating increased sensitivity to DNA-damaging agents including increased lethality by methylnitrosourea (MNU) and γ-rays, increased sister chromatid exchanges (SCE) and chromatid breaks by MNU and γ-rays in bone marrow, increased G2/M arrest of fibroblasts and increased apoptosis by MNU [121]. The rate of spontaneous SCE was 4- to 5-fold higher in *PARP -/-* than in *PARP +/+* mice. No difference was found in the frequency of spontaneous chromosome breakage.

Overexpression of Enzymes Involved in DNA Repair of Endogenous Damage

The elevated efficiency of repair of endogenous damage in mammalian cells is reflected by the low steady state levels of some endogenous lesions in human cells, in front of their number generated daily [19]. For example, while several thousands of AP sites are generated daily in a human cell, probably less than ten are present at each moment in the genome. The high efficiency of repair of AP sites can be easily observed *in vitro* [122] and *in vivo* [32] where these lesions are removed in relatively short times. Demple and coworkers [123] and Cappelli et al [305] have calculated that the amount of APE/HAP1 protein in mammalian cells varies from ~ 3.0×10^5 molecules/cell for primary human fibroblasts to ~ 7×10^6

molecules/cell for HeLa, T lymphoblast and CHO cells. These elevated levels may account for the speed with which AP sites are incised.

Unfortunately, not all endogenous lesions are repaired efficiently as AP sites. The only addition of the glycosylase step (i.e. repair of bases rather than base loss sites) involves a decrease in the repair replication capability of human cell extracts, the extent of which is varying according to the lesion investigated. For instance, U, one of the most readily repaired bases in mammalian cells, is repaired by normal cell extracts three-ten fold less efficiently than natural AP sites *in vitro*. In turn, 8-oxoG is repaired with lower efficiency with respect to U [125,305]. The same heterogeneity of repair efficacy applies to different oxidized lesions. Jaruga and Dizdaroglu [124] have determined the *in vivo* repair of eleven modified bases produced upon hydrogen peroxide treatment of human cells. In general, repair of purine lesions was slower than that of pyrimidine lesions and the most persistent adduct was 4,6-diamino-5-formamidopyrimidine (Fapy Ade) followed by 8-oxoG. The question arises whether the capacity of mammalian cells to repair endogenous lesions, in particular those showing marked miscoding properties, may be increased.

With some exceptions the enzymatic mechanisms of DNA repair are highly conserved in nature. As a consequence, the enzymes involved in these important functions are often able to work in heterologous cells [13,126,255]. Transfection of repair deficient cell lines with cloned bacterial or mammalian repair genes has been exploited powerfully in cloning strategies and might also be a tool to improve protection of eukaryotic cells from the genotoxic effects of exogenous and endogenous mutagens [127]. However, the concept that overexpression of DNA repair enzymes may invariably lead to enhanced resistance to endogenous or exogenous mutagenesis is certainly wrong. As it will come out from the following, imbalancing the cellular levels of DNA repair enzymes in mammalian cells may in fact result in some cases in genome instability [99,128].

DNA Glycosylases Removing Oxidized Bases

Transfection of the *E.coli NTH* gene into radiosensitive *Chinese hamster* cells resulted in increased resistance to hydrogen peroxide, an agent inducing a high level of endonuclease III-repairable damages. No substantial difference in γ-rays sensitivity and even increased sensitivity to the radiomimetic drug bleomycin were observed in the transfected cells [131].

The hOGG1 protein has been successfully expressed in human lung cells where it conferred resistance to toxicity of oxidizing agents [135] or BCNU [138]. hOGG1 expression could not yet protect *Chinese hamster* cells from oxidative mutagenesis [132].

The most effective heterologous enzyme that protects human cells from oxidative DNA damage is probably E.coli FPG [129,130]. This enzyme will be treated with special reference below (see Section "The FPG (MutM) protein" and following).

DNA Glycosylases Removing Methylated Bases

Overexpression in *Chinese hamster* ovary cells of the *ANPG* gene does not lead to enhanced resistance to the toxic and mutagenic effects of MMS and N-methyl-N'-nitro-N-nitrosoguanidine (MNNG) [133] and even may render cells more sensitive to alkylating agents [134]. In other studies, no increased resistance to the bifunctional alkylating agents bis-chloroethylnitrosourea and melphalan was observed [136]. Overexpression of ANPG in NIH3T3 cells and in a number of tumor cell lines leads to toxic effects [137]. It is likely that overexpression of the ANPG glycosylase causes an imbalance in BER, giving rise to an excess of AP sites and gapped DNA that would result in a substantial increase of SCEs and chromosomal aberrations as well as inhibition of DNA replication [134]. The latter hypothesis is supported by studies in *Saccharomyces cerevisiae.* In budding yeast, the mutagenic effect of overexpression of the *MAG1* gene, encoding the main 3meA DNA glycosylase activity was increased in the absence of APN1, the main AP endonuclease activity that covers > 97 % AP endonuclease activity in yeast [95,139]. This suggests that AP sites created by the excess MAG1 protein may be the causative intermediates: the glycosylase could act upon spontaneously produced altered bases and, most probably, normal purines [140], to produce mutagenic AP sites that, if left unrepaired as in *APN1* mutants, could eventually cause mutations during replication. The elevation of the spontaneous mutation rate was of approximately 600 fold [141], among the strongest mutator phenotypes observed in yeast. Comparison of the mutator effect of MAG1, which has a broad substrate specificity with that of *E.coli* 3-meA-DNA glycosylase I (TAG), which is highly specific for 3-meA and whose expression failed to generate a mutator phenotype, indicated that the increased spontaneous mutation rate was unlikely due to elevated removal of 3-meA endogenous lesions, but that AP sites were generated *via* removal of other, so far unidentified, bases [141]. It can be hypothesized that imbalances in certain glycosylases with broad substrate specificity may potentially lead to a number of AP sites exceeding the incision capacity of APE/HAP1, with a resulting increase of mutation frequency. Such kinds of imbalances may occur under physiological conditions as well, consequently to changes in gene expression or in the nuclear localization of the enzymes [142]. It is interesting (and puzzling) that different results can be obtained when overexpression of ANPG is achieved in the alkylation sensitive *Chinese hamster* mutant IRS1 [143] or in bacterial cells [136] that indeed acquire enhanced resistance to alkylating agents. It has been proposed to exploit the paradoxical phenomenon of sensitization of mammalian cells to alkylating agents after overexpression of ANPG to increase the sensitivity of tumoral cells to chemotherapeutic agents [137,144].

As mentioned above, overexpression of the *E.coli* TAG protein, a DNA glycosylase with narrow substrate specificity that only repairs 3-meA, does not generate a mutator phenotype in yeast. Klungland et al [145] have reported that overexpression of the *TAG* gene in mammalian cells may in fact increase resistance to the genotoxic effects of methylating agents such as MMS. *TAG* expression reduced the frequency of MMS- and EMS-induced mutations at the *HPRT* locus to about half the normal rate [146]. Another group was however unable to obtain the same results with 3T3 NIH murine fibroblasts [147].

AP Endonucleases

Consistent with the notion that APE/HAP1 is not rate-limiting in mammalian BER, in most overexpression studies no increased resistance to DNA damaging agents but rather increased genomic instability and even sensitization to bio-reductive drugs were observed. For instance, overexpression of APE/HAP1 protein has been achieved in the rat C6 glioma cell line. No increased resistance to γ-irradiation at high or low dose rates or to the methylating agent MMS could be observed thus indicating that increased levels of APE/HAP1 protein cannot improve in those cells the efficiency of repair of radiation or alkylation damage [148]. Another analysis of the modulation of resistance to DNA damaging agents by overexpression of the APE/HAP1 protein in CHO cells has been reported by Prieto-Alamo and Laval [149]. No increased resistance to the toxic effects of various agents inducing BER-processed damages (e.g. MMS, bleomycin, H_2O_2) was observed in the transfected cells. However these cells became more sensitive to drugs that require reduction to be activated. It has been demonstrated that this effect was linked to the increased redox activity associated to APE/HAP1 [149]. Sossou and coworkers have recently observed increased genomic instability after APE/HAP1 overexpression in XRCC1-deficient cells [152]. One exception to this genome instability effect was the germ cell tumoral line NT2 where APE1 overexpression conferred resistance to bleomycin and gamma-irradiation [153].

Tomicic et al [150], have shown that *Chinese hamster* cells become more resistant to DNA damaging agents such as MMS and H_2O_2 after transfection of yeast but not human AP endonuclease genes. Cells transfected with the yeast *APN1* gene express high levels of APN1 protein and become clearly resistant to the genotoxic effects of MMS and H_2O_2. Expression studies with the *Saccharomyces cerevisiae* APN1 activity to protect mammalian cells by DNA damaging agents were also reported by the group of W. Martin [151]. Yeast APN1 endonuclease was overexpressed in the alveolar epithelial A549 cell line by transduction with a MIEG3 retroviral vector encoding both enhanced green fluorescent protein and APN1. Cells which overexpressed APN1 showed increased resistance to the antitumor drug bleomycin. These results may indicate therefore that expression of the *APN1* cDNA in mammalian cells may result in increased resistance to certain oxidative and alkylating agents. The different outcome with respect to the works by Herring et al [148] and Prieto-Alamo and Laval [149], may be linked to the different characteristics of yeast and human AP endonuclease. As mentioned above, APE/HAP1 shows a very weak 3'-repair diesterase activity and evolved probably to act preferentially on natural AP sites rather than deoxyribose fragments located at DNA strand breaks. On the contrary, the yeast *APN1* gene encodes an AP endonuclease function that is homologous to E.coli endonuclease IV and, like this bacterial enzyme, is associated to a robust 3' repair diesterase activity. The capacity to trim 3' – blocking deoxyribose fragments may thus make the difference. In vitro studies have shown yet that pure APN1 protein supplemented to repair reactions does not stimulate DNA BER performed by mammalian cell extracts [218]. Additional cellular factors may cooperate with the 3' repair diesterase activty of APN1 to increase resistance to DNA damaging agents.

DNA Pol β

In comparison to APE/HAP1, a lower level of DNA pol β (5 x 10^4 molecules per cell) has been estimated in mammalian cells [154]. In view of this, can an increased expression of pol β lead to enhanced capacity to repair endogenous damage to DNA? Indeed, up-regulation of DNA pol β in mouse cell lines by agents inducing oxidative stress such as lipopolysaccharide correlates with enhanced protection to MMS, an effect that was not observed in cells deleted of the DNA pol β gene [155]. However, overexpression of pol β in human cells does not result in increased protection from endogenous or exogenous damage. On the contrary, it results in increased spontaneous mutagenesis and a highly mutagenic tolerance phenotype toward bifunctional DNA cross-linking anticancer drugs [156,215]. Sensitization to MMS, but not ultraviolet (UV) damage, has also been reported [157]. It has been hypothesized that excess pol β may perturb the well-defined functions of the other DNA polymerases and may introduce in the genome illegitimate deoxyribonucleotides or mutagenic base analogs produced by intracellular oxidation [156,215].

As mentioned under MOLECULAR MECHANISMS OF DNA REPAIR OF ENDOGENOUS DAMAGE, in addition to its polymerization function, DNA pol β contains a dRPase activity that excises dRP from DNA incised by hydrolytic AP endonucleases (Fig. 1). It has been demonstrated that the dRPase activity resides in a definite 8 kDa domain that retains its functionality when isolated from the rest of the protein [57]. Srivastava et al, [61] have shown that excision of dRP is the rate-limiting step in a reconstituted in vitro BER system and there are indications that this is also true in vivo [32]. Increasing the velocity of this step by overexpression of the whole pol β protein could be useless, given the above mentioned genome instability phenotype [156,215]. It is possible however that the overall efficiency of BER in mammalian cells may be augmented by overexpression of the isolated dRPase activity associated to the 8-kDa amino-terminal domain of pol β. The isolated domain efficiently removes 5'dRP from a preincised AP site and its catalytic mechanism as well as a detailed functional analysis of its protein sequence have been determined [158,159].

PARP

PARP was investigated in overexpression studies performed in rat tumor cells by Bernges and coworkers [160] and in *Chinese hamster* cells by Van Gool and coworkers [161]. Sensitization upon treatment with γ-radiation or MNNG was observed in several, but not all, subclones expressing human PARP. A protective effect against DNA damage was never observed. A slight reduction in MMS toxicity and MMS-induced chromosomal aberrations has been observed in *Chinese hamster* cells expressing increased levels of PARP [162]. Taken together the available evidence indicates that overexpression of PARP does not lead to significant protective effects against the toxicity and mutagenicity of DNA-damaging agents. A non physiological level of PARP may even be detrimental in mammalian cells.

XRCC1

Overexpression of XRCC1 in Cos-7 or HeLa cells dramatically decreases PARP activity *in vivo* [75]. Cells overproducing XRCC1 are impaired in poly(ADP-ribose) synthesis in response to H_2O_2-generated DNA breaks. Whether spontaneous or induced mutagenesis were affected in the transfected cells was not reported [75].

ATase

Overexpression of ATase in mammalian cells is not associated to increased genomic instability [163]. On the contrary, expression of ATase protects Mex$^-$ *Chinese hamster* cells from spontaneous G:C to A:T transitions, thus showing that ATase efficiently counteracts the effects of an endogenous source of O^6-meG that is normally responsible for those mutations [97]. No significant difference occurs in these cells in the number or type of deletions, frameshifts, multiple substitutions or insertions. Further, CHO cells that overexpress ATase (or the bacterial ADA protein) gain resistance to point mutations induced by MNU, N-ethyl-N-nitrosourea (ENU) and MNNG and to alkylation-induced SCEs and aberrations as compared to ATase deficient cells [166-172]. Similarly, transfer and overexpression of the bacterial or mammalian *ATase* gene in human cells results in cellular resistance to MNNG [164] and alkylnitrosoureas [165]. Expression of exogenous ATase can also protect cells from *in vitro* transformation [173]. The protective effects were dependent on the level of ATase expression, the agent used for alkylation and the phase of cell cycle progression. Overexpression of the *E.coli* ADA protein deleted for the methylphosphotriester repair activity led to similar results with respect to cytotoxicity and SCE induction, thus pointing out the limited significance of methylphosphotriesters with respect to those endpoints [174]. One case has been reported of sensitization of cells expressing an heterologous ATase activity: cells expressing the full lenght or the O^6-meG region, but not the alkylphosphotriesters region of the *E.coli* ADA protein were found to be more sensitive to the growth inhibitory effects of dibromoalkanes [175]. It has been hypothesized in this case that the cell sensitization may be related to inactivation of the ATase protein.

Targeted expression of the protein can be obtained in transgenic mice, thus giving insight on the role of this enzyme for preventing mutation and cancer [18,100,176]. According to the targeted tissue, ATase-overexpressing mice were protected against alkylation-induced skin carcinogenesis [177], liver carcinogenesis [178], colon carcinogenesis [179], lung carcinogenesis [180] and lymphomagenesis [181-185]. The direct demonstration that O^6meG adducts account for the carcinogenic effect of MNU and that ATase overexpression is selectively able to reduce those adducts under a carcinogenic threshold has been obtained [187,188].

Increased resistance to chemotherapeutic nitrosoureas can be achieved in mammalian cells by increased expression of bacterial or human ATase protein [189-191]. A gene therapy approach with the *ATase* gene has been developed to circumvent both the acute toxic and chronic carcinogenic effects of chemotherapy with alkylating agents [192]. In particular, early hematopoietic progenitors could be protected by retroviral-mediated transfer of the

ATase gene from toxicity of chemotherapeutic nitrosoureas [193-199] or temozolomide [200,201]. A mutant *ATase* gene resistant to O^6 benzylguanine was employed in most cases, thus allowing the clinical use of this sensitizing agent [202-204]. ATase-transduced marrow cells could selectively repopulate the marrow and improve the therapeutic index of treatments, allowing adoption of more intensive therapeutic schedules [205,206].

Other Gene Products Involved in Resistance to Oxidizing and Methylating Agents

As mentioned above, the recently characterized S3 ribosomal protein of *Drosophila* is endowed with different enzymatic activities [59]. In particular, overexpression of this protein may confer to eukaryotic cells an increased capacity to perform repair of 8-oxoG residues. Recent expression studies showed yet that this protein may be toxic when overexpressed in mammalian cells [219].

A lower basal oxidized state has been observed after overexpression of oxyradical scavenger proteins such as carbonic anhydrase III [207] or metallothionein [208]. In the latter case, a decrease of spontaneous mutagenesis was also observed [208]. Further, *Chinese hamster* ovary cells transfected with the human metallothionein IIA are protected from killing by alkylating agents but not ionizing radiation or bleomycin [209].

Resistance to methylating mutagens, but not to UV irradiation or crosslinking agents can be obtained in *Chinese hamster* cells by transfer of human genes that do not cause increased repair of O^6-meG, 3-meA or 7-meG [210]. One of these genes has been characterized: it encodes a protein partially identical to a lysosomal membrane glycoprotein that is able to confer a methylation–resistant phenotype to mammalian ATase-deficient cells. Interestingly, the transfectants also showed a suppressed tumor phenotype [211].

Finally, overexpression of the *E.coli* ALKB protein, to which no specific DNA repair activity but only regulatory functions have been assigned, protects human cells against killing by SN2 alkylating agents [212]. Thus, the *E.coli* ALKB protein can provide alkylation resistance *per se*, by an as yet unidentified mechanism.

The FPG (MUT M) Protein

Structural and Catalytic Features

Escherichia coli FPG is a globular monomer that possesses combined DNA glycosylase-AP lyase activities [220,221] (Figure 2; Table II). The nucleotide sequence of the gene is composed of 809 base pairs and codes for a polypeptide of 269 aminoacids with molecular weight 31 kDa. The overall protein structure comprises two domains connected by a hinge polypeptide [222]. The N-terminal domain contains eight β-strands forming a β-sandwich with two α-helices parallel to its edges (Figure 2). The C-terminal domain includes four α-helices, two of which form a helix-twoturn-helix motif, and two β-strands that form a β-hairpin zinc finger. The latter domain is the DNA binding one. *E. coli* FPG protein differs from hOGG1 under several aspects. While hOGG1 only excises from DNA 8-oxoG and FapyG, FPG fastly releases 8-oxoG, FapyG and FapyA with similar excision kinetics [223]. The ability to excise FapyA is thus a major substrate difference between the two enzymes. Further, in comparison to hOGG1, FPG has different paired base and sequence context effects that probably originate from differences in the architecture of the DNA binding and catalytic domains. Similarities also exist: e.g. two sequential phosphates 3' adjacent to the 8-oxoG lesions are crucial in both enzymes [224]. Defects in FPG cause an increase of GC – TA transversions, as a consequence of the high frequency of misincorporated A in front of 8-oxoG during DNA replication [225,226]. FPG efficiently releases ring-opened guanine or adenine, oxidized guanine and guanidinohydantoin (Gh) and spiroiminodihydantoin (Sp), two minor lesions resulting from one-electron oxidation of 8-oxoG [227,228] (Figure 3; Table II). It has been further reported that FPG recognizes and excises oxidized pyrimidines from oligonucleotides with a single lesion only [229,230], but such repair activity is less evident on DNA substrates containing multiple lesions [229,231]. Hence, it is likely that the enzyme main physiological substrates are limited to 8-oxoG, FapyG, FapyA and AP sites.

FPG preferentially acts on 8-oxoG paired with G, C, T but not A [232]. The AP site resulting from base removal is incised by an associated β, δ-elimination reaction which leaves a single-nucleotide gap in DNA flanked by 3'- and 5'-phosphoryl groups [233]. As aforementioned, catalysis is fast: for instance, the k_{cat}/K_m values on 8-oxoG and me-Fapy are 0.14 and 0.13 respectively, roughly 80-fold higher than in hOGG1 [234] (Table II). How this protein efficiently locates a single lesion embedded in a vast excess of DNA has been investigating in detail [235,236]. These studies exploited a Schiff base intermediate involving Pro-1 of the enzyme and C1' of the damaged nucleotide that forms early in the reaction sequence and can be reductively trapped by sodium borohydride, forming a stable covalent complex [237] (Figure 2). Site-specific mutagenesis studies revealed that His-89 and Arg-109 form part of a "reading head", a structural feature used by the enzyme to scan DNA for damage [238]. His-89 and Lys-217 help determine the specificity of FPG in recognizing the oxidatively damaged base, while Arg-108 provides specificity for bases positioned opposite the lesion [238]. Binding occurs by the absolutely conserved residues Lys-56, His-70, Asn-168 and Arg-258 that form hydrogen bonds to the phosphodiester backbone of DNA, which is sharply kinked at the lesion site.

Cell Culture Studies

FPG has been investigated as a protective tool from DNA damage in several studies. It was initially expressed in the *Chinese hamster* CHO and V79 cells by F. Laval [129]. The transfected cells expressed a formamidopyrimidine-releasing activity 30 to 40 fold over the basal level. Resistance to lethality induced by γ-rays and hydrogen peroxide was similar in control and transfected cells but the mutagenicity of γ-rays measured as induction of 6 thioguanine-resistant mutants was significantly reduced in FPG-expressing cells (Fig. 4A). Mutagenicity of ionizing radiations is partially linked to lesions that are substrates of FPG.

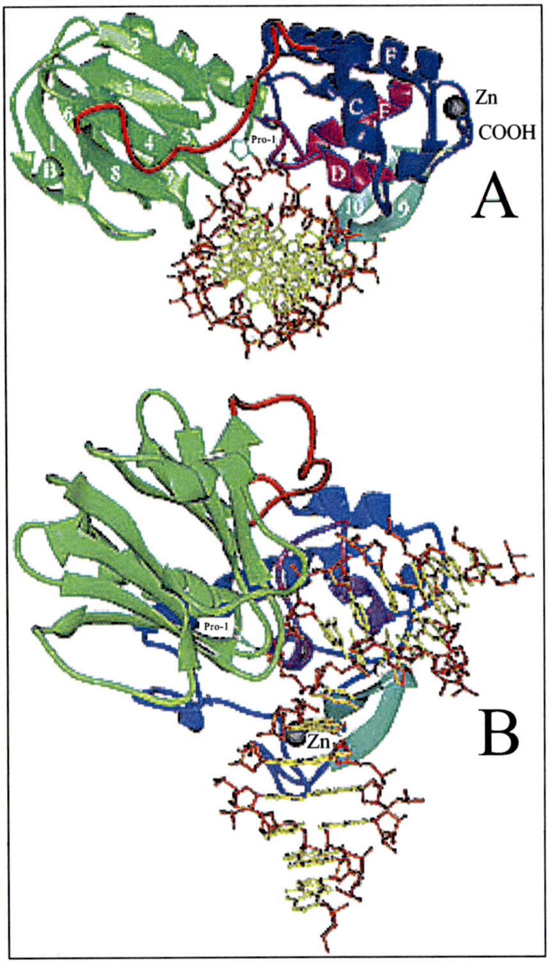

Figure 2. Overall structure of Fpg-DNA. A, view perpendicular to the long axis of Fpg, showing the DNA binding cleft. The N-terminal domain is shown in green, the interdomain hinge in red, and the C-terminal domain in blue, except for the helix-two turn-helix (purple) and zinc finger (cyan) motifs. The zinc atom is represented by a gray ball and the covalently linked Pro-1 is highlighted in cyan. The DNA backbone is colored orange and bases in yellow. B, view approximately orthogonal to A, highlighting the protein-DNA contact area and the DNA kink (from ref. [222] with permission).

Similar defending effects were subsequently observed on mutagenicity of aziridine (ethyleneimine), an alkylating agent produced in large amounts in industry [239] (Fig. 4B) and N,N',N''-triethylenethiophosphoramide (thioTEPA), an alkylating chemotherapeutic drug [240] (Fig. 4C).

In the latter study, the FPG expressing *Chinese hamster* cells were resistant to both lethal and mutagenic effects of thioTEPA and had lower number of replication blocking lesions, thus suggesting that FPG could be useful to protect bone marrow cells from thio-TEPA genotoxicity. This was demonstrated directly five years later by Kobune and coworkers [241] (see Preclinical models). A549 lung epithelial cells have been transduced with the recombinant retrovirus SF91 containing the *FPG* gene and expressing EGFP *via* an internal ribosomal entry site element. Those cells were protected from hyperoxia and hydrogen peroxide as evidenced by a decrease in DNA damage and a corresponding increase in cell survival [242]. FPG further protected the cells from DNA damage and toxicity induced by the alkylating drug BCNU, with higher efficacy than hOGG1 [138]. The broad damage recognition capability of FPG thus includes some BCNU-induced damages. A fusion protein EGFP-FPG has been recently expressed in the human bladder carcinoma T24 cell line [130]. The one nucleotide insertion, but not the PCNA-dependent BER pathway, for 8-oxoG and AP sites was significantly accelerated in FPG-expressing cells and potassium bromate-induced mutations at the Na^+/K^+ ATPase locus were reduced of one order of magnitude (Fig. 4D). Taken together, the available data indicate that supplementation of FPG to human cells protects from genotoxicity of a wide range of DNA-damaging agents.

Table II. The FPG (MutM) protein*.

Source	Escherichia coli
Shape and size	Globular monomer of 269 aa with a zinc-finger domain
Molecular weight	31 kDa
Catalytic activities	Combined DNA glycosylase-AP lyase (β,δ-elimination)
Substrate lesions	8-oxodG, 8-oxodeoxyguanosine; FapydG, guanosine with the imidazole ring opened; N^7-Me-FapydG, guanosine with the imidazole ring opened and a methyl group in N^7 position; FapydA, adenosine with the imidazole ring opened; Sp, spiroiminodihydantoin; Gh, guanidinohydantoin; 8-oxodA, 8-oxodeoxyadenosine; Z, oxazolone; DHU, dihydrouracil; 5OHC, 5-hydroxycytosine; N7-PM-FapydG, guanosine with the imidazole ring opened and a phosphoramide mustard group in N^7 position; N7-AFB1-FapydG, guanosine with the imidazole ring opened and an aflatoxin B1 group in N^7 position; C8-AF-irodG, imidazole ring-opened derivative of N-(deoxyguanosine-8-yl)-2-aminofluorene.
8-oxoG recognition requirements	8-oxoG paired with G,C,T but not A. Two sequential 3' phosphates
Antimutagenic activity	Prevention of GC –TA transversions
K_m (nM)	8-oxoG : 13 ± 4 N^7-Me-FapyG: 38 ± 9
V_{max} (nM/min)	8-oxoG : 2.7 ± 0.7 N^7-Me-FapyG : 7.7 ± 1.2
k_{cat} (min^{-1})	8-oxoG : 1.8 N^7-Me-FapyG : 5.1
k_{cat} / K_m	8-oxoG : 0.14 N^7-Me-FapyG : 0.13

*data from [220], [224-225], [228], [232-234] and references therein. Reproduced from ref. [255] with permission.

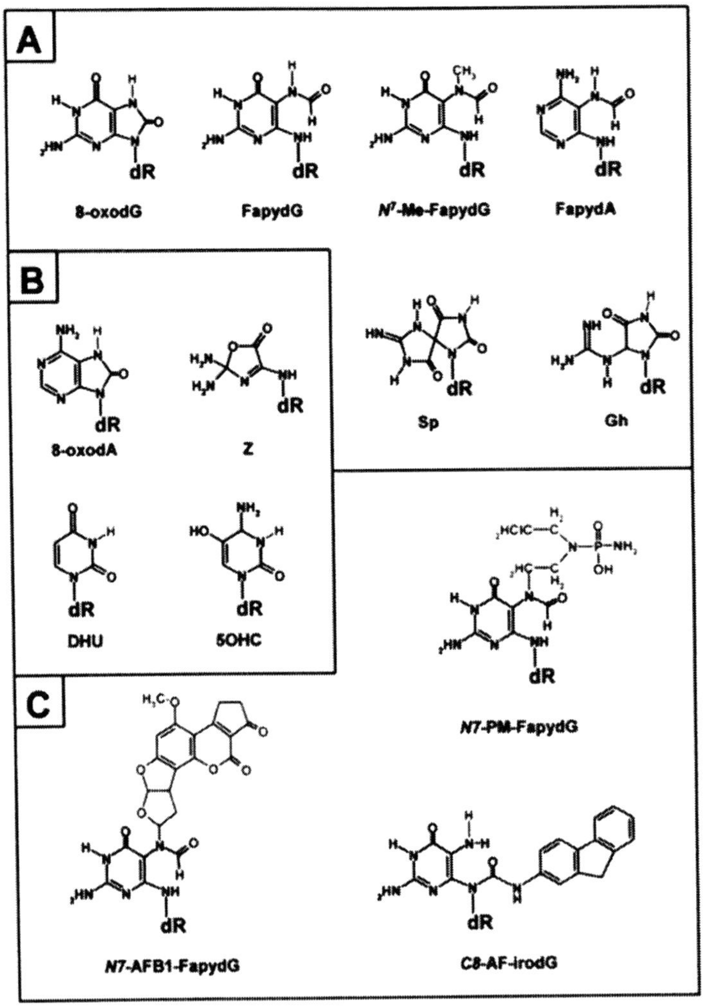

Figure 3. Substrates of Fpg. Substrates presented in A are more efficiently processed by Fpg than substrates shown in B, whereas substrates presented in C are poorly excisable by the enzyme. 8-oxodG, 8-oxodeoxyguanosine; FapydG, guanosine with the imidazole ring opened; N^7-Me-FapydG, guanosine with the imidazole ring opened and a methyl group in N^7 position; FapydA, adenosine with the imidazole ring opened; Sp, spiroiminodihydantoin; Gh, guanidinohydantoin; 8-oxodA, 8-oxodeoxyadenosine; Z, oxazolone; DHU, dihydrouracil; 5OHC, 5-hydroxycytosine; N7-PM-FapydG, guanosine with the imidazole ring opened and a phosphoramide mustard group in N^7 position; N7-AFB1-FapydG, guanosine with the imidazole ring opened and an aflatoxin B1 group in N^7 position; C8-AF-irodG, imidazole ring-opened derivative of N-(deoxyguanosine-8-yl)-2-aminofluorene (from ref. [228] with permission).

Enhancement of DNA repair and prevention of skin cancer has been achieved in normal and Xeroderma Pigmentosum mice highlighting the potential of gene therapy performed with DNA repair proteins [243,244] (Table III). Recombinant adenovirus carrying the human XPA gene was used for in vivo gene therapy in UVB-irradiated skin of XPA mice. Virus subcutaneous injection caused expression of the XPA protein in keratinocytes and protection of skin from deleterious effects, including late development of squamous cell carcinoma. Enhancement of DNA repair (NER in this case) may defend from cancer development.

Similar results have been obtained for XPC complementation group mice [245-247] (Table III). The application of topical liposomal encapsulated T4 endonuclease V (enzyme therapy) should also be quoted here [248,249]. This heterologous enzyme was shown to traverse the stratum corneum, reach the nuclei of living cells of the skin and enhance the repair of UV-induced cyclobutane pyrimidine dimers.

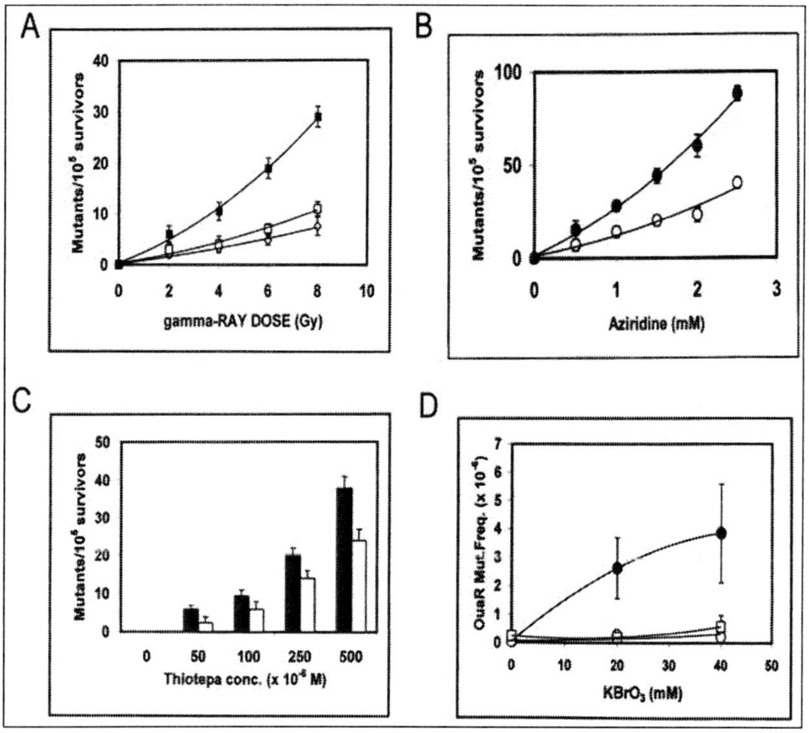

Figure 4. Antimutagenic effect of Fpg in mammalian cells. A. γ-rays mutagenicity in FPG-expressing CHO cells. Cells were exposed to γ-rays and then selected for 6-thioguanine resistance. Data are the means ± S.D. of three independent experiments. (■) Control cells. (□, ◊) Two independent clones of FPG-expressing CHO cells. B. Aziridine mutagenicity in FPG-expressing CHO cells. Cells were exposed to aziridine and then selected for 6-thioguanine resistance. Data are the means ± S.D. of three independent experiments. (●) Control cells. (○) FPG-expressing cells. C. Thiotepa mutagenicity in FPG-expressing CHO cells. Cells were exposed to Thiotepa and then selected for 6-thioguanine resistance. Data are the means ± S.D. of six independent experiments. Solid bars: control cells. Open bars: FPG-expressing CHO cells. D. $KBrO_3$ mutagenicity in FPG-expressing human cells. Cells were exposed to $KBrO_3$, and then selected for ouabain resistance. Data are the means ± S.E.M. of seven independent experiments. (●) Control cells. (○, □) Two independent clones of FPG-expressing human T24 cells (Data from ref. [129, 130, 239, 240]. Reproduced from ref. [255]with permission).

Preclinical Models

As hinted at above, enhanced DNA repair can be exploited to protect sensitive tissues from toxicity of chemotherapeutic agents. Mutations within the active site region of the ATase gene render the ATase protein resistant to inactivation by the inhibitor O^6 benzylguanine. Mutant ATase gene transfer into hematopoietic stem cells has been shown to selectively protect the marrow from the combination of an alkylating agent and O^6 benzylguanine, while at the same time sensitizing tumor cells [250]. Using this approach, a large animal study on dogs has been performed by Neff and coworkers [251] (Table III) who used both retroviral and lentiviral vectors.

Table III. Possible vectors for expression of DNA repair proteins in vivo.*

	Advantages	Disadvantages	Examples of expressed proteins	Ref.
Retroviral (other than lentiviral)	• High infectivity • Stable integration of transgene • Biosafety testing in animals	• Require cell division • Possible propagation of replication-competent particles • Silencing • Insertional mutagenesis	FPG APN1 XPC TTD/XPD	[241-242] [151] [245,247] [262]
Retroviral (lentiviral)	• High infectivity • Stable integration of transgene • Biosafety testing in animals (FIV) • Transduce both dividing and non-dividing cells • Accommodate large transgenes • Low host immune response • Possible mobilization.	• Possible propagation of replication-competent particles. • Silencing. • Insertional mutagenesis	FANCA FANCC ATase	[265] [251]
Adenoviral	• Affinity for lung epithelium • Easily produced at great titres • Transduce both dividing and non-dividing cells • Low risk of insertional mutagenesis	• Significant inflammation • Silencing	p53 XPA ANPG CFTR	[280,281] [244] [284] [286]
Adenoassociated viral	• Transduce both dividing and non-dividing cells • Little inflammation • Low risk of insertional mutagenesis	• Possible difficulties in production and in determination of transgene expression	FHIT Factor IX	[289] [292]
Herpes simplex	• Neurotropic	• Immunogenic and cytotoxic responses	Neurotrophin	[293]
Other				
Polyplexes	• Low inflammation • Low risk of adverse effects	• Low efficiency	p53	[302]
Lipoplexes	• Low inflammation • Low risk of adverse effects	• Low efficiency	alpha-1 antitripsin	[298]
Zinc-finger nucleases or integrase + DNA donor	• Specific sequence insertion	• Immunogenic responses? • Deletions and rearrangements?	IL2Rgamma (gene correction)	[299]

* from ref. [255] with permission.

Murine bone marrow cells have been infected with the aforementioned SF91 retroviral vector containing the FPG gene and expressing EGFP *via* an internal ribosomal entry site element [241] (Table III). FPG–transduced and EGFP+ bone marrow hematopoietic cells were resistant *in vitro* to thioTEPA-induced toxicity at multiple concentrations. Despite low transduction efficiency the peripheral blood leukocytes, haemoglobin and platelet counts of thioTEPA-treated FPG mice were significantly higher than treated control mice. Further, the bone marrow, spleen and thymic cellularity were significantly higher in FPG mice than control mice. Selection of FPG-expressing cells was demonstrated by an increase of EGFP-bright cells, thus indicating a preferential survival of FPG-expressing cells. Similar results were obtained with the major human DNA glycosylase for oxidative DNA damage, hOGG1. Expression of DNA glycosylases for oxidized purines may protect bone marrow cells from toxicity of thio-TEPA treatments, thus allowing higher chemotherapeutic regimens. No data are available concerning protection from oxidative mutagenesis nor have in vivo studies on animal cancer models been performed. In this regard, specific inbred strains of mice are susceptible to the development of spontaneous tumors in various organs (e.g. lungs) and may represent valuable tools (reviewed in [252]).

GENE PROPHYLAXIS BY DNA REPAIR FUNCTIONS

Gene therapy, the treatment of any disorder or pathophysiologic state on the basis of the transfer of genetic information has been thoroughly investigated for the treatment of a number of illnesses, for instance respiratory disorders such as cystic fibrosis and alpha1-antitrypsin deficiency [253]. Progress has been slower than anticipated but in some cases, gene transfer has been successfully achieved in patients.

Transfer of genetic information may be further used to elevate the level of protection of at-risk individuals, with prevention purposes. This concept can be described by the term “gene prophylaxis”. For instance, reinforcing the repair capacity for oxidative DNA damage of lung epithelia by expression of heterologous repair proteins, may afford protection in heavy smokers where the deleterious effects of ROS released in inflamed tissues multiplies the carcinogenic effects of tobacco [254,255]. FPG is just one possible candidate. But how safely and efficiently could DNA repair proteins be expressed in human tissues? We are still far from having a suited vector to this purpose. Major barriers still are the low number of expressing cells, even after use of virus-based vectors and the risks of insertional mutagenesis. Important improvements have been achieved yet as described below. For some types of vectors with which no DNA repair proteins have been studied, the survey will get out of the DNA repair field.

Viral Vectors

Viruses are naturally evolved vehicles which efficiently transfer their genes into host cells. This ability makes them convenient tools for development of vectors for delivery of therapeutic genes. Viral vectors vary widely under many aspects. The definition of the application range of a viral vector depends on a number of factors including packaging capacity, host range, cell or tissue-specific targeting, replication competency, genome integration, extent and duration of transgene expression [256-258]. Viral vectors can be divided into RNA and DNA viral vectors. The majority of RNA virus-based vectors have been derived from simple retroviruses like murine leukaemia virus. Some retroviral vectors have been specifically designed for gene transfer to airway epithelia [259]. The most commonly used DNA viral vectors have been derived from adenoviruses and adeno-associated viruses. Each class presents benefits and drawbacks (Table III) and a general-purpose vector will probably never be available. Further, the gene itself may be a problem, e.g. for excessive size [260] but this is not the case for most DNA repair proteins whose cDNAs can be easily accommodated in most vectors.

Retrovirus-Based Vectors

Retroviral vectors have been widely used for gene therapy purposes [261]. Those vectors can readily infect a number of organs. These systems have two components: the transfer vector carries all the cis-acting elements necessary for transgene expression and the efficient integration of viral DNA; the packaging cell line produces the trans-acting proteins (gag, pol, env) necessary for virion maturation. The most important advantage of retroviruses is their ability to transform their single stranded RNA genome into a double stranded DNA molecule that stably integrates into the target cell genome. For example, XPC and TTD/XPD cell lines have been permanently complemented using this kind of vector [262] (Table III). Phenotypical correction was obtained in both complementation groups with restoration of survival to UVC radiation to wild type levels. Initially, retroviruses were thought of as very efficient vectors that only required be collecting from a viral supernatant and applying to the desired cell. Four drawbacks soon appeared [263,264]: i) the need for cell division at the time of transduction initially limited their use to ex vivo approaches in which dividing cultured cells were genetically manipulated in culture prior to transplantation. Even the ex vivo approach often did not work: attempts to use retroviral vectors to cure Fanconi anemia (FA) patients proved ineffective given the impaired proliferation potential of human FA hematopoietic progenitors [265] (Table III). ii) Despite engineering of replication-incompetent viral particles by segregation of sequences required for transduction of the viral genome from sequences that encode replication proteins, recombination events could still produce at low frequency dangerous replication-competent viral particles [266]. iii) One additional obstacle was vector silencing that was more pronounced in stem cells, most desirable targets for therapeutic gene delivery [267]. iv) But the main disadvantage of retroviral vectors consists in the possibility of insertional mutagenesis during integration in the host genome. In a trial for X-linked severe combined immunodeficiency (X-SCID), important therapeutic effects were obtained with almost complete reconstitution of the immune system by receiving retrovirally transduced autologous CD34+ hematopoietic stem cells [268,269]. Gene transfer to stem cells resulted in substantial correction of the immune function and multi-lineage engrafment in patients with adenosine deaminase deficiency. However 2 out of 11 patients developed iatrogenic T-cell leukaemia probably due to activation of an oncogene [270]. Ongoing research deals with the above problems. i) and ii): the most important achievements derived from research on human immunodeficiency viruses, the acquired immune deficiency syndrome (AIDS) agents, that disclosed the potentials of lentiviruses as gene therapy vectors [271,272]. Lentiviruses have the unique feature among retroviruses of being able to infect and stably integrate their genes into the genome of both dividing and non dividing cells [257]. Although concerns remain, particularly with regards to propagation of replication-competent particles and insertional mutagenesis, those vectors are among the most likely tools to permit elevated and stable expression in vivo. For example, using lentiviral vectors, efficient genetic correction could be achieved in quiescent hematopoietic progenitors from FANCA and FANCC mice [265] (Table III). Resistance to DNA damaging agents was fully restored by lentiviral transduction, allowing for in vivo selection of the corrected cells with nonablative doses of cyclophosphamide. ATase-chemoprotection of allogeneic stem cells has been also obtained [251] (Table III). Lentivirus vectors can deliver and integrate > 8 kB of transgenic DNA into target cell genomes without inducing a host immune response against the transduced cells. Many primate and non-primate lentiviral vector systems have been described including those based on feline immunodeficiency virus (FIV), human immunodeficiency virus 1 (HIV-1) and 2 (HIV-2) and equine infectious anemia virus (EIAV). Most of them are replication incompetent, self inactivating or conditionally replicating (mobilizable) [273,274]. The design of FIV vectors adds the benefit that biosafety testing of vectors can occur in the natural hosts [275]. EIAV vectors are neurotropic, being able to infect neuronal tissues after remote delivery [276]. Lentiviral vectors are very efficient at transducing quiescent stem cells, a characteristic that is due to multiple karyophilic components of the lentiviral vector pre-integration complex. An important goal for the field remains derivation of improved clinical grade packaging/productive cell lines. iii) Cis-acting modifications have been introduced in order to reduce the probability of silencing. They include mutations of known retroviral silencer elements, addition of positive regulatory elements and insulator elements to protect the transgene from negative position effects. Other cis-acting elements currently under study control the clonal variability and the response to differentiation. Target cell recognition can be modulated through interventions on the surface glycoproteins of membrane-enveloped viruses [277]. Pseudotyping, the replacement of the native viral protein envelope with glycoprotein of a foreign enveloped virus, could expand the potential target cells for a certain vector or restrict entry to specific cells. For instance, the Ebola virus glycoprotein, for its evolutionary origin and the route of viral entry proved particularly useful for pseudotyping of viral particles used for gene therapy of lung disorders [278]. iv) Strategies to decrease risks of insertional mutagenesis and iatrogenic oncogenesis have seen important advancements recently by development of engineered zinc-finger nucleases promoting transgene integration at specific sites (see below: Non Viral Vectors).

Adenovirus-Based Vectors

Adenoviral vectors were first considered for lung gene therapy, for their natural affinity for lung epithelium and subsequently, they have been employed for gene therapy protocols of a number of additional tissues [279]. The life cycle does not normally involve integration into the host genome, rather they replicate as episomal elements in the nucleus of the host cells and consequently there is no risk of insertional mutagenesis. They can be easily grown at great titres (>10^{11} /ml) and, importantly, they can transduce the recombinant genome into non-dividing cells. However, results from studies in experimental animals and clinical trials have shown that adenovirus vectors can induce significant inflammation responses thus frequently reducing the beneficial effects of transgene expression. Dose-limiting inflammation effects were linked to the large amount of viral proteins delivered. Despite several modifications introduced in the viral backbone to circumvent it, the problem of cellular immunity remains with adenoviral vectors. Further, transgene expression in vivo tends to be transient with frequent promoter inactivation. Current studies explore adenoviral systems requiring single or widely spaced administrations, creation of a stealth vector, modification of immunogenic epitopes or tolerance induction.

Preclinical models for gene therapy of lung cancer by adenovirus-based vectors have been developed. Insertion of the p53 gene by the adenoviral vector Ad5CMV-p53 could arrest cancer cell growth and induce apoptosis and tumor regression in non small cell lung cancer [280,281] (Table III). Transgene expression appeared to occur even in case an immune response to the adenoviral vector was mounted. The apparent safety and the lack of adverse effects in adjacent normal tissues suggested an evaluation of intervention in the postoperative surgical adjuvant setting and in treating premalignancies [282,283]. Additional examples of transgenes transferred via adenoviruses in preclinical and cell culture studies were XPA and ANPG [244,284] (Table III). Much work has also been performed for introduction of the cystic fibrosis transmembrane conductance regulator (CFTR) gene in lung epithelia of Cystic Fibrosis (CF) patients [285,286] (Table III).

Adeno-Associated Viral Vectors

Adeno-associated viruses are non-pathogenic human parvoviruses, dependent on a helper virus, usually adenovirus, to proliferate. The wild type genome is a single stranded DNA molecule and they are capable of infecting both dividing and non dividing cells. In general, adeno-associated viral vectors appear safer and have superior duration profiles as compared to adenoviral vectors. Adeno-associated viral vectors are characterized by their variability in stimulating an immune response against the transgene product but, on average, this kind of vector causes little inflammation as compared to adenoviral vectors [287]. A number of new serotypes of adeno-associated viruses have demonstrated substantial promise in animal models [286,288]. One specific drawback of these vectors is that the level of messenger RNA expression may be difficult to ascertain. The small size of the wild type genome (4.7 kb) relative to some transgenes may not leave much space for vector specific sequences on which to base assays to distinguish endogenous from vector-expressed messenger RNA.

Adeno-associated viral vectors have been used as carriers of the tumor suppressor gene fragile histidine triad gene (FHIT – [289]) (Table III). The FHIT gene encompasses a human fragile site that is susceptible to environmental carcinogens. Mutations in FHIT have been observed in a number of neoplasms. Overall, roughly 60% of tumors showed absent or markedly reduced FHIT expression in cancer cells. Studies of preneoplastic lesions or early-stage cancer showed inactivated FHIT in 31% of cases. Preclinical studies showed that reintroduction of FHIT in cancer cell lines caused inhibition of in vitro cell growth or in vivo tumor development in 57% of cases [290,291]. Using adeno-associated viral vectors, oral introduction of FHIT could prevent tumor development in the gastrointestinal tract of heterozygous FHIT $^{+/-}$ mice that are prone to tumor development after carcinogen exposure. It has been proposed that reintroduction of FHIT may be useful not only for inhibition of tumor cell growth but also for prevention of carcinogen-induced tumor development [289].

Adeno-associated viral vectors have been further used for genetic correction of hemofilia B (Table III). Intramuscular injection of vectors to patients with severe hemofilia B resulted in evidence of Factor IX gene transfer to skin and muscle [292].

Other Viruses-Based Vectors

The Herper simplex virus type 1 is a human neurotropic virus whose genome is a linear double stranded DNA molecule of 152 kb. The ability of herpes simplex viruses to reside in neurons in latent state that does not appear to affect normal cellular physiology has sparked interest in them as potential vectors for gene therapy of neurological disorders. These and other viral vectors are currently employed to express neurotrophins in the injured spinal cord to promote regeneration [293] (Table III). Those vectors often elicit immunogenic and cytotoxic responses [294]. Alphaviruses such as Semliki Forest have been also used in some applications. Efficient tumor targeting has been reported in mice treated with intratumoral injections of naked or liposome-encapsulated alphavirus constructs [258,295].

Non-Viral Vectors

As aforementioned, viral vectors all induce an immunological response to some degree and may have safety risks (such as insertional mutagenesis and toxicity problems). The latter may be reduced (not eliminated) at the price of a much decreased transfer efficiency, through the use of non viral methods. Naked DNA (in the form of a plasmid) can be directly injected into the target tissue, or attached to gold particles that are bombarded within. This can result in a very low but prolonged expression in vivo. It has been used for development of some DNA vaccines. Lipid-based systems named DNA lipoplexes and polymer-based transfection particles named DNA polyplexes have undergone some evolution (Table III). Lipoplexes are essentially cationic liposomes. DNA spontaneously associates to the surface of cationic liposomes (by virtue of its charge) and is then incorporated into the cell. Lipids and DNA may form supramolecular lamellar structures that protect the construct from serum DNAse degradation [296]. Polymers used for polyplexes development include polyethylenimine shielded with polyethylene glycol [297] (see "Delivery mode"). These methods are less likely than viral systems to cause inflammation and iatrogenic disorders but their application will probably be limited to those cases where low levels of expression will be sufficient to achieve a therapeutic goal. Examples of gene therapy trials executed with non viral vectors include expression of alpha-1 antitrypsin in the nasal epithelium of deficient subjects and expression of the CFTR protein in CF patients [298].

Enzymes have been designed which can lead transgenes to specific locations in the chromosome thus promoting homologous recombination of a wild type sequence [gene correction [299,300] (Table III). In this approach, zinc finger proteins engineered to recognize a unique chromosomal site have been fused to a nuclease domain. A double strand break induced by the resulting zinc-finger nuclease could create specific sequence alterations by stimulating homologous recombination between the chromosome and an extrachromosomal DNA donor containing the right copy of the gene. A remarkable feature of this process is the frequency at which it occurred, even in the absence of any selection for corrected cells. Urnov et al [299] showed correction of 15-20% of mutated chromosomes in a cultured cell line with an X-linked severe combined immunodeficiency (SCID) mutation and around 5% correction in human T cells taken from a patient. But the additional and most promising aspect of this work is the possibility to drive transgenes to specific sites on the chromosome, thus avoiding random integration and the risk of insertional mutagenesis [301]. By a similar approach but using phage integrases rather than zinc finger proteins, site-specific insertions of transgenes have been also obtained by M. Calos to cure mouse models of hemophilia and L. Chen and S. Woo to cure a mouse version of phenylketonuria (reported in [264]).

Delivery Mode

Another challenge to the development of clinically feasible gene therapy is delivery mode. For instance, lung epithelia possess both physical and immunological barriers that can hinder effective transduction. In those cases, delivery of gene formulations via aerosol is a relatively recent field of investigation where some achievements have been obtained. Aerosol therapies with tumor suppressor gene constructs have shown the potential to shrink pulmonary metastases of selected histologies in experimental models. This approach took advantage of polyplexes technology: polyethyleneimine (PEI)-based formulations have proven stable during nebulisation and result in elevated transfection efficiency throughout the airways, as well as significant, although lower, levels of transfection throughout the lung parenchyma. Therapeutic responses have been obtained in several animal lung tumour models when PEI-based complexes of p53 and IL-12 genes were delivered [302,303] (Table III). Low toxicity and little immunostimulatory response characterized those attempts. Some survival benefits in selected patients with metastatic renal cell cancer have been also reported after treatment with aerosol delivery of tumor suppressor gene constructs [304].

Conclusions

Endogenous damage is probably the most important source of mutations in humans under physiological conditions. Dietary modulations, micronutrient supplementation and naturally or synthetically derived drugs are attractive chemopreventive approaches that may confer temporary protections, depending on absorption, stability, excretion rates and other pharmacokinetics parameters. Likewise gene therapy approaches to reduce the burden of UV damage in Xeroderma Pigmentosum [244], targeting the genes involved in DNA repair of endogenous damage may represent a novel mode to abate its mutagenic impact and confer permanent protection to human tissues. In this strategy, the following should be considered:

a) overexpression of some BER enzymes may be detrimental more than beneficial with respect to protection from spontaneous mutagenesis. This is particularly evident for ANPG and DNA pol β, whose imbalanced production with respect to the other major BER activities leads to sensitization of cells to DNA-damaging agents and a genome instability phenotype;
b) overexpression of ATase may be useful to reduce mutagenicity by endogenous O^6 meG lesions. Mammalian cells overexpressing this protein show a substantial reduction in some kinds of spontaneous mutations, as well as increased resistance to the toxic and mutagenic effects of alkylating agents. In no (but one) study in which ATase activity was enhanced, genome instability or sensitization of cells occurred;
c) FPG is probably the most promising tool against oxidative DNA damage. FPG is 80-fold faster than hOGG1 in repairing mutagenic oxidative lesions. It can be expressed in mammalian cells where it accelerates DNA repair and abates mutagenicity of a wide range of DNA damaging agents. Prophylaxis of oxidative damage and mutation could be achieved in lung epithelia and other tissues of at-risk individuals by expression of this protein.

Acknowledgements

This work was partially supported by grants of Compagnia di S. Paolo (Programma Oncologia) and Ministry of University and Research, Fondo Investimenti Ricerca Base (FIRB), Italy.

References

[1] Lindahl, T. (1993) Instability and decay of the primary structure of DNA. *Nature 362*, 709-715.
[2] Grollman, A.P. and Moriya, M. (1993) Mutagenesis by 8-oxoguanine: an enemy within. *Trends in Genetics, 9*, 246-249.
[3] Beckman, K.B. and Ames, B.N. (1997) Oxidative decay of DNA. *J. Biol. Chem., 272*, 19633-19636.
[4] Burcham, P.C. (1999) Internal hazards: baseline DNA damage by endogenous products of normal metabolism. *Mutat. Res., 443*, 11-36.
[5] Demple, B. and Harrison, L. (1994) Repair of oxidative damage to DNA: enzymology and biology. *Annu. Rev. Biochem., 63*, 915-948
[6] Satoh, M.S. and Lindahl, T. (1994) Enzymatic repair of oxidative DNA damage. *Cancer Res., 54*, 1899s-1901s.

[7] Seeberg, E., Eide, L. and Bjoras, M. (1995) The base excision repair pathway. *Trends Biochem. Sci., 20*, 391-397.

[8] Barzilay, G., Walker, L.J., Rothwell, D.G. and Hickson, I.D (1996) Role of the HAP1 protein in repair of oxidative DNA damage and regulation of transcription factors. *British J. Cancer, 74*, S145 - S150.

[9] 9. Wilson III D.M. and Thompson L.H. (1997) Life without DNA repair. *Proc. Natl. Acad. Sci. USA, 94*, 12754-12757.

[10] Girard, P.M. and Boiteux, S. (1997) Repair of oxidized DNA bases in the yeast Saccharomyces cerevisiae, *Biochimie, 79*, 559-566.

[11] Wallace, S.S. (1998) Enzymatic processing of radiation-induced free radical damage in DNA. *Radiat. Res., 150*, S60 – S79.

[12] Boiteux, S. and Radicella, J.P. (1999) Base excision repair of 8-hydroxyguanine protects DNA from endogenous oxidative stress. *Biochimie, 81*, 59-67.

[13] Frosina G. (2000) Overexpression of enzymes that repair endogenous damage to DNA, *Eur. J. Biochem, 267*, 2135-2149.

[14] Kolodner, R.D. (1995) Mismatch repair: mechanisms and relationship to cancer susceptibility. *Trends Biochem. Sci., 20*, 397-401.

[15] Jiricny, J. (1996) Mismatch repair and cancer. *Cancer Surv., 28*, 47-68

[16] Modrich, P. and Lahue, R. (1996) Mismatch repair in replication fidelity, genetic recombination and cancer biology. *Annu. Rev. Biochem., 65*, 101-133.

[17] Pegg, A.E. (1990) Mammalian O^6 alkylguanine DNA alkyltransferase: regulation and importance in response to alkylating carcinogenic and therapeutic agents. *Cancer Res., 50*, 6119- 6129.

[18] Sekiguchi, M., Nakabeppu, Y., Sakumi, K. and Tuzuki, T. (1996) DNA repair methyltransferase as a molecular device for preventing mutation and cancer. *J. Cancer Res. Clin. Oncol., 122*, 199-206.

[19] Kunkel, T.A. (1999) The high cost of living. *Trends in Genetics, 15*, 93-94.

[20] Gedik, C.M., Collins, A., ESCODD (European Standards Committee on Oxidative DNA damage) (2005) Establishing the background level of base oxidation in human lymphocyte DNA: results of an interlaboratory validation study. *FASEB J., 19*, 82-84.

[21] Wallace, S.S. (1988) AP endonucleases and DNA glycosylases that recognize oxidative DNA damage. *Environ. Mol. Mutagenesis, 12*, 431-477.

[22] Breimer, L.H. (1990) Molecular mechanisms of oxygen radical carcinogenesis and mutagenesis: the role of DNA base damage. *Molec. Carcinogenesis, 3*, 188-197.

[23] Hariharan, P.V., Achey, P.M. and Cerutti, P.A. (1977) Biological effect of thymine ring saturation in coliphage phiX174-DNA. *Radiat. Res., 69*, 375-378.

[24] Moran, E. and Wallace, S.S. (1985) The role of specific DNA base damages in the X-ray-induced inactivation of bacteriophage PM2. *Mutation Res., 146*, 229-241.

[25] Laspia, M.F. and Wallace, S.S. (1988) Excision repair of thymine glycols, urea residues, and apurinic sites in Escherichia coli. *J. Bacteriol., 170*, 3359-3366.

[26] Basu, A.K., Loechler, E.L., Leadon, S.A. and Essigmann, J.M. (1989) Genetic effects of thymine glycol: site specific mutagenesis and molecular modeling studies. *Proc. Natl. Acad. Sci., USA, 86*, 7677-7681.

[27] Lindahl, T., Satoh, M.S., Poirier, G.G. and Klungland, A. (1995) Post-translational modification of poly(ADP-ribose) polymerase induced by DNA strand breaks. *Trends Biochem. Sciences, 237*, 405-411.

[28] Burcham, P.C. (1998) Genotoxic lipid peroxidation products: their DNA damaging properties and role in formation of endogenous DNA addducts. *Mutagenesis, 13*, 287-305.

[29] Johnson, K.A., Mierzwa, M.L., Fink, S.P. and Marnett, L.J. (1999) MutS recognition of exocyclic DNA adducts that are endogenous products of lipid oxidation. *J. Biol. Chem., 274*, 27112-27118.

[30] Marnett, L.J. and Burcham, P.C. (1993) Endogenous DNA adducts: potential and paradox. *Chem. Res. Toxicol., 6*, 771-785.

[31] Kang, H., Konishi, C., Kuroki, T. and Huh, N. (1995) Detection of O^6-methylguanine, O^4-methylthymine and O^4-ethylthymine in human liver and peripheral blood leukocyte DNA. *Carcinogenesis, 16*, 1277-1280.

[32] Nakamura, J. and Swenberg, J.A. (1999) Endogenous apurinic/apyrimidinic sites in genomic DNA of mammalian tissues. *Cancer Res., 59*, 2522-2526.

[33] Cooper, D.N. and Youssoufian, H. (1988) The CpG dinucleotide and human genetic disease. Hum Genet., 78, 151-155.

[34] Lettieri, T. and Jiricny, J. (1997) Mammalian mismatch-specific DNA glycosylases, in: I.D. Hickson (Ed.) *Base Excision Repair of DNA damage*, Landes Bioscience, Austin, TX, pp. 45-66.

[35] Rideout, W.M. III, Coetzee, G.A., Olumi, A.F., and Jones, P.A. (1990) 5-methylcytosine as an endogenous mutagen in the human LDL receptor and p53 genes. *Science, 249*, 1288- 1290.

[36] Marra, G. and Boland, C. R. (1995) Hereditary nonpolyposis colorectal cancer: the syndrome, the genes and historical perspectives. *J. Natl. Cancer Inst., 87*, 1114-1125.

[37] Risinger, J.L., Umar, A., Glaab, W.E., Tindall, K.R., Kunkel, T.A. and Barrett, J. C. (1998) Single gene complementation of the hPMS2 defect in HEC-1-A endometrial carcinoma cells. *Cancer Res., 58*, 2978-2981.

[38] Lettieri, T., Marra, G., Aquilina, G., Bignami, M., Crompton, N.E.A., Palombo, F. and Jiricny, J. (1999) Effect of hMSH6 cDNA expression on the phenotype of mismatch repair-deficient colon cancer cell line HCT15. *Carcinogenesis, 20*, 373-382.

[39] Eckert, K.A. and Opresko, P.L. (1999) DNA polymerase mutagenic bypass and proofreading of endogenous DNA lesions. *Mutat. Res., 424*, 221-236.

[40] Xanthoudakis, S., Smeyne, R.J., Wallace, J.D. and Curran, T. (1996) The redox/DNA repair protein, Ref-1, is essential for early embryonic development in mice. *Proc. Natl. Acad. Sci. USA, 93*, 8919-8923.

[41] Sobol, R.W., Horton, J.K., Kuhn, R., Gu, H., Singhal, R.K., Prasad, R., Rajewsky, K. and Wilson, S.H. (1996) Requirement of mammalian DNA polymerase β in base-excision repair. *Nature, 379*: 183-186.

[42] Satoh, M.S., Jones, C.J., Wood, R.D. and Lindahl, T. (1993) DNA excision-repair defect of xeroderma pigmentosum prevents removal of a class of oxygen free radical-induced base lesions, *Proc. Natl. Acad. Sci. USA, 90*, 6335-6339.

[43] Neddermann, P. and Jiricny, J. (1993) The purification of a mismatch-specific thymine-DNA glycosylase from HeLa cells. *J. Biol. Chem., 268,* 21218-21224.

[44] Hataet, Z., Kow, Y.W., Purmal, A.A., Cunningham, R.P. and Wallace, S.S. (1994) New substrates for old enzymes. 5-hydrozy-2'-deoxycytidine and 5-hydroxy-2'-deoxyuridine are substrates for Escherichia coli endonuclease III and formamidopyrimidine DNA N-glycosylase, while 5-hydroxy-2'-deoxyuridine is a substrate for uracil DNA glycosylase. *J. Biol. Chem., 269,* 18814-18820.

[45] Zastawny, T.H., Doetsch, P.W. and Dizdaroglu, M. (1995) A novel activity of E.coli uracil DNA N-glycosylase: excision of isodialuric acid (5,6-dihydroxyuracil), a major product of oxidative DNA damage from DNA. *FEBS Lett., 364,* 255-258.

[46] Wyatt, M.D., Allan, J.M., Lau, A.Y., Ellenberger, T.E. and Samson, L.D. (1999) 3-methyladenine DNA glycosylases: structure, function and biological importance. *Bioessays, 21,* 668-676.

[47] Saparbaev, M., and Laval, J. (1994) Excision of hypoxanthine from DNA containing dIMP residues by the Escherichia coli, yeast, rat and human alkylpurine DNA glycosylases. *Proc. Natl. Acad. Sci. USA, 91,* 5873-5877.

[48] Saparbaev, M. Kleibl., K. and Laval, J. (1995) Escherichia coli, Saccharomyces cerevisiae, rat and human 3-methyladenine DNA glycosylases repair 1, N6-ethenoadenine when present in DNA. *Nucleic Acids Res., 23,* 3750-3755.

[49] Krokan, H.E., Standal, R. and Slupphaug, G. (1997) DNA glycosylases in the base excision repair of DNA. *Biochem. J., 325,* 1-16.

[50] Aspinwall, R., Rothwell, D.G., Roldan-Arjona, T., Anselmino, C., Ward, C.J., Cheadles, J.P., Sampson, J.R., Lindahl, T., Harris, P.C. and Hickson I. (1997) Cloning and characterization of a functional human homolog of Escherichia coli endonuclease III. *Proc. Natl. Acad. Sci. USA, 94,* 109-114.

[51] Klungland, A., Hoss, M., Gunz, D., Constantinou, A., Clarkson, S.G., Doetsch, P.W., Bolton, P.H., Wood, R.D. and Lindahl, T. (1999) Base excision repair of oxidative DNA damage activated by XPG protein. *Mol. Cell., 3,* 33-42.

[52] Grosch, S., Fritz, G. and Kaina, B. (1998) Apurinic endonuclease (Ref-1) is induced in mammalian cells by oxidative stress and involved in clastogenic adaptation. *Cancer Res., 58,* 4410-4416.

[53] Ramana, C.V., Boldogh, I., Izumi, T., and Mitra, S (1998) Activation of apurinic/apyrimidinic endonuclease in human cells by reactive oxygen species and its correlation with their adaptive response to genotoxicity of free radicals. *Proc. Natl. Acad. Sci., USA, 95,* 5061-5066.

[54] Yacoub, A., Kelley, M.R. and Deutsch, W.A. (1997) The DNA repair activity of human redox/repair protein APE/Ref-1 is inactivated by phosphorilation. *Cancer Res., 57,* 5457-5459.

[55] Xu, Y.J., Kim, E.Y. and Demple, B. (1998) Excision of C-4'-oxidized deoxyribose lesions from double-stranded DNA by human apurinic/apyrimidinic endonuclease (Ape1 protein) and DNA polymerase beta. *J. Biol. Chem., 273,* 28837-28844.

[56] Whitehouse, C.J., Taylor, R.M., Thistlethwaite, A., Zhang, H., Karimi-Busheri, F., Lasko, D.D., Weinfeld, M. and Caldecott, K.W. (2001) XRCC1 stimulates human

polynucleotide kinase activity at damaged DNA termini and accelerates DNA single-strand break repair. *Cell, 104*, 107-117.

[57] Matsumoto, Y. and Kim, K. (1995) Excision of deoxyribose phosphate residues by DNA polymerase β during DNA repair. *Science, 269*, 699-702.

[58] Wilson, S.H. (1998) Mammalian base excision repair and DNA polymerase beta. *Mutation Res., 407*, 203-215.

[59] Sandigursky, M., Yacoub, A., Kelley, M.R., Deutsch, W.A. and Franklin, W.A. (1997) The Drosophila ribosomal protein S3 contains a DNA deoxyribophosphodiesterase (dRpase) activity. *J. Biol. Chem, 272*, 17480-17484.

[60] Cappelli E., D'Osualdo A., Bogliolo M, Kelley M.R. and Frosina G. (2003) The Drosophila S3 ribosomal protein accelerates repair of 8 oxoguanine performed by human and mouse cell extracts. *Environ. Mol. Mutagenesis, 42*, 50-58.

[61] Srivastava, D. K., Vande Berg, B.J., Prasad, R., Molina, J.T., Beard, W.A., Tomkinson, A.E. and Wilson, S.H. (1998) Mammalian abasic site base excision repair. Identification of the reaction sequence and rate-determining steps. *J. Biol. Chem., 273*, 21203-21209.

[62] Dianov, G., Price, A. and Lindahl, T. (1992) Generation of single nucleotide repair patches following excision of uracil residues from DNA. *Mol. Cell. Biol., 12*, 1605-1612.

[63] Matsumoto, Y., Kim, K. and Bogenhagen, D.F. (1994) Proliferating cell nuclear antigen-dependent abasic site repair in Xenopus laevis oocytes: an alternative pathway of base excision DNA repair. *Mol. Cell. Biol., 14*, 6187-6197.

[64] Frosina G., Fortini, P., Rossi, O., Carrozzino, F., Raspaglio, G., Cox, L.S., Lane, D.P., Abbondandolo, A. and Dogliotti, E. (1996) Two pathways for base excision repair in mammalian cells, *J. Biol. Chem., 271*, 9573 - 9578.

[65] Fortini, P., Pascucci, B., Parlanti, E., Sobol, R.W., Wilson, S.H. and Dogliotti, E. (1998) Different DNA polymerases are involved in the short and long-patch base excision repair in mammalian cells, *Biochemistry, 37*, 3575- 3580.

[66] Dianov, G.L., Prasad, R., Wilson, S.H. and Bohr, V.A. (1999) Role of DNA polymerase beta in the excision step of long patch mammalian base excision repair, *J Biol. Chem., 274*, 13741- 13743.

[67] Fortini, P., Parlanti, E., Sidorkina, O.M., Laval, J. and Dogliotti, E. (1999) The type of DNA glycosylase determines the base excision repair pathway in mammalian cells. *J. Biol. Chem., 274*, 15230 - 15236.

[68] Klungland, A. and Lindahl, T. (1997) Second pathway for completion of human DNA base excision repair: reconstitution with purified proteins and requirement for DNAase IV (FEN1), *EMBO J., 16*, 3341-3348.

[69] Prasad, R., Singhal, R.K., Srivastava, D.K., Molina, J.T., Tomkinson, A.E. and Wilson, S.H. (1996) Specific interaction of DNA polymerase beta and DNA ligase I in a multiprotein base excision repair complex from bovine testis. *J. Biol. Chem., 271*, 16000-16007.

[70] Kubota, Y., Nash, R.A., Klungland, A., Schar, P., Barnes, D.E. and Lindahl, T. (1996) Reconstitution of DNA base excision repair with purified human proteins: interaction between DNA polymerase beta and the XRCC1 protein. *EMBO J., 15*, 6662-6670.

[71] Cappelli, E, Taylor, R., Cevasco, M., Abbondandolo, A., Caldecott, K. and Frosina, G. (1997) Involvement of XRCC 1 and DNA ligase III gene products in DNA base excision repair, *J. Biol. Chem. 272*, 23970-23975.

[72] Bennett, R.A., Wilson, D.M. 3rd, Wong, D. and Demple, B. (1997) Interaction of human apurinic endonuclease and DNA polymerase beta in the base excision repair pathway. *Proc. Natl. Acad. Sci. USA, 94*, 7166-7169.

[73] Nash, R.A., Caldecott, K.W., Barnes, D.E. and Lindahl, T. (1997) XRCC1 protein interacts with one of two distinct forms of DNA ligase III. *Biochemistry, 36*, 5207-5211.

[74] Caldecott, K.W., Aoufouchi, S., Johnson, O. and Shall, S. (1996) XRCC1 polypeptide interacts with DNA polymerase β and possibly poly (ADP-ribose) polymerase, and DNA ligase III is a novel molecular "nick-sensor" in vitro. *Nucleic Acids Res., 24*, 4387-4394.

[75] Masson, M., Niedergang, C., Schreiber, V., Muller, S., Menissier-deMurcia J. and de Murcia, G (1998) XRCC1 is specifically associated with poly (ADP-ribose) polymerase and negatively regulates its activity following DNA damage. *Mol. Cell Biol., 18*, 3563-3571.

[76] Pegg, A.E., Wiest, L., Foote, R.S., Mitra, S., and Perry, W. (1983) Purification and properties of O^6-methylguanine-DNA methyltransferase from rat liver. *J. Biol. Chem., 258*, 2327-2333.

[77] Harris, A.L., Karran, P. and Lindahl, T. (1983) O^6 methylguanine DNA methyltransferase of human lymphoid cells: structural and kinetic properties and absence in repair-deficient cells. *Cancer Res., 43*, 3247- 3252.

[78] Zak, P., Kleibl, K. and Laval, F. (1994) Repair of O6-methylguanine and O4-methylthymine by the human and rat O6-methylguanine-DNA methyltransferases. *J. Biol. Chem., 269*, 730-733.

[79] Laval, J., Jurado, J., Saparbaev, M. and Sidorkina, O. (1998) Antimutagenic role of base-excision repair enzymes upon free radical-induced DNA damage. *Mutat. Res., 402*, 93-102.

[80] Cabrera, M., Nghiem, Y. and Miller, J.H. (1988) mutM, a second mutator locus in Escherichia coli that generates GC – TA transversions. *J. Bacteriol., 170*, 5405-5407.

[81] Bessho, T., Tano, K, Kasai, H. and Nishimura, S. (1992) Deficiency of 8-hydroxyguanine DNA endonuclease activity and accumulation of the 8-hydroxyguanine in mutator mutant (mutM) of Escherichia coli. *Biochem. Biophys. Res. Commun., 188*, 372-378.

[82] Michaels, M.L., Cruz, C., Grollman, A.P. and Miller, J.H. (1992) Evidence that MutY and MutM combine to prevent mutations by an oxidatively damaged form of guanine in DNA. *Proc. Natl. Acad. Sci. USA, 89*, 7022-7025.

[83] Nghiem, Y., Cabrera, M., Cupples, C.G. and Miller, J.H. (1988) The mutY gene: a mutator locus in Escherichia coli that generates GC – TA transversions. *Proc. Natl. Acad. Sci. USA, 85*, 2709-2713.

[84] Michaels, M.L. and Miller, J.H. (1992) The GO system protects organisms from the mutagenic effect of the spontaneous lesion 8-hydroxyguanine (7,8-dihydro-8-oxoguanine). *J. Bacteriol., 174*, 6321-6325.

[85] Tajiri, T., Maki, H. and Sekiguchi, M. (1995) Functional cooperation of Mut T, mutM and MutY proteins in preventing mutations caused by spontaneous oxidation of guanine nucleotide in Escherichia coli. *Mutat. Res., 336*, 257-267.

[86] Jiang, D., Hatahet, Z., Blaisdell, J.O., Melamede, R.J. and Wallace, S.S. (1997) Escherichia coli endonuclease VIII: cloning, sequencing and overexpression of the nei structural gene and characterization of nei and nei nth mutants. *J. Bacteriol., 179*, 3773-3782.

[87] Saito, Y., Uraki, F., Nakajima, S., Asaeda, A., Ono, K., Kubo, K. and Yamamoto, K. (1997) Characterization of endonuclease III (nth) and endonuclease VIII (nei) mutants of Escherichia coli K-12. *J. Bacteriol., 179*, 3783-3785.

[88] Thomas, D., Scot, A.D., Barbey, R., Padula, M. and Boiteux, S. (1997) Inactivation of OGG1 increases the incidence of G.C – T.A transversions in Saccharomyces cerevisiae: evidence for endogenous oxidative damage to DNA in eukaryotic cells. *Mol. Gen. Genet., 254*, 171-178.

[89] Radicella, J.P., Dherin, C., Desmaze, C., Fox, M.S. and Boiteux, S. (1997) Cloning and characterization of hOGG1, a human homolog of the OGG1 gene of Saccharomyces cerevisiae. *Proc. Natl. Acad. Sci. USA, 94*, 8010-8015.

[90] Alseth, I., Eide, L., Pirovano, M., Rognes, T., Seeberg, E. and Bjoras, M. (1999) The Saccharomyces cerevisiae homologues of endonuclease III from Escherichia coli, Ntg1 and Ntg2, are both required for efficient repair of spontaneous and induced oxidative DNA damage in yeast. *Mol. Cell. Biol., 19*, 3779-3787.

[91] You, H.J., Swanson, R.L., Harrington, C., Corbett, A.H., Jinks-Robertson, S., Senturker, S., Wallace, S.S., Boiteux, S., Dizdaroglu, M. and Doetsch, P.W. (1999) Saccharomyces cerevisiae ntg1p and ntg2p: broad specificity N-glycosylases for the repair of oxidative DNA damage in the nucleus and mitochondria. *Biochemistry, 38*, 11298-11306.

[92] Rebeck, G.W. and Samson, L. (1991) Increased spontaneous mutation and alkylation sensitivity of Escherichia coli strains lacking the ogt O(6)-methylguanine DNA repair methyltransferase. *J. Bacteriol, 173*, 2068-2076.

[93] Mackay, W.J., Han, S. and Samson, L.D. (1994) DNA alkylation repair limits spontaneous base substitution mutations in Escherichia coli. *J. Bacteriol., 176*, 3224-3230.

[94] Taverna, P. and Sedgwick, B. (1996) Generation of an endogenous DNA methylating agent by nitrosation in Escherichia coli. *J. Bacteriol., 178*, 5105- 5111.

[95] Xiao, W, and Samson, L. (1993) In vivo evidence for endogenous DNA alkylation damage as a source of spontaneous mutation in eukaryotic cells. *Proc. Natl. Acad. Sci. USA, 90*, 2117-2121.

[96] Xiao, W. and Fontanie, T. (1995) Expression of the human MGMT O6-methylguanine DNA methyltransferase gene in a yeast alkylation-sensitive mutant: its effects on both exogenous and endogenous DNA alkylation damage. *Mutat. Res., 336*, 133-142.

[97] Aquilina, G., Biondo, R., Dogliotti, E., Meuth, M. and Bignami, M. (1992) Expression of the endogenous O^6-methylguanine-DNA methyltransferase protects Chinese hamster ovary cells from spontaneous G:C to A:T transitions. *Cancer Res., 52*, 6471-6475.

[98] Iwakuma, T., Sakumi, K., Nakatsuru, Y., Kawate, H., Igarashi, H., Shiraishi, A., Tsuzuki, T., Ishikawa, T. and Sekiguchi, M. (1997) High incidence of nitrosamine-induced tumorigenesis in mice lacking DNA repair methyltransferase. *Carcinogenesis, 18*, 1631-1635.

[99] Kaina, B., Fritz, G., Ochs, K., Haas, S., Grombacher, T., Dosch, J., Christmann, M., Lund, P., Gregel, C.M. and Becker, K. (1998) Transgenic systems in studies on genotoxicity of alkylating agents: critical lesions, thresholds and defense mechanisms. *Mutat. Res., 405*, 179-191.

[100] Sekiguchi, M., Sanada, M. (1999) Alkylation carcinogenesis in mice with altered levels of DNA repair methyltransferase. *Prog. Exp. Tumor Res., 35*, 25-36.

[101] Hang, B., Singer, B, Margison, G.P. and Elder, R.H. (1997) Targeted deletion of alkylpurine-DNA-N-glycosylase in mice eliminates repair of 1,N6-ethenoadenine and hypoxanthine but not of 3,N4-ethenocytosine or 8-oxoguanine. *Proc. Natl. Acad. Sci. USA, 94*, 12869-12874.

[102] Engelward, B.P., Dreslin, A., Christensen, J, Huszar, D., Kurahara, C. and Samson, L. (1996) Repair-deficient 3 methyladenine-DNA glycosylase homozygous mutant mouse cells have increased sensitivity to alkylation-induced chromosome damage and cell killing, *EMBO J., 15*, 945-952.

[103] Elder, R.H., Jansen, J.G., Weeks, R.J., Willington, M.A., Deans, B., Watson, A.J., Mynett, K.J., Bailey, J.A., Cooper, D.P., Rafferty, J.A., Heeran, M.C., Wijnhoven, S.W., van Zeeland, A.A. and Margison, G. P. (1998) Alkylpurine-DNA-N-glycosylase knockout mice show increased susceptibility to induction of mutations by methylmethane sulfonate. *Mol. Cell. Biol., 18*, 5828-5837.

[104] Engelward, B.P., Weeda, G., Wyatt, M.D., Broekhof, J.L., de Wit, J., Donker, I., Allan, J.M., Gold, B., Hoeijmakers, J.H. and Samson, L.D. (1997) Base excision repair deficient mice lacking the Aag alkyladenine DNA glycosylase. *Proc. Natl. Acad. Sci. USA, 94*, 13087-13092.

[105] Duncan, B.K. and Weiss, B. (1982) Specific mutator effects of ung (uracil-DNA glycosylase) mutations in Escherichia coli. *J. Bacteriol., 151*, 750-755.

[106] Olsen, L.C., Aasland, R., Krokan, H.E. and Helland, D.E. (1991) Human uracil-DNA glycosylase complements E.coli ung mutants. *Nucleic Acids Res., 19*, 4473- 4478.

[107] Chen, J.D. and Lacks, S.A. (1991) Role of uracil DNA-glycosylase in mutation avoidance by Streptococcus Pneumoniae. *J. Bacteriol., 173*, 283-290.

[108] Burgers, P.M.J. and Klein, M.B. (1986) Selection by genetic transformation of a Saccharomyces cerevisiae mutant defective for the nuclear uracil-DNA glycosylase. *J. Bacteriol., 166*, 905-913.

[109] Impellizzeri, K.J.., Anderson, B. and Burgers, P.M.J. (1991) The spectrum of spontaneous mutations in a Saccharomyces cerevisiae uracil-DNA glycosylase mutant limits the function of this enzyme to cytosine deamination repair. *J. Bacteriol., 173*, 6807-6810.

[110] Cunningham, R.P. and Weiss, B. (1985) Endonuclease III (nth) mutants of Escherichia coli. *Proc. Natl. Acad. Sci. USA, 82*, 474-478.

[111] Ramotar, D., Popoff, S.C., Gralla, E.B. and Demple, B. (1991) Cellular role of yeast Apn1 apurinic endonuclease/3'-diesterase: repair of oxidative and alkylation DNA damage and control of spontaneous mutation. *Mol. Cell. Biol., 11*, 4537-4544.

[112] Kunz, B.A., Henson, E.S., Roche, H., Ramotar, D., Nunoshiba, T. and Demple, B. (1994) Specificity of the mutator caused by deletion of the yeast structural gene (APN1) for the major apurinic endonuclease. *Proc. Natl. Acad. Sci. USA, 91*, 8165-8169.

[113] Masson, J.Y. and Ramotar, D. (1997) Normal processing of AP sites in Apn1-deficient Saccharomyces cerevisiae is restored by Escherichia coli genes expressing either exonuclease III or endonuclease III. *Mol. Microbiol., 24*, 711-721.

[114] Walker, L.J., Craig, R.B., Harris, A.L. and Hickson, I.D. (1994) A role for the human DNA repair enzyme HAP1 in cellular protection against DNA damaging agents and hypoxic stress. *Nucleic Acids Research, 22*, 4884-4889.

[115] Ono, Y., Futura, T., Ohmoto, T., Akiyama, K. and Seki, S. (1994) Stable expression in rat glioma cells of sense and antisense nucleic acids to a human multifunctional DNA repair enzyme APEX nuclease. *Mutation Res., 315*, 55-63.

[116] Gu, H., Marth, J.D., Orban, P.C., Mossmann, H. and Rajewsky, K. (1994) Deletion of a DNA polymerase β gene segment in T cells using cell type-specific gene targeting. *Science, 265*, 103-106.

[117] Zdzienicka, M.Z., van der Schans, G.P., Natarajan, A.T., Thompson, L.H., Neuteboom, I. and Simons, J.W.I.M. (1992) A Chinese hamster ovary cell mutant (EM-C11) with sensitivity to simple alkylating agents and a very high level of sister chromatid exchanges. *Mutagenesis, 7*, 265 - 269.

[118] Caldecott, K.W., McKeown, C.K., Tucker, J.D., Ljungquist, S. and Thompson, L.H. (1994) An interaction between the mammalian DNA repair protein XRCC1 and DNA ligase III. *Mol. Cell. Biol., 14*, 68 - 76.

[119] Caldecott, K.W., Tucker, J.D., Stanker, L.H. and Thompson, L.H. (1995) Characterization of the XRCC1-DNA ligase III complex in vitro and its absence from mutant hamster cells. *Nucleic Acids Res., 23*, 4836 - 4843.

[120] Tebbs, R.S., Flannery, M.L., Meneses, J.J., Hartmann, A., Tucker, J.D., Thompson, L.H., Cleaver, J.E. and Pedersen, R.A. (1999) Requirement for the Xrcc1 DNA base excision repair gene during early mouse development. *Dev. Biol., 208*, 513-529.

[121] Menissier-de Murcia, J., Niederganf, C., Trucco, C., Ricoul, M. , Dutrillaux, B.., Mark, M., Oliver, F.J., Masson, M., Dierich, A., LeMeur, M., Waltzinger, C., Chambon, P. and de Murcia, G. (1997) Requirement of poly(ADP-ribose)polymerase in recovery from DNA damage in mice and in cells. *Proc. Natl. Acad. Sci. USA, 94*, 7303-7307.

[122] Frosina, G., Fortini, P., Rossi, O., Carrozzino, F., Abbondandolo A. and Dogliotti, E. (1994) Repair of abasic sites by mammalian cell extracts. *Biochem. J., 304*, 699-705.

[123] Harrison, L., Galanopoulos, T., Ascione, A.G., Antoniades, H.N. and Demple, B. (1996) Regulated expression of APE apurinic endonuclease mRNA during wound healing in porcine epidermis, *Carcinogenesis, 17*, 377-381.

[124] Jaruga, P. and Dizdaroglu, M. (1996) Repair of products of oxidative DNA base damage in human cells. *Nucleic Acids Res., 24*, 1389-1394.

[125] Cappelli, E.., Degan, P. and Frosina, G. (2000) Comparative repair of the endogenous lesions 8-oxoguanine, uracil and abasic site by mammalian cell extracts: 8-oxoguanine is poorly repaired by human cell extracts. *Carcinogenesis, 21*, 1135-1141.

[126] Memisoglu, A. and Samson, L. (1996) DNA repair functions in heterologous cells. *Crit. Rev. Biochem. Mol. Biol., 31*, 405-447.

[127] Kaina, B., Fritz, G. and Coquerelle, T. (1991) Identification of human genes involved in repair and tolerance of DNA damage. *Radiat. Environ. Biophys., 30*, 1-19.

[128] Bramson, J. and Panasci, L.C. (1993) Effect of ERCC-1 overexpression on sensitivity of Chinese hamster ovary cells to DNA-damaging agents. *Cancer Res., 53*, 3237-3240.

[129] Laval, F. (1994) Expression of the E.coli fpg gene in mammalian cells reduces the mutagenicity of gamma-rays. *Nucleic Acids Res., 22*, 4943-4946.

[130] Ropolo, M., Geroldi, A., Degan, P., Andreotti, V., Zupo, S., Poggi, A., Reed, A., Kelley, M.R. and Frosina, G. (2006) Accelerated repair and reduced mutagenicity of oxidative DNA damage in human bladder cells expressing the E.coli FPG protein. *Int. J. Cancer, 118*, 1628-1634.

[131] Harrison, l., Skorvaga, M., Cunningham, R.P., Hendry, J.H. and Margison, G.P. (1992) Transfection of the Escherichia coli nth gene into radiosensitive Chinese hamster cells: effects on sensitivity to radiation, hydrogen peroxide, and bleomycin sulfate. *Radiat. Res., 132*, 30-39.

[132] Hollenbach, S., Dhenaut, A., Eckert, I., Radicella, J.P. and Epe, B. (1999) Overexpression of Ogg1 in mammalian cells: effects on induced and spontaneous oxidative DNA damage and mutagenesis. *Carcinogenesis, 20*, 1863-1868.

[133] Ibeanu, G., Hartenstein, B., Dunn, WC, Chang, L.Y., Hofmann, E., Coquerelle, T., Mitra, S. and Kaina, B. (1992) Overexpression of human DNA repair protein N-methylpurine-DNA glycosylase results in the increased removal of N-methylpurines in DNA without a concomitant increase in resistance to alkylating agents in Chinese hamster ovary cells. *Carcinogenesis, 13*, 1989-1995.

[134] Coquerelle, T, Dosch, J and Kaina, B. (1995) Overexpression of N-methylpurine-DNA glycosylase on Chinese hamster ovary cells renders them more sensitive to the production of chromosomal aberrrations by methylating agents – a case of imbalanced DNA repair. *Mutat. Res., 336*, 9-17.

[135] Wu, M., He, Y.H., Kobune, M., Xu, Y., Kelley, M.R. and Martin, W.J. 2nd (2002) Protection of human lung cells against hyperoxia using the DNA base excision repair genes hOGG1 and Fpg. *Am J. Respir. Crit. Care. Med., 166*, 192-199.

[136] Bramson, J., O'Connor, T. and Panasci, L. (1995) Effect of alkyl-N-purine DNA glycosylase overexpression on cellular resistance to bifunctional alkylating agents. *Biochem. Pharmacol., 50*, 39-44.

[137] Limp-Foster, M., Xu, Y., Williams, D.A. and Kelley, M.R. (1999) DNA base excision repair (BER) and gene transfer: use of the human N-methylpurine DNA glycosylase (MPG) to sensitize tumor cells to chemotherapy, *Env. Mol. Mutagenesis, 33*, s30, p. 40.

[138] He, Y-H, Xu, Y., Kobune, M., Wu, M., Kelley, M.R., Martin II, W.J. (2002) Escherichia coli FPG and human OGG1 reduce DNA damage and cytotoxicity by BCNU in human lung cells, *Am J. Physiol. Lung Cell. Mol. Physiol., 282*, L50-L55.

[139] Popoff, S.C., Spira, A.I., Johnson, A.W. and Demple, B. (1990) Yeast structural gene (APN1) for the major apurinic endonuclease: homology to Escherichia coli endonuclease IV. *Proc. Natl. Acad. Sci. USA, 87*, 4193-4197.

[140] Berdal, K.G., Johansen, R.F. and Seeberg, E. (1998) Release of normal bases from intact DNA by a native DNA repair enzyme. *EMBO J., 17*, 363-367.

[141] Glassner, B.J., Rasmussen, L.J., Najarian, M.T., Posnick, L.M. and Samson, L.D. (1998) Generation of a strong mutator phenotype in yeast by imbalanced base excision repair. *Proc. Natl. Acad. Sci. USA, 95*, 9997-10002.

[142] Glassner, B.J., Posnick, L.M. and Samson, L.D. (1998) The influence of DNA glycosylases on spontaneous mutation. *Mutation Res., 400*, 33-44.

[143] Habraken, Y. and Laval, F (1993) Increased resistance of the Chinese hamster mutant irs1 cells to monofunctional alkylating agents by transfection of the E.coli or mammalian N3-methyladenine-DNA-glycosylase genes. *Mutat. Res., 293*, 187-195.

[144] Rinne, M., Caldwell, D. and Kelley, M.R. (2004) Transient adenoviral N-methylpurine DNA glycosylase overexpression imparts chemotherapeutic sensitivity to human breast cancer cells. *Mol. Cancer Ther. 3*, 955-967.

[145] Klungland, A., Fairbairn, L., Watson, A.J., Margison, G.P. and Seeberg, E. (1992) Expresssion of the E.coli 3-methyladenine DNA glycosylase I gene in mammalian cells reduces the toxic and mutagenic effects of methylating agents. *EMBO J., 11*, 4439-4444.

[146] Klungland, A., Bjoras, M., Hoff, E. and Seeberg, E. (1994) Increased removal of 3-alkyladenine reduces the frequencies of hprt mutations induced by methyl- and ethylmethane sulfonate in Chinese hamster fibroblast cells. *Nucleic Acids Res., 22*, 1670-1674.

[147] Imperatori, L., Damia, G., Taverna, P., Garattini, E.., Citti , L.,, Boldrini, L. and D'Incalci, M. (1994) 3T3 NIH murine fibroblasts and B78 murine melanoma cells expressing the Escherichia coli N3-methyladenine-DNA glycosylase I do not become resistant to alkylating agents. *Carcinogenesis, 15*, 533 - 537.

[148] Herring, C.J., Deans, B., Elder, R.H., Rafferty, J.A., MacKinnon, J., Barzilay, G., Hickson, I.D., Hendry, J.H. and Margison, G.P. (1999) Expression levels of the DNA repair enzyme HAP1 do not correlate with the radiosensitivities of human or HAP1-transfected rat cell lines. *Br. J. Cancer, 80*, 940-945.

[149] Prieto-Alamo, M.J. and Laval, F. (1999) Overexpression of the human HAP1 protein sensitizes cells to the lethal effect of bioreductive drugs. *Carcinogenesis, 20*, 415-419.

[150] Tomicic, M., Eschbach, E. and Kaina, B, (1997) Expression of yeast but not human apurinic/apyrimidinic endonuclease renders Chinese hamster cells more resistant to DNA damaging agents. *Mutat. Res., 383*, 155-165.

[151] He, Y.H. 2nd, Wu, M., Kobune, M., Xu, Y., Kelley, M.R. and Martin, W.J. (2001) Expression of yeast apurinic/apyrimidinic endonuclease (APN1) protects lung epithelial cells from bleomycin toxicity. *Am. J. Respir. Cell Mol. Biol., 25*, 692-698.

[152] Sossou, M., Flohr-Beckhaus, C., Schulz, I., Daboussi, F., Epe, B. and Radicella, J.P. (2005) APE1 over expression in XRCC1-deficient cells complements the defective repair of oxidative single strand breaks but increases genomic instability. *Nucleic Acids Res. 33*, 298-306.

[153] Robertson, K.A., Bullock, H.A., Xu, Y., Tritt, R., Zimmermann, E., Ulbright, T.M., Foster, R.S., Einhorn, L.H. and Kelley, M.R. (2001) Altered expression of Ape/ref-1 in germ cell tumors and overexpression in NT2 cells confers resistance to bleomycin and radiation. *Cancer Res. 61*, 2220-2225.

[154] Horton, J.K., Srivastava, D.K., Zmudzka, B.Z. and Wilson, S.H. (1995) Strategic down-regulation of DNA polymerase beta by anti-sense RNA sensitizes mammalian cells to specific DNA damaging agents. *Nucleic Acids. Res., 23*, 3810-3815.

[155] Chen, K.H., Yakes, F.M., Srivastava, D.K., Singhal, R.K., Sobol, R.W., Horton, J.K., Van Houten, B. and Wilson, S.H. (1998) Up-regulation of base excision repair correlates with enhanced protection against a DNA damaging agent in mouse cell lines. *Nucleic Acids Res., 26,* 2001-2007.

[156] Canitrot, Y., Frechet, M., Servant, L., Cazayx, C. and Hoffmann, J.S. (1999) Overexpression of DNA polymerase beta: a genomic instability enhancer process. *FASEB J., 13*, 1107-1111.

[157] Clairmont, C.A. and Sweasy, J.B. (1996) Dominant negative rat DNA polymerase beta mutants interfere with base excision repair in Saccharomyces cerevisiae. *J. Bacteriol., 178*, 656-661.

[158] Prasad, R., Beard, W.A., Strauss, P.R. and Wilson, S.H. (1998) Human DNA polymerase beta deoxyribose phosphate lyase. Substrate specificity and catalytic mechanism. *J. Biol. Chem., 273*, 15263-15270.

[159] Prasad, R., Beard, W.A., Chyan, J.Y., Maciejewski, M.W., Mullen, G.P. and Wilson, S.H. (1998) Functional analysis of the amino-terminal 8-kDa domain of DNA polymerase beta as revealed by site-directed mutagenesis. DNA binding and 5'-deoxyribose phosphate lyase activities. *J. Biol. Chem., 273*, 11121-11126.

[160] Bernges, F., Burkle, A., Kupper, J.H. and Zeller, W.J. (1997) Functional overexpression of human poly(ADP-ribose) polymerase in transfected rat tumor cells. *Carcinogenesis, 18*, 663-668.

[161] Van Gool, L., Meyer, R., Tobiasch, E., Cziepluch, C., Jauniaux, J.C., Mincheva, A., lichter, P., Poirier, G.G., Burkle, A. and Kupper, J.H. (1997) Overexpression of human poly(ADP-ribose) polymerase in transfected hamster cells leads to increased poly(ADP-ribosyl)ation and cellular sensitization to gamma irradiation. *Eur. J. Biochem., 244*, 15-20.

[162] Fritz, G., Auer, B. and Kaina, B. (1994) Effect of transfection of human poly(ADP-ribose)polymerase in Chinese hamster cells on mutagen resistance. *Mutat. Res., 308*, 127-133.

[163] Maze, R., Kapur, R., Kelley, M.R., Hansen, W.K., Oh, S.Y. and Williams, D.A. (1997) Reversal of 1,3-bis(2-chloroethyl)-1-nitrosourea-induced severe immunodeficiency by transduction of murine long-lived hemopoietic progenitor cells using O^6-methylguanine DNA methyltransferase complementary DNA. *J. Immunol., 158*, 1006-1013.

[164] Ishizaki, K., Tsujimura, T., Yawata, H., Fujio, C., Nakabeppu, Y., Sekiguchi, M.. and Ikenaga, M. (1986) Transfer of the E.coli O6-methylguanine methyltransferase gene into repair-deficient human cells and restoration of cellular resistance to N-methyl-N'-nitro-N-nitrosoguanidine. *Mutat. Res., 166*, 135-141.

[165] Harris, L.C. and Margison, G.P. (1993) Expression in mammalian cells of the Escherichia coli O6 alkylguanine-DNA-alkyltransferase gene ogt reduces the toxicity of alkylnitrosoureas. *Br. J. Cancer, 67*, 1196-1202.

[166] Brennand, J. and Margison, G.P. (1986) Reduction of the toxicity and mutagenicity of alkylating agents in mammalian cells harboring the Escherichia coli alkyltransferase gene. *Proc. Natl. Acad. Sci. USA, 83*, 6292-6296.

[167] Brennand, J. and Margison, G.P. (1986) Expression in mammalian cells of a truncated Escherichia coli gene coding for O6-alkylguanine alkyltransferase reduces the toxic effects of alkylating agents. *Carcinogenesis, 7*, 2081-2084.

[168] White, G.R., Ockey, C.H., Brennand, J. and Margison, G.P. (1986) Chinese hamster cells harbouring the Escherichia coli O6-alkylguanine alkyltransferase gene are less susceptible to sister chromatid exchange induction and chromosome damage by methylating agents. *Carcinogenesis, 7*, 2077-2080.

[169] Bignami, M., Terlizzese, M., Zijno, A., Calcagnile, A., Frosina, G., Abbondandolo, A. and Dogliotti, E. (1987) Cytotoxicity, mutations and SCEs induced by methylating agents are reduced in CHO cells expressing an active mammalian O6 methylguanine-DNA methyltransferase gene. *Carcinogenesis, 8*, 1417-1421.

[170] Fox, M., Brennand, J. and Margison, G.P. (1987) Protection of Chinese hamster cells against the cytotoxic and mutagenic effects of alkylating agents by transfection of the Escherichia coli alkyltransferase gene and a truncated derivative. *Mutagenesis, 2*, 491-496.

[171] Fox, M. and Margison, G.P. (1988) Expression of an E.coli O6-alkylguanine DNA alkyltransferase gene in Chinese hamster cells protects against N-methyl and N-ethylnitrosourea induced reverse mutation at the hypoxanthine phosphoribosyltransferase locus. *Mutagenesis, 3*, 409-413.

[172] Kaina,B., Fritz, G., Mitra, S., and Coquerelle, T. (1991) Transfection and expression of human O6-methylguanine-DNA methyltransferase (MGMT) cDNA in Chinese hamster cells: the role of MGMT in protection against the genotoxic effects of alkylating agents. *Carcinogenesis, 12*, 1857-1867.

[173] Von Hofe E., Fairbairn, L. and Margison, G.P. (1992) Relationship between O6-alkylguanine-DNA alkyltransferase activity and N-methyl-N'-nitro-N-nitrosoguanidine-induced mutation, transformation, and cytotoxicity in C3H/10T1/2 cells expressing exogenous alkyltransferase genes. *Proc. Natl. Acad. Sci. USA, 89*, 11199-11203.

[174] Ishizaki, K., Tsujimura, T., Fujio, C., Zhang, Y.P., Yawata, H., Nakabeppu, Y., Sekiguchi, M. and Ikenaga, M. (1987) Expression of the truncated E.coli O6-methylguanine methyltransferase gene in repair-deficient human cells and restoration of cellular resistance to alkylating agents. *Mutat. Res., 184*, 121- 128.

[175] Abril, N. and Margison, G.P. (1999) Mammalian cells expressing Escherichia coli O6-alkylguanine-DNA alkyltransferases are hypersensitive to dibromoalkanes. *Chem Res. Toxicol., 12*, 544-551.

[176] Walter, C.A., Lu, J., Bhakta, M., Mitra, S., Dunn, W., Herbert, D.C., Weaker, F.J., Hoog, T., Garza, P., Adrian, G.S. et al (1993) Brain and liver targeted overexpression

of O6-methylguanine DNA methyltransferase in transgenic mice. *Carcinogenesis, 14*, 1537-1543.

[177] Becker, K., Dosch, J., Gregel, C.M., Martin, B.A. and Kaina, B. (1996) Targeted expression of human O(6)-methylguanine-DNA methyltransferase (MGMT) in transgenic mice protects against tumor initiation in two-stage skin carcinogenesis. *Cancer Res., 56*, 3244-3249.

[178] Gerson, S.L., Zaidi, N.H., Dumenco, L.L., Allay, E., Fan, C.Y., Liu, L. and O'Connor, P.J. (1994) Alkyltransferase transgenic mice: probes of chemical carcinogenesis. *Mutat. Res., 307*, 541-555.

[179] Zaidi, N.H., Pretlow, T.P., O'Riordan, M.A., Dumenco, L.L., Allay, E. and Gerson, S.L. (1995) Transgenic expression of human MGMT protects against azoxymethane-induced aberrant crypt foci and G to A mutations in the K-ras oncogene of mouse colon. *Carcinogenesis, 16*, 451-456.

[180] Liu, L., Qin, X. and Gerson, S.L. (1999) Reduced lung tumorigenesis in human methylguanine DNA-methyltransferase transgenic mice achieved by expression of transgene within the target cell. *Carcinogenesis, 20*, 279-284.

[181] Dumenco, L.L., Allay, E., Norton, K. and Gerson, S.L. (1993) The prevention of thymic lymphomas in transgenic mice by human O6-alkylguanine-DNA alkyltransferase. *Science*, 259, 219-222.

[182] Liu, L., Allay, E., Dumenco, L.L. and Gerson, S.L. (1994) Rapid repair of O6-methylguanine-DNA adducts protects transgenic mice from N-methylnitrosourea-induced thymic lymphomas. *Cancer Res.*, 54, 4648-4652.

[183] Allay, E., Reese, J.S., McGuire, E.A., Koc, O.N., Sedransk, N. and Gerson, S.L. (1997) Potentiation of lymphomagenesis by methylnitrosourea in mice transgenic for LMO1 is blocked by O6-alkylguanine DNA-alkyltransferase. *Oncogene, 15*, 2127-2132.

[184] Qin, X., Zhou, H., Liu, l. and Gerson, S.L. (1999) Transgenic expression of human MGMT blocks the hypersensitivity of PMS2-deficient mice to low dose MNU thymic lymphomagenesis. *Carcinogenesis, 20*, 1667-1673.

[185] Qin, X., Liu, L. and Gerson, S.L. (1999) Mice defective in the DNA mismatch gene PMS2 are hypersensitive to MNU induced thymic lymphoma and are partially protected by transgenic expression of human MGMT. *Oncogene, 18*, 4394-4400.

[186] Saporito, S.M., Gedenk, M. and Cunningham, R.P. (1989) Role of exonuclease III and endonuclease IV in repair of pyrimidine dimers initiated by bacteriophage T4 pyrimidine dimer-DNA glycosylase. *J. Bacteriol., 171*, 2542-2546.

[187] Dumenco, L.L., Arce, C., Norton, K., Yun, J., Wagner, T. and Gerson, S.L. (1991) Enhanced repair of O6-methylguanine DNA adducts in the liver of transgenic mice expressing the ada gene. *Cancer Res., 51*, 3391-3398.

[188] Allay, E., Veigl, M. and Gerson, S.L. (1999) Mice over-expressing human O6 alkylguanine-DNA alkyltransferase selectively reduce O6 methylguanine mediated carcinogenic mutations to threshold levels after N-methyl-N-nitrosourea. *Oncogene, 18*, 3783-3787.

[189] Dumenco, L.L., Warman, B., Hatzoglou, M., Lim, I.K., Abboud, S.L. and Gerson, S.L. (1989) Increase in nitrosourea resistance in mammalian cells by retrovirally mediated

gene transfer of bacterial O6-alkylguanine-DNA alkyltransferase. *Cancer Res., 49*, 6044-6051.

[190] Lim, I.K., Dumenco, L.L., Hatzoglou, M., Hanson, R.W. and Gerson, S.L. (1990) Increased drug resistance following retroviral gene transfer of a chimeric P-enolpyruvate carboxykinase (GTP)-bacterial O6-alkylguanine-DNA alkyltransferase gene into NRK cells. *Carcinogenesis, 11*, 737-743.

[191] Minnick, D.T., Gerson, S.L., Dumenco, L.L., Veigl, M.L. and Sedwick, W.D. (1993) Specificity of bischloroethylnitrosourea-induced mutation in a Chinese hamster ovary cell line transformed to express human O6-alkylguanine-DNA alkyltransferase. *Cancer Res., 53*, 997-1003.

[192] Rafferty, J.A., Hickson, I., Chinnasamy, N., Lashford, L.S., Margison, G.P., Dexter, T.M. and Fairbairn, L.J. (1996) Chemoprotection of normal tissues by transfer of drug resistance genes. *Cancer Metastasis Rev., 15*, 365-383.

[193] Allay, J.A., Dumenco, L.L., Koc, O.N., Liu, L. and Gerson, S.L. (1995) Retroviral transduction and expression of the human alkyltransferase cDNA provides nitrosourea resistance to hematopoietic cells. *Blood, 85*, 3342-3351.

[194] Moritz, T., Mackay, W., Glassner, B.J., Williams, D.A. and Samson, L. (1995) Retrovirus-mediated expression of a DNA repair protein in bone marrow protects hematopoietic cells from nitrosourea-induced toxicity in vitro and in vivo. *Cancer Res., 55*, 2608-2614.

[195] Reese, J.S., Koc, O.N., Lee, K.M., Liu, L., Allay, J.A., Phillips, W.P.Jr and Gerson, S.L. (1996) Retroviral transduction of a mutant methylguanine DNA methyltransferase gene into human CD34 cells confers resistance to O6-benzylguanine plus 1,3-bis (2-chloroethyl)-1-nitrosourea. *Proc. Natl. Acad. Sci. USA, 93*, 14088-14093.

[196] Jelinek, J., Fairbairn, L.J., Dexter, T.M., Rafferty, J.A., Stocking, C., Ostertag, W. And Margison, G.P. (1996) Long-term protection of hematopoiesis against the cytotoxic effects of multiple doses of nitrosourea by retrovirus-mediated expression of human O6-alkylguanine-DNA-alkyltransferase. *Blood, 87*, 1957-1961.

[197] Allay, J.A., Koc, O.N., Davis, B.M. and Gerson, S.L. (1996) Retroviral-mediated gene transduction of human alkyltransferase complementary DNA confers nitrosourea resistance to human hematopoietic progenitors. *Clin. Cancer Res., 2*, 1353-1359.

[198] Davis, B.M., Reese, J.S., Koc, O.N., Lee, K., Schupp, J.E. and Gerson, S.L. (1997) Selection for G156A O6-methylguanine DNA methyltransferase gene-transduced hematopoietic progenitors and protection from lethality in mice treated with O6-benzylguanine and 1,3-bis(2-chloroethyl)-1-nitrosourea. *Cancer Res., 57*, 5093-5099.

[199] Koc, O.N., Reese, J.S., Davis, B.M., liu, L., Majczenko, K.J. and Gerson, S.L. (1999) Delta MGMT-transduced bone marrow infusion increases tolerance to O6-benzylguanine and 1,3-bis(2-chloroethyl)-1-nitrosourea and allows intensive therapy of 1,3-bis(2-chloroethyl)-1-nitrosourea-resistant human colon cancer xenografts. *Hum. Gene Ther., 10*, 1021-1030.

[200] Jelinek, J., Kleibl, K, Dexter, T.M. and Margison, G.P. (1988) Transfection of murine multi-potent haemopoietic stem cells with an E.coli DNA alkyltransferase gene confers resistance to the toxic effects of alkylating agents. *Carcinogenesis, 9*, 81-87.

[201] Reese, J.S., Davis, B.M., Liu, L. and Gerson, S.L. (1999) Simultaneous protection of G156A methylguanine DNA methyltransferase gene-transduced hematopoietic progenitors and sensitization of tumor cells using O6-benzylguanine and temozolomide. *Clin. Cancer Res., 5*, 163-169.

[202] Hickson, I., Fairbairn, L.J., Chinnasamy, N., Dexter, T.M., Margison, G.P. and Rafferty, J.A. (1996) Protection of mammalian cells against chloroethylating agent toxicity by an O6-benzylguanine-resistant mutant of human O6-alkylguanine-DNA alkyltransferase. *Gene Therapy, 3*, 868-877.

[203] Hickson, I., Fairbairn, L.J., Chinnasamy, N., Lashford, I.S., Thatcher, N., Margison, g.p., Dexter, T.M. and Rafferty, J.A. (1998) Chemoprotective gene transfer I: transduction of human haemopoietic progenitors with O6-benzylguanine-resistant O6-alkylguanine-DNA alkyltransferase attenuates the toxic effects of O6-alkylating agents in vitro. *Gene Therapy, 5*, 835-841.

[204] Chinnasamy, N., Rafferty, J.A., Hickson, I., Lashford, L.S., Longhurst, S.J., Thatcher, N., Margison, G.P. Dexter, T.M. and Fairbairn, L.J. (1998) Chemoprotective gene transfer II: multilineage in vivo protection of haemopoiesis against the effects of an antitumor agent by expression of a mutant human O6-alkylguanine-DNA alkyltransferase. *Gene Therapy, 5*, 842-847.

[205] Maze, R., Carney, J.P., Kelley, M.R., Glassner, B.J., Williams, D.A. and Samson, L. (1996) Increasing DNA repair methyltransferase levels via bone marrow stem cell transduction rescues mice from the toxic effects of 1,3-bis(2-chloroethyl)-1-nitrosourea, a chemotherapeutic alkylating agent. *Proc. Natl. Acad. Sci., USA, 93*, 206-210.

[206] Koc, O.N., Reese, J.S., Szekely, E.M. and Gerson, S.L. (1999) Human long-term culture initiating cells are sensitive to benzylguanine and 1,3-bis(2-chloroethyl)-1-nitrosourea and protected after mutant (G156A) methylguanine methyltransferase gene transfer. *Cancer Gene Ther., 6*, 340-348.

[207] Raisanen, S.R., Lehenkari, P., Tasanen, M., Rahkila, P., Harkonen, P.L. and Vaananen, H.K. (1999) Carbonic anhydrase III protects cells from hydrogen peroxide-induced apoptosis. *FASEB J. 13*, 513-522.

[208] Rossman, T.G. and Goncharova, E.I. (1998) Spontaneous mutagenesis in mammalian cells is caused mainly by oxidative events and can be blocked by antioxidants and metallothionein. *Mutat. Res. 402*, 103-110.

[209] Kaina, B., Lohrer, H., Karin, M. and Herrlich, P. (1990) Overexpressed human metallothionein IIA gene protects Chinese hamster ovary cells from killing by alkylating agents. *Proc. Natl. Acad. Sci. USA, 87*, 2710-2714.

[210] Kaina, B., Van Zeeland, A.A., Backendorf, C., Thielmann, H.W. and Van de Putte, P. (1987) Transfer of human genes conferring resistance to methylating mutagens, but not to UV irradiation and cross-linking agents, into Chinese hamster ovary cells. *Mol Cell Biol, 7*, 2024-2030.

[211] Fritz, G., Dosch, J., Thielmann, H.W. and Kaina, B. (1993) Molecular and cellular characterization of Mex-/methylation-resistant phenotype. Gene and cDNA cloning, serum dependence, and tumor suppression of transfectant strains, *J. Biol. Chem., 268*, 21102-21112.

[212] Chen, B.J., Carroll, P. and Samson, L. (1994) The Escherichia coli AlkB protein protects human cells against alkylation-induced toxicity. *J. Bacteriol., 176*, 6255-6261.

[213] Jayaraman, L., Murthy, K.G.K., Zhu, C., Curran, T., Xanthoudakis, S. and Prives, C. (1997) Identification of redox/repair protein Ref-1 as a potent activator of p53. *Genes Dev., 11*, 558-570.

[214] Tsuzuki, T., Sakumi, K., Shiraishi, A., Kawate, H., Igarashi, H., Iwakuma, T., Tominaga, Y., Zhang, S., Shimizu, S., Ishikawa, T., Nakamura, K., Nakao, K., Katsuki, M. and Sekiguchi, M. (1996) Targeted disruption of the DNA repair methyltransferase gene renders mice hypersensitive to alkylating agent. *Carcinogenesis, 17*, 1215-1220.

[215] Canitrot, Y., Lautier, D., Laurent, G., Frechet, M., Ahmed, A., Turhan, A.G., Salles, B., Cazaux, C. and Hoffmann, J.S. (1999) Mutator phenotype of BCR—ABL transfected Ba/F3 cell lines and its association with enhanced expression of DNA polymerase β. *Oncogene, 18*, 2676-2680.

[216] Pascucci, B., Stucki, M., Jonsson, Z.O., Dogliotti, E. and Hubscher, U. (1999) Long patch base excision repair with purified human proteins. DNA ligase I as patch size mediator for DNA polymerases delta and epsilon. *J. Biol. Chem., 274*, 33696-33702.

[217] Lindahl, T. and Wood, R.D. (1999) Quality control by DNA repair. *Science, 286*, 1897-1905.

[218] Bogliolo, M., Cappelli, E., D'Osualdo, A., Rossi, O., Barbieri, O., Kelley, M.R. and Frosina, G. (2002) Effect of S. cerevisiae APN1 protein on mammalian DNA base excision repair, *Anticancer Res., 22*, 2797-2804.

[219] Ropolo, M., Geroldi, A., Rossi, O., Degan, P., Zupo, S., Poggi, A. and Frosina, G. (2004) Expression of the Drosophila melanogaster S3 ribosomal/repair protein in human cells. *Anticancer Res, 24*, 3811-3818.

[220] Boiteux, S., O'Connor, T.R. and Laval, J (1987) Formamidopyrimidine-DNA glycosylase of Escherichia coli: cloning and sequencing of the fpg structural gene and overproduction of the protein. *EMBO J., 6*, 3177-3183.

[221] O'Connor, T.R. and Laval, J. (1989) Physical association of the 2,6-diamino-4-hydroxy-5N-formamidopyrimidine-DNA glycosylase of Escherichia coli and an activity nicking DNA at apurinic/apyrimidinic sites. *Proc. Natl. Acad. Sci USA, 86*, 5222-5226.

[222] Gilboa, R., Zharkov, D.O., Golan, G., Fernandes, A.S., Gerchman, S.E., Matz, E., Kycia, J.H., Grollman, A.P. and Shoham, G. (2002) Structure of formamidopyrimidine-DNA glycosylase covalently complexed to DNA. *J. Biol. Chem., 277*, 19811-19816.

[223] Perlow-Poenhelt, R.A., Zharkov, D.O., Grollman, A.P. and Broyde, S. (2004) Substrate discrimination by formamidopyrimidine-DNA glycosylase: distinguishing interactions within the active site. *Biochemistry, 43*, 16092-16105.

[224] Rogacheva, M.V., Saparbaev, M.K., Afanasov, I.M. and Kuznetsova, S.A. (2005) Two sequential phosphates 3' adjacent to the 8-oxoguanine are crucial for lesion excision by E.coli fpg protein and 8-oxoguanine-DNA glycosylase. *Biochimie, in press.*

[225] Shibutani, S., Takeshita, M. and Grollmann, A.P. (1991) Insertion of specific bases during DNA synthesis past the oxidation-damaged base 8-oxodG. *Nature, 349*, 431-434.

[226] Avkin, S. and Livneh, Z. (2002) Efficiency, specificity and DNA polymerase-dependence of translesion replication across the oxidative DNA lesion 8-oxoguanine in human cells. *Mutat. Res., 510*, 81-90.

[227] Leipold, M.D., Muller, J.G., Burrows, C.J. and David, S.S. (2000) Removal of hydantoin products of 8-oxoguanine oxidation by the Escherichia coli DNA repair enzyme, FPG. *Biochemistry, 39*, 14984-14992.

[228] Coste, F., Ober, M., Carell, T., Boiteux, S., Zelwer, C. and Castaing, B. (2004) Structural basis for the recognition of the FapydG lesion (2,6-diamino-4-hydroxy-5-formamidopyrimidine) by formamidopyrimidine-DNA glycosylase. *J. Biol. Chem., 279*, 44074-44083.

[229] D'Ham,C., Romieu, A., Jaquinod, M., Gasparutto, D. and Cadet, J.(1999) Excision of 5,6-dihydroxy-5,6-dihydrothymine, 5,6-dihydrothymine, and 5-hydroxycytosine from defined sequence oligonucleotides by Escherichia coli endonuclease III and Fpg proteins: kinetic and mechanistic aspects. *Biochemistry, 38*, 3335-3344.

[230] Fromme, J.C., and Verdine, G.L. (2003) DNA lesion recognition by the bacterial repair enzyme MutM. *J. Biol. Chem., 278*, 51543-51548.

[231] Cadet, J., Bourdat, A.G., D'Ham, C., Duarte, V., Gasparutto, D., Romieu, A. and Ravanat, J.L. (2000) Oxidative base damage to DNA: specificity of base excision repair enzymes. *Mutat. Res. 462*, 121-128.

[232] Tchou, J., Kasai, H., Shibutani, S., Chung, M.H., Laval, J., Grollman, A.P. and Nishimura, S. (1991) 8-oxoguanine (8-hydroxyguanine) DNA glycosylase and its substrate specificity. *Proc. Natl. Acad. Sci. USA, 88*, 4690-4694.

[233] Bailly, V., Verly, W.G., O'Connor, T.R. and Laval, J. (1989) Mechanism of DNA strand nicking at apurinic apyrimidinic sites by Escherichia coli (formamidopyrimidine) DNA glycosylase. *Biochem J., 262*, 581-589.

[234] Asagoshi, K., Yamada, T., Terato, H., Ohyama, Y., Monden, Y., Arai, T., Nishimura, S., Aburatani, H., Lindahl, T. and Ide, H. (2000) Distinct repair activities of human 7,8-dihydro-8-oxoguanine DNA glycosylase and formamidopyrimidine DNA glycosylase for formamidopyrimidine and 7,8-dihydro-8-oxoguanine. *J. Biol. Chem. 275*, 4956-4964.

[235] Zharkov, D.O., Shoham, G. and Grollman, A.P. (2003) Structural characterization of the Fpg family of DNA glycosylases. 2, 839-862.

[236] Fromme, J.C., Banerjee, A. and Verdine, G.L. (2004) DNA glycosylases: recognition and catalysis. *Curr. Opin. Struct. Biol., 14*, 43-49.

[237] Zharkov, D.O., Rieger, R.A., Iden, C.R. and Grollman, A.P. (1997) NH2-terminal proline acts as a nucleophile in the glycosylase/AP-lyase reaction catalyzed by Escherichia coli formamidopyrimidine-DNA glycosylase (Fpg) protein. *J. Biol. Chem., 272*, 5335-5341.

[238] Zaika, E.I., Perlow, R.A., Matz, E., Broyde, S., Gilboa, R., Grollman, A.P. and Zharkov, D.O. (2004) Substrate discrimination by formamidopyrimidine-DNA glycosylase – A mutational analysis. *J. Biol. Chem., 279*, 4849-4861.

[239] Cussac, C. and Laval, F. (1996) Reduction of the toxicity and mutagenicity of aziridine in mammalian cells harboring the Escherichia coli fpg gene. *Nucleic Acids Res., 24*, 1742-1746.

[240] Gill, R.D., Cussac, C., Souhami, R.L. and Laval, F. (1996) Increased resistance to N,N',N''-triethylenethiophosphoramide (thioTEPA) in cells expressing the Escherichia coli formamidopyrimidine-DNA glycosylase. *Cancer Res., 56*, 3721-3724.

[241] Kobune, M., Xu, Y., Baum, C., Kelley, M.R. and Williams, D.A. (2001) Retrovirus-mediated expression of the base excision repair proteins formamidopyrimidine DNA glycosylase or human oxoguanine DNA glycosylase, protects hematopoietic cells from N, N', N'' – triethylenethiophosphoramide (thioTEPA)-induced toxicity in vitro and in vivo. *Cancer Res., 61*, 5116-5125.

[242] Wu, M., He, Y.H., Kobune, M., Xu, Y., Kelley, M.R. and Martin, W.J. 2nd (2002) Protection of human lung cells against hyperoxia using the DNA base excision repair genes hOGG1 and Fpg. *Am J. Respir. Crit. Care. Med., 166*, 192-199.

[243] Jans, J., Schul, W., Sert, Y. G., Rijksen, Y., Rebel, H., Eker, A.P., Nakajima, S., van Steeg, H., de Gruijl, F.R., Yasui, A., Hoeijmakers, J.H. and van der Horst, G.T. (2005) Powerful skin cancer protection by a CPD-photolyase transgene. *Curr. Biol., 15*, 105-115.

[244] Marchetto, M.C., Muotri, A.R., Burns, D.K., Friedberg, E.C. and Menck, C.F. (2004) Gene transduction in skin cells: preventing cancer in xeroderma pigmentosum mice. *Proc. Natl. Acad. Sci., 0101*, 17759-17764.

[245] Arnaudeau-Begard, C., Brellier, F., Chevallier-Lagente, O., Hoeijmakers, J., Bernerd, F., Sarasin, A., Magnaldo, T. (2003) Genetic correction of DNA repair-deficient/cancer-prone xeroderma pigmentosum group C keratinocytes. *Hum. Gene Ther., 14*, 983-996.

[246] Magnaldo, T. (2004) Xeroderma pigmentosum: from genetics to hopes and realities of cutaneous gene therapy. *Expert Opin. Biol. Ther., 4*, 169-179.

[247] Magnaldo, T. and Sarasin, A. (2004) Xeroderma pigmentosum: from symptoms and genetics to gene-based skin therapy. *Cells Tissues Organs., 177,* 189-198.

[248] Yarosh, D.B. (2002) Enhanced DNA repair of cyclobutane pyrimidine dimers changes the biological response to UV-B radiation. *Mutat. Res., 509*, 221-226.

[249] Norgauer, J., Idzko, M., Panther, E., Hellstern, O. and Herouy, Y. (2003) Xeroderma pigmentosum. *Eur. J. Dermatol., 13*, 4-9.

[250] Gerson, S.L. (2002) Clinical relevance of MGMT in the treatment of cancer. *J. Clin. Oncol., 20*, 2388-2399.

[251] Neff, T., Horn, P.A., Peterson, L.J., Thomasson, B.M., Thompson, J., Williams, D.A., Schmidt, M., georges, G.E., von Kalle, C. and Kiem, H-P. (2003) Methylguanine methyltransferase-mediated in vivo selection and chemoprotection of allogeneic stem cells in a large-animal model. *J. Clin. Invest., 112*, 1581-1588

[252] Jackson, E.L., and Jacks, MMHCC Lung Cancer Organ Site. *http://emice.nci.nih.gov/emice/mouse_models/organ_models/lung_models/murinecancer/murinemodels#table.*

[253] Ferrari, S., Griesenbach, U., Geddes, D.M. and Alton, E. (2003) Immunological hurdles to lung gene therapy. *Clin. Exp. Immunol., 132*, 1-8.

[254] Speina, E., Zielinska, M., Barbin, A., Gackowski, D., Kowalewski, J., Graziewicz, M.A., Siedlecki, J.A., Olinski, R. and Tudek, B. (2003) Decreased repair activities of

1,N(6)-ethenoadenine and 3,N (4)-ethenocytosine in lung adenocarcinoma patients. *Cancer Res., 63*, 4351-4357.

[255] Frosina G. (2006) Prophylaxis of oxidative DNA damage by formamidopyrimidine DNA glycosylase, *Int. J. Cancer, 119*, 1-7.

[256] Gould, D.J. and Favorov, P. (2003) Vectors for the treatment of autoimmune disease. *Gene Ther., 10*, 912-927.

[257] Barker, E. and Pianelles, V. (2003) Vectors derived from the human immunodeficiency virus, HIV-1. *Front Biosci., 8*, d491-d510.

[258] Lundstrom, K. (2003) Latest developments in viral vectors for gene therapy. *Trends Biotechnol. 21*, 117-122.

[259] Wang, G., Sinn, P.L. and McCray, P.B. Jr. (2000) Development of retroviral vectors for gene transfer to airway epithelia. *Curr. Opin. Mol. Ther., 2*, 497-506.

[260] Takeda, S. and Miyagoe-Suzuki, Y. (2001) Gene therapy for muscular dystrophies: current status and future prospects. *BioDrugs, 15*, 635-644.

[261] Brenner, S and Malech, H.L. (2003) Current developments in the design of onco-retrovirus and lentivirus vector systems for hematopoietic cell gene therapy. *Biochim Biophys. Acta, 1640*, 1-24.

[262] Zhou N.Y., Bates, S.E., Bouziane, M., Stary, A., Sarasin, A. and O'Connor, T.R. (2003) Efficient repair of cyclobutane pyrimidine dimers at mutational hot spots is restored in complemented xeroderma pigmentosum group C and trichothiodystrophy/xeroderma pigmentosum group D cells. *J. Mol. Biol., 332*, 337-351

[263] Pages, J.C. and Bru, T. (2004) Toolbox for retrovectorologists. *J. Gene Med., 6 Suppl. 1*, S67-S82.

[264] Kaiser, J. (2005) Retroviral vectors: a double-edged sword. *Science. 308*. 1735-1736.

[265] Galimi, F., Noll, M., Kanazawa, Y., Lax, T., Chen, C., Grompe, M. And Verma, I.M. (2002) Gene therapy of Fanconi anemia: preclinical efficacy using lentiviral vectors. *Blood, 100*, 2732-2736.

[266] Fuller, M. and Anson, D.S. (2004) Can the use of HIV-1 derived gene transfer vectors for clinical applications be justified? *Curr. Gene Ther., 4*, 65-77.

[267] Pannell, D. and Ellis, J. (2001) Silencing of gene expression: implications for design of retrovirus vectors. *Rev. Med. Virol., 11*, 205-217.

[268] Aiuti, A., Slavin, S., Aker, M., Ficara, F., Deola, S., Mortellaro, A., Morecki, S., Andolfi, G., Tabucchi, A., Carlucci, F., Marinello, E., Cattaneo, F., Vai, S., servida, P., miniero, R., roncarolo, M.G. and Bordignon, C. (2002) Correction of ADA-SCID by stem cell gene therapy combined with nonmyeloablative conditioning. *Science, 296*, 2410-2413.

[269] Hacein-Bey-Abina, S., Le Deist, F., Carlier, F., Bouneaud, C., Hue, C., De Villartay, J-P., Thrashier, A.J., Wulfraat, N., Sorensen, R., Dupuis-Girod, S., Fischer, A., and Cavazzana-Calvo M. (2002) Sustained correction of X-linked severe combined immunodeficicncy by ex vivo gene therapy. *N. Engl. J. Med., 346*, 1185-1193.

[270] Hacein-Bey-Abina, S., von Kalle, C., Schmidt, M. and Le Deist, F. (2003) A serious adverse event after successful gene therapy for X-linked severe combined immunodeficiency. *N. Engl. J. Med., 348*, 255.

[271] Cockrell, A.S. and Kafri, T. (2003) HIV-1 vectors: fulfillment of expectations, further advancements, and still a way to go. *Curr HIV Res., 1*, 49-439.

[272] Anson, D.S. (2004) The use of retroviral vectors for gene therapy- what are the risks? A review of retroviral pathogenesis and its relevance to retroviral vector-mediated gene delivery. *Genetic Vaccines Therapy, 2*, 9.

[273] Logan, A.C., Lutzko, C. and Kohn, D.B. (2002) Advances in lentiviral vector design for gene-modification of hematopoietic stem cells. Curr. *Opin. Biotechnol., 13*, 429-436.

[274] Morris, K.V. and Rossi, J.J. (2004) Anti-HIV-1 gene expressing lentiviral vectors as an adjunctive therapy for HIV-1 infection. *Curr HIV Research, 2*, 185-191.

[275] Poeschla, E.M. (2003) Non-primate lentiviral vectors. *Curr. Opin. Mol. Ther., 5*, 529-540.

[276] Azzouz, M. and Mazarakis, N. (2004) Non-primate EIAV-based lentiviral vectors as gene delivery systems for neuron diseases. *Curr. Gene Therapy, 4*, 277-286.

[277] Sandrin, V., Russell, S.J. and Cosset, F.L. (2003) Targeting retroviral and lentiviral vectors. Curr. Top. Microbiol. *Immunol., 281*, 137-178.

[278] Sanders, D.A. (2004) Ebola virus glycoproteins: guidance devices for targeting gene therapy vectors. *Expert Opin. Biol. Ther., 4*, 329-336.

[279] Kanerva, A. and Hemminki, A. (2005) Adenoviruses for treatment of cancer. *Ann. Med., 37*, 33-43.

[280] Fujiwara T., Kataoka, M. and Tanaka, N. (2000) Adenovirus-madiated p53 gene therapy for human cancer. *Mol. Urol., 4*, 51-54.

[281] Moon, C., Oh, Y. and Roth, J.A. (2003) Current status of gene therapy for lung cancer and head and neck cancer. *Clin. Cancer Res., 9*, 5055-5067.

[282] Merritt, J.A., Roth, J.A. and Logothetis, C.J. (2001) Clinical evaluation of adenoviral-mediated p53 gene transfer: review of INGN 201 studies. *Semin Oncol., 28*, 105-114.

[283] Roth, J.A., Grammer, S.F., Swisher, S.G., Komaki, R., Nemunaitis, J., Merritt, J., Fujiwara, T. and Meyn, R.E. Jr (2001) Gene therapy approaches for the management of non-small cell lung cancer. *Semin. Oncol., 28*, 50-56.

[284] Rinne, M., Caldwell, D. and Kelley, M.R. (2004) Transient adenoviral N-methylpurine DNA glycosylase overexpression imparts chemotherapeutic sensitivity to human breast cancer cells. *Mol. Cancer Ther., 3*, 955-967.

[285] Cockett, M. I. (1999) Technology evaluation: cystic fibrosis therapy, Genzyme. *Curr. Opin. Mol. Ther. 1*, 279-283.

[286] Flotte, T.R. and Laube, B.L. (2001) Gene therapy in cystic fibrosis. *Chest, 120*, 124S-131S.

[287] Jooss, K. and Chirmule, N. (2003) Immunity to adenovirus and adeno-associated viral vectors: implications for gene therapy. *Gene Ther., 10*, 955-963.

[288] Driskell, R.A. and Engelhardt, J.F. (2003) Current status of gene therapy for inherited lung diseases. *Annu. Rev. Physiol., 65*, 585-612.

[289] Dumon, K.R., Ishii, H., Fong, L.Y.Y., Zanesi, N., Fidanza, V., Mancini, R., Vecchione, A., Baffa, R., Trapasso, F., During, M.J., Huebner, K. and Croce, C.M. (2001) FHIT gene therapy prevents tumor development in Fhit-deficient mice. *Proc. Natl. Acad. Sci USA, 98*, 3346-3351.

[290] Ishii, H., Dumon, K.R., Vecchione A., Fong, L.Y., Baffa, R., Huebner, K. and Croce, C.M. (2001) Potential cancer therapy with the fragile histidine triad gene: review of the preclinical studies. *JAMA, 286*, 2441-2449.

[291] Roz, L., Gramegna, M., Ishii, H., Croce, C.M. and Sozzi, G. (2002) Restoration of fragile histidine triad (FHIT) expression induces apoptosis and suppresses tumorigenicity in lung and cervical cancer cell lines. *Proc. Natl. Acad. Sci., 99*, 3615-3620.

[292] Herzog, R.W. and Arruda, V.R. (2003) Update on gene therapy for hereditary haematological disorders. *Expert Rev. Cardiovasc. Ther. 1*, 215-232.

[293] Hendriks, W.T., Ruitenberg, M.J., Blits, B., Boer, G.J. and Verhaagen, J. (2004) Viral vector-mediated gene transfer of neurotrophins to promote regeneration of the injured spinal cord. *Prog. Brain Res., 146*, 451-476.

[294] Blits, B., Boer, G.J. and Verhaagen, J. (2002) Pharmacological, cell and gene therapy strategies to promote spinal cord regeneration. *Cell Transplant., 11*, 593-613.

[295] Lundstrom, K. (2003) Alphaviruses vectors for vaccine production and gene therapy. Expert Rev. *Vaccines, 2*, 447-459.

[296] Tranchant, I., Thompson, B., Nicolazzi, C., Mignet, N. And Scherman, D. (2004) Physicochemical optimisation of plasmid delivery by cationic lipids. *J. Gene Med., 6* Suppl 1, S24-S35.

[297] Ogris, M., and Wagner, E. (2002) Tumor-targeted gene transfer with DNA polyplexes. *Somat. Cell Mol. Genet., 27*, 85-95.

[298] Stecenko, A.A. and Brigham, K. L. (2003) Gene therapy progress and prospects: alpha-1antitrypsin. *Gene Ther., 10*, 95-99.

[299] Urnov, F.D., Miller, J.C., Lee, Y.L., Beausejour, C.M., Rock, J.M., Augustus, S., Jamieson, A.C., Porteus, M.H., Gregory, P.D. and Holmes, M.C. (2005) Highly efficient endogenous human gene correction using designed zinc-finger nucleases. *Nature, 435*, 646-651.

[300] Alwin, S., Gere, M.B., Guhl, E., Effertz, K., Barbas, C.F. 3rd, Segal, D.J., Weitzman, M.D. and Cathomen, T. (2005) Custom Zinc-Finger Nucleases for use in human cells. *Mol Ther., in press.*

[301] High, K.A. (2005) The moving finger. *Nature, 435*, 577-578.

[302] Densmore, C.L., Kleinerman, E.S., Gautam, A., Jia, S.F., Xu, B., Worth, L.L., Waldrep, J.C., fung, Y.K., T'Ang, A. and Knight, V. (2001) Growth suppression of established human osteosarcoma lung metastases in mice by aerosol gene therapy with PEI-p53 complexes. *Cancer Gene Ther., 8*, 619-627.

[303] Jia, S-F., Worth, L.L., Densmore, C.L., Xu, B., Duan, X. and Kleinerman, E.S. (2003) Aerosol gene therapy with PEI:IL-12 eradicates osteosarcoma lung metastases. *Clinical Cancer Res., 9*, 3462-3468.

[304] Rao, R.D., Markovic, S.N. and Anderson, P.M. (2003) Aerosol therapy for malignancy involving the lungs. *Curr. Cancer Drug Targets, 3*, 239-250

[305] Cappelli, E., Hazra, T., Hill, J.W., Slupphaug, G., Bogliolo M. and Frosina, G. (2001) Rates of base excision repair are not solely dependent on levels of initiating enzymes. *Carcinogenesis, 22*, 387-393

In: New Research on DNA Repair
Editor: Breehn R. Landseer, pp. 167-190

ISBN 1-60021-385-5

Chapter VI

TRANSLESION DNA SYNTHESIS-A FUNDAMENTAL DNA DAMAGE TOLERANCE MECHANISM

Errol C. Friedberg* and Lisa D. McDaniel

Laboratory of Molecular Pathology, Department of Pathology, University of Texas Southwestern Medical Center at Dallas, Dallas, TX 75390-9072.

ABSTRACT

Translesion DNA synthesis (TLS) is a fundamental biological process whereby arrested DNA replication caused by the presence of base damage that cannot be readily bypassed by replicative DNA polymerases, is relieved by an alternative DNA replication that utilizes DNA polymerases that are specialized for DNA synthesis past such lesions. Multiple such polymerases have been identified in all life forms. In particular, vertebrate cells contain no less than 10 such specialized enzymes. This article reviews our present knowledge of these specialized DNA polymerases in general and their specific role in TLS.

Keywords: Bypass polymerase, dCMP Transferase, Lesion Bypass, Xeroderma Pigmentosum Varient

INTRODUCTION

The genome of living cells is constantly subjected to damage associated with exposure to numerous environmental agents and to chemicals and ionic species generated during normal metabolism [1]. In particular, free radicals generated during metabolic reactions are typically

---

* Correspondence concerning this article should be addressed to Errol C. Friedbeg, Department of Pathology, U.T. Southwestern Medical Center at Dallas, 5323 Harry Hines Blvd., Dallas, TX 75390-9702.

highly reactive with DNA. Indeed, the evolution of a genetic code based on deoxyribonucleic acid, a macromolecule that is chemically unstable and that reacts with multiple chemical and physical agents, required the parallel evolution of a diverse and complex series of biological responses to DNA damage in order to preserve genetic integrity. These biological responses include (but are not limited to) multiple strategies that evolved specifically for coping with the potentially lethal effects of arrested DNA replication and/or the transcription of essential genes. They are generically referred to as *DNA damage tolerance* mechanisms because they mitigate the lethal potential of DNA damage without removing the offending lesion(s) from the genome. This review focuses on a particular mode of DNA damage tolerance during which arrested or stalled DNA replication is relieved by DNA synthesis directly across sites of base damage, a process called *translesion DNA synthesis* (TLS).

TRANSLESION DNA SYNTHESIS (TLS)

Historical Perspectives

The pioneering studies of Evelyn Witkin revealed that exposure of bacteria (e.g., *E. coli*) to DNA-damaging agents such as ultraviolet (UV) light results in a dose-dependent increase in mutations [2-5]. Studies aimed at understanding the molecular basis of DNA damage-induced mutagenesis were in due course extended to other prokaryotes and to lower and higher eukaryotes, with the same general result. The genetic requirements for observing DNA damage-dependent mutagenesis in *E. coli* implicated several genes, notably ones called *umuC*, *umuD* and *dinB* [6]. (It was subsequently shown that *umuC* and *umuD* are in a single operon [7].)

The late Harrison (Hatch) Echols and his colleagues worked toward understanding the specific functions of the proteins encoded by these genes [8-11]. They established systems for monitoring the replicative bypass of base damage *in vitro* and demonstrated a requirement for the UmuC/D proteins to observe TLS. These and other observations prompted a model for TLS in which these proteins were postulated to interact with the DNA replication machinery in a manner that relaxed the stringent fidelity of the replicative DNA polymerase(s), thereby facilitating their direct involvement in DNA synthesis across damaged DNA templates. They further hypothesized that TLS supported by the relaxed fidelity of the replicative DNA polymerases is highly error-prone, providing a satisfying explanation for the requirement of genes such as *umuC* and *umuD* for UV radiation-induced mutagenesis in *E. coli* [10].

In the course of experiments carried out after Echols's untimely death it was determined by several investigators that UmuC protein is itself a *bone fide* DNA polymerase that supports the error-prone replicative bypass of various types of base damage independent of the presence of the replicative polymerases [12]. At about the same time independent studies demonstrated that DinB protein is also a low fidelity, error-prone DNA polymerase [13]. More detailed characterization of both UmuC and DinB proteins demonstrated that they have extremely low fidelity [11-13], even when copying undamaged DNA, are devoid of 5'->3' proofreading exonuclease activity, and are weakly processive [13].

The low fidelity of these DNA polymerases when copying undamaged DNA templates is directly related to their ability to support the incorporation of nucleotides opposite sites of base damage. Both properties derive from structural determinants in the catalytic site that distinguish these enzymes from high fidelity DNA polymerases [14-16]. In particular, all of these enzymes are endowed with a more open active site, which can accommodate the distorted or bulky structures of many damaged bases (thus promoting TLS), as well as mispaired nucleotides (thus promoting error-prone copying undamaged DNA). Additionally, some of these low fidelity DNA polymerases are endowed with novel structural domains [17].

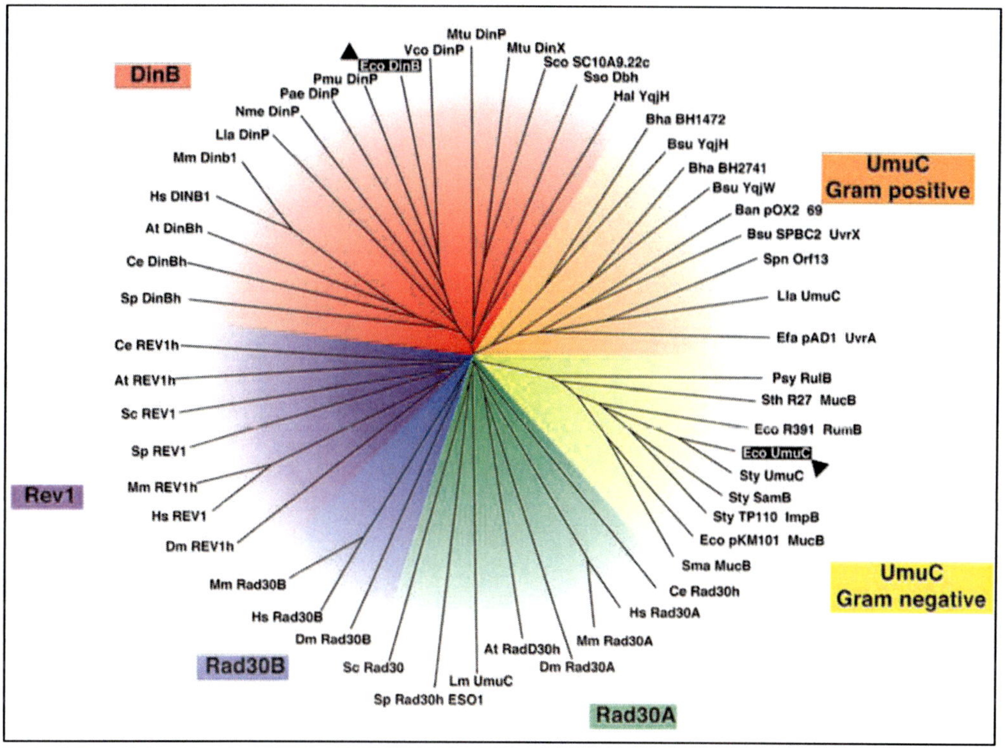

Figure 1. The Y-family of specialized DNA polymerases. This unrooted phylogenetic tree shows evolutionary relationships between proteins that are all believed to be specialized DNA polymerases, in both bacteria and eukaryotic organisms. The Y-family comprises 6 subfamilies designated by various color codes. The UmuC sub-families are identified exclusively in bacteria and the Rev1, Rad30A and Rad30B subfamilies are exclusively eukaryotic. Members of the DinB sub-family span the entire evolutionary spectrum. (Adapted from [1] with permission from the ASM Press).

The observation that *E. coli* is endowed with two low fidelity DNA polymerases (reviewed in [18]) that are devoid of proofreading capacity, prompted searches for structural and functional homologs in eukaryotes. Facilitated by the availability of the nucleotide sequences of multiple eukaryotic genomes, structural orthologs of UmuC and DinB proteins were identified *in silico* across the entire animal kingdom. Careful comparisons of amino acid sequence identity and similarity showed that these polymerases belong to a new superfamily of proteins (the Y family), distinct from the A, B and X families of DNA polymerases previously identified [17]. The Y family is comprised of 5 sub-families designated UmuC, DinB, REV1, RAD30A and RAD30B, after prominent members of each sub-family (Fig. 1). The RAD30A subgroup is now called Polη and the RAD30B subgroup is called Polι. The UmuC subfamily is only represented in the prokaryotic world and the REV1, RAD30A and RAD30B subfamilies are restricted to eukaryotes (Fig. 1). The DinB (Polκ) subfamily is evolutionarily conserved from bacteria to man [19]. However, overall, DinB protein (DNA polymerase V) in *E. coli* is considerably smaller than its evolutionarily conserved Polκ vertebrate counterpart and is devoid of specific functional domains present in Polκ.

Whereas all members of the Y family thus far examined are *bone fide* DNA template-directed DNA polymerases, albeit highly inaccurate ones, members of the REV1 subfamily are more appropriately designated as dCMP transferases. When copying DNA all members of the REV1 subfamily support the incorporation of one (or two) dCMP moieties, regardless of DNA template composition [20]. This feature derives from unique structural determinants in the protein despite the fact that dCMP incorporation is DNA template directed [21].

None of the low-fidelity DNA polymerases that are devoid of 3'->5' proofreading exonuclease activity have been implicated in replicative DNA synthesis. Rather, all of those studied to-date have been shown to subserve highly specialized functions, including, not limited to TLS. We therefore designate this entire class of enzymes as *specialized DNA polymerases*. The remainder of this review is devoted to a consideration of specialized DNA polymerases in eukaryotes, with a particular emphasis on those in mammalian cells.

Specialized DNA Polymerases in Eukaryotic Cells

In contrast to prokaryotes such as *E. coli*, which are endowed with just two DNA polymerases that are specialized for TLS (Table 1), lower eukaryotes contain more such enzymes (Table 1). As shown in Table 1, different specialized polymerases are represented in different lower eukaryotes, but all lower eukaryotes possess at least three such enzymes, one of which, REV1, is invariant. Vertebrates are endowed with at least 10 specialized DNA polymerases (Table 1).

Table 1. Identified Bypass Polymerase Proteins.

Bacterial	S. cerevisiae	C. elegans	D. melanogaster	Vertebrate enzyme	Gene (H.s./M.m.)	Protein
	REV1	REV-1	Rev1	Rev1	REV1L/Rev1l	REV1L
	pol η (RAD30)	POLH-1	DNApol-eta	polη	POLH/Polh	POLH
pol IV (DinB)		POLK-1		Polκ	POLK/Polk	POLK
			Poli (Rad30B)	Polι	POLI/Poli	POLI
pol V (umuC)	POL4			Polλ	POLL/Poll	POLL
				Polμ	POLM/Polm	POLM
				Polβ	POLB/Polb	POLB
	Pol ζ (REV3/REV7)	Y37B11A.2 (REV3)	Polz (Rev3 or Mus205/Rev7)	Polζ	REV3L/Rev3l	REV3L
		POLQ-1	polq (Mus308)	Polθ	POLQ/Polq	POLQ
				Polν	POLN/Poln	POLN

TLS catalyzed by specialized DNA polymerases is readily demonstrated by incubating a purified enzyme of interest with the required cofactors (divalent metal ions and the four usual deoxyribonucleoside triphosphates) and an oligonucleotide primer-template carrying a single defined type of base damage placed in a known location in the template DNA [22-25]. Simple primer extension affords the ability to both monitor TLS (by observing extension beyond the lesion) and the fidelity of this bypass (by sequencing the extended primer strand) (Fig. 2).

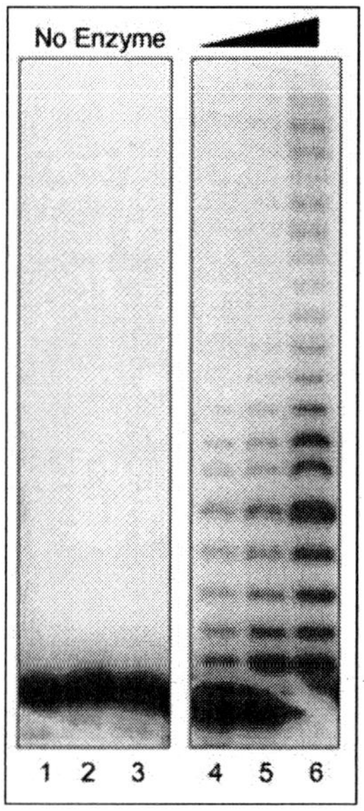

Figure 2. Replicative bypass of sites of base damage (TLS) by specialized DNA polymerases can be demonstrated *in vitro* by primer extension, using a short oligonucleotide primer annealed to a radiolabeled damage-containing template. The left panel shows the electrophoretic mobility of the primer-template in the absence of added enzyme. In the presence of an appropriate DNA polymerase the primer is extended to the end of the template as a function of increasing enzyme concentration. The ladder effect reflects the very low processivity of the polymerase, in this case DNA polymerase kappa (Polκ) from mouse cells.

This simple assay has demonstrated that specialized polymerases are able to support the replicative bypass of a diverse array of base damage that arrests normal DNA replication in cells (Table 2). The structural diversity of lesions bypassed by some individual polymerases *in vitro* is confounding, since these lesions bear little if any structural similarity. This caveat prompts the caution that as is the case with many reductionist *in vitro* systems, the qualitative and quantitative results of primer extension must be interpreted with care, especially if the ratio of enzyme to substrate is high. Even some high fidelity polymerases can be "pushed" to synthesize DNA across sites of base damage in vitro. Additionally, as discussed fully later, in intact vertebrate cells most if not all specialized DNA polymerases interact with other proteins involved in TLS, and some of them undergo specific post-translational modification that modify the dynamics and range of these interactions [26,27]. Thus, it is useful to interpret the results of primer extension studies in the context of established genetic and biological information about the enzymes under study.

Table 2. Damaging Agent. Bypassed by TLS Polymerases.

E. coli	*S. Cerevisae*	*H. sapiens*
	Rev1	**REV1L**
	abasic site[e] [78]	8-oxoguanine [79]
	N(2)-G[e] [80]	benzo(a)pyrene [79]
		1,N6-ethenoadenine [79]
	Pol eta (RAD30)	**POLH**
	CPD [28]	AAF [81]
	8-oxoguanine [82]	psoralen crosslinks [83]
	abasic site [84]	CPD [85]
	G[8-5]C intrastrand cross-link [86]	oxaliplatin and cisplatin [32]
		abasic site [31]
		8-oxoguanine [87]
		benzo(a)pyrene [87]
		cyclodeoxyadenosine [88]
		dG-N(2)-3MeE[f] [89]
		dA-N(6)-3MeE[g] [89]
		4-OHEN-dC[i] [90]
		N2-methyl(Me)G [91]
		N2-ethyl(Et)G [91]
		N2-isobutyl(Ib)G [91]
		N2-benzyl(Bz)G [91]
		N2-CH2(2-naphthyl)G [91]
		N2-isopropyl G [92]
		N6-isopropyl A [92]
		T-U CPD [93]
		1,N(6)-Ethanoadenine [94]
Pol IV (DinB)		**POLK**
UV lesions [95]		Abasic sites [51]
benzo(a)pyrene[d] [96]		AAF[51]
abasic site [97]		8-oxoguanine [30]
		benzo(a)pyrene [30]
		thymine glycol [22]
		dG-N(2)-3MeE[f] [89]
		dA-N(6)-3MeE[g] [89]
		4-OHEN-dC[i] [90]
		T-U CPD [93]
		POLI
		6-4PP [98]
		abasic sites [98]
		CPD [99] not [100]
		8-oxoguanine [101]
		5-hydroxycytosine [101]
		benzo(a)pyrene [102]
		uracil[101]
		5-hydroxyuracil [101]
		5,6-dihydrouracil [101]
		N2-isopropyl G [92]
		N6-isopropyl A [92]
		T-U CPD [93]

(Table 2. Continued)

E. coli	*S. Cerevisae*	*H. sapiens*
Pol V	**PolIV**	**POLL**
CPD[a] [103]	UV lesions [104]	abasic sites [61]
6-4PP[b] [103]	abasic site [105]	Cisplatin [61]
abasic site [106]		
AAF[c] [96]		
benzo(a)pyrene[d] [96]		
		POLM
		8-oxoguanine [107]
		abasic site [107]
		1,N(6)-ethenoadenine [107]
		AAF [107]
		benzo[a]pyrene [107]
		CPD [107]
		oxaliplatin and cisplatin [108]
		POLB
		AAF [109]
		Cisplatin [110]
		Styrene oxide [111]
		Propanodeoxyguanosine [112]
		O6-methylguanine [113]
		CPD [114]
		6-4PP [114]
		1,N(6)-Ethanoadenine [94]
	Rev 3/ Pol zeta	**REV3**
	CPD [115]	Cisplatin [116]
	6-4PP [117]	benzo[a]pyrene [118]
	AAF [117]	
	thymine glycol [119]	
	abasic site[e] [78]	
	N(2)-G[e] [80]	
	T-U CPD [93]	
		POLQ
		apurinic site [120]
		POLN
		Inferred from D. melanogaster
		Nitrogen mustard [121]
		Hexamethylphosphoramide [118] hexamethylmelamine [118]
		N-ethyl-N-nitrosourea [118]

[a]thymine-thymine cis-syn cyclobutane dimmer

[b]6-4 thymine-thymine pyrimidine-pyrimidone photoproduct

[c]2-acetylaminofluorene

[d]both pol IV and V are required

[e]both are REV1 and Pol zeta are required

[f]N(2)-[3-methoxyestra-1,3,5(10)-trien-6-yl]-2'-deoxyguanosine

[g]N(6)-[3-methoxyestra-1,3,5(10)-trien-6-yl]-2'-deoxyadenosine

[h]gamma-hydroxy-1,N(2)-propano-2'deoxyguanosine DNA minor-groove adduct

The genetic tractability of lower eukaryotes such as the budding yeast *S. cerevisiae*, the fission yeast *S. pombe*, the worm *C. elegans* and the fruitfully *Drosophila melanogaster*, has facilitated the elaboration of extensive genetic frameworks that invite complementary biochemical studies. However, as noted in Table 1, none of these organisms embrace the complete spectrum of specialized DNA polymerases observed in higher organisms such as mammals. Additionally, as noted later, there are indications that TLS in at least some of these lower eukaryotes may be qualitatively different from that in vertebrates. Genetic information about specialized DNA polymerases in higher eukaryotes is more restricted. Nevertheless, the study of several human hereditary diseases, as well as the generation of mouse strains defective in genes that encode some specialized DNA polymerases have provided important insights into the biological relevance of some of these enzymes.

The Molecular Biology of TLS in Vertebrates

The Variant Form of Xeroderma Pigmentosum (XP) and DNA Polymerase Eta (POLH or Polη)

Xeroderma pigmentosum (XP) is a multigene human hereditary disorder usually characterized by defective nucleotide excision repair (NER), a DNA repair process during which various types of base damage that arrest DNA replication are removed from the genome. However, some individuals with this disorder are proficient in NER but are defective for a specialized DNA polymerase called POLH, encoded by the *POLH/XPV* gene.

In vitro primer extension assays have demonstrated that POLH can support the bypass of multiple types of base damage [28-30]. To reiterate the cautions stated above about *in vitro* primer extension, the biological relevance of these observations is not obvious, since the phenotypes of the XP variant disease are believed to relate primarily, if not exclusively, to an inability to properly respond to sunlight-induced DNA damage. Thus a presumably relevant biochemical observation is that POLH catalyzes TLS past thymine-thymine cyclobutane pyrimidine dimers (T<>T CPD) generated in skin cells exposed to UV light from the sun [28]. In so doing POLH accurately incorporates the base adenine opposite T<>T lesions. In fact, detailed kinetic studies indicate that the enzyme has a preference for incorporating AA opposite these lesions compared to TT in undamaged DNA, suggesting that the enzyme evolved specifically to fulfill this role. The enzyme can presumably also accurately bypass C<>T and C<>C dimers in DNA. However, this has not yet been formally demonstrated because C in pyrimidine dimers is very labile, readily converting to U by deamination. Despite the fact that like all specialized DNA polymerases, POLH is distinctly error-prone compared to the high fidelity of replicative DNA polymerases, its intrinsic fidelity when copying T<>T in DNA, coupled with the fact that bypass of this lesion requires the incorporation of just two nucleotides, provides for mainly accurate TLS past such lesions in vertebrate cells [31,32].

Collectively, these genetic and biochemical observations have led to the following hypothesis about the generation of the so-called variant form of XP. In normal individuals the combined outcome of proficient NER and functional POLH maintains a limited mutational burden in cells exposed to UV radiation. In contrast, in the absence of functional POLH cells deploy some other mechanism for tolerating T<>T CPD in DNA, likely by error-prone TLS catalyzed by one (or more) other specialized polymerases, or by some different error-prone process. While this response mitigates the potential for cell death in the face of persistent replicative arrest, enhanced cell survival is achieved at the cost of a significantly increased frequency of mutations, ultimately leading to skin cancer and the other dermatological abnormalities associated with XP.

The XP variant story has lent considerable credence to the notion that POLH evolved principally, if not exclusively, to support TLS past T<>T (and presumably other types of *cis-syn*) CPD, a potent and common source of UV radiation-induced replication arrest. Extrapolating this hypothesis to other specialized DNA polymerases leads to the suggestion that the multiplicity of such enzymes in vertebrates reflects the selective pressures of multiple types of naturally occurring base damage in DNA during biological evolution. To validate this hypothesis it might be useful to systematically survey the ability of multiple specialized DNA polymerases to accurately bypass multiple different known types of naturally occurring base damage that arrest DNA replication.

However, there are indications that some specialized DNA polymerases may have been selected for primary functions other than TLS. In particular, the reduced fidelity of these enzymes has apparently facilitated their participation in somatic hypermutation in the immune system (see later discussion). Additionally, some of them have been implicated in DNA repair processes such as base excision repair (BER) and nucleotide excision repair (NER) [33-36]. Hence, though all mammalian specialized DNA polymerases thus far examined can indeed support TLS by primer extension *in vitro*, not all of them may necessarily function in this capacity *in vivo*. And those that do may not support the replicative bypass of the entire spectrum of base damage observed during TLS *in vitro*. Since lower eukaryotes are not endowed with immune functions, perhaps it is valid to conclude that all lower eukaryotic members of this class of enzymes reflect a primary, if not an exclusive role in TLS in vertebrates. As shown in Table 1, only 5 of the 10 specialized DNA polymerases identified in vertebrates are represented in at least two lower eukaryotes.

In summary, the paucity of genetic models that clearly correlate with biochemical studies limits the general validity of the notion that the multiplicity of specialized DNA polymerases in vertebrates reflects the imperative to bypass multiple types of naturally occurring base damage during biological evolution.

Mouse Strains Defective in Genes that Encode Specialized DNA Polymerases

The facility with which genes can be inactivated in mouse embryonic stem cells offered the hope that mouse mutants would be highly informative with respect to the biological function of specialized DNA polymerases in mammals. Mouse strains defective in the *Polh* [37,38], *Polk* [39,40], *Poli* [41], *Poll* [42,43], *Polm* [43], *Polq* [44] or *Rev1l* [45] genes have been generated. Like human XP variant patients, *Polh*-defective mice manifest a pronounced predisposition to UV radiation-induced skin cancer [38]. It remains to be established what phenotypes (if any) transpire in these mice when they are exposed to DNA-damaging agents other than UVC radiation, which are also bypassed by POLH *in vitro* [37]. Mouse strains defective in genes that encode the specialized DNA polymerases POLI, POLL and POLM do not manifest phenotypes indicative of a preferred substrate for TLS by these enzymes. However, some of these mice manifest phenotypes indicative of altered somatic hypermutation (SH) in immunoglobulin genes [37,44,46] (see later discussion). Recent studies on double mutant mice defective in both the Polκ and *Poli* genes shows a modest increased UV sensitivity of cells relative to those defective in just the *Polk* gene as well as a slightly increased susceptibility to UV radiation-induced skin cancer (F. Hanaoka, personal communication). This observation is significant because it has been suggested that in the absence of POLK, POLI may be the specialized DNA polymerase that primarily, if not exclusively, contributes to error-prone TLS past CPD and hence to skin cancer [47]. If this were the case one would expect that in double mutant *Polk/Poli* mice skin cancer might be alleviated rather than aggravated.

Polk Mutant Mice

Two strains of *Polk*$^{-/-}$ mice have been independently generated by conventional gene knockout technology [39,40]. Extensive immunological characterization of one of these mouse strains revealed no evidence of a role for POLK in SH in immunoglobulin genes in the mouse. Hence, at the present time POLK is considered to be an enzyme that evolved primarily (if not exclusively) for a role in TLS past a particular type or class of naturally occurring base damage(s).

Phenotypes in Polk Mutant Mice

Both strains of *Polk*-defective mice are viable and manifest no obvious phenotypes. However, cells in culture derived for these animals display increased sensitivity to killing by UV light [39,40] and following exposure to the polycyclic aromatic compound benzo[a]pyrene (B[a]P) [48-50], a well known DNA damaging-agent that can be covalently adducted to guanine in DNA. Several independent studies have failed to demonstrate the bypass of TLS past CPD or [6-4] photoproducts (PP, the major photoproducts in UV irradiated DNA) by in vitro primer extension [51]. Assuming that this result reflects the inability of this bypass reaction in living cells, it is not obvious why the absence of this enzyme renders cells more sensitive to UV light. However, several formal possible explanations have been entertained.

1. As discussed later, there are indications that TLS in eukaryotes may sometimes (possibly always) involve two specialized DNA polymerases, one of which catalyzes the incorporation of nucleotides directly opposite offending template base lesions, while a second specialized polymerase supports extension of the primer beyond such sites. Several *in vitro* studies have implicated Polκ in the latter role [19,22,52]. If such a role is indeed served during TLS *in vivo*, it provides a possible explanation

for the increased sensitivity of Polk-defective cells to UV radiation. However, the direct demonstration of such a role is beyond the capacity of presently established *in vitro* systems for TLS.

2. Conceivably, one or more types of base damage generated by UV radiation other than CDP or [6-4]PP is a "cognate" (preferred) substrate for TLS by Polκ. In this regard, the purified enzyme has been shown to efficiently bypass thymine glycol lesions in DNA, (one of several types of single nucleotide damage promoted by exposure to UV radiation), and in so doing to mainly accurately incorporate A opposite thymine glycol [22]. However, it is difficult to evaluate the biological relevance of this observation. Notably, thymine glycol is readily generated in DNA by perturbations that generate oxidative damage to DNA. But *Polk*$^{-/-}$ cells are not abnormally sensitive to killing by oxidative stress (I. Boldogh, S. Mitra, C. Guo and E. C. Friedberg, unpublished observations).
3. Recent studies have demonstrated that *Polk* mutant cells are at least partially defective in the repair synthesis step of the removal of CPD and [6-4]PP by conventional nucleotide excision repair (NER) [34]. This result requires careful evaluation since specialized DNA polymerases have not been previously implicated in NER. Nor is such implication intuitively compelling, since these enzymes are notoriously error-prone when copying undamaged DNA and their primary involvement in repair synthesis during NER would be expected to generate mutations, whereas NER is considered to be an error-free process. Additionally, a general involvement of Polk in NER predicts that *Polk* mutant cells should be hypersensitive to killing following exposure of all DNA-damaging agents that generate lesions the repair of which requires operational NER. As discussed in the ensuing paragraphs, *Polk*$^{-/-}$ cells are indeed hypersensitive to several bulky polycyclic hydrocarbon adducts in DNA. However, there are alternative (and more cogent) explanations for these observations besides defective NER. Negative cell survival experiments may be difficult to come by since they are frequently not reported in the literature. Recent studies have shown that chicken DT40 cells defective in the Polk gene are slightly more sensitive to killing by the nonfunctional alkylating agent methylmethane sulfonate (MMS) [53].

The biological significance of the sensitivity of *Polk*$^{-/-}$ cells to B[a]P is also not unequivocally resolved. B[a]P may have been naturally generated by fires during early evolution. However, such occurrences are unlikely to have been sufficiently widespread or prolonged to drive biological selection during evolution. Thus the polycyclic ring structure of B[a]P prompts one to consider that it may mimic one or more naturally occurring polycyclic compounds that also bind to guanine in DNA. In this regard steroids come readily to mind. While there is no definitive proof that steroid molecules bound to DNA constitute the preferred substrate for POLK, a number of indirect observations certainly support this notion.

1. In wild-type mouse embryos aged about 15.5 days the adrenal cortex (a major site of adrenal steroid biosynthesis) reveals very high levels of *Polk* mRNA expression by *in situ* hybridization. Indeed, the adrenal cortex is the only site of *Polk* gene

expression that can be identified by this technique in embryonic mice. This expression pattern persists in adult mice, in which the adrenal cortex is not the sole tissue in which *Polk* mRNA can be detected [54].

2. Overexpression of *Poli*, *Polh* or *Poll* mRNA is not observed in the adrenal cortex of the mouse. Hence enhanced expression of a specialized DNA polymerase in adrenal cortical cells is apparently unique to POLK [54].
3. Chicken DT40 cells deleted for the *Polk* gene are abnormally sensitive to killing and to chromosomal aberrations associated with exposure of cells to tamoxifen or estrogen derivatives [53].
4. Mouse breedings that involve *Polk*$^{-/-}$ females result in reduced litter sizes, suggestive of a subtle fertility defect in these animals. Additionally, many *Polk*$^{-/-}$ males manifest subtle age-dependent fertility problems (L.D. McDaniel and E. C. Friedberg, unpublished observations)
5. In vitro, purified POLK efficiently and accurately bypasses estrogen adducts in DNA [55].

These observations notwithstanding, *Polk*$^{-/-}$ mice do not manifest adrenal insufficiency under standard laboratory conditions. However, this observation must be qualified by the fact that cells in the adrenal cortex do not normally replicate their DNA. Thus, it might be informative to subject *Polk*$^{-/-}$ mice to stresses that promote adrenal cortical hyperplasia. The obligate DNA replication associated with this cellular proliferation may be accompanied by error-prone TLS past steroid adducts, prompting an increased mutational burden in adrenal cortical cells, or by arrested replication, prompting death of adrenal cortical cells. Both of these outcomes may (or may not) result in phenotypically detectable adrenal cortical insufficiency.

Other Phenotypes Related to the *Polk* Mutation

While *Polk*$^{-/-}$ mice do not manifest obvious phenotypes that are directly linked to the absence of Polκ enzymatic activity, several prominent phenotypes observed in a cohort of Polk mice inbred for multiple generations suggest that loss of Polk activity may promote a mutator state in the male germline.

Male and female *Polk*$^{+/-}$ mice were inbred in a non-pathogen-free (conventional) mouse facility at the University of Texas Southwestern Medical Center for 8 generations of pups. Typical diabetes insipidus (DI) was observed in some offspring as early as the second generation. Diabetes insipidus is distinct from the more common and unrelated disease diabetes mellitus. It is characterized by extensive loss of water and electrolytes due to failure to reabsorb liquid in the kidneys as a result of insufficient antidiuretic hormone (ADH) (vasopressin). However, the DI phenotype did not segregate with a particular *Polk* genotype, i.e., DI mice were either *Polk*$^{-/-}$, *Polk*$^{+/-}$ or *Polk*$^{+/+}$ in approximately equal numbers and the disease was inherited as an autosomal recessive.

Collectively, these observations led to the hypothesis that *Polk*$^{-/-}$ male (and possibly also female) mice suffer spontaneous mutations in germline cells. (While this mutator phenotype most obviously derives from the absence of wild type POLK protein, later discussion will address the possible presence of multiple POLK isoforms in the mouse testis. Hence, loss of any one or more of these isoforms may promote a mutator phenotype.) Assuming that such spontaneous mutations are generated randomly in the genome, one might expect that DI arose in some mice by a chance founder mutation in a gene that governs some aspect of vasopressin activity. Three genes designated *AVPR2*, *AQP2* and *AVP*, all of which are required for normal vasopressin metabolism, have been implicated in hereditary DI in humans. However, no mutations were detected in the open reading frames of the *Aqp2* or *Avp* genes in several DI mice examined (L.D. McDaniel and E. C. Friedberg, unpublished observations). The *Avpr2* gene is sex-linked and was thus not examined.

While we have not formally eliminated the possibility of functionally inactivating mutations in non-coding regions of these genes, these findings weaken the mutator hypothesis as presented here. However, DI can also occur in a non-hereditary form, usually from loss of vasopressin production in the supraoptic and paraventricular nuclei in the brain due to multiple causes, including autoimmune destruction of cells in these nuclei. Consistent with this pathogenesis, recent studies have demonstrated an obvious loss of cells in the supraoptic and paraventricular nuclei in several DI mice in the Polk cohort (J. Richardson, L.D. McDaniel and E. C. Friedberg, unpublished observations). Hence, it is conceivable that among the random mutations generated in this cohort, one (or more) mutations caused an autoimmune state that is reflected in the phenotypes observed, notably acquired DI. Obviously the stochastic nature of the mutations predicted in *Polk* mutant mice precludes reproducing this experimental result at will. In support of the notion of a chance founder mutation in a gene(s) for autoimmune disease in the *Polk* cohort with DI, several mice in this cohort manifested other disease states. These include uncharacterized neuromuscular disturbances and skeletal abnormalities, and typical vitiligo. The latter condition in humans has a known autoimmune basis.

The *Polk* Gene is highly Expressed in the Male Germline

Besides the adrenal cortex, the *Polk* gene is highly expressed in the seminiferous tubules of the testis in adult mice. Furthermore, northern analysis reveals that in addition to a single transcript of ~4.4 kb present in all cells, a group of smaller transcripts is observed uniquely in the testis [54]. Among this group of transcripts (about a dozen distinct transcripts have been observed by RT-PCR), at least 5 are in-frame and could potentially encode multiple POLK isoforms in the testis, each of which may be endowed with unique functions. At this time the presence of distinct protein isoforms in the testis of mice has not been formally demonstrated.

The Molecular Mechanism of Translesion DNA Synthesis in Vertebrates

It is intuitively compelling that any mechanism of TLS must involve switching various DNA polymerases at the arrested replication fork. Multiple such switches are likely. (i) The arrested replication machinery must be replaced by at least one specialized DNA polymerase that can effect replication past the offending lesion, incorporating at least a single nucleotide. We define this DNA synthesis mode as *TLS insertion*. (ii) As indicated earlier, there are indications that TLS past at least some, and possibly all lesions, requires a second specialized DNA polymerase to extend the primer terminus generated during TLS insertion; so-called *TLS extension*. This may serve the function of advancing the primer terminus to a position sufficiently downstream of the base damage that when the high fidelity replication machinery reengages, the associated proofreading exonucleases cannot remove nucleotides that were incorporated during TLS. (iii) Productive reengagement of the replicative machinery presumably requires a third DNA polymerase switch.

Recent studies have begun to reveal biochemical events that are apparently required for TLS in mammalian cells. These involve both specific protein-protein interactions and specific posttranslational modifications of proteins.

Protein-Protein Interactions and Posttranslational Modifications Associated with TLS

Several Y family DNA polymerases have been shown to interact with one another *in vitro* and *in vivo* in the absence of deliberate exposure of cells to DNA damage. Notably, POLH binds to POLI [56], and REV1L protein binds POLK, POLH, POLI and the REV7 subunit of Polζ through the same highly conserved C-terminal domain [57]. Indeed, when REV1L is incubated in the presence of both POLK and increasing concentrations of REV7, the latter protein competes with POLK for binding to REV1L. POLI [58], POLH [59], POLK [60], POLL [61], POL ζ [62] and REV1L [63] have also been shown to interact with PCNA. Specific PCNA-interacting peptide (PIP) domains in POLI, POLH and POLK, and a BRCT domain in REV1L are required for these interactions. The functional significance of the interactions between these various polymerases and PCNA in the absence of deliberate exposure of cells to DNA damage is not clear. There are no experimental indications that specialized DNA polymerases are involved in normal high fidelity DNA replication. Conceivably their association with PCNA facilitates their tracking with the replication machinery, providing functional access to sites of stalled replication when appropriate.

When cells are exposed to specific DNA-damaging agents new biochemical events transpire. For example, when cells are exposed to UV light PCNA undergoes monoubiquitylation at a specific lysine residue (Lys164) [64]. This posttranslational modification requires an E3 ubiquitin ligase encoded by the *RAD6* (*UBE2A/UBE2B*) and *RAD18* genes. Monoubiquitylated PCNA binds more avidly to POLH, POLI and REV1L protein than does unmodified PCNA, suggesting that monoubiquitylation of PCNA is a key event during TLS [65].

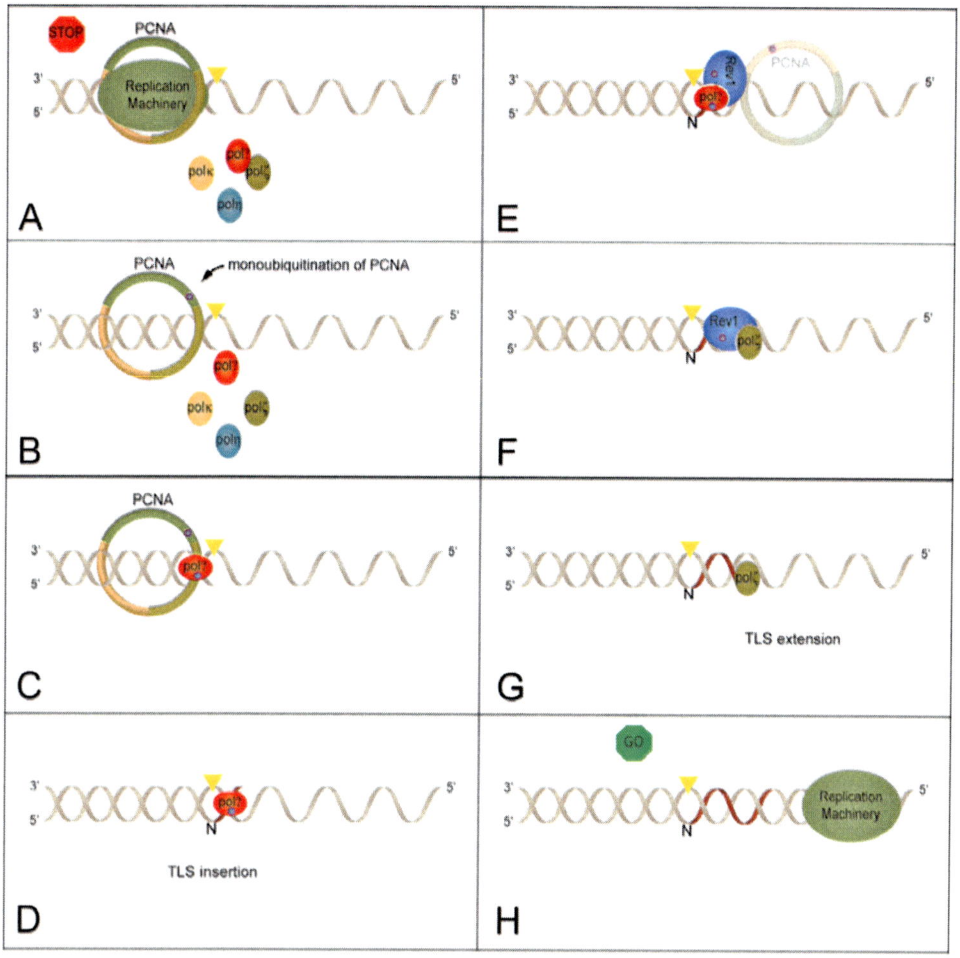

Figure 3. A model for translesion DNA synthesis in a vertebrate cell based on existing information. (A) The replicative machinery of a cell (containing among other things, high-fidelity DNA polymerases and PCNA) is arrested at a site of base damage in the template strand (inverted yellow triangle). Several specialized DNA polymerases (polκ, polη, polζ and pol?) are shown in the vicinity. However, the precise physical relationship of these enzymes to the arrested replication machinery is not known. (B) Some feature(s) of the arrested replication results in monoubqcuitylation of PCNA, allowing its interaction with a specialized DNA polymerase (pol?), which is now positioned at the primer-template (C), the first of several DNA polymerases during TLS. It remains to be determined how this particular enzyme is selected among the numerous other specialized polymerases. (D) Incorporation of just one or two nucleotides allows bypass of the template base damage, a process called *TLS insertion*. (E) The model postulates that pol? (possibly in a monoubuitylated state) is then recruited to interact with the C-terminal region of Rev1 protein. This interaction may also involve monoubiquitylated PCNA. (F) These interactions allow for a second polymerase switch, between pol? and polζ. (G) The second specialized DNA polymerase now extends the primer (*TLS extension*) to a location sufficiently downstream of the site of base damage such that when the replication machinery reengages in high fidelity DNA synthesis (H), nucleotides incorporated during the processes of TLS insertion and TLS extension are not removed by the proof-reading exonucleases intrinsic to the high fidelity polymerases. Reengagement of the replicative machinery at the primer-template requires a third polymerase switch.

Besides PCNA, the Y-family polymerases POLH, POLI, POLK and REV1L can themselves become monoubiquitylated [26,66]. The requisite E3 ligase(s) for these events have not been identified. However, all these polymerases contain conserved ubiquitin-binding motifs (UBMs or UBZ's) that are required for binding ubiquitin (and presumably for monoubiquitylation) *in vitro* [26]. These domains are also required for robust interaction with monoubiquitylated PCNA, suggesting that the robust polymerase-PCNA interactions may transpire between monoubiquitylated specialized polymerases and monoubiquitylated PCNA. Such a scenario affords the possibility that reversible monoubiquitylation of one or more of these proteins (involving specific deubiquitylating enzymes) regulate DNA polymerase switching during TLS. A model for DNA polymerase switching in mammalian cells based on the observations detailed above is presented in Fig. 3.

Interactions between REV1 and various specialized DNA polymerases has not been reported in lower eukaryotes, nor is the amino acid sequence in the C-terminal region of REV1 as well conserved as in vertebrates. Negative results *in vitro* must obviously be cautiously interpreted. For instance, REV1 may indeed interact with specialized polymerases *in vivo*, but through one or more intermediary proteins rather than directly, as is apparently the case in vertebrate cells. Nonetheless, is entirely possible that TLS in lower eukaryotes is mechanistically distinct from that in higher organisms, in which case the precise role of REV1 protein in TLS in organisms such as yeast and Drosophila may require reevaluation.

Potential Immunological Functions for Specialized DNA Polymerases

The discovery of specialized DNA polymerases has yielded significant inroads into our understanding of somatic hypermutation (SH) and class switching (CS) in the immune system. After exposure to antigen, B-lymphocytes expressing IgM with affinity to the antigen proliferate and genomic rearrangement results in antibody class switching. During this time the variable region experiences a higher mutation rate described as SH [67]. It is known that activation-induced cytosine deaminase (AICDA, formerly AID) protein is important in this process [68,69] promoting deamination of cytosine to uracil [70-72]. Since the deamination of cytosine is the means by with mutations are introduced, C>G mutations should predominate. However, A>T mutation rate is similar to the C>G mutation rate [73]. There is now evidence that mismatch repair proteins and TLS polymerases are involved in this process (reviewed in [74]). Mouse models with mutated DNA polymerases have all been tested for SH rates and mutations in *Polh* [37], *Rev1l* [75], and *Polq* [44] also have defective SH. In addition, mouse B cell lines with reduced expression of *Rev3l* manifest reduced SH [76,77]. It is not known how specific specialized DNA polymerases are recruited to the variable region of immunoglobulin genes.

Conclusion

The relief of arrested DNA replication through translesion DNA synthesis (TLS) catalyzed by multiple specialized DNA polymerases is a fundamental DNA damage tolerance mechanism. This process helps mitigate cell death associated with prolonged S phase arrest, and we presume that this outcome was a primary selective force during the evolution of this class of proteins. However, while the relaxed fidelity of DNA synthesis that is intrinsic to these enzymes facilitates the bypass of arresting lesions in DNA, this property also promotes frequent incorporation of incorrect nucleotides, leading to mutations. TLS of spontaneous base damage may thus be a primary mechanism of spontaneous mutagenesis in eukaryotic cells.

Much remains to be learned about the mechanism of TLS, especially the events that determine the multiple polymerase switching events that are intrinsic to the process. The establishment of a cell-free system that accurately mirrors TLS *in vivo* would likely contribute to our understanding of TLS, but is not a trivial undertaking. Meanwhile, meticulous dissection of the genetic requirements for TLS will facilitate biochemical studies. Finally, it is cogent to recall that aside from their roles in TLS, there is increasing evidence that specialized DNA polymerases participate in other biological events that require DNA synthesis, notably the process of somatic hypermutation.

REFERENCES

[1] Friedberg, E.C., et al., *DNA Repair And Mutagenesis*. 2nd Edition ed. 2005, Washington D.C.: ASM Press. 1118.

[2] Witkin, E.M., Inherited Differences in Sensitivity to Radiation in Escherichia Coli. *Proc Natl Acad Sci U S A*, 1946. 32(3): p. 59-68.

[3] Witkin, E.M., Radiation-induced mutations and their repair. *Science*, 1966. 152(727): p. 1345-53.

[4] Witkin, E.M., Mutation and the repair of radiation damage in bacteria. *Radiat Res*, 1966: p. Suppl 6:30-53.

[5] Witkin, E.M., The mutability toward ultraviolet light of recombination-deficient strains of Escherichia coli. *Mutat Res*, 1969. 8(1): p. 9-14.

[6] Kato, T. and Shinoura, Y., Isolation and characterization of mutants of Escherichia coli deficient in induction of mutations by ultraviolet light. *Mol Gen Genet*, 1977. 156(2): p. 121-31.

[7] Elledge, S.J. and Walker, G.C., Proteins required for ultraviolet light and chemical mutagenesis. Identification of the products of the umuC locus of Escherichia coli. *J Mol Biol*, 1983. 164(2): p. 175-92.

[8] Burckhardt, S.E., et al., UmuD mutagenesis protein of Escherichia coli: overproduction, purification, and cleavage by RecA. *Proc Natl Acad Sci U S A*, 1988. 85(6): p. 1811-5.

[9] Petit, M.A., et al., Sequential folding of UmuC by the Hsp70 and Hsp60 chaperone complexes of Escherichia coli. *J Biol Chem*, 1994. 269(38): p. 23824-9.

[10] Rajagopalan, M., et al., Activity of the purified mutagenesis proteins UmuC, UmuD', and RecA in replicative bypass of an abasic DNA lesion by DNA polymerase III. *Proc Natl Acad Sci U S A*, 1992. 89(22): p. 10777-81.

[11] Woodgate, R., et al., UmuC mutagenesis protein of Escherichia coli: purification and interaction with UmuD and UmuD'. *Proc Natl Acad Sci U S A*, 1989. 86(19): p. 7301-5.

[12] Reuven, N.B., et al., The mutagenesis protein UmuC is a DNA polymerase activated by UmuD', RecA, and SSB and is specialized for translesion replication. *J Biol Chem*, 1999. 274(45): p. 31763-6.

[13] Wagner, J., et al., The dinB gene encodes a novel E. coli DNA polymerase, DNA pol IV, involved in mutagenesis. *Mol Cell*, 1999. 4(2): p. 281-6.

[14] Ling, H., et al., Crystal structure of a Y-family DNA polymerase in action: a mechanism for error-prone and lesion-bypass replication. *Cell*, 2001. 107(1): p. 91-102.

[15] Silvian, L.F., et al., Crystal structure of a DinB family error-prone DNA polymerase from Sulfolobus solfataricus. *Nat Struct Biol*, 2001. 8(11): p. 984-9.

[16] Zhou, B.L., Pata, J.D., and Steitz, T.A., Crystal structure of a DinB lesion bypass DNA polymerase catalytic fragment reveals a classic polymerase catalytic domain. *Mol Cell*, 2001. 8(2): p. 427-37.

[17] Friedberg, E.C., Wagner, R., and Radman, M., Specialized DNA polymerases, cellular survival, and the genesis of mutations. *Science*, 2002. 296(5573): p. 1627-30.

[18] Fuchs, R.P., Fujii, S., and Wagner, J., Properties and functions of Escherichia coli: Pol IV and Pol V. *Adv Protein Chem*, 2004. 69: p. 229-64.

[19] Gerlach, V.L., et al., Purification and characterization of pol kappa, a DNA polymerase encoded by the human DINB1 gene. *J Biol Chem*, 2001. 276(1): p. 92-8.

[20] Lin, W., et al., The human REV1 gene codes for a DNA template-dependent dCMP transferase. *Nucleic Acids Res*, 1999. 27(22): p. 4468-75.

[21] Nair, D.T., et al., Rev1 employs a novel mechanism of DNA synthesis using a protein template. *Science*, 2005. 309(5744): p. 2219-22.

[22] Fischhaber, P.L., et al., Human DNA polymerase kappa bypasses and extends beyond thymine glycols during translesion synthesis in vitro, preferentially incorporating correct nucleotides. *J Biol Chem*, 2002. 277(40): p. 37604-11.

[23] Suzuki, N., et al., Translesional synthesis past acetylaminofluorene-derived DNA adducts catalyzed by human DNA polymerase kappa and Escherichia coli DNA polymerase IV. *Biochemistry*, 2001. 40(50): p. 15176-83.

[24] Vaisman, A. and Chaney, S.G., The efficiency and fidelity of translesion synthesis past cisplatin and oxaliplatin GpG adducts by human DNA polymerase beta. *J Biol Chem*, 2000. 275(17): p. 13017-25.

[25] Burnouf, D.Y., Miturski, R., and Fuchs, R.P., Sequence context modulation of translesion synthesis at a single N-2-acetylaminofluorene adduct located within a mutation hot spot. *Chem Res Toxicol*, 1999. 12(2): p. 144-50.

[26] Bienko, M., et al., Ubiquitin-binding domains in Y-family polymerases regulate translesion synthesis. *Science*, 2005. 310(5755): p. 1821-4.

[27] Sutton, M.D., Narumi, I., and Walker, G.C., Posttranslational modification of the umuD-encoded subunit of Escherichia coli DNA polymerase V regulates its interactions with the beta processivity clamp. *Proc Natl Acad Sci U S A*, 2002. 99(8): p. 5307-12.

[28] Johnson, R.E., Prakash, S., and Prakash, L., Efficient bypass of a thymine-thymine dimer by yeast DNA polymerase, Poleta. *Science*, 1999. 283(5404): p. 1001-4.

[29] Yuan, F., et al., Specificity of DNA lesion bypass by the yeast DNA polymerase eta. *J Biol Chem,* 2000. 275(11): p. 8233-9.

[30] Zhang, Y., et al., Error-free and error-prone lesion bypass by human DNA polymerase kappa in vitro. *Nucleic Acids Res*, 2000. 28(21): p. 4138-46.

[31] Masutani, C., et al., Mechanisms of accurate translesion synthesis by human DNA polymerase eta. *Embo J*, 2000. 19(12): p. 3100-9.

[32] Vaisman, A., et al., Efficient translesion replication past oxaliplatin and cisplatin GpG adducts by human DNA polymerase eta. *Biochemistry*, 2000. 39(16): p. 4575-80.

[33] Boiteux, S. and Guillet, M., Abasic sites in DNA: repair and biological consequences in Saccharomyces cerevisiae. *DNA Repair (Amst),* 2004. 3(1): p. 1-12.

[34] Ogi, T. and Lehmann, A.R., The Y-family DNA polymerase kappa (pol kappa) functions in mammalian nucleotide-excision repair. *Nature Cell Biology*, 2006. 8: p. 640-642.

[35] Otterlei, M., et al., Repair of chromosomal abasic sites in vivo involves at least three different repair pathways. *Embo J,* 2000. 19(20): p. 5542-51.

[36] Wu, X., et al., Critical roles for polymerase zeta in cellular tolerance to nitric oxide-induced DNA damage. *Cancer Res*, 2006. 66(2): p. 748-54.

[37] Delbos, F., et al., Contribution of DNA polymerase eta to immunoglobulin gene hypermutation in the mouse. *J Exp Med*, 2005. 201(8): p. 1191-6.

[38] Lin, Q., et al., Increased susceptibility to UV-induced skin carcinogenesis in polymerase eta-deficient mice. *Cancer Res*, 2006. 66(1): p. 87-94.

[39] Shimizu, T., et al., The absence of DNA polymerase kappa does not affect somatic hypermutation of the mouse immunoglobulin heavy chain gene. *Immunol Lett*, 2003. 86(3): p. 265-70.

[40] Schenten, D., et al., DNA polymerase kappa deficiency does not affect somatic hypermutation in mice. *Eur J Immunol*, 2002. 32(11): p. 3152-60.

[41] Mcdonald, J.P., et al., 129-derived strains of mice are deficient in DNA polymerase iota and have normal immunoglobulin hypermutation. *J Exp Med*, 2003. 198(4): p. 635-43.

[42] Kobayashi, Y., et al., Hydrocephalus, situs inversus, chronic sinusitis, and male infertility in DNA polymerase lambda-deficient mice: possible implication for the pathogenesis of immotile cilia syndrome. *Mol Cell Biol*, 2002. 22(8): p. 2769-76.

[43] Bertocci, B., et al., Cutting edge: DNA polymerases mu and lambda are dispensable for Ig gene hypermutation. *J Immunol*, 2002. 168(8): p. 3702-6.

[44] Masuda, K., et al., DNA polymerase theta contributes to the generation of C/G mutations during somatic hypermutation of Ig genes. *Proc Natl Acad Sci U S A*, 2005. 102(39): p. 13986-91.

[45] Jansen, J.G., et al., The BRCT domain of mammalian Rev1 is involved in regulating DNA translesion synthesis. *Nucleic Acids Res*, 2005. 33(1): p. 356-65.

[46] Faili, A., et al., DNA polymerase eta is involved in hypermutation occurring during immunoglobulin class switch recombination. *J Exp Med*, 2004. 199(2): p. 265-70.

[47] Vaisman, A., et al., Sequence context-dependent replication of DNA templates containing UV-induced lesions by human DNA polymerase iota. *DNA Repair (Amst),* 2003. 2(9): p. 991-1006.

[48] Bi, X., et al., DNA polymerase Kappa is specifically required for recovery from the benzo[a]pyrene-Di-hydrodiol epoxide (BPDE)-induced S-phase checkpoint. *J Biol Chem*, 2005.

[49] Ogi, T., et al., Polkappa protects mammalian cells against the lethal and mutagenic effects of benzo[a]pyrene. *Proc Natl Acad Sci U S A*, 2002. 99(24): p. 15548-53.

[50] Zhang, Y., et al., Activities of human DNA polymerase kappa in response to the major benzo[a]pyrene DNA adduct: error-free lesion bypass and extension synthesis from opposite the lesion. *DNA Repair (Amst)*, 2002. 1(7): p. 559-69.

[51] Ohashi, E., et al., Error-prone bypass of certain DNA lesions by the human DNA polymerase kappa. *Genes Dev*, 2000. 14(13): p. 1589-94.

[52] Ohashi, E., et al., Fidelity and processivity of DNA synthesis by DNA polymerase kappa, the product of the human DINB1 gene. *J Biol Chem*, 2000. 275(50): p. 39678-84.

[53] Takenaka, K., et al., Involvement of vertebrate Polkappa in translesion DNA synthesis across DNA monoalkylation damage. *J Biol Chem*, 2006. 281(4): p. 2000-4.

[54] Velasco-Miguel, S., et al., Constitutive and regulated expression of the mouse Dinb (Polkappa) gene encoding DNA polymerase kappa. *DNA Repair (Amst)*, 2003. 2(1): p. 91-106.

[55] Shibutani, S., Suzuki, N., and Grollman, A.P., Mechanism of frameshift (deletion) generated by acetylaminofluorene-derived DNA adducts in vitro. *Biochemistry*, 2004. 43(50): p. 15929-35.

[56] Kannouche, P., et al., Localization of DNA polymerases eta and iota to the replication machinery is tightly co-ordinated in human cells. *Embo J*, 2002. 21(22): p. 6246-56.

[57] Guo, C., et al., Mouse Rev1 protein interacts with multiple DNA polymerases involved in translesion DNA synthesis. *Embo J*, 2003. 22(24): p. 6621-30.

[58] Vidal, A.E., et al., Proliferating cell nuclear antigen-dependent coordination of the biological functions of human DNA polymerase iota. *J Biol Chem*, 2004. 279(46): p. 48360-8.

[59] Haracska, L., et al., Physical and functional interactions of human DNA polymerase eta with PCNA. *Mol Cell Biol*, 2001. 21(21): p. 7199-206.

[60] Haracska, L., et al., Stimulation of DNA synthesis activity of human DNA polymerase kappa by PCNA. *Mol Cell Biol*, 2002. 22(3): p. 784-91.

[61] Maga, G., et al., Human DNA polymerase lambda functionally and physically interacts with proliferating cell nuclear antigen in normal and translesion DNA synthesis. *J Biol Chem*, 2002. 277(50): p. 48434-40.

[62] Garg, P., et al., Proliferating cell nuclear antigen promotes translesion synthesis by DNA polymerase zeta. *J Biol Chem*, 2005. 280(25): p. 23446-50.

[63] Ross, A.L., Simpson, L.J., and Sale, J.E., Vertebrate DNA damage tolerance requires the C-terminus but not BRCT or transferase domains of REV1. *Nucleic Acids Res*, 2005. 33(4): p. 1280-9.

[64] Haracska, L., et al., Opposing effects of ubiquitin conjugation and SUMO modification of PCNA on replicational bypass of DNA lesions in Saccharomyces cerevisiae. *Mol Cell Biol*, 2004. 24(10): p. 4267-74.

[65] Haracska, L., et al., Ubiquitylation of yeast proliferating cell nuclear antigen and its implications for translesion DNA synthesis. *Proc Natl Acad Sci U S A*, 2006. 103(17): p. 6477-82.

[66] Guo, C., et al., Multiple PolK (POLK) transcripts in mammalian testis. *DNA Repair (Amst)*, 2005. 4(3): p. 397-402.

[67] Rajewsky, K., Clonal selection and learning in the antibody system. *Nature*, 1996. 381(6585): p. 751-8.

[68] Muramatsu, M., et al., Class switch recombination and hypermutation require activation-induced cytidine deaminase (AID), a potential RNA editing enzyme. *Cell*, 2000. 102(5): p. 553-63.

[69] Yoshikawa, K., et al., AID enzyme-induced hypermutation in an actively transcribed gene in fibroblasts. *Science,* 2002. 296(5575): p. 2033-6.

[70] Chaudhuri, J., et al., Transcription-targeted DNA deamination by the AID antibody diversification enzyme. *Nature*, 2003. 422(6933): p. 726-30.

[71] Petersen-Mahrt, S.K., Harris, R.S., and Neuberger, M.S., AID mutates E. coli suggesting a DNA deamination mechanism for antibody diversification. *Nature*, 2002. 418(6893): p. 99-103.

[72] Pham, P., et al., Processive AID-catalysed cytosine deamination on single-stranded DNA simulates somatic hypermutation. *Nature*, 2003. 424(6944): p. 103-7.

[73] Diaz, M. and Flajnik, M.F., Evolution of somatic hypermutation and gene conversion in adaptive immunity. *Immunol Rev*, 1998. 162: p. 13-24.

[74] Diaz, M. and Lawrence, C., An update on the role of translesion synthesis DNA polymerases in Ig hypermutation. *Trends Immunol*, 2005. 26(4): p. 215-20.

[75] Jansen, J.G., et al., Strand-biased defect in C/G transversions in hypermutating immunoglobulin genes in Rev1-deficient mice. *J Exp Med*, 2006. 203(2): p. 319-23.

[76] Diaz, M., et al., Decreased frequency of somatic hypermutation and impaired affinity maturation but intact germinal center formation in mice expressing antisense RNA to DNA polymerase zeta. *J Immunol*, 2001. 167(1): p. 327-35.

[77] Zan, H., et al., The translesion DNA polymerase zeta plays a major role in Ig and bcl-6 somatic hypermutation. *Immunity*, 2001. 14(5): p. 643-53.

[78] Kow, Y.W., et al., Mutagenic effects of abasic and oxidized abasic lesions in Saccharomyces cerevisiae. *Nucleic Acids Res*, 2005. 33(19): p. 6196-202.

[79] Zhang, Y., et al., Response of human REV1 to different DNA damage: preferential dCMP insertion opposite the lesion. *Nucleic Acids Res*, 2002. 30(7): p. 1630-8.

[80] Washington, M.T., et al., Efficient and error-free replication past a minor-groove N2-guanine adduct by the sequential action of yeast Rev1 and DNA polymerase zeta. *Mol Cell Biol*, 2004. 24(16): p. 6900-6.

[81] Cordonnier, A.M., Lehmann, A.R., and Fuchs, R.P., Impaired translesion synthesis in xeroderma pigmentosum variant extracts. *Mol Cell Biol*, 1999. 19(3): p. 2206-11.

[82] Kozmin, S.G., et al., Roles of Saccharomyces cerevisiae DNA polymerases Poleta and Polzeta in response to irradiation by simulated sunlight. *Nucleic Acids Res*, 2003. 31(15): p. 4541-52.

[83] Raha, M., et al., Mutagenesis by third-strand-directed psoralen adducts in repair-deficient human cells: high frequency and altered spectrum in a xeroderma pigmentosum variant. *Proc Natl Acad Sci U S A*, 1996. 93(7): p. 2941-6.

[84] Zhao, B., et al., Role of DNA polymerase eta in the bypass of abasic sites in yeast cells. *Nucleic Acids Res*, 2004. 32(13): p. 3984-94.

[85] Cordonnier, A.M. and Fuchs, R.P., Replication of damaged DNA: molecular defect in xeroderma pigmentosum variant cells. *Mutat Res*, 1999. 435(2): p. 111-9.

[86] Gu, C. and Wang, Y., LC-MS/MS identification and yeast polymerase eta bypass of a novel gamma-irradiation-induced intrastrand cross-link lesion G[8-5]C. *Biochemistry*, 2004. 43(21): p. 6745-50.

[87] Zhang, Y., et al., Error-prone lesion bypass by human DNA polymerase eta. *Nucleic Acids Res*, 2000. 28(23): p. 4717-24.

[88] Kuraoka, I., et al., Oxygen free radical damage to DNA. Translesion synthesis by human DNA polymerase eta and resistance to exonuclease action at cyclopurine deoxynucleoside residues. *J Biol Chem*, 2001. 276(52): p. 49283-8.

[89] Suzuki, N., et al., Translesion synthesis past estrogen-derived DNA adducts by human DNA polymerases eta and kappa. *Biochemistry*, 2004. 43(20): p. 6304-11.

[90] Suzuki, N., et al., Translesion synthesis past equine estrogen-derived 2'-deoxycytidine DNA adducts by human DNA polymerases eta and kappa. *Biochemistry*, 2004. 43(35): p. 11312-20.

[91] Choi, J.Y. and Guengerich, F.P., Adduct size limits efficient and error-free bypass across bulky N2-guanine DNA lesions by human DNA polymerase eta. *J Mol Biol*, 2005. 352(1): p. 72-90.

[92] Perrino, F.W., et al., Polymerization past the N2-isopropylguanine and the N6-isopropyladenine DNA lesions with the translesion synthesis DNA polymerases eta and iota and the replicative DNA polymerase alpha. *Chem Res Toxicol*, 2005. 18(9): p. 1451-61.

[93] Vaisman, A., et al., DNA polymerase iota-dependent translesion replication of uracil containing cyclobutane pyrimidine dimers. *DNA Repair (Amst)*, 2006. 5(2): p. 210-8.

[94] Hang, B., et al., Miscoding properties of 1,N6-ethanoadenine, a DNA adduct derived from reaction with the antitumor agent 1,3-bis(2-chloroethyl)-1-nitrosourea. *Mutat Res*, 2003. 531(1-2): p. 191-203.

[95] Roush, A.A., et al., Deletion of the Saccharomyces cerevisiae gene RAD30 encoding an Escherichia coli DinB homolog confers UV radiation sensitivity and altered mutability. *Mol Gen Genet*, 1998. 257(6): p. 686-92.

[96] Napolitano, R., et al., All three SOS-inducible DNA polymerases (Pol II, Pol IV and Pol V) are involved in induced mutagenesis. *Embo J*, 2000. 19(22): p. 6259-65.

[97] Maor-Shoshani, A., et al., Analysis of translesion replication across an abasic site by DNA polymerase IV of Escherichia coli. *DNA Repair (Amst)*, 2003. 2(11): p. 1227-38.

[98] Johnson, R.E., et al., Eukaryotic polymerases iota and zeta act sequentially to bypass DNA lesions. *Nature*, 2000. 406(6799): p. 1015-9.

[99] Tissier, A., et al., Misinsertion and bypass of thymine-thymine dimers by human DNA polymerase iota. *Embo J*, 2000. 19(19): p. 5259-66.

[100] Zhang, Y., et al., Response of human DNA polymerase iota to DNA lesions. *Nucleic Acids Res*, 2001. 29(4): p. 928-35.

[101] Vaisman, A. and Woodgate, R., Unique misinsertion specificity of poliota may decrease the mutagenic potential of deaminated cytosines. *Embo J*, 2001. 20(22): p. 6520-9.

[102] Rechkoblit, O., et al., trans-Lesion synthesis past bulky benzo[a]pyrene diol epoxide N2-dG and N6-dA lesions catalyzed by DNA bypass polymerases. *J Biol Chem*, 2002. 277(34): p. 30488-94.

[103] Tang, M., et al., Roles of E. coli DNA polymerases IV and V in lesion-targeted and untargeted SOS mutagenesis. *Nature*, 2000. 404(6781): p. 1014-8.

[104] Budd, M.E. and Campbell, J.L., Purification and enzymatic and functional characterization of DNA polymerase beta-like enzyme, POL4. expressed during yeast meiosis. *Methods Enzymol*, 1995. 262: p. 108-30.

[105] Bebenek, K., et al., Biochemical properties of Saccharomyces cerevisiae DNA polymerase IV. *J Biol Chem*, 2005. 280(20): p. 20051-8.

[106] Tang, M., et al., UmuD'(2)C is an error-prone DNA polymerase, Escherichia coli pol V. *Proc Natl Acad Sci U S A*, 1999. 96(16): p. 8919-24.

[107] Zhang, Y., et al., Lesion bypass activities of human DNA polymerase mu. *J Biol Chem*, 2002. 277(46): p. 44582-7.

[108] Havener, J.M., et al., Translesion synthesis past platinum DNA adducts by human DNA polymerase mu. *Biochemistry*, 2003. 42(6): p. 1777-88.

[109] Rabkin, S.D. and Strauss, B.S., A role for DNA polymerase in the specificity of nucleotide incorporation opposite N-acetyl-2-aminofluorene adducts. *J Mol Biol*, 1984. 178(3): p. 569-94.

[110] Hoffmann, J.S., et al., DNA polymerase beta bypasses in vitro a single d(GpG)-cisplatin adduct placed on codon 13 of the HRAS gene. *Proc Natl Acad Sci U S A*, 1995. 92(12): p. 5356-60.

[111] Latham, G.J., et al., The efficiency of translesion synthesis past single styrene oxide DNA adducts in vitro is polymerase-specific. *Chem Res Toxicol*, 1995. 8(3): p. 422-30.

[112] Hashim, M.F., Schnetz-Boutaud, N., and Marnett, L.J., Replication of template-primers containing propanodeoxyguanosine by DNA polymerase beta. Induction of base pair substitution and frameshift mutations by template slippage and deoxynucleoside triphosphate stabilization. *J Biol Chem*, 1997. 272(32): p. 20205-12.

[113] Singh, J. and Snow, E.T., Chromium (III) decreases the fidelity of human DNA polymerase beta. *Biochemistry*, 1998. 37(26): p. 9371-8.

[114] Servant, L., et al., A role for DNA polymerase beta in mutagenic UV lesion bypass. *J Biol Chem*, 2002. 277(51): p. 50046-53.

[115] Nelson, J.R., Lawrence, C.W., and Hinkle, D.C., Thymine-thymine dimer bypass by yeast DNA polymerase zeta. *Science*, 1996. 272(5268): p. 1646-9.

[116] Chaney, S.G. and Vaisman, A., Specificity of platinum-DNA adduct repair. *J Inorg Biochem*, 1999. 77(1-2): p. 71-81.

[117] Guo, D., et al., Translesion synthesis by yeast DNA polymerase zeta from templates containing lesions of ultraviolet radiation and acetylaminofluorene. *Nucleic Acids Res*, 2001. 29(13): p. 2875-83.

[118] Li, Z., et al., hREV3 is essential for error-prone translesion synthesis past UV or benzo[a]pyrene diol epoxide-induced DNA lesions in human fibroblasts. *Mutat Res*, 2002. 510(1-2): p. 71-80.
[119] Johnson, R.E., et al., Yeast DNA polymerase zeta (zeta) is essential for error-free replication past thymine glycol. *Genes Dev*, 2003. 17(1): p. 77-87.
[120] Seki, M., et al., High-efficiency bypass of DNA damage by human DNA polymerase Q. *Embo J*, 2004. 23(22): p. 4484-94.
[121] Boyd, J.B. and Harris, P.V., Mutants partially defective in excision repair at five autosomal loci in Drosophila melanogaster. *Chromosoma*, 1981. 82(2): p. 249-57.

In: New Research on DNA Repair
Editor: Breehn R. Landseer, pp. 191-217

ISBN 1-60021-385-5

Chapter VII

LONG-TERM RNA INTERFERENCE: TOWARD A SET OF ISOGENEIC HUMAN CELLS DEFICIENT IN DNA REPAIR GENES

D.S.F. Biard and J.F. Angulo*

Commissariat à l'Energie Atomique. Laboratoire de Génétique de la Radiosensibilité, Département de Radiobiologie et de Radiopathologie, Direction des Sciences du Vivant, Fontenay-aux-Roses 92265, France

ABSTRACT

The functional characterization of mammalian radiosensitivity genes has been greatly facilitated by using rodent cells derived from mutants sensitive to either ultraviolet or ionizing radiation and has led to the molecular identification of two important groups of genes called ERCC and XRCC. Numerous genetic studies have used cells from patients with hereditary disorders. However, the genetic heterogeneity of human samples remains one of the greatest impediments to the identification and characterization of human radiosensitivity genes. Over the last two years, RNA interference (RNAi) has emerged as a promising technology in biological science which has broad applications, in particular in the field of reverse genetics. RNAi is the process by which double-strand RNA (dsRNA) silences gene expression by inducing the sequence-specific degradation of the complementary mRNA. RNAi is becoming a state-of-the-art method whereby researchers can knock down the expression of a specific gene in order to untangle its biological functions. We have developed new pEBVsiRNA plasmids to impose stable gene silencing in human cells. These plasmids, which remain as independent replicons in transfected human cells, greatly enhance RNAi by allowing efficient, specific and very long-term gene silencing, even after more than 500 days in culture. We successfully silenced several genes involved in the nucleotide excision repair pathway (e.g. *XPA*, *XPC*, *hHR23A or hHR23B),* in the non-homologous end-joining

* Correspondence concerning this article should be addressed to Denis S.F. Biard.CEA DSV DRR, Laboratoire de Génétique de la Radiosensibilité, BP 6, 92265 Fontenay aux Roses, France. E-mail: denis.biard@cea.fr.

pathway (e.g. *DNA-PKcs* and *XRCC4)* and in the homologous recombination repair pathway (e.g. *Rad52).* Other genes involved in signaling and other types of recovery after irradiation (e.g. *P53*, *MRE11*, *Rad50*, *NBS1*) were also targeted. Stable knock-down clones were propagated in culture and monitored for several months. They displayed the expected DNA repair deficiency several months after transfection. At present, we are constructing more than 140 vectors to target 30 genes. We are using these cells to unravel the complex regulatory relationships between the different DNA repair pathways. For the first time, we have created a set of stable isogeneic human cells deficient in DNA repair genes without the use of viral vehicle.

Keywords: siRNA, shRNA, gene silencing, EBV vector, NER, NHEJ, DNA repair

INTRODUCTION

Cell genomes are threatened by a plethora of environmental genotoxic agents (ultraviolet light, ionizing radiation, hazardous chemicals) and endogenous reactive chemical species (by-products of cellular metabolism). The cellular response to DNA damage encompasses different coordinated pathways, including damage prevention (e.g. detoxification metabolism), damage recognition (signal-transduction pathways with both *sensor* and *transducer* proteins), and ultimately the response to the DNA damage *per se* (cell cycle checkpoints, DNA repair, apoptosis) (Zhang et al., 2006). For many years, a great challenge has been to uncover the different proteins involved in the response to DNA damage. A large body of evidence demonstrates that impaired responses to genotoxic insults could be involved in different human hereditary disorders and in cancer development. The critical role of DNA repair proteins in the hypersensitivity of cancer patients to radiation therapy has also been highlighted. The challenge is to create relevant cellular models to elucidate the damage-response mechanisms and their roles in the early stages of human cancer development.

Overview of Rodent Models for Studying DNA Repair Genes

To identify and characterize DNA repair genes, different cellular models have been developed, in particular genotoxic-sensitive rodent cell lines. For instance, mutants of CHO cells that are defective in the NER pathway provided the first tool for isolating human DNA repair genes. The genetic complementation of UV-sensitive rodent mutants has allowed the characterization of genes involved in DNA repair. These genes were termed "Excision Repair Cross Complementing" (ERCC) genes because they allowed the retrieval of the resistance phenotype (Hoeijmakers, 1993). The first member of this group, the ERCC1 gene (homologous to the yeast RAD10) was identified in 1986 for its complementation of CHO UV20 mutant cells (van Duin et al., 1986). Because of the intimate involvement of these genes in human cancer, we describe ERCC genes in the next section.

In the course of characterizing DNA repair genes, other studies have examined genes involved in double-strand break (DSB) repair (DSBR). DSBs are deadly damage which must be efficiently repaired in order to safeguard genomic integrity. The use of mutant mice and hamster cells (CHO, V79) exhibiting hypersensitivity to ionizing radiation or to radiomimetic agents led to the identification of 11 X-ray cross complementing (*XRCC*) groups (for review Kraakman-van der Zwet et al., 2002; Thacker and Zdzienicka, 2003; Thacker and Zdzienicka, 2004). *XRCC1*, the first mammalian gene shown to affect cellular sensitivity to ionizing radiation, was cloned in 1990 and its function investigated (Thompson et al., 1990). Because ionizing radiation causes a wide range of DNA lesions, no single repair process can deal with all kinds of damage. As a consequence, the XRCC genes were found to belong to different DNA repair pathways such as BER (*Base Excision Repair*), HR (*Homologous Recombination*), or NHEJ (*Non-Homologous End Joining*). XRCC1 participates in BER; XRCC2 and XRCC3 (two members of the RecA/Rad51-related protein family), XRCC11 (BRCA2/FANCD1) and XRCC9 (FANCG) belongs to HR pathway; XRCC4, XRCC5 (Ku80), XRCC6 (Ku70) and XRCC7 (DNAPKcs) are building blocks of NHEJ, while XRCC8 (with AT and NBS1-like characteristics) could be involved in DNA damage signaling pathways.

Another interesting biological source of rodent "mutants" was obtained after targeted gene inactivation: "knock-out" mouse models. This approach allowed the accumulation of a number of mouse mutant strains covering all aspects of DNA repair and cellular responses to DNA damage (Friedberg and Meira, 2004). In particular, these studies outlined the close connection between the different enzymatic machineries coordinating the DNA repair pathways and other DNA transactions such as DNA replication, transcription, cell cycle progression and apoptosis.

New Human Cell Models

The exclusive source of DNA repair deficient cells available in a human context arises from hereditary radiosensitive disorders and cancer cells (Shiloh, 2003). These cells from patients have been used to demonstrate that mutations in genes encoding proteins involved in different DNA repair pathways, such as NER (*Nucleotide Excision Repair*), MMR (*MisMatch Repair*), NHEJ and BER, may cause chromosome and genetic instability, telomere defects, and cancer predisposition.

Inherited defects in the MMR (mainly due to mutated *MLH1*, *MSH2*, *MSH6* or *PMS2* genes) were implicated in the most prevalent cancer syndrome in humans, hereditary nonpolyposis colon cancer (HNPCC) (Peltomaki, 2003). The MMR is required to remove single-base mismatches and insertion-deletion loops that arise during DNA replication.

The NER pathway is involved in the repair of bulky, helix-distorting DNA lesions such as those arising after exposure to ultraviolet (UV) radiation or certain chemical treatments. Three human syndromes have been extensively characterized and exhibit a defect in the NER pathway: xeroderma pigmentosum (XP), Cockayne's syndrome (CS), and trichothiodystrophy (TTD). Some mutations in ERCC genes were identified to be components of the NER (Hoeijmakers, 1993). In XP patients, mutations affecting 8 genes were identified: *XPA*, *XPB*

(*ERCC3*), *XPC*, *XPD* (*ERCC2*), *XPE* (*DDB1*), XPF (ERCC4), XPG (*ERCC5*) and a "variant" form. In CS patients, 5 complementation groups were identified, *CSA* (*ERCC8*), *CSB* (*ERCC6*), *XPB*, *XPD* and *XPG*, and in TTD patients 3 complementation groups were found: *XPB*, *XPD* and *TTDA* (human TFB5 ortholog) (Giglia-Mari et al., 2004; Gillet and Scharer, 2006). The first human genetic disorder attributed to a BER defect was identified by Tassan and collaborators in a family with multiple colorectal adenoma and carcinoma predisposition, displaying a biallelic germline mutation in the *MUTYH* gene, the human homolog of the *E. coli mutY* gene encoding a DNA glycosylase involved in the repair of oxidative damage (Al-Tassan et al., 2002). Mutations in some NHEJ components were also identified, such as in the *ligase IV* gene (LIG4 syndrome), in *ARTEMIS* (RS-SCID or *human radiosensitive severe combined immunodeficiency syndrome*), in *DNA-PKcs* (M059J glioblastoma cells), and in *CERNUNNOS* (Callebaut et al., 2006). Chromosomal instability associated with impaired RecQ helicase function is associated with Bloom syndrome (BS), Werner syndrome (WS) and Rothmund-Thompson syndrome (RTS) (Duker, 2002; O'Driscoll et al., 2004; O'Driscoll and Jeggo, 2006).

Besides *p53* gene mutation, which is the most frequently encountered genetic event in human cancers, mutations are also found in genes coding for proteins which act as *sensor* proteins for DNA lesions and/or chromatin alterations, or *transducer* proteins for conveying the damage signal to downstream *effector* proteins. Hence, such hereditary mutations were found in ataxia-telangiectasia (AT; ATM protein), ATR-Seckel syndrome (ATR protein), AT-like disease (ATLD; Mre11 protein), Nijmegen breakage syndrome (NBS; Nbs1 protein), and Fanconi anemia (FAN-A, C, D1, D2, E, F, G proteins).

These examples emphasize the need for outstanding biological tools to study DNA repair genes. However, major constraints restrict this approach due to the limited number of human deficient cells and the large genetic variation in the human population. Moreover, only a few hereditary radiosensitive disorders have been identified so far (about 15 to 20). Therefore, it becomes necessary to establish a set of isogeneic human cell lines in which one specific gene is silenced by the RNA interference method. The challenge is to develop a new tool to obtain cells where one specific gene is silenced for a very long time (one year and more). As a consequence, this new tool will increase the short-, middle-, and long-term efficiency of RNAi in different cell types. For this purpose, we have developed and characterized replicative Epstein Barr-derived siRNA vectors (pEBVsiRNA) for efficient, specific and very long-term gene silencing in human cells (Biard et al., 2005). This approach will help us to untangle the interwoven DNA repair pathways in isogeneic human cells.

RNA Interference, the State-of-the Art

RNA interference (RNAi) is a coordinated cellular pathway that induces the silencing of specific endogenous and exogenous genes by the expression of small RNA in which one strand is complementary (perfectly or not) to a target mRNA (Fire et al., 1998; Zamore et al., 2000). During this process, long double-stranded RNAs (dsRNAs) are converted into 21 to 27 nt RNA molecules such as small interfering RNAs (siRNAs) by the RNA silencing machinery (Elbashir et al., 2001; Tomari and Zamore, 2005). In particular, the type III endoribonuclease Dicer generates siRNAs (McManus and Sharp, 2002; Schwarz et al., 2003). The antisense strand of the duplex siRNA guides the RISC (*RNA-induced silencing complex*) to the homologous mRNA, where a RISC-associated endoribonuclease (Argonaute 2) cleaves the target mRNA between the 10 and 11 nt from the 5'-phosphate of the guide strand, which results in the degradation of the target mRNA (Hammond et al., 2000). During the RISC assembly, the sense ("passenger") strand of the siRNA is cleaved according to the same rules established for the siRNA-guided cleavage of a target RNA (Leuschner et al., 2006).

Because RNAi is an endogenous natural pathway, its power as a gene-silencing tool is likely to be orders of magnitude greater than other approaches, like anti-sense RNA. In human cells, transfection of siRNA duplexes is very effective in the field of reverse genetics as shown by a plethora of articles that describe the relevance of the RNAi in abrogating gene expression. The crucial step in RNAi-imposed gene silencing remains the unfailing recognition of the cognate mRNA by the siRNA sequence. Many empirical guidelines for the design of powerful siRNAs have been developed (Mittal, 2004). Tuschl and colleagues described the initial design paradigm for synthesis of 21 nt siRNA, which efficiently suppresses gene expression without inducing the interferon pathway (Elbashir et al., 2001; Tuschl, 2001). More rational designs have been developed, based on a better understanding of RNAi biochemistry. In particular, the thermodynamic profile of the siRNA duplex termini greatly influences the incorporation of the antisense strand of an siRNA into the RISC to efficiently trigger the cleavage of the sense mRNA target (Khvorova et al., 2003; Schwarz et al., 2003). Considerations for generating effective and specific siRNAs have been listed (Mittal, 2004). We adopted the algorithm developed by Sonnhammer and collaborators (Chalk et al., 2004), which allowed us to successfully silence more than one dozen genes over several months of culture (and more), as described below.

RNA Interference in Human Cultured Cells

Most studies describing RNA interference in human cultured cells are based on transient transfections of siRNA duplexes or vectors. The cost and transient nature of siRNA duplexes greatly hamper this approach. In 2002, a pioneer plasmid allowing transient knock-down of one exogenous gene and one endogenous gene was described by Agami and collaborators (Brummelkamp et al., 2002). This plasmid, termed pSUPER, carries the H1 RNA polymerase III promoter (RNA pol III) to drive the transcription of short hairpin RNA (shRNA), giving rise to siRNA-like molecules *in vivo*. RNA pol III promoters are widely used to direct the expression of shRNAs because they are active in all cell types and efficiently direct the synthesis of small, non-coding transcripts (which are structurally close to shRNAs) that contain well-defined ends. This breakthrough afforded the opportunity to develop new strategies for imposing a more stable expression of RNAi in human cells.

In parallel, virus-expressing siRNA constructs were developed based on integrative retroviruses (e.g. lentiviruses) or episomal adenoviruses. However, adenovirus-based gene transfer initiates a host immune response which rapidly destroys adenoviral vector particles *in vivo*. These unwanted side effects depend on vector design, titer, host, and transgene (Fish and Kruithof, 2004). Besides, while human immunodeficiency virus type 1 (HIV-1)-derived lentiviruses are elegant tools for *in vivo* assays and transduction of cultured cells (including non-dividing cells and stem cells), major drawbacks have to be solved, concerning long-term efficiency, safety, and ethical concerns (Romano et al., 2003). The expected long-term transgene expression is due to the insertion of the proviral DNA into chromosomal DNA, which results in insertional mutagenesis, which has already been demonstrated in normal or malignant hematopoietic cells (Woods et al., 2003). Beside, the uncontrollable variation in positions of the proviral DNA integration might contribute to transcriptional silencing (mediated by DNA methylation and chromatin modification) that may compromise long-term transgene expression (Ellis, 2005; He et al., 2005). It appears that improved vector design will be the most straightforward way of increasing the safety of lentivirus vectors (Connolly, 2002; Romano et al., 2003).

In the strategy based on siRNA-expressing plasmids, authors always use integrative ones carrying (or not) a selection marker. In cells, most of these vectors have lost their siRNA cassette under selection pressure, which considerably hampers the establishment of stable cell populations. Hence, only a few stable clones are characterized for a reduced period of time in culture (Butz et al., 2003; Laposa et al., 2003; Paddison et al., 2002). Besides, a high copy number of siRNA-based plasmid per cell (or an elevated viral titer of lentivirus-mediated siRNA expression) may saturate the RNAi machinery (e.g. the RISC complex) and trigger unwanted side effects (Anonymous, 2003). Interestingly, siRNA-induced toxicity (through *apoptosis*) occurs in an siRNA dose-dependent manner and in a target-independent fashion (Fedorov et al., 2006). As a consequence, over-expression of siRNA through either siRNA duplexes or integrative (viral or not) vectors may lead to erroneous conclusions in transient experiments. To overcome these shortcomings, we harnessed the efficiency of the siRNA approach with Epstein Barr virus-derived (EBV) plasmids in order to establish cell lines carrying a few plasmid copies of per cell and expressing a given shRNA for a long time (> 500 days). To demonstrate the usefulness of this approach, we have constructed more than 140 pEBVsiRNA expression vectors targeting several genes involved in different DNA repair or signaling pathways. At present, about 12 genes are silenced for a long period of time, while 18 others are still under investigation.

Main Feature of EBV-Based Plasmids

pEBV-based plasmids have been used for many years to efficiently modify human cell genotypes (Biard et al., 1992; Calos, 1996). These plasmids contain the latent replication origin of EBV virus (*OriP*) and allow the expression of the viral transactivator EBV Nuclear Antigen-1 (EBNA-1). The minimal replicator sequence *OriP* is a 1.8 kbp fragment that consists of two clusters of cognate EBNA-1-binding sites: the family of repeats (FR) and the dyad symmetry (DS or DYAD) element (Aiyar et al., 1998; Lupton and Levine, 1985; Reisman et al., 1985). FR is a cluster of 20 binding sites for EBNA-1 which acts as an EBNA-1-dependent transcription enhancer and is responsible for the stable segregation and maintenance of EBV episomes (Wysokenski and Yates, 1989). The DS element contains four EBNA-1 binding sites and is the site at or near which initiation of DNA replication occurs (Collins and Medveczky, 2002). EBNA-1 tethers EBV episomes to metaphase chromosomes, providing the basis for their nuclear retention and their successful segregation at mitosis (Kapoor and Frappier, 2003; Marechal et al., 1999). In the presence of both EBNA-1 and *oriP*, EBV vectors persist in cell nuclei as stable episomes even after a long period of culture (Reisman et al., 1985). In human stem cells, pEBV plasmids are maintained as episomal DNA for several months in the presence of a drug selection marker and a gradual loss (2-4% per cell doubling) was observed after selection pressure withdrawal (Zhang et al., 2003). Interestingly, the presence of EBNA-1 does not interfere with the human ES cell properties or differentiation.

EBV DNA replication is semi-conservative and is coordinated with genomic replication (Dhar et al., 2001). Host cellular factors assist the EBV replication, such as Orc2, which tightly interacts with EBNA-1, suggesting that EBNA-1 may function as a recruiting factor for ORC to *OriP* (Chaudhuri et al., 2001; Schepers et al., 2001). The sequential binding and release of replication initiation proteins to *OriP* in different phases of the cell cycle follow kinetics similar to those of their association with global chromatin (Ritzi et al., 2003). Prereplicative-complex components such as Mcm2p-Mcm7p proteins and Orc1p are regulated in a cell-cycle dependent fashion, whereas other Orc subunits and EBNA-1 are constantly bound. These interactions restrict EBV replication to exactly one single round during each cell cycle, ensuring a low vector copy number per cell and high stability. Furthermore, *OriP* is flanked by nucleosomes which are subject to cell cycle-dependent chromatin remodeling and histone modifications. Histone H3 acetylation of DS-bound nucleosomes decreases during late G1, which coincides with nucleosome remodeling and Mcm3 loading, preceding the onset of DNA replication (Zhou et al., 2005). Therefore, oriP appears to be regulated like a human endogenous origin of DNA replication.

For more than two decades, pEBV vectors have been widely used in *in vitro* and *in vivo* experiments. The high stability of these vectors allows the incorporation of large inserts such as elements of the bacterial artificial chromosome (BAC). In 1999, Wade-Martins et al. demonstrated that a 117 kb episome could be retained at approximately 2 copies per cell and shuttled unrearranged from the human cells into bacterial cells, even after 15 months of continuous cell growth (Wade-Martins et al., 1999). Furthermore, pEBV vectors appear to be relevant in mouse cells, thus opening the opportunity to develop mouse models. While preliminary studies suggested that EBV replicons failed to replicate in mouse embryonic fibroblasts (MEFs), Einav and collaborators have observed that pEBV vectors are maintained unrearranged in MEFs six weeks after transfection (Einav et al., 2003). In the absence of selection pressure, these episomes persist with a half-life of 7–14 days. In addition, several *in vivo* approaches have also been attempted after injection through the vascular route. Strikingly, efficient transgene expression was observed in mouse lung endothelial cells 3 weeks after injection (Zhang et al., 2003).

In these conditions, it seemed relevant to us to assess the potential of pEBV vectors for imposing stable and long-term gene silencing in human cells. Because these vectors behave like endogenous transcription units, it was expected that they would tightly regulate siRNA expression over a long period of time.

pEBVsiRNA Vectors

We have reported the long-term silencing of both *XPA* and *XPC* genes in HeLa cells (Biard et al., 2005). XPA^{KD} or XPC^{KD} (*KD = knock-down*) syngeneic clones described in this article are still in culture after more than 500 days. Importantly, they are maintaining a near-complete gene silencing phenotype demonstrating that we are able to mimic the characteristics of cells from a human cancer-prone syndrome (XP) for an extended period in culture.

We have improved the cloning efficiency of pEBVsiRNA vectors by creating pEBVsiRNA-LacZ' (termed pBD751) (Figure 1). This plasmid allows blue/white screening to select bacteria colonies carrying a recombinant pEBVsiRNA plasmid that incorporates the DNA coding for an shRNA. As a consequence, shRNA cloning is nearly 100% efficient, thus reducing the time of selection and characterization of recombinant plasmids. For each shRNA sequence, we need to select only three "white" colonies for further characterization and purification, and one of them is currently being sequenced. More than 80% of recombinant pEBVsiRNA vectors contain the shRNA sequence (*without mutation*). The cloning procedure proposed here is easy, cheap and rapid since in about 7 days it is possible to construct the desired plasmid and to test it in human cells. Since the pBD751 plasmid carries hygromycin B resistance, we have also constructed its puromycin counterpart (pBD899). Puromycin resistance allows efficient selection within 24 h.

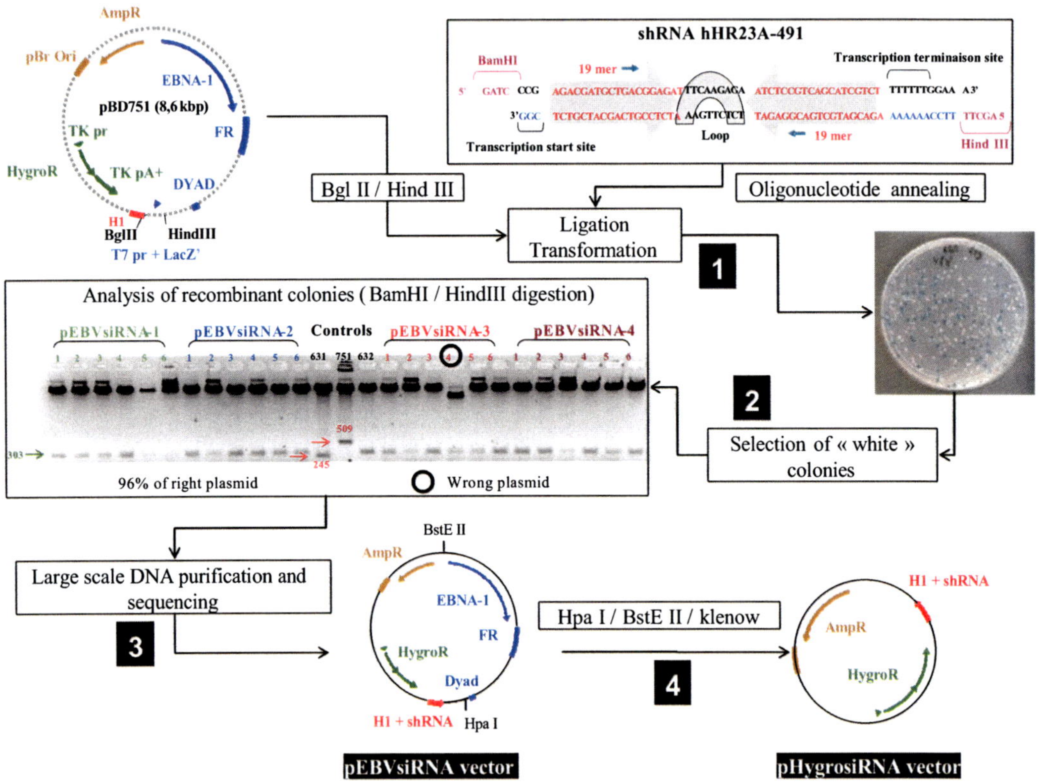

Figure 1. Scheme for cloning pEBVsiRNA vectors.

For each gene, 2 to 4 vectors are assessed for both short- and long-term silencing (several months). Most of our shRNA sequences are designed with the help of the siSearch website developed by Chalk, Wahlestedt and Sonnhammer (Chalk et al., 2004; Chalk et al., 2005). All siRNA sequences target the open reading frame. After transfection of human cells (e.g. HeLa), 12 clones per gene are propagated in culture and analyzed. Usually, different sets of transfection experiments are carried out with each vector. For long-term experiments, only one "*validated*" vector is used. Some vectors depicted here are described in Table 1. All our vectors carry the siRNA cassette in a reverse position as described elsewhere (Biard et al., 2005).

Table 1. Efficient pEBVsiRNA vectors for long-term silencing of *KIN17, XPC, XPA, XRCC4, DNA-PKcs, Rad50* and *hHR23B* genes. Hygromycin B and puromycin vectors are represented. Each siRNA sequence and its position in the cDNA are mentioned.

Gene	siRNA name	pEBVsiRNA	siRNA sequence / Position
KIN17 (O60870)	**siKIN17-180**	**pBD674** (pEBVsiK180)	5'-TCCTCAGCAGTTTATGGAT-3'
		pBD901 (pEBVsiK180-Puro)	180 to 198 nucleotides
	siKIN17-663	**pBD632** (pEBVsiK663)	5'-GTCAAGTACTCTGGGACCG-3'
		pBD900 (pEBVsiK663-Puro)	663 to 681 nucleotides
XPC (Q01831)	**siXPC-267**	**pBD634** (pEBVsiXPC267)	5'-GGATGAAGCCCTCAGCGAT-3'
		pBD894 (pEBVsiXPC267-Puro)	267 to 285 nucleotides
XPA (P23025)	**siXPA-587**	**pBD695** (pEBVsiXPA587)	5' GTCAAGAAGCATTAGAAGA 3'
		pBD895 (pEBVsiXPA587-Puro)	587 to 605 nucleotides
XRCC4 (Q13426)	**siXRCC4-674**	**pBD694** (pEBVsiXRCC4-674)	5' GAGATCCAGTCTATGATGA-3'
		pBD902 (pEBVsiXRCC4-674-Puro)	674 to 692 nucleotides
Ku70 (P12956)	**siKu70-156**	**pBD699** (pEBV-siKu70-156)	5'-GAGTGAAGATGAGTTGACA-3'
		pBD903 (pEBV-siKu70-156-Puro)	156 to 174 nucleotides
DNA-PKcs (P78527)	**siDNAPK-5980**	**pBD743** (pEBVsiDNAPK-5980)	5'-GTTGAGGTTCCTATGGAAA-3'
		pBD904 (pEBVsiDNAPK-5980-Puro)	5980 to 5998 nucleotides
hRad50 (Q92878)	**Rad50-274**	**pBD872** (pEBVsiRad50-274)	5'-GAACTTATAGCTGTGCAAA-3'
		pBD908 (pEBVsiRad50-274-Puro)	274 to 292 nucleotides
hHR23B (P54727)	**siR23B-1123**	**pBD804** p(EBVsiR23B-1123)	5'-GCATTAGGATTTCCTGAAG-3'
		pBD896 (pEBVsiR23B-1123-Puro)	1123 to 1141 nucleotides

Silencing of Genes Belonging to the NER Pathway

To assess the capacity of human cells to silence a gene for a long period, we target different proteins involved in the DNA damage recognition step of the NER pathway, such as XPA, XPC, hHR23A and hHR23B. NER is the more versatile DNA repair process and removes diverse DNA lesions, including UV-induced photoproducts. In mammals, NER requires at least 20 polypeptides comprising damage recognition and/or structure distortion factors (XPA, XPC-hHR23B, XPE, replication protein A (RPA), and the transcription factor TFIIH, strand-separating helicases to create an open pre-incision complex (TFIIH containing XPB and XPD DNA helicases), two structure-specific endonucleases (ERCC1-XPF and XPG), and the enzymes needed for gap filling (DNA polymerase δ/ε, PCNA, RFC, and

RPA). NER is divided into two subpathways which differ from their DNA recognition step: the *transcription-coupled repair* (TCR) and the *global genome repair* (GGR) (Friedberg, 2001). The initial event inducing TCR is a lesion which blocks the progression of the RNA polymerase II. GGR is involved in repairing DNA lesions on non-transcribed genes or non-transcribed strands of active genes.

In GGR, the recognition of the distortion is performed by XPC associated with hHR23B (Rad4/Rad23 yeast homologs). Although the precise functions of hHR23A&B in NER remain unknown, two interpretations have been proposed: (i) hHR23A&B may stabilize the XPC protein by protecting it from 26*S* proteasome-dependent protein degradation (Lommel et al., 2002), and (ii) hHR23B may be a key event in the displacement of XPC from damaged DNA by the XPA/RPA complex (You et al., 2003). In fact, XPA, in conjunction with RPA, monitors DNA bending and unwinding to verify the damage-specific localization of repair complexes or control their correct three-dimensional assembly (Missura et al., 2001; Rademakers et al., 2003). Mutations in hHR23B are not known to be associated with human disorders, but mutations in the Rad23 yeast counterpart lead to UV sensitivity. XPC also interacts with the dimer DDB1/DDB2 (XPE group) and centrin 2 (Okuda et al., 2004; Yang et al., 2006; You et al., 2003).

In our preliminary study performed with pEBVsiRNA vectors directed against XPA and XPC genes, we isolated different clones displaying very low (nearly undetectable) amounts of either XPA or XPC protein (Biard et al., 2005). We also focused our attention on the $_{\text{HSA}}$KIN17 (*Homo sapiens* KIN17) gene which encodes a nuclear zinc finger protein involved in DNA replication (Biard et al., 2003; Miccoli et al., 2005), RNA processing (Pinon-Lataillade et al., 2004), and in the cellular response to DNA damage (for review Angulo et al., 2005).

We have compared the efficiency of replicative pEBVsiRNA vectors with integrative ones, such as those found in the literature. For that, the EBV cartridge from the pEBVsiRNA vector was removed in order to obtain an integrative siRNA vector, termed pHygro-siRNA, as shown in Figure 1. Fourteen days after transfection of RKO cells, clonogenic cell growth assays revealed a higher number of proliferating clones with pEBV-siRNA vectors in comparison with integrative pHygro-siRNA ones. This is due to the nuclear retention of pEBV vector which enhances the plating efficiency of transfected cells in the presence of selection pressure. Besides, most of these RKO clones down-regulate the gene of interest. We observed that nearly 100% of transfected cells are silenced for *KIN17* (pBD632 vectors), *XPA* (pBD695), *hHR23A* (pBD804) or *Rad50* gene (pBD872) two weeks after transfection and hygromycin B selection of either RKO or HeLa cells ((Biard et al., 2005) and data not shown).

The long-term silencing of *KIN17* gene in RKO cells was monitored by Western blotting (figure 2A), immunocytochemical staining (figure 2B), or RT-PCR (figures 2C&D). As expected, the reduced KIN17 protein level was correlated with a significant decrease of *KIN17* mRNA content, even after 88 days in culture. In HeLa cells, among 4 efficient pEBVsiKIN17 vectors, only one leads to the selection of stable clones; the others trigger cell death. Since the kin17 protein seems to be an essential factor, the residual level of this protein (less than 10% of the control) may be required to allow cell survival.

As a critical point in the RNAi approach, we evaluated the *loss of function* associated with the loss of the targeted protein. For XPA and XPC deficiencies, we observed a dramatic decrease in DNA repair capacity of XPA^{KD} and XPC^{KD} cells after UVC irradiation, as evidenced by impaired unscheduled DNA synthesis (UDS). This impaired UDS in response to UVC was correlated with an enhanced sensitivity to UVC of both XPA^{KD} and XPC^{KD} HeLa clones (Biard et al., 2005).

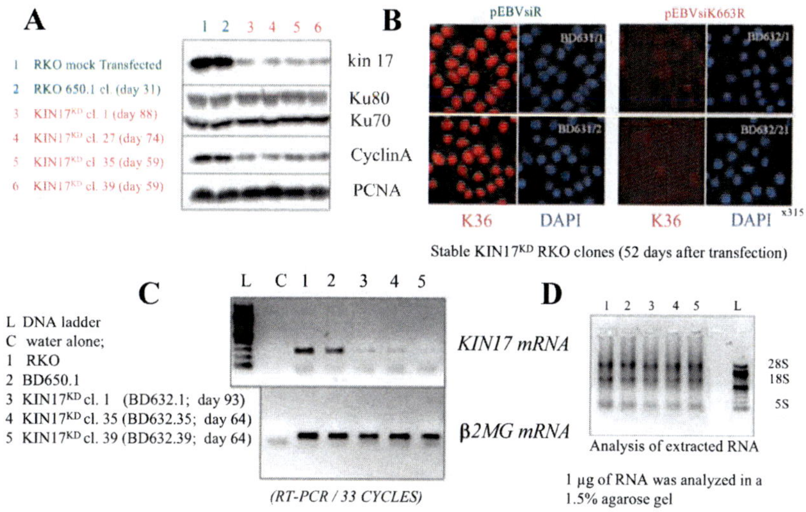

Figure 2. Long-term silencing of *KIN17* gene in colorectal carcinoma RKO cells.

Silencing of Genes Belonging to other DNA Repair Pathways (e.g. NHEJ)

These results prompted us to construct more than 140 pEBVsiRNA vectors targeting about 30 genes; some of them are depicted in figure 3. Our aim was to target genes belonging to the main pre- and post-replicative DNA repair pathways. Although NHEJ can function throughout the cell cycle, it predominates in the G0-G1 phases. In contrast, HR occurs predominantly in the S and G_2 phases of the cell cycle. Besides, while most DNA damage is removed by the different aforementioned DNA repair processes, some damage is not repaired, leading to stalled DNA replication forks. TLS (*translesion synthesis*) is a DNA damage tolerance mechanism that allows DNA replication across the lesion. Dysfunction in the TLS leads to cancer, as observed in XP *variant* cells (XP-V). XP-V cells are proficient in NER but are impaired in lesion bypass associated with DNA replication on damaged templates. XP-V cells display a mutation in the translesional polymerase pol η (eta) gene (low fidelity Y family polymerases) (for review Andreassen et al., 2006).

To establish NHEJ-deficient cells, we have silenced Ku70, XRCC4 and DNA-PKcs. The non-homologous DNA end joining pathway (NHEJ) includes at least seven factors: Ku70, Ku80, DNA-PKcs, Artemis, XRCC4, and DNA ligase IV (Lig IV) and recently Cernunnos (Callebaut et al., 2006). The Ku70/Ku80 heterodimer binds to DNA ends thus protecting the DNA from exonuclease digestion. Following DNA binding, Ku70/Ku80 recruit the ATM-related kinase DNA-PKcs. DNA-PKcs phosphorylates, associates with and activates the Artemis endonuclease which is required to resolve DNA hairpin structures. One of the targets of DNA-PKcs is the XRCC4 protein, which associates with DNA-Ligase IV (Lig IV). XRCC4 / Lig IV complex is responsible for the final ligation step (for review Buck et al., 2006a).

We have designed plasmids to silence the *Ku70* gene (e.g. pBD699 plasmid), resulting in undetectable Ku70 a few days after transfection of HeLa cells. Unfortunately, all our attempts to propagate Ku70^{KD} HeLa cells resulted in cellular death after 10 to 15 days of culture (in the presence of hygromycin B). Hence, Ku70 appears to be an essential factor for cell survival in human cells. In contrast, Ku70 knock-out mice are viable, but display defects in DSBR, growth retardation, and impaired B cell development. Ku$^{-/-}$ MEFs are sensitive to ionizing radiation, with premature senescence and accumulation of non-dividing cells in culture (for review Friedberg and Meira, 2004).

We successfully established XRCC4^{KD} (pBD694 plasmid) and DNA-PKcs^{KD} (pBD743 plasmid) cells for a very long period of time.

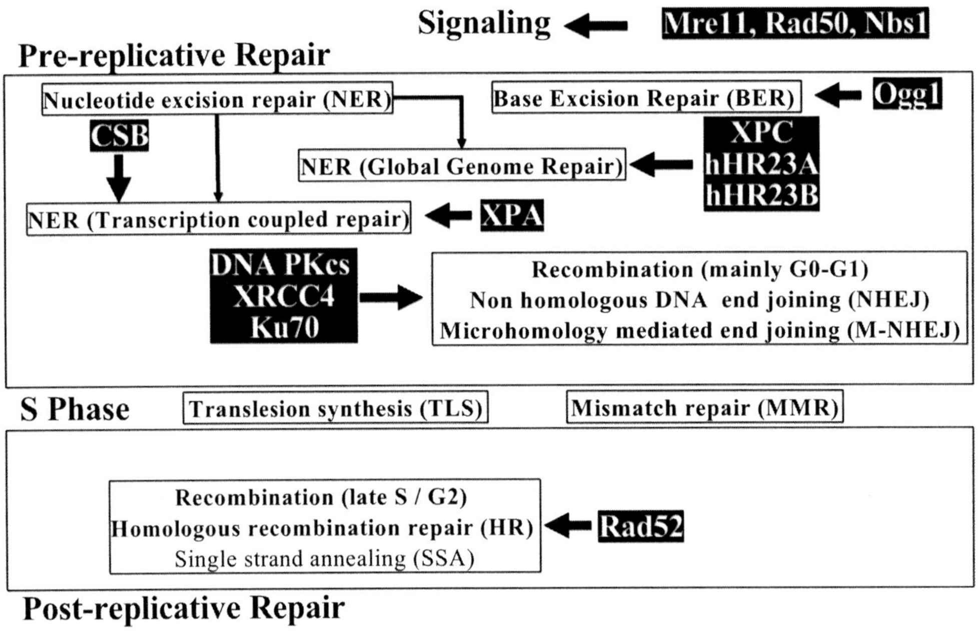

Figure 3. Genes targeted with pEBVsiRNA vectors for long-term silencing in HeLa cells.

The protein content of knock-down HeLa clones was constantly monitored, as shown in figure 4, where XPA (15 and 436 days after transfection), XPC (day 97), hHR23B (day 179), KIN17 (day 15), XRCC4 (15 and 107 days), Rad50 (day 73) or Rad52 (day 62) were silenced. In Western blot analysis, PCNA (*proliferating cell nuclear antigen*) protein was used as loading control. These results confirm not only the efficiency of gene silencing but also its specificity. Interestingly, serial steps of freezing / thawing did not modify the silenced status of knock-down clones.

The establishment of XRCC4^{KD} and DNA-PKcs^{KD} clones demonstrates that the dramatic reduction in either XRCC4 or DNA-PKcs protein levels triggers a failure in DSB rejoining activity, as evidenced by an *in vitro* NHEJ assay (figure 5; Despras et al. 2006, submitted). In parallel, these cells are more sensitive to ionizing radiation and DSB-inducing agents (e.g. neocarzinostatin, etoposide). The establishment of XRCC4^{KD} HeLa clones is interesting because no human cell line lacking this protein has been identified to date. Only Artemis, Lig IV and Cernunnos are known to be V(D)J / NHEJ factors deficient in human heritable diseases. While null mutations in *ARTEMIS* result in a human radiosensitive syndrome (RS-SCID), hypomorphic mutations of *Lig IV* were found in six patients and mutations of *CERNUNNOS* in five patients (Buck et al., 2006a; Buck et al., 2006b; Callebaut et al., 2006). Only one naturally occurring DNA-PKcs human mutant has been found in M059J glioblastoma cells (Allalunis-Turner et al., 1993; Lees-Miller et al., 1995). Strikingly, M059J cells also exhibit a reduced ATM protein level.

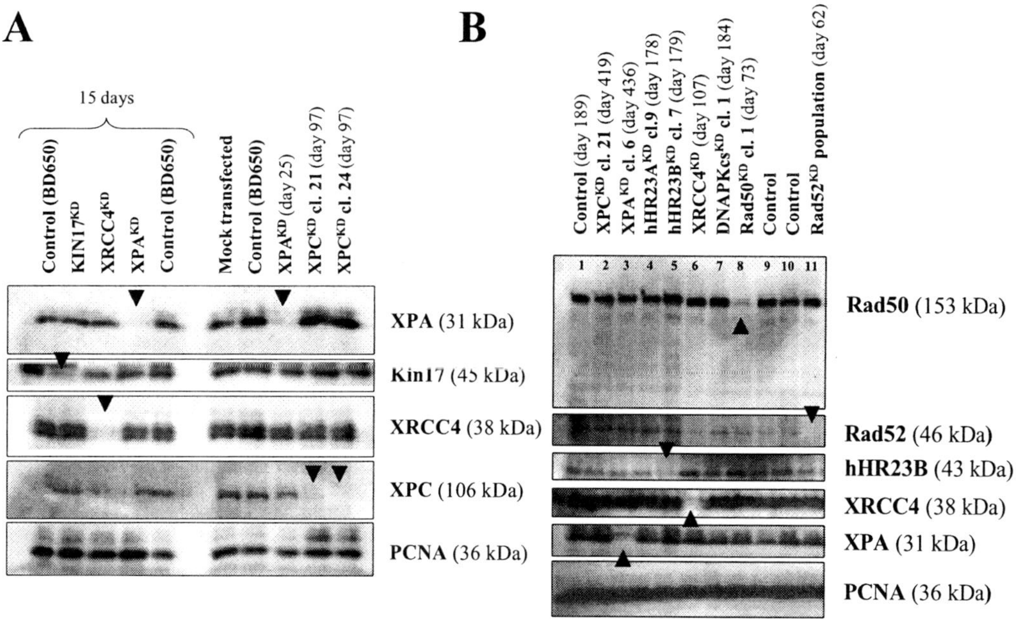

Figure 4. Long-term gene silencing in HeLa cells.

Cells deficient in genes involved in DSBR pathways are sensitive to agents that induce the formation of topoisomerases I and II cleavage complexes. Topoisomerases are essential proteins in the control of the superhelical tension and DNA interlacing associated with DNA replication, recombination, and transcription. While a covalent linkage between topoisomerase I and DNA is formed at the 3'-end of one strand of the DNA, cleavage complexes of topoisomerase II with DNA are encountered at the 5′-end of both strands of the DNA duplex (Connelly and Leach, 2004). We treated XRCC4^{KD} and DNA-PKcs^{KD} cells with etoposide (VP16), a well-known topoisomerase type II inhibitor, to evaluate their sensitivity to DSB-inducing agents. We monitored the appearance of γH2AX foci as a sensor of DSB formation because the earliest detectable cellular event in radiation-induced DNA damage is the phosphorylation of the histone variant H2AX (γH2AX). Phosphorylated H2AX stretches over large regions of the chromosome adjacent to DNA breaks, creating huge intranuclear foci. These foci concentrate repair factors within the vicinity of DNA DSBs.

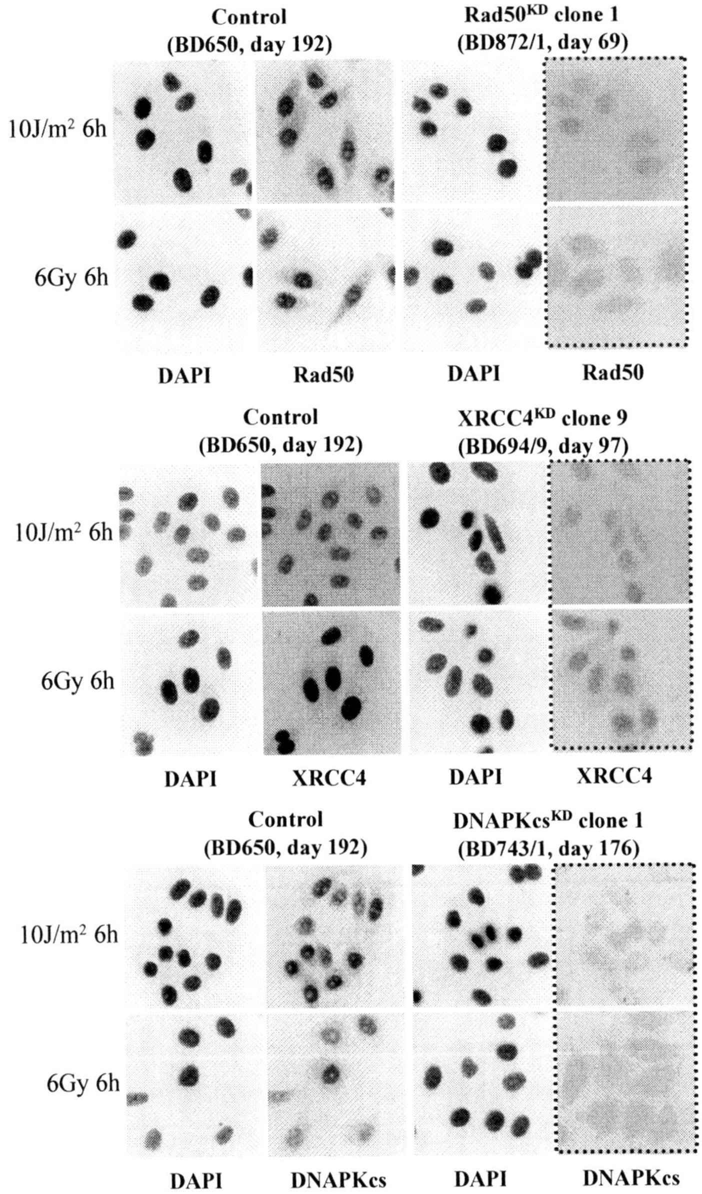

Figure 5. $Rad50^{KD}$, $XRCC4^{KD}$ and $DNAPKcs^{KD}$ HeLa clones 6 h after UVC or γ irradiation.

Endogenous γH2AX protein level depends on the cell type used. Elevated γH2AX levels are observed in S-G2 phases as the consequence of DSB formation during DNA replication (Banath et al., 2004). We show that histone H2AX is efficiently phosphorylated in control, $Rad50^{KD}$, XPC^{KD}, DNA-$PKcs^{KD}$ or $XRCC4^{KD}$ cells 4 h after etoposide treatment (figure 6). Therefore, no significant differences are detected in our DNA repair-deficient cells in regard to γH2AX recruitment. As a consequence, DNA-PKcs does not interfere with H2AX phosphorylation after etoposide treatment of HeLa cells, and other protein kinases of the phosphoinositide (PI)-3 family may be involved, such as ATM or ATR. Ongoing experiments are designed to silence these genes.

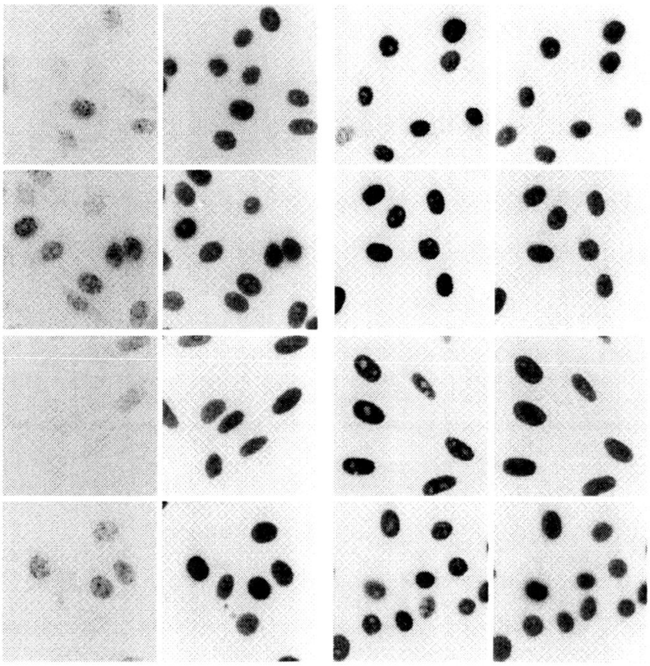

Figure 6. Phosphorylation of histone H2AX 4 h after etoposide (VP16; 25 μM) treatment in different knock-down HeLa clones.

A high level of γH2AX staining was still detected in control and $RAD50^{KD}$ cells 24 h after treatment, suggesting the presence of irreversible damage. However, control and $RAD50^{KD}$ cells failed to show acute sensitivity to etoposide 24 h after treatment (figure 7). $XRCC4^{KD}$, $DNA\text{-}PK_{CS}{}^{KD}$, and surprisingly XPC^{KD} cells appear to be rapidly sensitive to VP16, as evidenced by the appearance of cells with apoptotic features (figure 7). It seems unexpected that XPC^{KD} cells did not tolerate high dose of VP16, and we attempt to explain this observation in a recent study (Despras et al, 2006, submitted).

We sought to assess the effects of different genotoxic injuries, such as UVC (10 J/m^2), γ rays (6 Gy) or etoposide (25 μM), on cell cycle progression. Knocked-down cells were recovered 24 h after treatment, and analyzed by flow cytometry. We only used doses that did not induce lethality in control cells at the time of the analysis. In control HeLa cells, UVC irradiation (10 J/m^2) induced strong G2-M phase arrest (52%) as compared to mock irradiated cells (17%), suggesting the accumulation of damaged cells or the synchronization of cells after the treatment (figure 8). Hence, in NER-proficient cells, UVC lesions are removed without hampering progression of DNA replication. Ionizing radiation triggers a slight G1 phase arrest (60% as compared to 55% in nonirradiated cells), without detectable G2 phase blockage, which is in agreement with previously reported results (Banath et al., 2004). This could be due to the cell-doubling time of these cells (about 20 h) and to the dose used. In fact, γ ray-induced G2 arrest is observed with higher doses. Lastly, etoposide (VP16) leads to an early S-phase blockage as already mentioned in other cell lines (Biard et al., 2002).

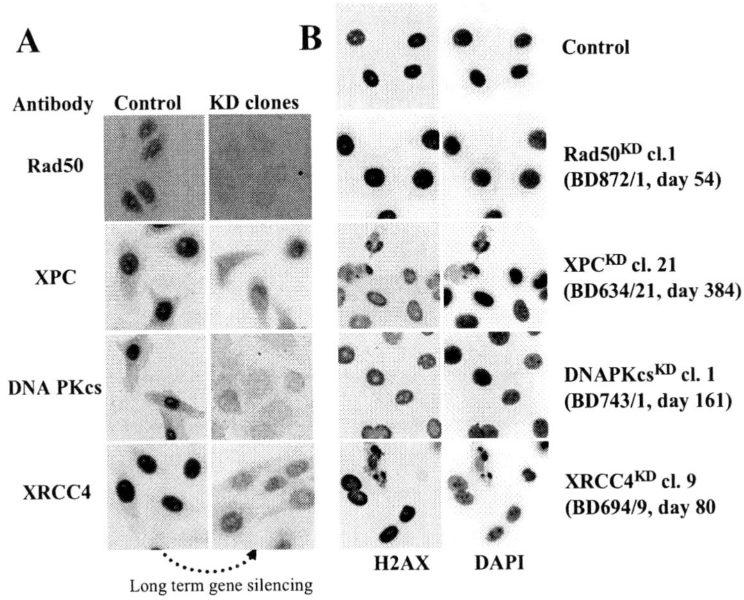

Figure 7. Treatment of knock-down HeLa clones with etoposide (VP16; 25 μM for 24 h).

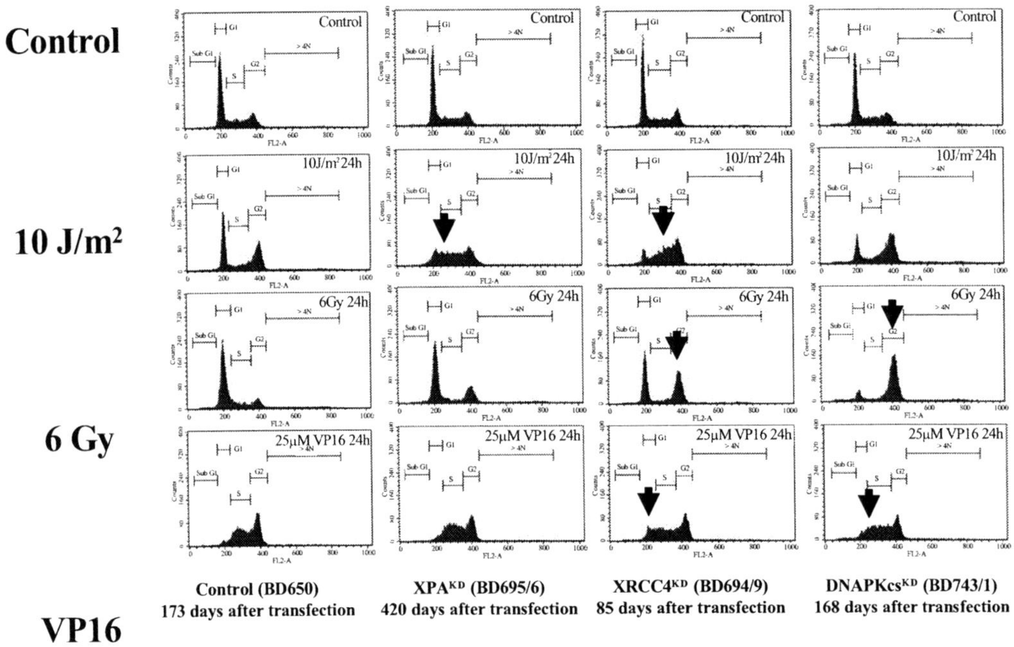

Figure 8. Treatment of XPA^{KD}, XRCC4^{KD} and DNAPKcs^{KD} HeLa clones with UVC (10 J/m^2), γ rays (6 Gy) or etoposide (VP16; 25 μM for 24 h). One day later, cells were collected by trypsinization, washed and fixed in 75% ethanol at 4°C for at least 24 h. Cells were stained with propidium iodide (4 μg/ml) in the presence of RNase (10 μg/ml) for at least 30 min. Stained cells were analyzed on a FACScalibur (Becton Dickinson, Franclin Lakes, NJ) using CellQuest software. 10,000 cells gated as single cells using FL2A/FL2W scatter were analyzed.

After UVC irradiation, highly sensitive XPA^{KD} cells were arrested in early and middle S phase suggesting that remaining UVC lesions block DNA replication. As expected, UVC lesions are not removed in NER-deficient cells, which is the typical feature of XP cells. In contrast, no significant variation was observed after γ ray irradiation and etoposide treatment in XPA^{KD} cells. DNA-PKcs^{KD} and XRCC4^{KD} cells, which are deficient in the major DSB-repair mechanism (NHEJ), display a clear G2-M arrest 24 h after ionizing radiation. In particular, 76% of DNA-PKcs^{KD} cells are blocked in the G2-M phase 24 h after 6 Gy, in comparison to 45% and 12% for XRCC4^{KD} and control cells, respectively. This accumulation of damaged cells argues for intolerance to DSB-inducing agents in both DNA-PKcs^{KD} and XRCC4^{KD} cells. The fact that etoposide induces a persistent arrest during the whole S phase in NHEJ-deficient cells (DNA-PKcs^{KD} and XRCC4^{KD}) suggests the presence of unrepaired (or unrepairable) DNA damage.

Comparison between these knock-down clones highlights the interwoven feature of DNA repair pathways since we observed that XRCC4^{KD} cells elicited an impaired cell cycle arrest after acute UVC irradiation (10 J/m^2 for 24 h), which is found in neither control nor DNA-PKcs^{KD} cells. This suggests that XRCC4^{KD} cells are, to some extent, sensitive to elevated UVC doses. Further experiments will help to unravel this response of XRCC4^{KD} cells to UVC.

Other Cell Lines

HeLa and RKO cells have been extensively characterized for long-term gene silencing. In parallel, we have silenced *XPA*, *XPC* and *KIN17* in two immortalized human fibroblasts (e.g. MRC5 V2) and established stable clones (figure 9). We have also performed successful silencing in colorectal carcinoma HCT 116 cells and breast carcinoma MCF-7 cells (data not shown).

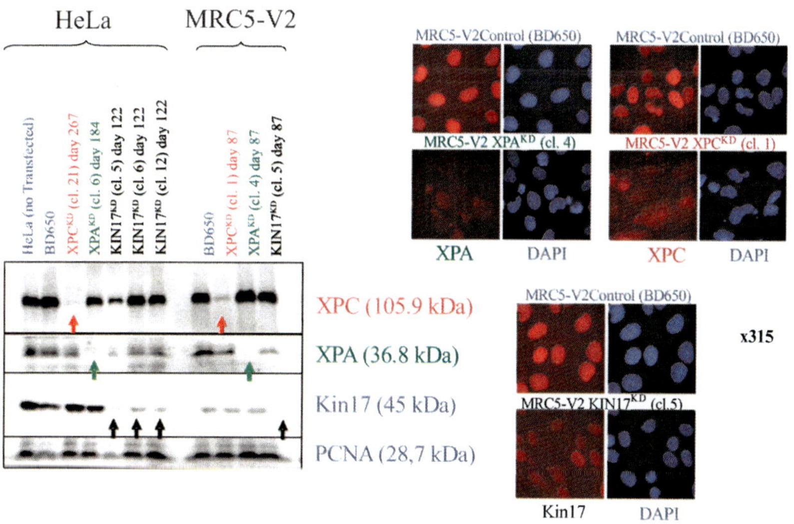

Figure 9 : Gene silencing in MCR5-V2 human SV40-immortalized cells (87 days after transfection).

Figure 9. Gene silencing in MCR5-V2 human SV40-immortalized cells (87 days after transfection).

The characterization of these cells showed that choice of cell line is important because some cells might be more suitable than others to particular studies. On the other hand, phenotypes may vary from one cell type to another and different cell lines have to be analyzed in parallel.

CONCLUSION

The long-term silencing of specific genes in isogeneic cultured human cells opens new areas of research, in particular in the field of genetic diseases and cancer. Evolution has retained highly sophisticated, interwoven DNA repair systems that prevent most of the damage inflicted on the genome. To unravel the connections between these pathways it is now possible to obtain genetically homogenous biological samples. We present here a new cellular model displaying a common genetic background. We show that replicative pEBVsiRNA plasmids greatly improve long-term gene silencing in different cell lines. We demonstrate that replicative pEBV vectors help to circumvent the shortcomings due to (i) the overexpression of siRNA molecules in transient and long-term assays, and to (ii) the random integration into the genome subsequent to the loss of plasmids. For the first time, we are able to compare defects in DNA repair pathways imposed by gene silencing over a long period of time.

The main properties of our pEBV-based siRNA vectors are the following: **(i)** transient gene silencing efficiencies reach a level similar to that observed with siRNA duplexes; **(ii)** more than 95% of cells are silenced two weeks after transfection (in comparison with an integrative plasmid giving rise to less than 50% of silenced cells); **(iii)** most isolated clones are silenced (e.g. 100% for XPA); **(iv)** in the selected clones, the protein of interest is still undetectable more than 500 days after transfection (XPA); **(v)** the phenotype associated with the loss of the targeted protein is maintained over the whole period, and **(vi)** the silencing is observed in different cell lines, such as HeLa, RKO, HCT 116, MCF-7, MRC5-V2, immortalized human fibroblasts (data not shown). A recent study reinforces our strategy by demonstrating that pEBV vector could effectively enhance the RNA interference efficiency in human stem cells (hES) (Ren et al., 2006). Abrogation of DNA repair genes in human stem cells could offer new opportunities because genetic modifications of stem cells have great therapeutic potential.

Our results also emphasize that a weak polymerase III promoter, such as H1, is sufficient to drive stable and long-term expression of a shRNA sequence. Many of the unwanted side effects observed transiently after transfection of siRNA duplexes or integrative plasmids, as well as after transduction of lentivirus vectors (at elevated viral titer), may stem from the large amount of siRNA molecules in transfected cells which saturate enzymatic complexes involved in gene silencing (Fish and Kruithof, 2004). As recently mentioned, unwanted side effects (such as toxicity) disappear with very low concentrations of siRNA duplexes (below 1 nM) (Fedorov et al., 2006). Therefore, it seems unnecessary to employ a strong polymerase II promoter (e.g. IE CMV promoter) to transcribe shRNA sequences except when tissue-specific expression is required.

After having performed numerous experiments on more than 20 genes involved in DNA repair pathways, we have noticed that there are significant differences in transient (*2-3 days after transfection*) and medium-term (*one month*) experiments, in comparison with long-term experiments (*several months or more than one year*). While most of our pEBVsiRNA vectors appeared to be very efficient in transient experiments, the fate of transfected cells after several weeks in culture is varies. We have observed four cases, depending on the target gene:

1. The reduction of protein level does not significantly change HeLa cell growth and viability even several months after transfection. This is the case for *XPA*, *hHR23A* or *MRE11* gene silencing. "Silenced" clones are easily obtained, and in almost all the protein of interest is reduced by more than 95% in comparison to controls. This suggests that these proteins are neither essential nor limiting for the recipient cells. Besides, we may observe a selective growth advantage in some cases a few weeks after transfection (e.g. *XPA* silencing in HeLa). The emerging question is whether the residual protein is sufficient to ensure its biological functions. A recent study indicates that XPA becomes a limiting factor for NER and consequently for UV sensitivity when reduced to less than 10% of that present in a normal cell (Koberle et al., 2006). In our hands, XPA is nearly undetectable in XPA^{KD} cells, and this is correlates with impaired UDS and enhanced UVC sensitivity.

2. Target genes contribute to cell viability even at a low level (limiting factors for some specific pathways). As a consequence, knock-down clones will be difficult to establish (e.g. *XRCC4* or *KIN17*). All knock-down clones maintain a residual amount of the protein of interest (about 10 to 15% of the controls) to allow cell viability, otherwise they die. The question is whether we could analyze the loss-of-function. For *XRCC4*, its remaining protein level in $XRCC4^{KD}$ cells is sufficient for survival but irrelevant for performing accurate joining activities (Despras et al, 2006, submitted). This suggests that XRCC4 is a limiting factor for NHEJ activity and certainly in other processes. In contrast, for *KIN17*, isolated clones obtained to date seem to have compensated for their deficiency.

3. Most silenced cells rapidly die after transfection, and few of them are able to be propagated in culture after an "*adaptive period*". However, established knock-down clones produced nearly undetectable levels of the protein of interest. This is the case for many genes, such as *XPC*, *hHR23B*, *DNA-PKcs*, *Rad50* or *NBS1*. For *DNA PK*, once expanded (30 to 40 days later), $DNA\text{-}PKcs^{KD}$ cells fail to express detectable DNA-PKcs protein; they display a total loss of the endogenous DNA-PKcs kinase activity and exhibit impaired NHEJ activity. This demonstrates the entire loss of function in these $DNA\text{-}PKcs^{KD}$ cells. For XPC and hHR23B, we observed a similar behavior with dramatic disadvantages and cell lethality a few weeks after transfection. XPC^{KD} and $hHR23B^{KD}$ clones were stably obtained only after several weeks of culture. All of these clones remained stable thereafter. For XPC^{KD} cells, the loss of function is maintained for the whole period of culture.

4. Ultimately, the targeted protein is essential for cell life and its shutting down entails cell death 7 to 10 days after transfection (and selection). This is the case for Ku70, SAF-A, and centrin 2. For these proteins, inducible vectors have to be constructed.

The maintenance for several months (or more) of the gene silencing imposed by pEBVsiRNA plasmids affords the opportunity to *validate different siRNA sequences* for an unfailing and specific gene silencing. siRNA validation should be a prerequisite before beginning time-consuming experiments, such as *in vivo* approaches or assays on human stem cells. Importantly, transient assays may mask the real effects of gene silencing. For instance, the establishment of some stable knock-down clones can trigger an elevated rate of cell death a few days after transfection (and selection), because the targeted protein is strongly required for cell survival (e.g. *DNA-PKcs*, *XRCC4*, *NBS1*, *KIN17*, and so on). This raises an essential question: what is the meaning of experiments where freshly "silenced" cells (*transient experiments*) are treated with genotoxic agents (physical or chemical)? However, in the long-term experiments, we do not exclude the possibility of skews. It must keep in mind that the suppression of gene expression over a long period may provoke compensatory cellular responses during an "*adaptive period*". During this period, cellular metabolism may compensate for the decrease in protein concentration, particularly if the protein plays an important role in the cell. This adaptation period may be accompanied by changes that mask the true biological consequences of specific gene silencing. However, this is also true during the multistage process leading to tumorigenesis, where a normal cell encounters serial genetic changes, including initiation, clonal expansion, pre-malignant lesions, and malignant progression, before acquiring a tumoral phenotype. This adaptation is also present in other strategies to switch off gene expression, such as in "knock-out" mouse models where clonal cell expansion is required. It is therefore necessary to constantly characterize the expected phenotype in knock-down cells during the culture, and to compare short- and long-term experiments. We have demonstrated that long-term silencing of genes of either the NER or NHEJ pathways mediated by pEBVsiRNA vectors in HeLa cells entails the expected phenotypic modifications.

At present, the establishment of knock-down clones for different DNA repair and sensor / transducing pathways may help us to answer some exciting questions. (i) Firstly, we are seeking to unravel the possible connection between DNA repair pathways as well as different subpathways (e.g. TCR *versus* GRR). In particular, we are interested in the existence of (*at least*) two NHEJ complementary subpathways with different kinetics, the high-speed classical D-NHEJ (for *DNA-PK and XRCC4-dependent*) and a low-speed B-NHEJ (for *Backup*; also termed *microhohomology-based end-joining pathway*), both of which contribute to the major rescue of IR-induced DSBs in higher eukaryotes (Iliakis et al., 2004; Verkaik et al., 2002). $XRCC4^{KD}$ and $DNA\text{-}PKcs^{KD}$ cells demonstrate that the *in vitro* NHEJ activity is tightly dependent on the protein level of both XRCC4 and DNA-PKcs. While most of the building blocks of the B-NHEJ remain unknown, DNA ligase III (Lig III) could be one of these components (Wang et al., 2005). Vectors directed against Lig III will help to explore this issue. (ii) Secondly, we focus our attention on the systems that act to repair DNA molecules containing covalently bound proteins (e.g. after topoisomerase inhibitor treatment) (Connelly and Leach, 2004). The contributions of HR *versus* NHEJ pathways have to be determined. (iii) Thirdly, another exciting field of investigation is the sensor / transducer pathway which is critical for genomic stability. The MRN complex (MRE11, Rad50 and NBS1) is critical for the intracellular signaling pathway activated after DNA damage. The literature reveals that MRN acts as a DNA damage sensor, continuously localized at sites of unrepaired DNA damage. Nbs1 binds to γH2AX in the vicinity of DSBs and recruits Rad50 and Mre11. Mre11 has nuclease and DNA-unwinding activities, and a homodimer of Rad50 could serve as a bridge to hold together the broken ends of DSBs and prevent extensive degradation (for review Zhou et al., 2006). Targeting of MRN components is essential to understand the role of the MRN complex in the NHEJ and HR pathways. While the requirement of the MRN complex in the NHEJ pathway remains controversial (Di Virgilio and Gautier, 2005), MRN plays a role in the activation of cell cycle checkpoint induced by ATM and in an early step of HR (for review Kim et al., 2005). *MRE11*, *Rad50* and *NBS1* are essential genes in vertebrates which limit experimental work. Experiments are usually performed with cells from patients and display hypomorphic mutations in either *MRE11* (ATLD disease) or *NBS1* (NBS syndrome). The latter cells may maintain several biological activities associated with the MRN complex. Interestingly, no naturally occurring human Rad50-deficient cells have been found. Our preliminary experiments revealed an efficient silencing of the *Rad50* gene two weeks after transfection, which is maintained with a similar efficiency after 120 days in culture (Figure 4). Beside $Rad50^{KD}$ cells, we have constructed pEBVsiRNA expression vectors directed against *MRE11* and *NBS1* genes. *NBS1* gene silencing triggers major growth disadvantages, with dramatic cell death. This suggests an essential role of the Nbs1 protein in HeLa cells. It is noteworthy that null mutation of NBS1 is lethal in the mouse. However, we have now isolated $NBS1^{KD}$ clones displaying an undetectable level of Nbs1. We are currently establishing $MRE11^{KD}$ HeLa clones.

The combination of RNAi technology with pEBV-derived vectors offers an exceptional opportunity to create a set of stable knock-down clones covering various fields of biology. We could mimic human diseases in a comparable genetic background. This approach may also have several applications in drug screening and in the first steps to test new therapies After characterizing our set of knock-down HeLa clones, we aim to transfect these pEBVsiRNA vectors in primary human cell lines. We wish to analyze the consequences (*genetic stability*) of long-term silencing of DNA repair genes in non-tumor cells.

Acknowledgements

We would like to thank Dr Germain Rousselet for his comments on the manuscript. This work was entirely done in the LGR laboratory. In accordance with the quality management program of our department, this work is reported in custom laboratory books numbered CEA/DRR/SRBF/LGR/ 0760, 3090, 3191, 4658, 4659 4660, 6093 and 6094. We thank "Electricité de France" for its support (contract no. 8702).

References

Aiyar, A., Tyree, C. and Sugden, B. (1998) The plasmid replicon of EBV consists of multiple *cis*-acting elements that facilitate DNA synthesis by the cell and a viral maintenance element. *EMBO J.*, **17**, 6394-6403.

Allalunis-Turner, M.J., Barron, G.M., Day, R.S., 3rd, Dobler, K.D. and Mirzayans, R. (1993) Isolation of two cell lines from a human malignant glioma specimen differing in sensitivity to radiation and chemotherapeutic drugs. *Radiat Res*, **134**, 349-354.

Al-Tassan, N., Chmiel, N.H., Maynard, J., Fleming, N., Livingston, A.L., Williams, G.T., Hodges, A.K., Davies, D.R., David, S.S., Sampson, J.R. and Cheadle, J.P. (2002)

Inherited variants of MYH associated with somatic G:C-->T:A mutations in colorectal tumors. *Nat Genet*, **30**, 227-232.

Andreassen, P.R., Ho, G.P. and D'Andrea, A.D. (2006) DNA damage responses and their many interactions with the replication fork. *Carcinogenesis*, **27**, 883-892.

Angulo, J.F., Mauffrey, P., Pinon-Lataillade, G., Miccoli, L. and Biard, D.S.F. (2005) Putative roles of kin17, a mammalian protein binding curved DNA, in transcription. In Takashi, E. (ed.), *DNA Conformation and Transcription.* Landes Biosciences / Eurekah.com, Vol. chapter 6, pp. 75-89.

Anonymous. (2003) Whither RNAi ? *Nat Cell Biol*, **5**, 489-490.

Banath, J.P., Macphail, S.H. and Olive, P.L. (2004) Radiation sensitivity, H2AX phosphorylation, and kinetics of repair of DNA strand breaks in irradiated cervical cancer cell lines. *Cancer Res*, **64**, 7144-7149.

Biard, D.S., Cordier, A. and Sarasin, A. (1992) Establishment of a human cell line for the detection of demethylating agents. *Exp Cell Res*, **200**, 263-271.

Biard, D.S., Despras, E., Sarasin, A. and Angulo, J.F. (2005) Development of new EBV-based vectors for stable expression of small interfering RNA to mimick human syndromes: application to NER gene silencing. *Mol Cancer Res*, **3**, 519-529.

Biard, D.S., Miccoli, L., Despras, E., Frobert, Y., Creminon, C. and Angulo, J.F. (2002) Ionizing radiation triggers chromatin-bound kin17 complex formation in human cells. *J Biol Chem*, **277**, 19156-19165.

Biard, D.S., Miccoli, L., Despras, E., Harper, F., Pichard, E., Creminon, C. and Angulo, J.F. (2003) Participation of kin17 protein in replication factories and in other DNA transactions mediated by high molecular weight nuclear complexes. *Mol Cancer Res*, **1**, 519-531.

Brummelkamp, T.R., Bernards, R. and Agami, R. (2002) A system for stable expression of short interfering RNAs in mammalian cells. *Science*, **296**, 550-553.

Buck, D., Malivert, L., de Chasseval, R., Barraud, A., Fondaneche, M.C., Sanal, O., Plebani, A., Stephan, J.L., Hufnagel, M., le Deist, F., Fischer, A., Durandy, A., de Villartay, J.P. and Revy, P. (2006a) Cernunnos, a novel nonhomologous end-joining factor, is mutated in human immunodeficiency with microcephaly. *Cell*, **124**, 287-299.

Buck, D., Moshous, D., de Chasseval, R., Ma, Y., le Deist, F., Cavazzana-Calvo, M., Fischer, A., Casanova, J.L., Lieber, M.R. and de Villartay, J.P. (2006b) Severe combined immunodeficiency and microcephaly in siblings with hypomorphic mutations in DNA ligase IV. *Eur J Immunol*, **36**, 224-235.

Butz, K., Ristriani, T., Hengstermann, A., Denk, C., Scheffner, M. and Hoppe-Seyler, F. (2003) siRNA targeting of the viral E6 oncogene efficiently kills human papillomavirus-positive cancer cells. *Oncogene*, **22**, 5938-5945.

Callebaut, I., Malivert, L., Fischer, A., Mornon, J.P., Revy, P. and de Villartay, J.P. (2006) Cernunnos interacts with the XRCC4/DNA-ligase IV complex and is homologous to the yeast nonhomologous end-joining factor NEJ1. *J Biol Chem.*

Calos, M.P. (1996) The potential of extrachromosomal replicating vectors for gene therapy. *Trends Genet*, **12**, 463-466.

Chalk, A.M., Wahlestedt, C. and Sonnhammer, E.L. (2004) Improved and automated prediction of effective siRNA. *Biochem Biophys Res Commun*, **319**, 264-274.

Chalk, A.M., Warfinge, R.E., Georgii-Hemming, P. and Sonnhammer, E.L. (2005) siRNAdb: a database of siRNA sequences. *Nucleic Acids Res*, **33**, D131-134.

Chaudhuri, B., Xu, H., Todorov, I., Dutta, A. and Yates, J.L. (2001) Human DNA replication initiation factors, ORC and MCM, associate with oriP of Epstein-Barr virus. *Proc Natl Acad Sci U S A*, **98**, 10085-10089.

Collins, C.M. and Medveczky, P.G. (2002) Genetic requirements for the episomal maintenance of oncogenic herpesvirus genomes. *Adv Cancer Res*, **84**, 155-174.

Connelly, J.C. and Leach, D.R. (2004) Repair of DNA covalently linked to protein. *Mol Cell*, **13**, 307-316.

Connolly, J.B. (2002) Lentiviruses in gene therapy clinical research. *Gene Ther*, **9**, 1730-1734.

Dhar, S.K., Yoshida, K., Machida, Y., Khaira, P., Chaudhuri, B., Wohlschlegel, J.A., Leffak, M., Yates, J. and Dutta, A. (2001) Replication from oriP of Epstein-Barr virus requires human ORC and is inhibited by geminin. *Cell*, **106**, 287-296.

Di Virgilio, M. and Gautier, J. (2005) Repair of double-strand breaks by nonhomologous end joining in the absence of Mre11. *J Cell Biol*, **171**, 765-771.

Duker, N.J. (2002) Chromosome breakage syndromes and cancer. *Am J Med Genet*, **115**, 125-129.

Einav, Y., Shistik, E., Shenfeld, M., Simons, A.H., Melton, D.W. and Canaani, D. (2003) Replication and episomal maintenance of Epstein-Barr virus-based vectors in mouse embryonal fibroblasts enable synthetic lethality screens. *Mol Cancer Ther*, **2**, 1121-1128.

Elbashir, S.M., Harborth, J., Lendeckel, W., Yalcin, A., Weber, K. and Tuschl, T. (2001) Duplexes of 21-nucleotide RNAs mediate RNA interference in cultured mammalian cells. *Nature*, **411**, 494-498.

Ellis, J. (2005) Silencing and variegation of gammaretrovirus and lentivirus vectors. *Hum Gene Ther*, **16**, 1241-1246.

Fedorov, Y., Anderson, E.M., Birmingham, A., Reynolds, A., Karpilow, J., Robinson, K., Leake, D., Marshall, W.S. and Khvorova, A. (2006) Off-target effects by siRNA can induce toxic phenotype. *Rna*.

Fire, A., Xu, S., Montgomery, M.K., Kostas, S.A., Driver, S.E. and Mello, C.C. (1998) Potent and specific genetic interference by double-stranded RNA in Caenorhabditis elegans. *Nature*, **391**, 806-811.

Fish, R.J. and Kruithof, E.K. (2004) Short-term cytotoxic effects and long-term instability of RNAi delivered using lentiviral vectors. *BMC Mol Biol*, **5**, 9.

Friedberg, E.C. (2001) How nucleotide excision repair protects against cancer. *Nat Rev Cancer*, **1**, 22-33.

Friedberg, E.C. and Meira, L.B. (2004) Database of mouse strains carrying targeted mutations in genes affecting biological responses to DNA damage (Version 6). *DNA Repair (Amst)*, **3**, 1617-1638.

Giglia-Mari, G., Coin, F., Ranish, J.A., Hoogstraten, D., Theil, A., Wijgers, N., Jaspers, N.G., Raams, A., Argentini, M., van der Spek, P.J., Botta, E., Stefanini, M., Egly, J.M., Aebersold, R., Hoeijmakers, J.H. and Vermeulen, W. (2004) A new, tenth subunit of TFIIH is responsible for the DNA repair syndrome trichothiodystrophy group A. *Nat Genet*, **36**, 714-719.

Gillet, L.C. and Scharer, O.D. (2006) Molecular mechanisms of mammalian global genome nucleotide excision repair. *Chem Rev*, **106**, 253-276.

Hammond, S.M., Bernstein, E., Beach, D. and Hannon, G.J. (2000) An RNA-directed nuclease mediates post-transcriptional gene silencing in Drosophila cells. *Nature*, **404**, 293-296.

He, J., Yang, Q. and Chang, L.J. (2005) Dynamic DNA methylation and histone modifications contribute to lentiviral transgene silencing in murine embryonic carcinoma cells. *J Virol*, **79**, 13497-13508.

Hoeijmakers, J.H. (1993) Nucleotide excision repair. II: From yeast to mammals. *Trends Genet*, **9**, 211-217.

Iliakis, G., Wang, H., Perrault, A.R., Boecker, W., Rosidi, B., Windhofer, F., Wu, W., Guan, J., Terzoudi, G. and Pantelias, G. (2004) Mechanisms of DNA double strand break repair and chromosome aberration formation. *Cytogenet Genome Res*, **104**, 14-20.

Kapoor, P. and Frappier, L. (2003) EBNA1 partitions Epstein-Barr virus plasmids in yeast cells by attaching to human EBNA1-binding protein 2 on mitotic chromosomes. *J Virol*, **77**, 6946-6956.

Khvorova, A., Reynolds, A. and Jayasena, S.D. (2003) Functional siRNAs and miRNAs exhibit strand bias. *Cell*, **115**, 209-216.

Kim, J.S., Krasieva, T.B., Kurumizaka, H., Chen, D.J., Taylor, A.M. and Yokomori, K. (2005) Independent and sequential recruitment of NHEJ and HR factors to DNA damage sites in mammalian cells. *J Cell Biol*, **170**, 341-347.

Koberle, B., Roginskaya, V. and Wood, R.D. (2006) XPA protein as a limiting factor for nucleotide excision repair and UV sensitivity in human cells. *DNA Repair (Amst)*, **5**, 641-648.

Kraakman-van der Zwet, M., Overkamp, W.J., van Lange, R.E., Essers, J., van Duijn-Goedhart, A., Wiggers, I., Swaminathan, S., van Buul, P.P., Errami, A., Tan, R.T., Jaspers, N.G., Sharan, S.K., Kanaar, R. and Zdzienicka, M.Z. (2002) Brca2 (XRCC11) deficiency results in radioresistant DNA synthesis and a higher frequency of spontaneous deletions. *Mol Cell Biol*, **22**, 669-679.

Laposa, R.R., Feeney, L. and Cleaver, J.E. (2003) Recapitulation of the cellular xeroderma pigmentosum-variant phenotypes using short interfering RNA for DNA polymerase H. *Cancer Res*, **63**, 3909-3912.

Lees-Miller, S.P., Godbout, R., Chan, D.W., Weinfeld, M., Day, R.S., 3rd, Barron, G.M. and Allalunis-Turner, J. (1995) Absence of p350 subunit of DNA-activated protein kinase from a radiosensitive human cell line. *Science*, **267**, 1183-1185.

Leuschner, P.J., Ameres, S.L., Kueng, S. and Martinez, J. (2006) Cleavage of the siRNA passenger strand during RISC assembly in human cells. *EMBO Rep*.

Lommel, L., Ortolan, T., Chen, L., Madura, K. and Sweder, K.S. (2002) Proteolysis of a nucleotide excision repair protein by the 26 S proteasome. *Curr Genet*, **42**, 9-20.

Lupton, S. and Levine, A.J. (1985) Mapping genetic elements of Epstein-Barr virus that facilate extrachromosomic persistence of Epstein-Barr virus-derived plasmids in human cells. *Mol. Cell. Biol.*, **5**, 2533-2542.

Marechal, V., Dehee, A., Chikhi-Brachet, R., Piolot, T., Coppey-Moisan, M. and Nicolas, J.C. (1999) Mapping EBNA-1 domains involved in binding to metaphase chromosomes. *J Virol*, **73**, 4385-4392.

McManus, M.T. and Sharp, P.A. (2002) Gene silencing in mammals by small interfering RNAs. *Nat Rev Genet*, **3**, 737-747.

Miccoli, L., Frouin, I., Novac, O., Di Paola, D., Harper, F., Zannis-Hadjopoulos, M., Maga, G., Biard, D.S.F. and Angulo, J.F. (2005) The Human Stress-Activated Protein kin17 Belongs to the Multiprotein DNA Replication Complex and Associates in Vivo with Mammalian Replication Origins. *Mol Cell Biol*, **in press**.

Missura, M., Buterin, T., Hindges, R., Hubscher, U., Kasparkova, J., Brabec, V. and Naegeli, H. (2001) Double-check probing of DNA bending and unwinding by XPA-RPA: an architectural function in DNA repair. *Embo J*, **20**, 3554-3564.

Mittal, V. (2004) Improving the efficiency of RNA interference in mammals. *Nat Rev Genet*, **5**, 355-365.

O'Driscoll, M., Gennery, A.R., Seidel, J., Concannon, P. and Jeggo, P.A. (2004) An overview of three new disorders associated with genetic instability: LIG4 syndrome, RS-SCID and ATR-Seckel syndrome. *DNA Repair (Amst)*, **3**, 1227-1235.

O'Driscoll, M. and Jeggo, P.A. (2006) The role of double-strand break repair - insights from human genetics. *Nat Rev Genet*, **7**, 45-54.

Okuda, Y., Nishi, R., Ng, J.M., Vermeulen, W., van der Horst, G.T., Mori, T., Hoeijmakers, J.H., Hanaoka, F. and Sugasawa, K. (2004) Relative levels of the two mammalian Rad23 homologs determine composition and stability of the xeroderma pigmentosum group C protein complex. *DNA Repair (Amst)*, **3**, 1285-1295.

Paddison, P.J., Caudy, A.A., Bernstein, E., Hannon, G.J. and Conklin, D.S. (2002) Short hairpin RNAs (shRNAs) induce sequence-specific silencing in mammalian cells. *Genes Dev*, **16**, 948-958.

Peltomaki, P. (2003) Role of DNA mismatch repair defects in the pathogenesis of human cancer. *J Clin Oncol*, **21**, 1174-1179.

Pinon-Lataillade, G., Masson, C., Bernardino-Sgherri, J., Henriot, V., Mauffrey, P., Frobert, Y., Araneda, S. and Angulo, J.F. (2004) KIN17 encodes an RNA-binding protein and is expressed during mouse spermatogenesis. *J Cell Sci*, **117**, 3691-3702.

Rademakers, S., Volker, M., Hoogstraten, D., Nigg, A.L., Mone, M.J., Van Zeeland, A.A., Hoeijmakers, J.H., Houtsmuller, A.B. and Vermeulen, W. (2003) Xeroderma pigmentosum group A protein loads as a separate factor onto DNA lesions. *Mol Cell Biol*, **23**, 5755-5767.

Reisman, D., Yates, J. and Sugden, B. (1985) A putative origin of replication of plasmids derived from Epstein-Barr virus is composed of two cis-acting components. *Mol. Cell. Biol.*, **5**, 1822-1832.

Ren, C., Zhao, M., Yang, X., Li, D., Jiang, X., Wang, L., Shan, W., Yang, H., Zhou, L., Zhou, W. and Zhang, H. (2006) Establishment and Applications of Epstein-Barr Virus-based Episomal Vectors in Human Embryonic Stem Cells. *Stem Cells*.

Ritzi, M., Tillack, K., Gerhardt, J., Ott, E., Humme, S., Kremmer, E., Hammerschmidt, W. and Schepers, A. (2003) Complex protein-DNA dynamics at the latent origin of DNA replication of Epstein-Barr virus. *J Cell Sci*, **116**, 3971-3984.

Romano, G., Claudio, P.P., Tonini, T. and Giordano, A. (2003) Human immunodeficiency virus type 1 (HIV-1) derived vectors: safety considerations and controversy over therapeutic applications. *Eur J Dermatol*, **13**, 424-429.

Schepers, A., Ritzi, M., Bousset, K., Kremmer, E., Yates, J.L., Harwood, J., Diffley, J.F. and Hammerschmidt, W. (2001) Human origin recognition complex binds to the region of the latent origin of DNA replication of Epstein-Barr virus. *Embo J*, **20**, 4588-4602.

Schwarz, D.S., Hutvagner, G., Du, T., Xu, Z., Aronin, N. and Zamore, P.D. (2003) Asymmetry in the assembly of the RNAi enzyme complex. *Cell*, **115**, 199-208.

Shiloh, Y. (2003) ATM and related protein kinases: safeguarding genome integrity. *Nat Rev Cancer*, **3**, 155-168.

Thacker, J. and Zdzienicka, M.Z. (2003) The mammalian XRCC genes: their roles in DNA repair and genetic stability. *DNA Repair (Amst)*, **2**, 655-672.

Thacker, J. and Zdzienicka, M.Z. (2004) The XRCC genes: expanding roles in DNA double-strand break repair. *DNA Repair (Amst)*, **3**, 1081-1090.

Thompson, L.H., Brookman, K.W., Jones, N.J., Allen, S.A. and Carrano, A.V. (1990) Molecular cloning of the human XRCC1 gene, which corrects defective DNA strand break repair and sister chromatid exchange. *Mol Cell Biol*, **10**, 6160-6171.

Tomari, Y. and Zamore, P.D. (2005) Perspective: machines for RNAi. *Genes Dev*, **19**, 517-529.

Tuschl, T. (2001) RNA interference and small interfering RNAs. *Chembiochem*, **2**, 239-245.

van Duin, M., de Wit, J., Odijk, H., Westerveld, A., Yasui, A., Koken, H.M., Hoeijmakers, J.H. and Bootsma, D. (1986) Molecular characterization of the human excision repair gene ERCC-1: cDNA cloning and amino acid homology with the yeast DNA repair gene RAD10. *Cell*, **44**, 913-923.

Verkaik, N.S., Esveldt-van Lange, R.E., van Heemst, D., Bruggenwirth, H.T., Hoeijmakers, J.H., Zdzienicka, M.Z. and van Gent, D.C. (2002) Different types of V(D)J recombination and end-joining defects in DNA double-strand break repair mutant mammalian cells. *Eur J Immunol*, **32**, 701-709.

Wade-Martins, R., Frampton, J. and James, M. (1999) Long-term stability of large insert genomic DNA episomal shuttle vectors in human cells
10.1093/nar/27.7.1674. *Nucl. Acids Res.*, **27**, 1674-1682.

Wang, H., Rosidi, B., Perrault, R., Wang, M., Zhang, L., Windhofer, F. and Iliakis, G. (2005) DNA ligase III as a candidate component of backup pathways of nonhomologous end joining. *Cancer Res*, **65**, 4020-4030.

Woods, N.B., Muessig, A., Schmidt, M., Flygare, J., Olsson, K., Salmon, P., Trono, D., von Kalle, C. and Karlsson, S. (2003) Lentiviral vector transduction of NOD/SCID repopulating cells results in multiple vector integrations per transduced cell: risk of insertional mutagenesis. *Blood*, **101**, 1284-1289.

Wysokenski, D.A. and Yates, J.L. (1989) Multiple EBNA1-binding sites are required to form an EBNA1-dependent enhancer and to activate a minimal replicative origin within oriP of Epstein-Barr virus. *J Virol*, **63**, 2657-2666.

Yang, A., Miron, S., Mouawad, L., Duchambon, P., Blouquit, Y. and Craescu, C.T. (2006) Flexibility and plasticity of human centrin 2 binding to the xeroderma pigmentosum group C protein (XPC) from nuclear excision repair. *Biochemistry*, **45**, 3653-3663.

You, J.S., Wang, M. and Lee, S.H. (2003) Biochemical analysis of the damage recognition process in nucleotide excision repair. *J Biol Chem*, **278**, 7476-7485.

Zamore, P.D., Tuschl, T., Sharp, P.A. and Bartel, D.P. (2000) RNAi: double-stranded RNA directs the ATP-dependent cleavage of mRNA at 21 to 23 nucleotide intervals. *Cell*, **101**, 25-33.

Zhang, J., Wilson, A., Alber, S., Ma, Z., Tang, Z.L., Satoh, E., Mazda, O., Watkins, S., Huang, L., Pitt, B. and Li, S. (2003) Prolonged gene expression in mouse lung endothelial cells following transfection with Epstein-Barr virus-based episomal plasmid. *Gene Ther*, **10**, 822-826.

Zhang, Y., Zhou, J. and Lim, C.U. (2006) The role of NBS1 in DNA double strand break repair, telomere stability, and cell cycle checkpoint control. *Cell Res*, **16**, 45-54.

Zhou, J., Chau, C.M., Deng, Z., Shiekhattar, R., Spindler, M.P., Schepers, A. and Lieberman, P.M. (2005) Cell cycle regulation of chromatin at an origin of DNA replication. *Embo J*, **24**, 1406-1417.

Zhou, J., Lim, C.U., Li, J.J., Cai, L. and Zhang, Y. (2006) The role of NBS1 in the modulation of PIKK family proteins ATM and ATR in the cellular response to DNA damage. *Cancer Lett*.

In: New Research on DNA Repair
Editor: Breehn R. Landseer, pp. 219-249

ISBN 1-60021-385-5

Chapter VIII

PECULIARITIES OF BASE EXCISION REPAIR IN LOACH EMBRYOGENESIS

Natalia Sharova[*]

N.K. Koltsov Institute of Developmental Biology, Russian Academy of Sciences, Vavilova St, 26, Moscow 119334, Russia.

ABSTRACT

Base excision repair (BER) is one of the DNA repair types, by means of which most of damages in DNA structure are removed. BER is performed through short patch or long patch pathways, the choice of which depends on the kind of the DNA damage and the level of the repair DNA polymerases in a cell. In this work, we were interested in the elucidation of mechanisms of the repair DNA polymerases action in embryogenesis of teleost fish *Misgurnus fossilis* (loach). We identified repair DNA polymerase β (pol β) and DNA polymerase δ (pol δ) in loach eggs and embryos. The pol β activity was extremely low in eggs and embryos (7.2×10^{-4} U per an egg or 3.6 U per gram of protein) up to gastrula stage. The pol δ activity 55 times exceeded the pol β activity at the same period of development, that is why pol δ became the object of our close attention. Repair properties of pol δ were investigated with the use of the model substrates – DNA duplexes containing gaps, the length of which (1-9 nucleotides, nt) corresponded with the length of gaps forming in BER. Pol δ displayed unexpected properties upon the interaction with such duplexes: this enzyme processively filled the gaps of 5 nt and 9 nt in the absence of processivity factors (PCNA, RFC, and ATP), while it elongated primers on single stranded DNA templates in a distributive manner, as it was expected. The mechanism of the unusual processivity of pol δ was studied by electrophoretic mobility shift assay. Pol δ was shown to interact with both ends of 5 nt and 9 nt gaps, demonstrating more affinity to the latter. This interaction stabilizes the complex between pol δ and DNA duplex, allowing pol δ to reveal the processivity.

[*] Correspondence concerning this article should be addressed to Natalia Sharova, N.K. Koltsov Institute of Developmental Biology, Russian Academy of Sciences, Vavilova St, 26, Moscow 119334, Russia; fax: (7-495)135-8012, e-mail: nsharova@proxima.idb.ac.ru.

To confirm the participation of pol δ in BER *in vivo*, after the beginning of cell division, embryos were affected by superoxide radicals in the xanthine – xanthine oxidase system generating mainly single strand breaks in DNA, which are repaired by long patch pathway. In response to this influence, the activity of pol δ (but not replicative pol α) increased by 370% in embryos at the middle gastrula stage. Thus, pol δ is the main repair DNA polymerase in loach embryos and this enzyme seems to be able to carry out long patch pathway of BER by earlier unknown mechanism, which is important for extreme conditions at low levels of PCNA and PCNA-independent repair DNA polymerases in a cell.

INTRODUCTION

Most damages in DNA structure (modified and damaged bases, normal and abnormal AP-sites, and single strand breaks) are removed by means of base excision repair (BER) (Sharova, 2005; Sweasy et al., 2006). This explains why this type of repair is the most studied. There are two pathways of BER, short patch BER and long patch BER. The former involves excision of only one (damaged) nucleotide, whereas the latter involves excision of a longer DNA stretch, which may include from 2 to 10 nucleotides (Wood, Shivji, 1997; Pascucci et al., 1999; Dogliotti et al., 2001).

What factors determine selection of the BER pathway? First of all, it is the type of DNA damage. Repair of modified bases, such as 8-oxoG, uracil, and also normal AP-sites usually involves short patch BER (Dogliotti et al., 2001). Abnormal AP-sites (reduced or oxidized) and single strand breaks are repaired via long patch BER (Klungland, Lindahl, 1997; Wilson, Thompson, 1997; Sekelsky et al., 1998). Removal of hypoxanthine, the product of adenine deamination, may involve either pathway of BER. The cellular level of DNA polymerases is another crucial factor influencing selection of the BER pathway. For example, wild type mouse embryonal fibroblast cell lines preferentially repair AP-sites via short patch BER (in 80% of cases), whereas cells with a damaged gene encoding DNA polymerase β repair AP-sites only via the long patch pathway (Dogliotti et al., 2001).

Short Patch BER

The first step of the short patch BER includes recognition and removal of modified base by corresponding DNA N-glycosylase. This step leaves deoxyribophosphate, a structure similar to an AP-site. The next step includes nicking the sugar-phosphate backbone by AP-endonuclease type II at the 5'-end of this intermediate and removal of the 5'-terminal residue of deoxyribose-phosphate. The resulting gap (of one nucleotide) is correctly filled in by DNA polymerase and the break is linked by DNA ligase (Wood, Shivji, 1997). Good evidence exists that DNA polymerase β of low processivity is involved into short patch BER. This was proved by: 1) studies of uracil repair in mammalian cell extracts in the presence of various DNA polymerase inhibitors (Dianov et al., 1992); 2) fractionation of cell extracts resulted in separation of activities repairing AP-sites or uracil (Matsumoto et al., 1994; Singhal et al., 1995; Nealon et al., 1996); 3) BER sensitivity to antibodies against DNA polymerase β

(Wiebauer, Jiricny, 1990; Singhal et al., 1995); 4) reconstitution of repair processes using purified enzymes (Kubota et al., 1996); 5) mutation analysis (Sobol et al., 1996). DNA polymerase β plays a dual role in short patch BER: it removes the 5'-terminal deoxyribose-phosphate residue of the AP-site via β-elimination due to AP-lyase (5'-deoxyribose-phosphatase) activity (Matsumoto, Kim, 1995) and fills the gap of one nucleotide due to DNA polymerase activity. It was shown that short patch repair of DNA containing uracil *in vitro* may involve recently discovered DNA polymerase λ. Besides DNA polymerase activity, this enzyme exhibits 5'-deoxyribose-phosphate lyase (5'-dRp-lyase) but not AP-lyase activity (Garcia-Diaz et al., 2001). This activity of DNA polymerase λ removes nucleotide residues containing modified bases (but not AP-sites as DNA polymerase β does). It is possible that in the case of necessity the cell can employ both types of DNA polymerases for both types of short patch BER.

After the filling of one nucleotide gap by DNA polymerase β or λ, short patch BER is completed with ligation of the break by DNA ligase I (Prasad et al., 1996) or complex of protein XRCCl and DNA ligase III (Caldecott et al., 1995; Cappelli et al., 1997). Protein XRCCl lacks catalytic activity, but it binds to repair proteins such as DNA polymerase β and poly(ADP-ribose) polymerase (PARP) and promotes involvement of these proteins into BER. What is the putative role of PARP in short patch repair? Perhaps PARP is a sensor recognizing DNA nicks (Wilson, Thompson, 1997). Short patch BER requires coordinative interactions between all protein components of this process. Undoubtedly, such concerted action significantly facilitates the repair process. Interaction between DNA polymerase β and AP-endonuclease and between DNA polymerase β and DNA ligase I has also been demonstrated (Wilson, Thompson, 1997; Tomkinson et al., 2001). Kinetic studies proved that AP-endonuclease accelerates removal of 5'-terminal deoxyribose-5'-phosphate by DNA polymerase β (Bennett et al., 1997).

Long Patch BER

Certain kinds of DNA damage (such as reduced and oxidized AP-sites, single strand breaks) can be repaired only by long patch BER. In some cases repair of modified bases and normal AP-sites also involves long patch BER. Repair of single strand breaks obviously begins with PARP (Satoh, Lindahl, 1992). The authors proposed the following model for functioning of this protein. PARP recognizes a break and binds to DNA at the damaged site. This activates synthesis of poly(ADP-ribose) at a specific site of PARP; thus PARP undergoes automodification. The latter decreases PARP affinity to DNA and the damaged site becomes accessible to repair enzymes, which are obviously directed by PARP. Regeneration of the modified PARP includes degradation of poly(ADP-ribose) by a specific glycohydrolase.

Modified base long patch BER (as well as short patch BER) begins with corresponding DNA N-glycosylase, and repair of the AP-site involves AP-endonuclease (Wilson, Thompson, 1997). It is suggested that these enzymes also act as factors recognizing damages. After DNA nicking by AP-endonuclease, FEN1 nuclease removes from 2 to 10 nucleotides from the 5'-side of the AP-site (Klungland, Lindahl, 1997; Wood, Shivji, 1997). For repair of reduced and oxidized AP-sites this is the only possible way because abnormal AP-sites are not substrates for 5'-phosphodiesterase activity of DNA polymerase β. Why are modified AP-sites not cleaved by DNA polymerase β as in the case of short patch BER? A mechanism for interaction between DNA polymerase β and oxidized AP-site has been proposed (DeMott et al., 2002). Oxidized AP-site appears as the result of effects of oxidative stress or radiation. Free radicals attack C1 carbon atom and this results in formation of 2-deoxyribonolactone (dL). Although AP-endonuclease could nick dL-containing DNA, subsequent dL excision by DNA polymerase β was strongly inhibited compared with normal AP-sites. This inhibition was accompanied by accumulation of crosslinks between DNA and DNA polymerase β. Such crosslinks were not detected in the case of unmodified AP-sites, but they were formed during reduction by sodium borohydride. Crosslinking of protein with dL involved Lys72 forming a Schiff base with the Cl aldehyde during removal of an unmodified AP-site. In the presence of dL, attack of Cl lactone by Lys72 occurs slower and apparently produces an amide compound. This explains why oxidized AP-sites cannot be removed via β-elimination by DNA polymerase β in the process of short patch BER. They are removed by FEN1 in long patch BER.

FEN1 is a nuclease exhibiting dual activity. It acts as: 1) 5'→3'-exonuclease cleaving mononucleotides and short oligonucleotides from the 5'-end of one strand of DNA duplex (Lindahl et al., 1969); 2) endonuclease cleaving single strand “branches”, known as flap structures, possessing free 5'-end in DNA duplex (Harrington, Lieber, 1994). The name FEN1 is an abbreviation reflecting the presence of both exo- and endonuclease activities: 5(**f**ive)'-**e**xo-**n**uclease and **f**lap-**e**ndo-**n**uclease (Wu et al., 1996). Do exo- and endonuclease activities of FEN1 reflect different catalytic properties or does this enzyme exhibit the same activity with different substrates (double strand DNA with nick containing phosphorylated 5'-end and flap structure with free 5'-end which is not necessarily phosphorylated)? Certain evidence exists that the second suggestion is correct (Wu et al., 1996). The 5'-OH end of a 5'-flap structure is a substrate, which binds to FEN1, but 5'-OH end within a nick in a DNA duplex is not equivalent to that. Authors suggest that electrostatic pulsation of 5'-terminal phosphate within nick (FEN1 substrate) may create “breathing” 5'-end, which may acquire the configuration of a pseudo flap structure. It is also known that a flap one nucleotide in length is a substrate effectively utilized by FEN1 (Harrington, Lieber, 1994). Anyway, both double stranded DNA and single strand flap structures are the substrates for FEN1. This suggests the existence of two mechanisms employing FEN1 for removal of damaged DNA sites: 1) FEN1 forms gaps in DNA duplex which are filled by DNA polymerases; 2) FEN1 removes flap structures formed during preliminary DNA synthesis with strand displacement.

Which DNA polymerases are involved in DNA synthesis in long patch BER? Experiments with model systems reconstituting long patch repair of AP-sites using purified human cell proteins revealed the involvement of DNA polymerases δ and ε in reparative DNA synthesis (Matsumoto et al., 1999; Pascucci et al., 1999). These DNA polymerases exhibit high processivity in the presence of PCNA (proliferating cell nuclear antigen); this protein retains polymerases on a DNA molecule until they synthesize a strand of requested length. Maximal length of newly synthesized strand reaches 7 (Matsumoto et al., 1999) or 10 (Pascucci et al., 1999) nucleotides. In most cases, the length of newly synthesized DNA fragment does not exceed 2 (Matsumoto et al., 1999) or 2-4 (Pascucci et al., 1999) nucleotides.

Certain evidence exists that DNA polymerase β may be involved in long patch repair of AP-sites (Klungland, Lindahl, 1997) and uracil containing DNA (Horton et al., 2000) as the enzyme synthesizing DNA. In long patch BER, DNA polymerase β synthesizes DNA fragments of 2 to 6 nucleotides in length (Klungland, Lindahl, 1997) in spite of low processivity of this enzyme on primer-templates and insensitivity to stimulation by PCNA. Involvement of DNA polymerase β becomes possible due to stabilization of enzyme-DNA complex by binding of the catalytic deoxyribose-phosphate lyase center of this enzyme to 5'-phosphate limiting a single strand gap (Matsumoto, Kim, 1995; Prasad et al., 1998). Recent data suggest that newly discovered DNA polymerase λ may also fill small gaps (up to 5 nucleotides) in processive manner and possibly participates in reparative DNA synthesis during long patch BER (Garsia-Diaz et al., 2002). It is possible that DNA polymerase λ employs the same mechanism filling gaps as DNA polymerase β. Long patch BER is finished with the linking of DNA fragments with DNA ligase I.

Recently it was shown that DNA ligase III and XRCC1 regulate the switch between short patch and long patch BER of damaged DNA bases (Petermann et al., 2006). XRCC1 stimulates DNA polymerase β strand displacement activity under conditions of ATP shortage and thus promotes long patch BER. If sufficient ATP is available, ligation by ligase III prevents strand displacement, leading to short patch BER.

By analogy with short patch BER, regulation of long patch BER involves coordinated protein-protein interactions. For example, AP-endonuclease may interact with FEN1 (Dianova et al., 2001). This interaction activates by 1.5-2- fold removal of reduced 5'-terminal AP-site by FEN1. Interestingly, PCNA can interact with all enzymes involved in long patch BER. Besides the activating effect of PCNA on DNA polymerases δ and ε, PCNA can also stimulate FEN1 nuclease (Li et al., 1995; Wu et al., 1996; Klungland, Lindahl, 1997; Jonsson et al., 1998; Gary et al., 1999; Gomes, Burgers, 2000; Tom et al., 2000) and bind AP-endonuclease (Dianova et al., 2001), DNA-ligase I (Levin et al., 1997; Jonsson et al., 1998; Matsumoto, 2001; Tom et al., 2001), DNA N-glycosylases (Otterlei et al., 1999; Matsumoto, 2001), DNA polymerase β (Kedar et al., 2002), and DNA polymerase λ (Maga et al., 2002). It was also demonstrated that PCNA binding to DNA polymerase δ, FEN1, DNA-ligase I, and two DNA N-glycosylases (UNG2 and MYH) involves the same consensus sequence found in these proteins (Matsumoto, 2001). It is suggested that PCNA plays a role of molecular adaptor facilitating interaction of these enzymes with sites of DNA repair. However, it should be noted that an activating effect of PCNA on some of these enzymes (AP-endonuclease, DNA N-glycosylase, DNA polymerases β and λ) has not been found.

In spite of the successful study of BER for last years, the problem touching peculiarities of BER mechanisms in embryogenesis remained outside of interest of investigators. At the same time, embryogenesis requires the concentrated control of DNA structure and it is very important for the study of DNA repair. So, we investigated the peculiarities of functioning of DNA polymerases, which are the key enzymes of BER, in eggs and embryos of teleost fish loach (*Misgurnus fossilis*).

MATERIALS AND METHODS

Mature loach eggs were obtained as described by Neifakh (Neifakh, 1959).

Enzymes and Reagents

Human DNA polymerase ε, calf thymus DNA polymerase δ and rabbit polyclonal antibodies against calf thymus DNA polymerase δ were provided by D. Mozzherin (State University of New York at Stony Brook). DNA polymerase α was purified from mature loach eggs as previously described (Sharova et al., 1993). Antibodies against rabbit IgG conjugated with peroxidase were purchased from Sigma (USA). Polynucleotide kinase of phage T4 was from Boehringer (Germany).

Fractionation of DNA Polymerases

Buffers

Buffer A: 50 mM potassium phosphate buffer, pH 7.5, containing 10% glycerol, 5 mM 2-mercaptoethanol, 1 mM EDTA, 10 mM $Na_2S_2O_5$, 1 mM phenylmethylsulfonyl fluoride (PMSF), 1 μM leupeptin, pepstatin (1 μg/ml), and aprotinin (1 μg/ml). Buffer B: 10 mM potassium phosphate buffer, pH 7.5, containing 10% glycerol, 5 mM 2-mercaptoethanol, 1 mM EDTA, 10 mM $Na_2S_2O_3$, 0.2 mM PMSF, 1 μM leupeptin, pepstatin (1 μg/ml), and aprotinin (1 μg/ml). Buffer C is identical to buffer B except that concentration of glycerol is increased to 20%. Buffer D: 10 mM potassium phosphate buffer, pH 7.5, containing 50% glycerol, 5 mM 2-mercaptoethanol, 1 mM EDTA, 10 mM $Na_2S_2O_5$, 0.2 mM PMSF, 1 μM leupeptin, pepstatin (1 μg/ml), and aprotinin (1 μg/ml). Buffer E: 200 mM potassium phosphate buffer, pH 7.5, containing 5 mM 2-mercaptoethanol, 1 mM EDTA, 10 mM $Na_2S_2O_5$, 0.2 mM PMSF, 1 μM leupeptin, pepstatin (1 μg/ml), and aprotinin (1 μg/ml).

Chromatography of Extract from Eggs on DEAE-Toyopearl Column

All procedures were carried out at 0-4°C. The extract was obtained by centrifugation of eggs using an SW-28 rotor (Beckman, USA) at 20,000 rpm for 30 min as described by Minin et al. (Minin et al., 1987). The extract (4.5 ml) was suspended in buffer A in at 1 : 3 ratio and centrifuged at 105,000g for 1 h. Buffer B was added to supernatant to potassium phosphate concentration of 20 mM. The solution was placed on the column with DEAE-Toyopearl (1.6 x 10 cm) equilibrated with buffer B. The rate of elution was 20 ml/h. The column was washed with 40 ml of buffer C and eluted with buffer C in which the concentration of potassium phosphate increased linearly to 0.4 M over a period of 10 h. The rate of elution was 20 ml/h. In fractions obtained, we determined the activity of DNA polymerases and the content of immunoreactive proteins by Western blotting using polyclonal antibodies against calf thymus DNA polymerase δ.

Purification of DNA Polymerases β and δ

DNA polymerases were isolated from 12 ml of extract of eggs. The extract was obtained as described above. The supernatant was placed on a column with phosphocellulose P-ll (2.5 x 15 cm) equilibrated with buffer B at the rate of 50 ml/h. The column was washed with 350 ml of buffer B containing 0.23 M KCl and eluted with 0.5 M KCl in buffer B. The active fractions were combined and dialyzed against buffer C. Then the preparation was placed on a column with DEAE-Toyopearl (1.6 x 8.5 cm) equilibrated with buffer C at the rate of 20 ml/h. The column was washed with 120 ml of buffer C containing 20 mM potassium phosphate and eluted with buffer C in which the concentration of potassium phosphate increased linearly from 20 to 90 mM over a period of 6 h. The rate of elution was 20 ml/h. Fractions possessing DNA polymerase activity sensitive to aphidicolin but resistant to N2-(p-n-butylphenyl)-2'-deoxyguanosine 5'-triphosphate (BuPdGTP) were combined. Dry ammonium sulfate was added to the preparation by small portions with stirring until they completely dissolved. The final concentration of $(NH_4)_2SO_4$ was 0.5 M. Then the preparation was placed on a column with phenyl-Sepharose (1 x 3.5 cm) equilibrated with buffer C containing 0.5 M $(NH_4)_2SO_4$ at the rate of 10 ml/h. The column was washed with 6 ml of buffer C containing 0.5 M $(NH_4)_2SO_4$ and then eluted at the rate of 8 ml/h it two stages (each over a period of 2 h): first by 0.5-0 M gradient of $(NH_4)_2SO_4$ in buffer C and then by 0-1% gradient of Triton X-100 in buffer C. DNA polymerase β was not bound to phenyl-Sepharose and was separated when the preparation was loaded in 0.5 M $(NH_4)_2SO_4$. DNA polymerase δ was eluted in two fractions: the first fraction (δ1) corresponded to 0.37-0.10 M interval of $(NH_4)_2SO_4$ and the second fraction (δ2) was obtained in 0.6-0.9% interval of Triton X-100. The fractions of DNA polymerases β, δ1 and δ2 were dialyzed separately against buffer C and placed on identical columns with single-stranded DNA-cellulose (0.8 x 1.7 cm) equilibrated with buffer C with the rate of 5 ml/h. The columns were washed with 2.5 ml of buffer C and eluted with a 0-0.8 M gradient of KCl in buffer C over a period of 4 h with the rate of 5 ml/h. DNA polymerase β was eluted from DNA-cellulose at relatively high ionic strength (0.4-0.6 M KCl), while DNA polymerases δ1 and δ2 were eluted with 0.05-0.2 M KCl. The active fractions of DNA polymerases δ1 and δ2 were dialyzed against buffer D and stored at -20°C. The active fractions of DNA polymerase β were concentrated using polyethylene glycol (M_r 40,000) to a volume of 0.1-0.2 ml. This preparation was dialyzed against 10% glycerol in buffer E for 4 h and then applied onto 4.9 ml of a linear gradient of 15 to 30% glycerol in buffer E for further centrifugation. Centrifugation was carried out using rotor SW 50.1 (Beckman, USA) at 48,000 rpm and 2°C for 26 h. Catalase (11.3S), aldolase (8.3S), and BSA (4,3S) at 0.5 mg of each were used as standards in separate tubes. The gradients were fractionated from the bottom of the tubes and the obtained fractions were investigated.

Immunoblotting

The proteins were separated by electrophoresis in 8% polyacrylamide gel in the presence of SDS. Then the semi-dry transfer of the proteins from the gel to nitrocellulose membrane was carried out. The membrane was incubated at room temperature for 1 h in buffer TNT (10 mM Tris-HCl buffer, pH 7.5, 150 mM NaCl and 0.1 % Tween-20) containing 20% serum of cattle and polyclonal antibodies against calf thymus DNA polymerase δ in dilution 1 : 1000. Then the membrane was washed several times by buffer TNT and incubated for 1 h in buffer TNT containing 20% serum of cattle and antibodies against rabbit IgG conjugated with peroxidase in dilution 1 : 5000. After washing in buffer TNT the membrane was placed in 50 mM Tris-HCl buffer, pH 7.5, containing diaminobenzidine (0.1 mg/ml) and 30% H_2O_2 until the coloring appears.

DNA Substrates

To prepare DNA substrates, the following oligonucleotides were used: 17-mer (TGCCGGGATCATAGAAG), 21-mer (AGAGTGCCGGGATCATAGAAG), 25-mer (ACACAGAGTGCCGGGATCATAGAAG), 29-mer (AGCCACACAGAGTGCCGGGATC ATAGAAG), standard primer 21-mer (AAGCGGAGTGTATGTGCAGTG), primer with unpaired 3'- tail (17+4)-mer (AAGCGGAGTGTATGTGCCTGA), and template 51-mer (CTTCTATGATCCCGGCACTCTGTGTGGCTACACTGCACATACACTCCGCTT). Oligonucleotides (primers 21-mer, (17+4)-mer, or template 51-mer) were labeled at the 5'-terminus with [γ-^{32}P]ATP using phage T4 polynucleotide kinase. Labeled primers were hybridized with the template at a molar ratio of 1 : 1.2. Labeled template 51-mer was hybridized with the primers at a ratio of 1 : 1.5. To obtain duplexes with gaps of one nucleotide (*21 : 51 : 29-mer), five nucleotides (*21 : 51 : 25-mer), nine nucleotides (*21 : 51 : 21-mer), and thirteen nucleotides (*21 : 51 : 17-mer), the labeled primer *21-mer was hybridized with the template 51-mer and the corresponding oligonucleotides at a ratio of 1 : 2 : 2.5. Duplexes with the primer (17+4), in which four residues at the 3'-end are not complementary to the template, were obtained in a similar way. 5'-[^{32}P]dT_{10} was hybridized with poly (dA) at a ratio 1:10.

Elongation of Primers by DNA Polymerase δ

The ability of DNA polymerase δ for processive synthesis was determined in 20-μl samples containing 30 mM Tris-HCl, pH 7.5, 6 mM $MgCl_2$, BSA (100 ug/ml), 2 mM dithiothreitol, 0.42 pmol of ^{32}P-labeled primer-template complex *21 : 51-mer (or of the duplex with the gap of five nucleotides (*21 : 51 : 25-mer) or nine nucleotides (*21 : 51 : 21-mer)), and dATP, dGTP, dCTP, and TTP (50 μM each). The reaction was started by adding 0.015 units of DNA polymerase δ to the reaction medium at 30°C. After 1, 2, and 4 min of incubation, 5-μl aliquots were taken and mixed with 7 μl of stop solution containing formamide and the leader dyes. The preparations were analyzed by electrophoresis in polyacrylamide gel (10 or 12.5% polyacrylamide) in the presence of 8 M urea. The ability of DNA polymerase δ for strand displacement synthesis was determined in reaction medium (20 μl) that contained 0.42 pmol of gapped ^{32}P-labeled duplexes and the other components listed above. Aliquots of the reaction mixture were taken at 5, 12, and 30 min after the reaction was initiated by adding 0.15 units of DNA polymerase δ.

Binding of DNA polymerase δ to DNA Substrates

DNA polymerase δ (0.004-0.06 units) was incubated with 0.005-0.04 pmol of ^{32}P-labeled DNA substrate in 10 μl of buffer containing 10 mM Tris-HCl, pH 8.0, BSA (100 μg/ ml), and 2 mM dithiothreitol at 22-23°C for 10-15 min. The samples were analyzed by electrophoresis in polyacrylamide gel (60 x 100 x 0.75 mm) containing 6% acrylamide (acrylamide/bisacrylamide, 60 : 1) prepared in 20 mM HEPES, pH 8.0, and 0.1 mM EDTA. Electrophoresis was performed at 80 V in buffer containing 20 mM HEPES, pH 8.0, 0.1 mM EDTA, and 2 mM thioglycolic acid at room temperature. Then the gels were transferred onto DE81 paper and dried under vacuum.

Determination of DNA Polymerase δ Activity in the Gel

DNA polymerase δ (0.03 units) was incubated with 0.04 pmol of the duplex 21:51: 21-mer (nine-nucleotide gap) or the duplex 21 : 51 : 17-mer (thirteen-nucleotide gap) in 10 μl of the reaction mixture described above at 4°C for 20 min. Electrophoresis was performed at 4°C in buffer containing 20 mM HEPES, pH 8.0, 0.1 mM EDTA, and 2 mM thioglycolic acid. The gel was incubated in 15 ml of the buffer containing 10 mM Tris-HCl, pH 7.5, 8 mM $MgCl_2$, 2 mM dithiothreitol, BSA (100 μg/ml), dGTP, dCTP, TTP (5 μM each), and 50 pmol of [α-^{32}P]dATP (750 Ci/mmol) at 30°C for 10 min. Then the gel was washed with 5% TCA supplemented with 1% sodium pyrophosphate, transferred onto chromatographic paper, and dried under vacuum.

Determination of DNA Polymerase Activity *in vitro*

The enzymatic activity of DNA polymerases was measured by incorporation of [^{3}H]dNMP into acid-insoluble product as described previously (Sharova et al., 1992). The reaction mixture (50 μl) contained 10 mM Tris-HCl (pH 7.5), 50 μg/ml DNA activated by DNAse, 8 mM $MgCl_2$, 200 μg/ml BSA, 2 mM dithiothreitol, 25 μM dATP, 25 μM dCTP, 25 μM dGTP, 10 μM TTP, and 20 μCi/ml [^{3}H]TTP. Incorporation of 1 nmol of dNMP into activated DNA during 1 h incubation was taken as the DNA polymerase activity unit.

Exonuclease Assay

Labeled oligonucleotides (3′-[^{3}H]dC$_1$)dT$_{100}$, (3′-[^{3}H]dT$_1$)dT$_{100}$, and 5′-[^{32}P]dT$_{100}$ were prepared as described earlier (Sharova et al., 1995). The oligonucleotides were annealed with a twofold quantity of poly(dA). Reaction mixtures (100 μl) contained labeled dT$_{100}$:poly(dA) (0.025:0.05 optical units), 20 mM Tris-HCl, pH 7.5, 10 mM $MgCl_2$, BSA (100 μg/ml), and 1 mM dithiothreitol. Reaction was started by addition of DNA polymerase (0.07 units) and carried out at 30°C for different incubation times. For some experiments 3′-[^{3}H]poly(dT)

(0.02 optical units, 7000-9000 counts/min) was used as DNA polymerase β or δ substrate. Radioactivity of the acid-insoluble fraction was determined.

Induction of the Oxidative Stress

After the beginning of cell division, the embryos were incubated for 20 min in 15 mM sodium carbonate buffer (pH 10.2) containing 0.3 x 10^{-4} M xanthine, 0.3 x 10^{-4} M EDTA, and 0.1 µl/ml xanthine oxidase (Sigma, grade III). Control embryos were incubated in the same medium minus xanthine oxidase. At certain stages of development, 2000 embryos were collected and prepared as described earlier (Peskin et al., 1997).

Other Methods

To study changes of the DNA polymerases β and δ activities in loach embryogenesis 2000 embryos at different stages of development were collected, homogenized in buffer A and centrifuged at 105,000g for 1 h; proteins of supernatant were chromatographied on DEAE-Toyopearl and phenyl-Sepharose columns successively (as described above). The activity of Cu,Zn-superoxide dismutase was measured as described earlier (Peskin et al., 1997). Protein concentration was determined spectrophotometrically. Salt concentration in fractions obtained in the chromatographic experiments was measured using a conductometer.

Results and Discussion

DNA Polymerase β in Loach Embryogenesis

Isolation and Characterization of DNA Polymerase β

In order to characterize DNA polymerase β we had to obtain highly purified preparation of this enzyme. The strategy for DNA polymerase β purification is described in "Materials and Methods". In the first steps of the purification (up to DEAE-Toyopearl chromatography), loach DNA polymerases were eluted together. When the preparation was subjected to DEAE-Toyopearl chromatography, DNA polymerase α (that was sensitive to BuPdGTP) was separated from other DNA polymerases that were resistant to BuPdGTP. After the phenyl-Sepharose chromatography the preparation of DNA polymerases, resistant to BuPdGTP, was divided into three fractions. A small portion of polymerase activity (about 2%) of this preparation was not bound to the hydrophobic sorbent when loaded in 0.5 M $(NH_4)_2SO_4$. This fraction contained DNA polymerase β that was identified by its resistance to N-ethylmaleimide and sensitivity to d_2TTP. Other fractions were bound to the sorbent and were eluted by 0.5-0 M gradient of $(NH_4)_2SO_4$ and then by 0-1% gradient of Triton X-100. Both of them were highly sensitive to N-ethylmaleimide. The DNA polymerase β fraction was subjected to single-stranded DNA-cellulose chromatography. All these procedures allowed to obtain the highly purified preparation of DNA polymerase β with the specific activity 30,000 units/mg of protein and to investigate its properties.

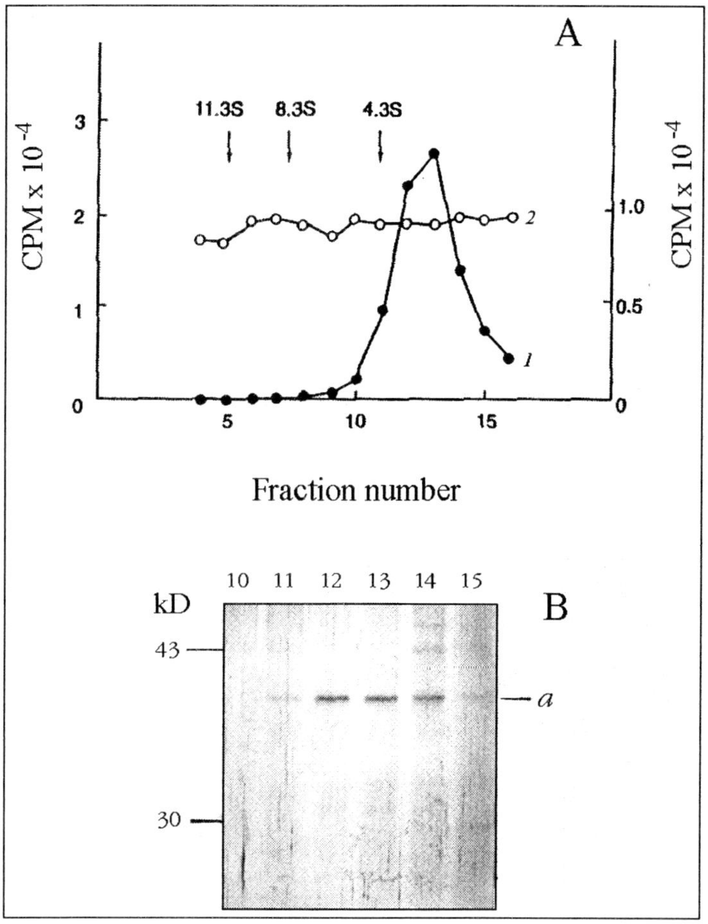

Fig. 1. (A) Sedimentation of a preparation of loach DNA polymerase β in 15-30% glycerol gradient: (*1*) DNA polymerase activity measured by the incorporation of [^{3}H]dNMP into activated DNA (left ordinate axis); (*2*) 3′→5′-exonuclease activity measured by hydrolysis of 3′-[^{3}H]poly(dT) (right ordinate axis). Standard deviation is less than 2%. (B) Electrophoresis of protein fractions in 8% polyacrylamide gel in the presence of SDS. 1/7 part of each active fraction obtained by sedimentation in glycerol gradient was used per sample. Sedimentation standards: catalase (11.3S), aldolase (8.3S), and BSA (4.3S). Standards for electrophoresis: ovalbumin (43 kD), and carbonic anhydrase (30 kD); *a* is polypeptide with apparent molecular mass of 39 kD (the gel was stained with silver).

SDS-polyacrylamide (8.5%) gel electrophoresis of DNA polymerase β fractions, obtained by sedimentation in glycerol gradient, revealed one major polypeptide with apparent molecular mass 39 kD. Its amount was proportional to the DNA polymerase activity (Fig. 1). The sedimentation constant of DNA polymerase β is 3S. DNA polymerase β is free from 3′→5′-exonuclease (Fig. 1).

The processivity of DNA polymerase β was determined by addition of increasing amounts of poly(dA)·dT_{10} (10:1) to a given amount of the enzyme in order to saturate the reaction with the primer-template complex (Fig. 2). The processivity of loach DNA polymerase β (lanes 8-10) was compared with that of human DNA polymerase ε (lanes 2-4) and loach DNA polymerase α (lanes 5-7). Highly processive DNA polymerase ε synthesizes DNA chains up to 100 bp on poly(dA) template. DNA polymerase α with moderate processivity synthesizes DNA fragments up to 15 bp. DNA polymerase β adds on average one nucleotide to the primer and displays the low processivity.

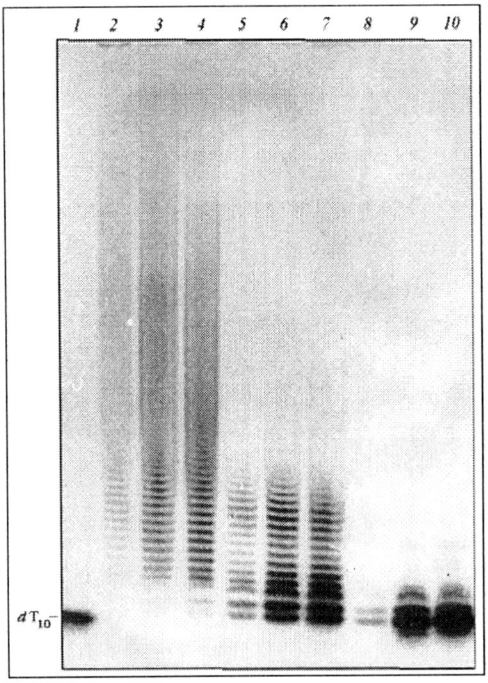

Fig. 2. Processivity of human DNA polymerase ε (lanes *2-4*), loach DNA polymerase α (lanes *5-7*), and loach DNA polymerase β (lanes *8-10*) on poly(dA)·dT_{10} (10:1) complex at concentrations of 10 (lanes *2, 5, 8*), 50 (lanes *3, 6, 9*), and 250 µg/ml (lanes *4, 7, 10*). Each of DNA polymerases was added in amount of 0.005 units. Autoradiography after polyacrylamide gel (10%) electrophoresis of reaction products in the presence of 7 M urea. 5′-[^{32}P]dT_{10} was used as a marker (*1*).

Thus, the properties of loach DNA polymerase β, namely the molecular mass of 39 kD, the sedimentation constant of 3S, the lack of 3′→5′-exonuclease activity, the low processivity, and the unique sensitivity to inhibitors are typical for this type of DNA polymerases of higher eukaryotes.

Dynamics of DNA Polymerase β Activity in Loach Embryogenesis

The peculiarity of loach eggs and embryos up to gastrula stage is extremely low activity of DNA polymerase β (7.2×10^{-4} unit per an egg or 3.6 units per gram of protein). At the same time, the DNA polymerase β activity in eggs of axolotl almost 100 times exceeds that in loach eggs (Carre et al., 1981). The DNA polymerase β activity increases during loach embryogenesis, but not strongly: it is twice more at the middle gastrula stage and it is three times more at the beginning of organogenesis in comparison with the blastula stage. Indirectly, the deficit of the repair DNA polymerase β activity in loach embryos indicates the presence of another enzymatic system, capable to perform BER. That is why we began to look for another repair DNA polymerase in loach eggs and embryos.

DNA Polymerase δ in Loach Embryogenesis

Identification of DNA Polymerase δ in Extracts of Loach Eggs

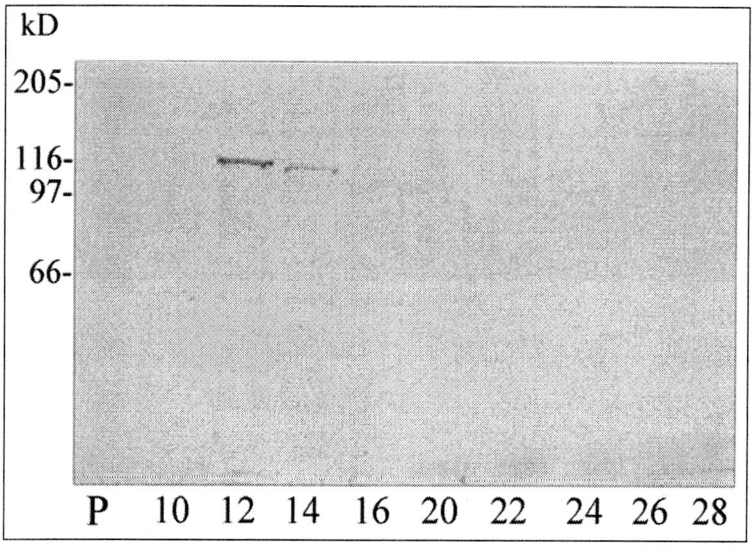

Fig. 3. Western blot of proteins from fractions of loach eggs extract. Fractions were obtained by chromatography on a column with DEAE-Toyopearl. Proteins were separated by electrophoresis in 8% polyacrylamide gel in the presence of SDS. Polyclonal antibodies against calf thymus DNA polymerase δ were used as primary antibodies. Numbers of fractions used in the experiments are shown in the horizontal axis (P are the proteins that were not bound with the ion exchanger). Standards for electrophoresis: myosin (205 kD), β-galactosidase (116 kD), phosphorylase b (97 kD), BSA (66 kD).

Extracts of loach eggs were fractionated using column chromatography on ion-exchanger DEAE-Toyopearl. Proteins were eluted with a 0-0.4 M gradient of potassium phosphate buffer. The activity of DNA polymerases in the fractions was determined using activated DNA as a template-primer in the absence and in the presence of BuPdGTP, a selective inhibitor of DNA polymerase α. The main DNA polymerase activity in extracts corresponded to DNA polymerase α. This enzyme was eluted from the ion exchanger at concentration of potassium phosphate higher than 100 mM and was sensitive to the inhibiting action of aphidicolin and BuPdGTP. DNA polymerase α of loach was isolated and characterized by us previously (Sharova et al., 1993; Sharova, Mikhailov, 1995). In fractions eluted in the interval from 30 to 80 mM (fractions 9-14), minor DNA polymerase activity sensitive to aphidicolin but resistant to BuPdGTP was found. This activity might correspond to DNA polymerases δ and ε and probably other unidentified enzymes. To identify DNA polymerase δ in fractions obtained with a gradient, the corresponding proteins were analyzed by electrophoresis in polyacrylamide gel in the presence of SDS followed by Western blotting with the use of polyclonal antibodies against calf thymus DNA polymerase δ (Fig. 3).

Fractions 12 and 14 showed an immunoreactive polypeptide with molecular mass of 120-130 kD. There were no other immunoreactive polypeptides in these fractions. (Special experiments showed that fractions 1-4 and 17-19 lacked immunoreactive polypeptides.) The molecular mass of 120-130 kD corresponds to the molecular mass of the catalytic subunit of DNA polymerases δ from various sources (Hindges, Hubscher, 1997). Thus, the immunoreactive polypeptide with molecular mass of 120-130 kD is probably the catalytic subunit of DNA polymerase δ of loach. We use the fact that chromatography on DEAE-Toyopearl results in separation of DNA polymerases δ and α for elaboration of a preparative method of isolation of DNA polymerase δ from loach eggs.

Isolation of DNA Polymerase δ

To obtain the preparation of DNA polymerase δ of loach in the preparative quantities, column chromatography of extract of unfertilized eggs was carried out sequentially on phosphocellulose, DEAE-Toyopearl, phenyl-Sepharose, and single-stranded DNA-cellulose as described in "Materials and Methods". In some experiments additional purification of the enzyme preparations was performed by centrifugation in glycerol gradients. Chromatography on phosphocellulose yielded a combined preparation of DNA polymerases. DNA polymerase α was separated on a column with DEAE-Toyopearl. DNA polymerase β was not bound to phenyl-Sepharose and was separated when the preparation was placed on this sorbent (see above). Chromatography on phenyl-Sepharose yielded additional fractionation of the preparation of DNA polymerase δ. Part of the DNA polymerase activity was eluted from phenyl-Sepharose when the concentration of $(NH_4)_2SO_4$ decreased from 0.5 to 0 M (fraction δ1), whereas the main activity was eluted by 0-1% gradient of nonionic surfactant Triton X-100, (fraction δ2). Then chromatography of both fractions was carried out on single-stranded DNA-cellulose. Western blotting with the use of antibodies against calf thymus DNA polymerase δ showed that fractions obtained by chromatography on DNA-cellulose and possessing the DNA polymerase activity contained immunoreactive polypeptides with molecular mass of 120 or 130 kD (Fig. 4). The apparent molecular mass of the immunoreactive polypeptide in fraction δ1 was ~ 120 kD (lanes *1* and *3*). The position of this polypeptide in the gel is coincident with that for a minor fraction in the preparation of calf thymus DNA polymerase δ that is probably a product of proteolysis of the catalytic subunit of the enzyme with molecular mass of ~ 125 kD (lane *4)*. After centrifugation of the preparation δ1 in the gradient of glycerol and rechromatography on phenyl-Sepharose, the amount of the polypeptide with molecular mass 120 kD in fractions determined by silver staining correlated with the DNA polymerase activity (data not presented). These results suggest that the immunoreactive polypeptide with molecular mass of 120 kD found in the preparation δ1 is the catalytic subunit of DNA polymerase δ. The apparent molecular mass of the immunoreactive polypeptide in fraction δ2 exceeds that for fraction δ1 and was found to be 130 kD (Fig. 4, lane *2)*. Thus, fraction δ1 may be formed by limited hydrolysis of fraction δ2. The specific DNA polymerase activity of the purified preparations with respect to activated DNA was 1000 units per mg of protein for fraction δ1 and 400 units per mg of protein for fraction δ2. The degree of purification in relation to the homogenate of eggs was 4000-5000-fold.

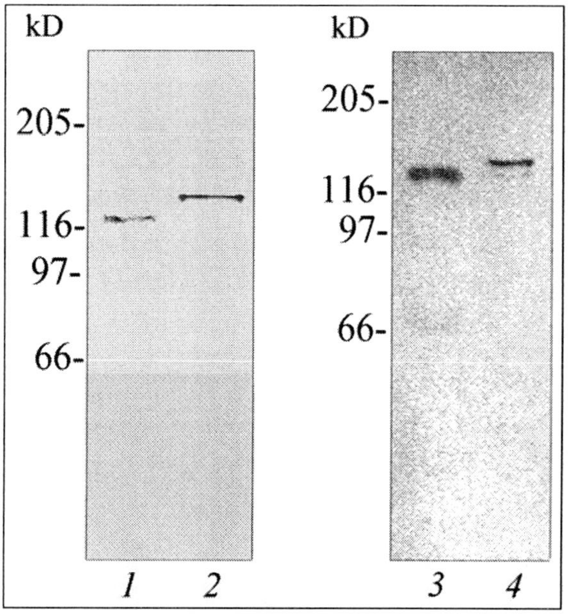

Fig. 4. Western blot of the preparations of loach DNA polymerase δ obtained by chromatography on columns with single-stranded DNA-cellulose. Proteins were separated by electrophoresis in 8% polyacrylamide gel in the presence of SDS. Polyclonal antibodies against calf thymus DNA polymerase δ were used as primary antibodies. The following preparations of DNA polymerases were used per sample: 0.15 units of δ1 (lane *1*), 0.5 units of δ2 (lane *2*), 2.5 units of δ1 (lane *3*), and 0.08 units of calf thymus DNA polymerase δ (lane *4*). Standards for electrophoresis: myosin (205 kD), β-galactosidase (116 kD), phosphorylase b (97 kD), BSA (66 kD).

Characteristic of DNA Polymerase δ

Apart from the DNA polymerase activity, the preparations of DNA polymerase δ possessed 3'→5'-exonuclease activity specific to single-stranded DNA. This activity was readily revealed at the terminal stages of purification after the separation of admixed nucleases. When carrying out chromatography of the preparations of δ1 and δ2 on single-stranded DNA-cellulose and sedimentation in gradients of glycerol, the 3'→5'-exonuclease activity in fractions obtained correlated with the DNA polymerase activity. Fig. 5 shows results obtained for fractionation of δ1 on the column with single-stranded DNA-cellulose. Both the enzymatic activities (polymerase and exonuclease activities) were eluted from the column at 0.2 M KCl as a single homogeneous peak. The preparations of loach DNA polymerase δ did not reveal 5'→3'-exonuclease activity. When incubated with double-stranded DNA, 3'→5'-exonuclease in the DNA polymerase δ preparations removed efficiently the unpaired 3'-end nucleotide residues but did not hydrolyze (or poorly hydrolyzed) the paired 3'-end nucleotide residues (Fig. 6). Thus, the 3'→5'-exonuclease activity observed is evidently connected with the proofreading function of DNA polymerase δ.

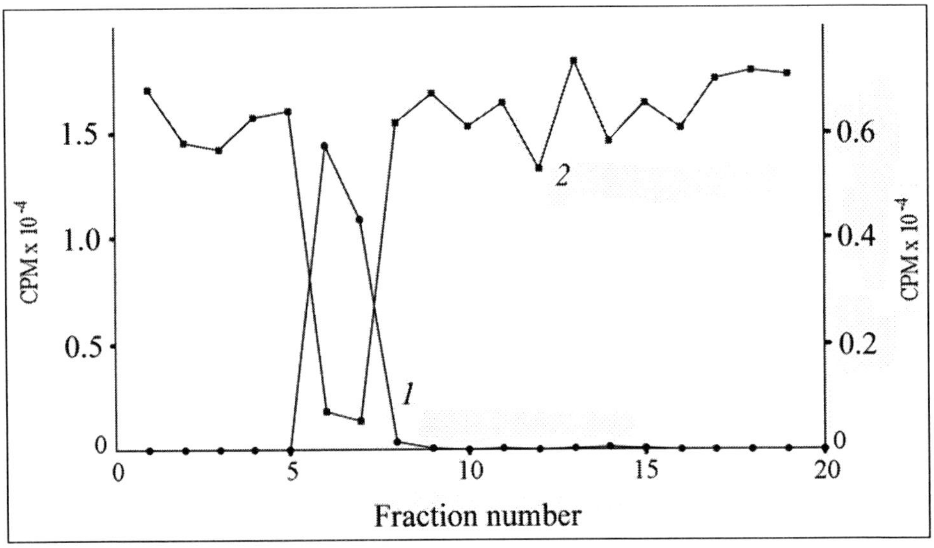

Fig. 5. Chromatography of a preparation of loach DNA polymerase δ1 on a column with single-stranded DNA-cellulose: (*1*) DNA polymerase activity measured by the incorporation of [^{3}H]dNMP into activated DNA (left ordinate axis); (*2*) 3′→5′-exonuclease activity measured by hydrolysis of 3′-[^{3}H]poly(dT) (right ordinate axis). Standard deviation is less than 2%.

In the course of the nucleotidyltransferase reaction, loach DNA polymerase δ (like DNA polymerase β) exerted distributive elongation of primers on single-stranded templates. The main product of the reaction catalyzed by fraction δ1 with excess of the primer-template complex dT_{10}: poly(dA) was primers extended by 1-2 nucleotides. In the case of fraction δ2, the elongation was analogous.

Thus, we have characterized DNA polymerase δ of loach eggs using inhibitor analysis as well as on the basis of physicochemical, enzymatic, and immunochemical properties:

1. the DNA polymerase activity of the enzyme is sensitive to the inhibiting action of aphidicolin and N-ethylmaleimide but resistant to BuPdGTP and d_2TTP;
2. the DNA polymerase activity correlates with the presence of a polypeptide with molecular mass of 120-130 kD that interacts with polyclonal antibodies against calf thymus DNA polymerase δ;
3. apart from DNA polymerase activity, the enzyme possesses 3'→5'-exonuclease activity specific to single-stranded DNA;
4. the enzyme catalyzes distributive synthesis of DNA in the absence of auxiliary factors (PCNA, replication factor C, and ATP).

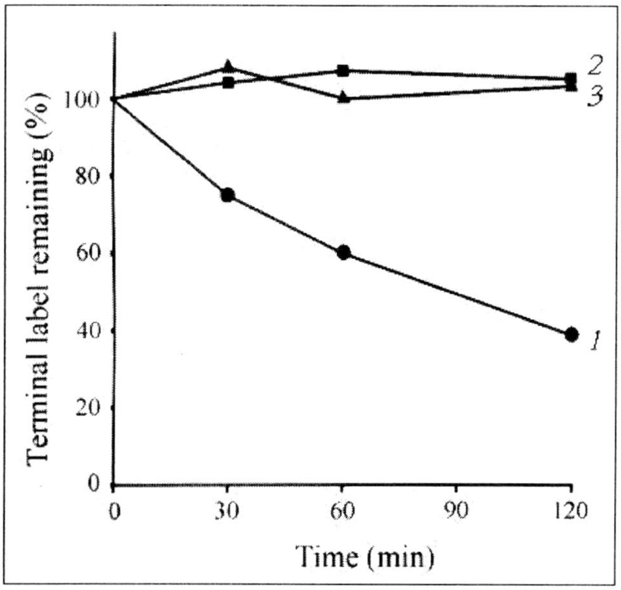

Fig. 6. Exonucleolytic hydrolysis of unpaired 3′-terminal residues by loach DNA polymerase δ. Reaction mixtures (100 μl) contained: (*1*) (3′-[^{3}H]dC$_1$)dT$_{100}$ annealed to poly(dA) (0.025:0.05 optical units), (*2*) (3′-[^{3}H]dT$_1$)dT$_{100}$ annealed to poly(dA) (0.025:0.05 optical units) or (*3*) 5′-[^{32}P]dT$_{100}$ annealed to poly(dA) (0.025:0.05 optical units), and other components (see "Materials and methods"). Reactions were started by the addition of DNA polymerase δ1 (0.07 units) and carried out at 30°C. At time points 0, 30, 60, and 120 min, 20-μl portions were removed and processed to determine the acid-insoluble radioactivity. The initial radioactivity of the probes corresponding respectively to 2600 (*1*), 6000 (*2*), and 2100 cpm (*3*) was taken as 100%. Standard deviation is less than 2%.

When purifying DNA polymerase δ by liquid chromatography, heterogeneity was observed, the latter resulting in the separation of the enzyme into fractions δ1 and δ2 with phenyl-Sepharose. The different molecular masses of the immunoreactive polypeptides, namely 120 kD in fraction δ1 and 130 kD in fraction δ2, suggest that the reason for the appearance of fraction δ1 may be limited proteolysis of the original form of the enzyme. However, a question is raised as to whether limited proteolysis accompanies the procedure of purification of the enzyme or proteolysis is connected with physiological processing of the enzyme in eggs. It should be noted that eggs contain a considerable store of the enzymes and structural proteins providing a rapid increase in growth of the embryo. The form in which DNA polymerase δ is stored remains unknown.

Identification of DNA polymerase δ in loach eggs and the preparation of partially purified enzyme allowed us to study the possible role of DNA polymerase δ in DNA repair in experiments with model DNA substrates simulating repair intermediates.

DNA Polymerase δ-Mediated Elongation

Repair properties of DNA polymerase δ were investigated with the use of DNA duplexes containing gaps, the length of which corresponded with the length of gaps forming in BER. A set of oligonucleotide substrates was constructed. Using hybridization of oligonucleotides comprised of 17-29 nucleotides with the 51-nucleotide template, we obtained duplexes containing single-strand gaps of one, five, nine, and thirteen nucleotides. Standard primer containing 21 nucleotides was radiolabeled at the 5'-terminus with radioactive phosphorus. Elongation of this primer by DNA polymerase δ in standard primer-template complex *21 : 51-mer and in gapped duplexes was examined.

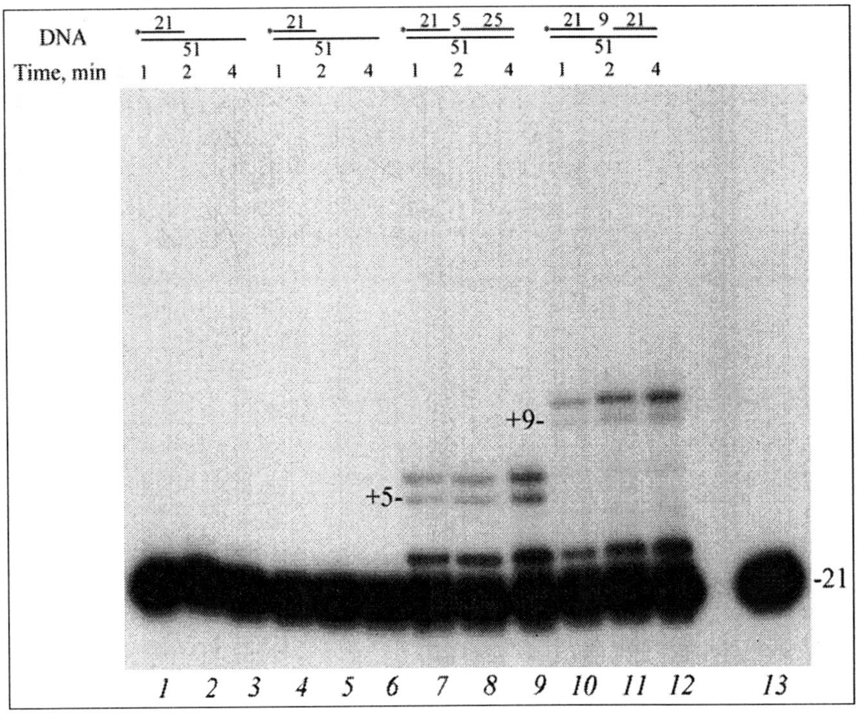

Fig. 7. DNA polymerase δ1-catalyzed polymerization of the primer 5′-[^{32}P]21-mer in the primer-template complex *21 : 51-mer (lanes *1-6*), duplex *21 : 51 : 25-mer with a five-nucleotide gap (lanes *7-9*), and duplex *21 : 51 : 21-mer with a nine-nucleotide gap (lanes *10-12*). DNA polymerase δ1 was added in the amount of 0.015 units. The reaction was carried out for 1, 2, and 4 min in a complete synthesis system (lanes *1-3* and *7-12*) and in medium devoid of dCTP (lanes *4-6*). The reaction products were analyzed by electrophoresis in 10% polyacrylamide gel in the presence of 8 M urea. Lane *13* shows migration of the primer 5′-[^{32}P]21-mer.

In the first series, we assayed the ability of loach DNA polymerase δ in processive synthesis during elongation of primers in the primer-template complex *21 : 51-mer and in partial duplexes with gaps of five and nine nucleotides (duplexes *21 : 51 : 25-mer and *21 : 51 : 21-mer, respectively). The amount of the enzyme, concentration of the DNA substrate, and duration of the reaction were selected so that the reaction products evolved from a single act of binding of the enzyme to the DNA substrate (Fig. 7). DNA polymerase δ displayed low activity in standard primer-template complex, and the major reaction products were the primers elongated by 1 or 2 nucleotides, which under these conditions were masked by the excess of the original radiolabeled primer (lanes *1-3)*. This is indicative of a distributive mode of synthesis performed by DNA polymerase δ. Elongation observed in the absence of dCTP served as a control so as to provide termination of the synthesis at the first residue dGMP in the template (the fourth nucleotide of the template at the 3'-end). Exclusion of dCTP from the reaction mixture did not change the pattern of distribution of the elongation products in the gel (lanes *4-6),* thus confirming the nonprocessive mode of elongation performed by DNA polymerase δ. However, when DNA duplex had a single-strand gap of 5 or 9 nucleotides, i.e., in the presence of oligonucleotide restricting the single-strand region in the template, elongation activity of the enzyme sharply increased. As a result, the enzyme gained the capacity for filling the gap in a processive mode (lanes 7-*12*). Besides the primers elongated by 1 to 2 nucleotides, the major reaction products were primers elongated by 5 or 6 and 9 or 10 nucleotides in the case of duplexes with gaps of 5 and 9 nucleotides, respectively. Therefore, the enzyme was able to fill all gap in a processive mode and to terminate the synthesis near the 5'-terminal residue restricting this gap.

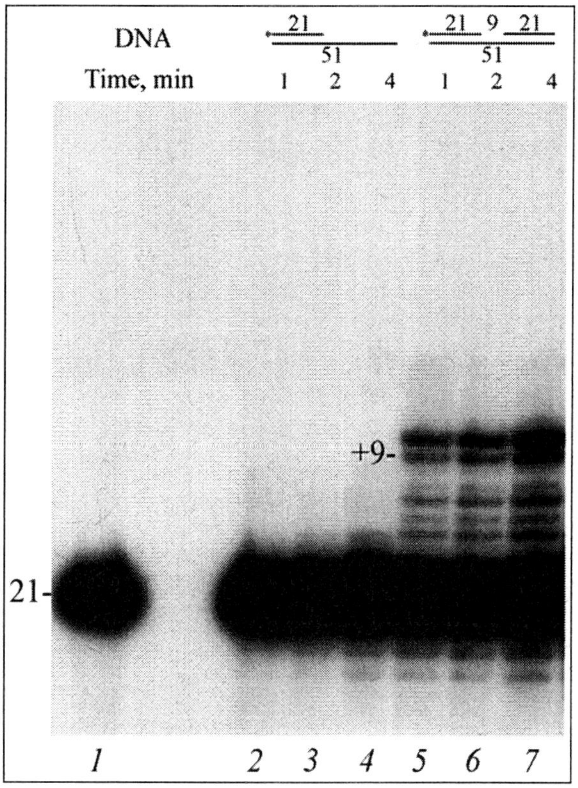

Fig. 8. DNA polymerase δ2-catalyzed polymerization of the primer 5′-[^{32}P]21-mer in the primer-template complex *21 : 51-mer (lanes *2-4*) and duplex *21 : 51 : 21-mer with a nine-nucleotide gap (lanes *5-7*). DNA polymerase δ2 was added in the amount of 0.015 units. The reaction was carried out for 1, 2, and 4 min. The reaction products were analyzed by electrophoresis in 12.5% polyacrylamide gel in the presence of 8 M urea. Lane *1* shows migration of the primer 5′-[^{32}P]21-mer.

The results shown in Fig. 7 were obtained with fraction δ1. It was of interest to find out whether the mode of elongation described for the most purified fraction of DNA polymerase δ1 is characteristic of the other fractions of the enzyme. Fig. 8 shows the results obtained in fraction δ2. In this experiment we used the standard primer-template complex *21 : 51-mer and the duplex *21 : 51 : 21-mer with a nine-nucleotide gap. We found that in the primer-template complex fraction δ2 performed synthesis in a distributive mode, whereas in the DNA duplex it filled the gap predominantly in the processive mode. Thus, similarly to fraction δ1, fraction δ2 can change from distributive elongation of primers in long patch single-strand regions in the template to processive synthesis when the single-strand region of the template is restricted by the proximal 5'-terminus of the oligonucleotide.

The results of the experiments led us to the following conclusions on the mode of elongation performed by loach egg DNA polymerase δ. First, the enzyme performs distributive elongation of primers in the standard primer-template complexes when the template is not restricted by the 5'-terminal nucleotide residues hybridized with the template. Second, the enzyme displays substantially higher elongation activity in DNA duplexes with short single-strand gaps (five to nine nucleotides) than in standard primer-template complexes and is capable of processive synthesis over the whole gap region.

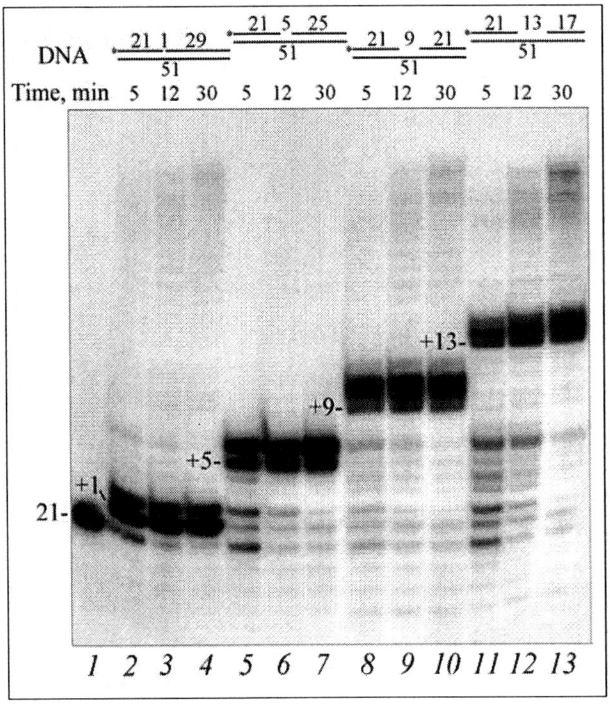

Fig. 9. DNA polymerase δ1-catalyzed polymerization of the primer 5'-[^{32}P]21-mer in duplexes with gaps of one nucleotide (*21 : 51 : 29-mer) (lanes *2-4*), five nucleotides (*21 : 51 : 25-mer) (lanes *5-7*), nine nucleotides (*21 : 51 : 21-mer) (lanes *8-10*), and thirteen nucleotides (*21 : 51 : 17-mer) (lanes *11-13*). DNA polymerase δ1 was added in the amount of 0.15 units. The reaction was carried out for 5, 12, and 30 min. The reaction products were analyzed by electrophoresis in 10% polyacrylamide gel in the presence of 8 M urea. Lane *1* shows migration of the primer 5'-[^{32}P]21-mer.

During processive filling short gaps in the duplexes, DNA polymerase δ can terminate synthesis at the 5'-terminal nucleotide of the fragment restricting the single-strand gap, thus displacing this residue from binding to the complementary nucleotide of the template (Figs. 7 and 8). However, in the aforementioned experiments elongation was performed in the presence of excess primer-template complexes so as to evaluate the real processivity of the enzyme. It remained unclear whether DNA polymerase δ can perform strand displacement DNA synthesis during long-term incubation under conditions that allow for multiple binding of the enzyme to the DNA substrate. To assess the ability of DNA polymerase δ for strand displacement DNA synthesis, we performed an experiment in which excess DNA polymerase δ1 was incubated for 30 min in reaction medium containing DNA duplexes with gaps of one, five, nine, and thirteen nucleotides (Fig. 9). It appeared that the labeled primers were utilized almost completely during the first five minutes of incubation. Irrespective of the gap size, the final reaction products were a mixture of oligonucleotides corresponding to the complete filling of the gap and an additional incorporation of usually one nucleotide. The absence of visible changes in the composition of elongation products during a 30-min incubation is indicative of a futile cycle when the enzyme adds one nucleotide to the oligonucleotide filling the gap and then removes this residue due to its 3'→5'-exonuclease activity. The pattern of distribution of the reaction products reflects the balance between the reactions of synthesis and exonuclease- mediated hydrolysis. On the whole, the experiment demonstrates the inability of DNA polymerase δ to perform DNA synthesis with displacement of the DNA strand that restricts the single-strand gap.

Binding of DNA Polymerase δ to DNA

The above-mentioned change of DNA polymerase δ from distributive synthesis on single-strand templates to processive synthesis when filling the gaps in DNA duplexes implies an interaction of the enzyme with the 5'-terminus restricting the gap. The possibility of DNA polymerase δ binding to the 5'-terminus restricting the gap was studied by electrophoretic mobility shift assay. Labeled DNA substrates used before for elongation study were incubated with DNA polymerase δ. Then the complexes formed by the enzyme and labeled DNA were analyzed by native electrophoresis in polyacrylamide gel. The occurrence of radiolabeled complexes with lower mobility than the original DNA substrate is indicative of a stable binding of this substrate to the enzyme. At first we compared DNA polymerase δ binding to the standard primer-template complex *21 : 51-mer and duplexes with gaps of one, five, nine, and thirteen nucleotides (Fig. 10). DNA polymerase δ exerted low affinity for the primer-template complex *21 : 51-mer and duplex with one-nucleotide gap. The affinity of the enzyme for five-nucleotide duplex was higher. Binding of the enzyme substantially increased with increase of the gap to 9 nucleotides and then somewhat decreased again as the gap grew to thirteen nucleotides. Thus, DNA polymerase δ exhibits much higher affinity for DNA duplexes with gaps of approximately ten nucleotides compared to standard template-primer complexes and duplexes with short gaps of one to five nucleotides.

In a control experiment, we obtained a formal confirmation that the discovered complexes of labeled DNA substrates with the enzyme contain functionally active DNA polymerase δ. Duplexes containing gaps of nine and thirteen nucleotides were used as DNA probes. In the first and second series, the enzyme was bound to duplexes labeled with radioactive phosphorus or unlabeled duplexes, respectively. To determine the position of the complexes of the enzyme with DNA in the gel, the radioactive portion of the gel after electrophoresis was exposed to a Roentgen film (Fig. 11, lanes *1-4*). To reveal DNA polymerase δ activity *in situ,* the portion of the gel containing unlabeled DNA probes was incubated in reaction mixture for DNA synthesis which contained [α-^{32}P] dATP (lanes *5-8*). Complexes exerting DNA polymerase δ activity were present in the samples that contained DNA substrates and the enzyme (lanes *6* and *7*). A coincidence between the position of functional complexes of DNA polymerase δ with DNA substrates in the gel and the position of the complexes of radiolabeled probes with the enzyme was observed (lanes *2* and *4*).

Therefore, the previously discovered complexes of DNA probes with the enzyme are in fact complexes of DNA with DNA polymerase δ.

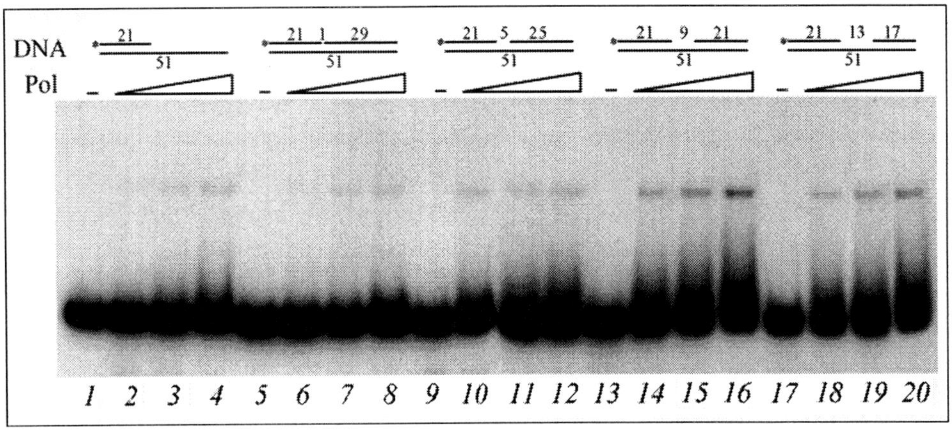

Fig. 10. Binding of DNA polymerase δ1 to the primer-template complex *21 : 51-mer (lanes *1-4*) and duplexes with gaps of one nucleotide (*21 : 51 : 29-mer) (lanes *5-8*), five nucleotides (*21 : 51 : 25-mer) (lanes *9-12*), nine nucleotides (*21 : 51 : 21-mer) (lanes *13-16*), and thirteen nucleotides (*21 : 51 : 17-mer) (lanes *17-20*). The samples (10 μl) contained 0.01 pmol of ^{32}P-labeled DNA substrates and 0.004 (lanes *2*, *6*, *10*, *14*, and *18*), 0.01 (lanes *3*, *7*, *11*, *15*, and *19*) or 0.03 units of DNA polymerase δ1 (lanes *4*, *8*, *12*, *16*, and *20*). Lanes *1*, *5*, *9*, *13*, and *17* contain no enzyme. The enzyme was incubated with the DNA substrates at room temperature for 15 min. The reaction products were analyzed by electrophoresis in 6% polyacrylamide gel.

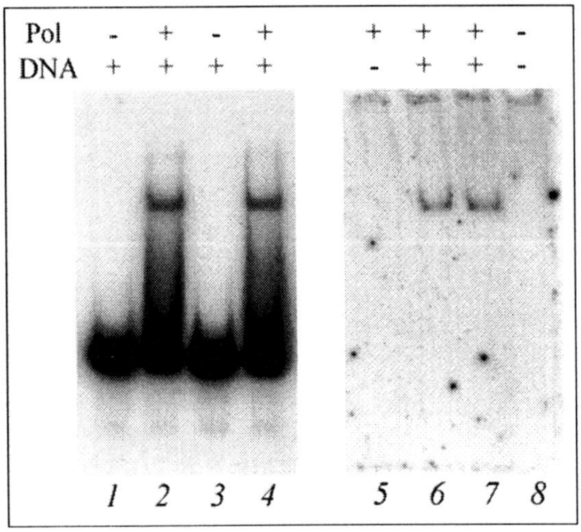

Fig. 11. *In situ* activity of DNA polymerase δ on gapped DNA following mobility shift electrophoresis. Reaction mixtures (10 μl) containing 0.04 pmol unlabeled partial duplexes 21 :51 :21-mer (9-nt gap) (lane *6*) and 21 :51 :17-mer (13-nt gap) (lane *7*) or 0.01 pmol of ^{32}P-labeled partial duplexes *21 : 51 : 21-mer (9-nt gap) (lanes *1*, *2*) and *21 : 51 : 17-mer (13-nt gap) (lanes *3*, *4*) and other components (see "Materials and methods") were incubated for 15 min on ice in the absence of DNA polymerase δ (lanes *1*, *3*, *8*) or in the presence of 0.03 units of DNA polymerase δ1 (lanes *2*, *4-7*). After electrophoresis in 6% polyacrilamide gel at 4°C, the gel region with lanes *1-4* was dried, while the other part with lanes *5-8* was placed in a reaction mixture with [α-^{32}P]dATP for an *in situ* assay of DNA polymerase activity as described in "Materials and methods".

The fact that DNA polymerase δ possesses 3'→5'-exonuclease activity specific towards single-strand DNA implies the possibility of binding of the enzyme to unpaired 3'-termini in DNA. To evaluate the affinity of DNA polymerase δ for unpaired DNA regions, we synthesized partial duplexes containing four unpaired nucleotide residues at the 3'-terminus of the primer comprised of 21 nucleotides. Binding of DNA polymerase δ to duplexes containing the standard primer and those with unpaired 3'-tails was compared by electrophoretic mobility shift assay as described above. As shown in Fig. 12, DNA polymerase δ displays high affinity for the substrates with unpaired 3'-tails both in standard template-primer complexes and duplexes with single-strand gaps. Tight binding of DNA polymerase δ to the unpaired 3'-tails suggests that the enzyme also exhibits high activity in similar structures in damaged DNA in cell nuclei.

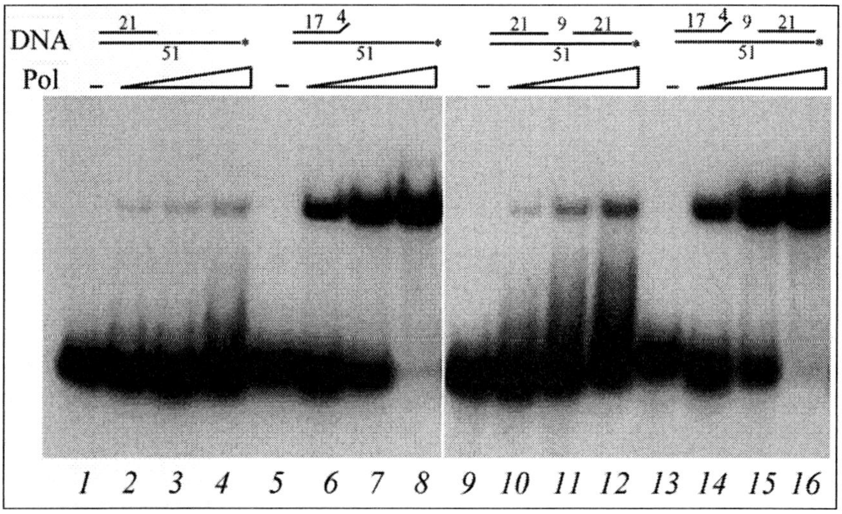

Fig. 12. Binding of DNA polymerase δ to primer-template complexes containing primers with unpaired 3′-terminal tails. The samples (10 μl) contained 0.005 pmol of ^{32}P-labeled DNA substrates: 21 : *51-mer (lanes *1-4*), (17+4) : *51-mer (lanes *5-8*), 21 : *51 :21-mer (lanes *9-12*), and (17+4) : *51 :21-mer (lanes *13-16*). DNA polymerase δ1 was added in the amount of 0.004 (lanes *2, 6, 10,* and *14*), 0.01 (lanes *3, 7, 11,* and *15*) or 0.03 units (lanes *4, 8, 12,* and *16*). Lanes *1, 5, 9,* and *13* contain no enzyme. The enzyme was incubated with the DNA substrates for 15 min at room temperature. The reaction products were analyzed by electrophoresis in 6% polyacrylamide gel.

Because the DNA polymerase δ preparation was heterogeneous, it was of interest to determine whether characteristics of the enzyme binding to DNA substrates described for the most purified fraction δ1 can be applied to the less purified fraction δ2. Incubation of fraction δ2 with labeled DNA probes yielded complexes that had similar (although somewhat lower) mobility in the gel as fraction δ1 (Fig. 13). Thus, the mode of DNA binding by fraction δ2 is similar to that of fraction δ1. Besides the major area of slowly migrating complexes, fraction δ2 contained a minor area with higher mobility that coincided with the mobility of the complexes formed by δ1 (lanes 7 and *8*). With regard for different molecular weights of immunoreactive polypeptides contained in fractions δ1 and δ2 (120 and 130 kD, respectively), it can be suggested that the occurrence of close but different in mobility complexes of the enzyme with DNA probes can result from partial proteolysis of the catalytic subunit of DNA polymerase δ. The results suggest that partial proteolysis does not significantly change the mode of binding of the enzyme to DNA substrates and the mode of elongation of primers during DNA synthesis.

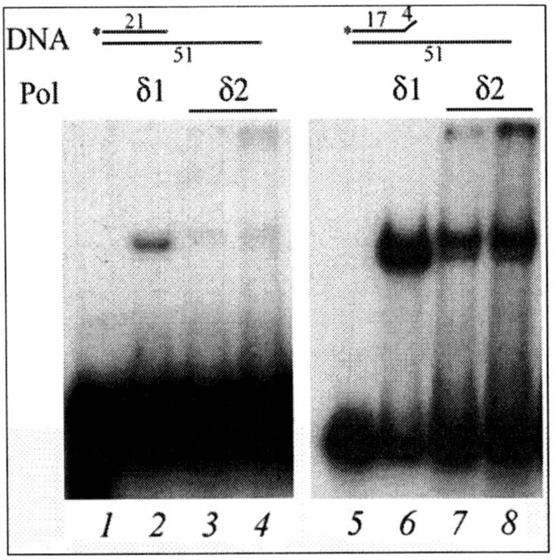

Fig. 13. Comparison of electrophoretic mobility of the complexes of DNA substrates with DNA polymerases δ1 and δ2. The samples (10 μl) contained 0.01 pmol of ^{32}P-labeled DNA substrates: *21 : 51-mer (lanes *1-4*) or *(17+4) : 51-mer (lanes *5-8*). DNA polymerase δ1 was added in the amount of 0.017 units (lanes *2* and *6*), DNA polymerase δ2 was added in the amount of 0.006 (lane *3*), 0.015 (lanes *4* and *7*), and 0.045 units (lane *8*). Lanes *1* and *5* contain no enzyme. The enzyme was incubated with DNA substrates at room temperature for 15 min. The reaction products were analyzed by electrophoresis in 6% polyacrylamide gel.

Thus, in this work we for the first time showed that DNA polymerase δ is capable of processive filling of gaps in DNA in the absence of any processivity factors. Apparently, the enzyme can interact with the 5'-terminal region of a DNA strand restricting the single-strand gap, and this interaction stabilizes the complex of the enzyme with the template-primer. A similar phenomenon was described earlier for DNA polymerase β (Prasad et al., 1994). In this case binding of the enzyme to the 5'-terminal nucleotide restricting the gap is due to the presence of a deoxyribophosphate lyase (dRp- or AP-lyase) site in the DNA polymerase β molecule (Matsumoto, Kim, 1995; Piersen et al., 1996; Matsumoto et al., 1998). However, data on AP-lyase activity in DNA polymerase δ are absent.

A low affinity of loach DNA polymerase δ for standard template-primers is consistent with data published in the literature. It was shown that the formation of a tight complex between the enzyme and DNA requires a structural “clamp” consisting of three PCNA molecules, which is formed by the RF-C protein complex in the presence of ATP (Hindges, Hubscher, 1997; Waga, Stillman, 1998). An addition of oligonucleotide restricting the template region to the template-primer increased the amount of the enzyme bound in the complex with DNA. This increase was insignificant in the case of short gaps of one to five nucleotides but relatively pronounced in the case of nine-nucleotide gaps. Further increase in the gap size from nine to thirteen nucleotides resulted in some decrease in binding. Thus, the optimal distance between the 5'- and 3'-termini restricting the gap that provides for DNA polymerase δ binding should be about ten nucleotides. Such a distance corresponds to the size of gaps which are formed during long patch BER (Sharova, 2005). The results of experiments on binding of the enzyme to duplexes with single-strand gaps account for the ability of DNA polymerase δ to fill gaps in DNA duplexes in a processive mode. When filling the gap, DNA polymerase δ interacts not only with the primer and template, but also with the proximal 5'-terminus downstream from the gap. Binding to the 5'-terminus additionally stabilizes the productive complex of the enzyme with the DNA substrate and confines the enzyme in this complex after a successive addition of a nucleotide residue to the growing primer, thus allowing the enzyme to perform processive polymerization. In the absence of proximal 5'-terminus downstream from the gap, binding of DNA polymerase δ with the template-primer is insufficient for the formation of the productive complex. For this reason, the enzyme dissociates from the DNA substrate after the addition of a successive nucleotide residue to the primer. In this case, polymerization proceeds according to the distributive mechanism. During processive synthesis, loach DNA polymerase δ filled the gap completely and either terminated synthesis directly before the 5'-terminal nucleotide downstream from the gap or added one more nucleotide to the growing strand. Even when present in excessive amount, DNA polymerase δ did not perform strand displacement DNA synthesis. Thus, DNA polymerase δ meets the requirements for an enzyme that fills gaps during DNA repair.

Tight binding of DNA polymerase δ to duplexes containing unpaired 3'-tails in primers possibly reflects the participation of the exonuclease site of the enzyme in DNA substrate binding. In this binding type, the 5'-terminus downstream from the gap does not significantly affect the affinity of the enzyme for DNA substrate. High affinity of DNA polymerase δ for duplexes containing unpaired 3'-termini implies that DNA with such structural lesions represents an effective substrate for DNA polymerase δ. Thus, unpaired 3'-termini and single-strand gaps, which are conventional intermediates during DNA repair, can serve as targets for repair activity of DNA polymerase δ in the cell nucleus. It is believed that participation of DNA polymerase δ in DNA excision repair is associated with the PCNA-dependent pathway of restoring intact DNA structure (Sharova, 2005). The processive mode of gap filling by DNA polymerase δ in the absence of processivity factors, as described in this work, suggests that this enzyme can participate particularly in the long patch BER.

Dynamics of DNA Polymerase δ Activity in Loach Embryogenesis

In early loach embryos, the specific DNA polymerase δ activity was 200 units per gram of protein up to the late gastrula stage. At the beginning of organogenesis it 2.5 times increased. Thus, the DNA polymerase δ activity 30-55 times exceeded the DNA polymerase β activity in early loach embryogenesis and seems to be the main repair DNA polymerase of loach embryos.

Effect of Oxidative Stress on DNA Polymerases in Loach Embryogenesis

It is commonly known that factors damaging DNA induce enzymes and other proteins which repair these damages. So, to confirm the participation of DNA polymerase δ in BER *in vivo*, after the beginning of cell division, embryos were affected by superoxide radicals in the xanthine – xanthine oxidase system generating mainly single strand breaks in DNA, which are repaired by long patch pathway. The oxidative stress development was controlled by the measurement of Cu,Zn-superoxide dismutase activity. In our experiments, the treatment of loach embryos with the xanthine – xanthine oxidase system led to the increase of Cu,Zn-superoxide dismutase activity by 30% at the stage of middle gastrula.

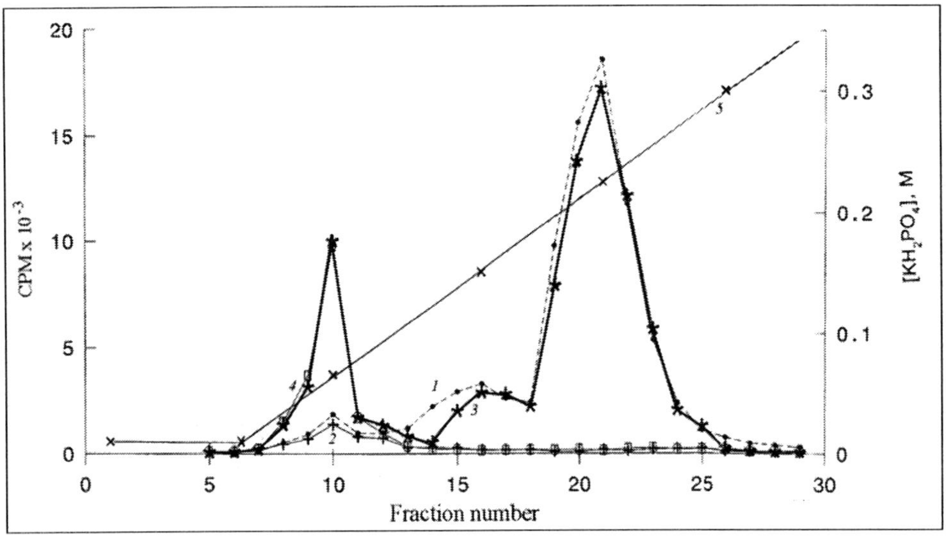

Fig. 14. Effect of xanthine – xanthine oxidase system on the DNA polymerase activity of loach embryos at the middle gastrula stage: (*1*) control; (*2*) in the presence of 3 μM BuPdGTP; (*3*) after oxidative stress; (*4*) in the presence of 3 μM BuPdGTP after oxidative stress; (*5*) gradient of salt concentration. Standard deviation is less than 2%.

At this stage, the oxidative stress changed the profile of the DNA polymerase activity on a DEAE-Toyopearl column (Fig. 14). The DNA polymerase activity was eluted from the DEAE-Toyopearl column in three peaks. The first peak contained DNA polymerases β and δ. The second and third peaks contained the replicative DNA polymerase α, which in loach embryos was represented by two forms, *viz.* α1 and α2 (Sharova et al., 1992). DNA polymerases α1 and α2 in the eluted fractions were identified with the specific inhibitor BuPdGTP (3 μM). The oxidative stress produced no activating effect on the replicative DNA polymerases α1 and α2, but significantly increased the activity of the first peak. Then DNA polymerases β and δ were separated chromatographically on a phenyl-Sepharose column. Before this procedure, the preparation of DNA polymerases β and δ was dialyzed against 0.5 M $(NH_4)_2SO_4$ solution in buffer C. After dialysis, the resulting preparation was applied to the column equilibrated with buffer C containing 0.5 M $(NH_4)_2SO_4$. DNA polymerase δ was bonded to the sorbent and it was eluted with buffer C containing 1 % Triton X-100. DNA polymerase β did not bind with the sorbent. The DNA polymerase activity was assayed as described above, but pH of the reaction mixture for DNA polymerase β was maintained at the value 8.3. Before the DNA polymerases β and δ activities were determined, the preparations were dialyzed against buffer C. Thus, the chromatography of DNA polymerases on phenyl-Sepharose revealed that the oxidative stress increased the activities of DNA polymerase β and

DNA polymerase δ by 140 and 370%, respectively. This result confirms the participation of these DNA polymerases in DNA repair.

Conclusion

This work discovered that loach eggs and embryos possess an unusual system of BER.

First of all, the activity of loach repair DNA polymerase β is very low in early embryogenesis up to the stage of organogenesis. At the same time DNA polymerase β is an enzyme capable to realize both short patch and long patch pathways of damaged bases repair. It is clear that the deficit of this enzyme in embryonic cells may lead to sorrowful consequences, if these cells have lack for another enzyme able to perform BER.

The enzyme that substitutes DNA polymerase β in loach eggs and embryos is DNA polymerase δ which activity 30-55 times exceeds the activity of DNA polymerase β in early embryogenesis. Besides the replicative function, DNA polymerase δ carries out different types of DNA repair including BER. Literature data indicate that in different eukaryots, DNA polymerase δ performs processive synthesis of DNA only in the presence of the replicative factors (PCNA, RFC and ATP). RFC (exhibiting ATPase activity) binds to a DNA site limiting a gap from the 5′-end and threads PCNA onto DNA duplex due to transient opening of PCNA ring in an ATP-dependent manner. The resultant complex binds DNA polymerase δ which further effectively fills a gap. Alternatively, in the absence of processivity factors, loach DNA polymerase δ displays unexpected high processivity upon the filling of 5-nt and 9-nt gaps due to the interaction with both ends of the gaps. This property of DNA polymerase δ is very important for extreme conditions at low levels of PCNA and PCNA-independent repair DNA polymerases in a cell.

At the same time, it remains unclear whether PCNA-independent repair synthesis of DNA by DNA polymerase δ is a peculiarity of long patch BER in loach cells or it is one else DNA repair pathway in cells of another eukaryots too. This issue is very actual since the PCNA-independent pathway of damaged bases repair is usually related to DNA polymerase β which is discovered in vertebrates but absent in insects, plants, yeast and so on. The problem indicated is a challenge for further investigators of DNA repair.

Acknowledgements

The author is grateful to V. Mikhailov and A. Peskin for assistance and discussion of the results. This study was supported by the Russian Foundation for Basic Research (grants 00-04-49183, 03-04-49127).

References

Bennett, R.A.O., Wilson, D.M. III, Wong, D. and Demple, B. (1997). Interaction of human apurinic endonuclease and DNA polymerase β in the base excision repair pathway. Proc. *Natl. Acad. Sci. USA, 94*, 7166-7169.

Caldecott, K.W., Tucker, J.D., Stanker, L.H. and Thompson, L.H. (1995). Characterization of the XRCC1-DNA ligase III complex in vitro and its absence from mutant hamster cells. *Nucleic Acids Res., 23*, 4836-4843.

Cappelli, E., Taylor, R., Cevasco, M, Abbondandolo, A., Caldecott, K. and Frosina, G. (1997). Involvement of XRCC1 and DNA ligase III gene products in DNA base excision repair. *J. Biol. Chem., 272*, 23970-23975.

Carre, D., Signoret, J., Lefresne, J. and David, J.C. (1981). Enzymes involved in DNA replication in the axolotl. Analysis of the forms and activities of DNA polymerase and DNA ligase during development. *Develop. Biol., 87*, 114-125.

DeMott, M.S., Beyret, E., Wong, D., Bales, B.C., Hwang, J.T., Greenberg, M.M. and Demple, B. (2002). Covalent trapping of human DNA polymerase beta by the oxidative DNA lesion 2-deoxyribonolactone. *J. Biol. Chem., 277*, 7637-7640.

Dianov, G., Price, A. and Lindahl, T. (1992). Generation of single-nucleotide repair patches following excision of uracil residues from DNA. *Mol. Cell. Biol., 12*, 1605-1612.

Dianova, I.I., Bohr, V.A. and Dianov, G.L. (2001). Interaction of human AP endonuclease 1 with flap endonuclease 1 and proliferating cell nuclear antigen involved in long-patch base excision repair. *Biochemistry, 40*, 12639-12644.

Dogliotti, E., Fortini, P., Pascucci, B. and Parlanti, E. (2001). The mechanism of switching among multiple BER pathways. *Progr. Nucleic Acids Res. Mol. Biol., 68*, 3-27.

Garcia-Diaz, M., Bebenek, K., Kunkel, T.A. and Blanco, L. (2001). Identidication of an intrinsic 5′-deoxyribose-5-phospdate lyase activity in human DNA polymerase λ. A possible role in base excision repair. *J. Biol. Chem., 276*, 34659-34663.

Garcia-Diaz, M., Bebenek, K., Sabariegos, R., Domiguez, O., Rodriguez, J., Kirchhoff, T., Garcia-Palomero, E., Picher, A.J., Juarez, R., Ruiz, J.F., Kunkel, T.A. and Blanco, L.(2002). DNA polymerase lambda, a novel DNA repair enzyme in human cells. *J. Biol. Chem., 277*, 13184-13191.

Gary, R., Kim, K., Cornelius, H.L., Park, M.S. and Matsumoto, Y. (1999). Proliferating cell nuclear antigen facilitates excision in long-patch base excision repair. *J. Biol. Chem., 274*, 4354-4363.

Gomes, X.V. and Burgers, P.M.J. (2000). Two modes of FEN1 binding to PCNA regulated by DNA. *EMBO J., 19*, 3811-3821.

Harrington, J.J. and Lieber, M.R. (1994). Functional domains within FEN-1 and Rad2 define a family of structure-specific endonucleases - implications for nucleotide excision repair. *Genes Dev., 8*, 1344-1355.

Hindges, R. and Hubscher, U. (1997). DNA polymerase δ, an essential enzyme for DNA transactions. *Biol. Chem., 378*, 345-362.

Horton, J.K., Prasad, R., Hou, E. and Wilson, S.H. (2000). Protection against methylation-induced cytotoxicity by DNA polymerase beta-dependent long patch base excision repair. *J. Biol. Chem., 275*, 2211-2218.

Jonsson, Z.O., Hindges, R. and Hubsdher, U. (1998). Regulation of DNA replication and repair proteins through interaction with the front side of proliferating cell nuclear antigen. *EMBO J., 17*, 2412-2425.

Kedar, P.S., Kim, S.J., Robertson, A., Hou, E., Prasad, R., Horton, J.K., Lewis, M. and Wilson, S.H. (2002). Direct interaction between mammalian DNA polymerase beta and proliferating cell nuclear antigen. *J. Biol. Chem., 277*, 31115-31123.

Klungland, A. and Lindahl, T. (1997). Second pathway for completion of human DNA base excision-repair: reconstitution with purified proteins and requirement for DNase IV (FEN1). *EMBO J., 16*, 3341-3348.

Kubota, Y., Nash, R., Klungland, A., Schar, P., Barnes, D. and Lindahl, T. (1996). Reconstitution of DNA base excision repair with purified human proteins: interaction between DNA polymerase β and the XRCC1 protein. *EMBO J., 15*, 6662-6670.

Levin, D.S., Bai, W., Yao, N., O'Donnell, M. and Tomkinson, A.E. (1997). An interacyion between DNA ligase I and proliferating cell nuclear antigen: Implications for Okazaki fragment synthesis and joining. *Proc. Natl. Acad. Sci. USA, 94*, 12863-12868.

Li, X., Li, J., Harrington, J., Lieber, M.R. and Burgers, P.M. (1995). Lagging strand DNA synthesis at the eukaryotic replication fork involved binding and stimulation of FEN-1 by proliferating cell nuclear antigen. *J. Biol. Chem., 270*, 22109-22112.

Lindahl, T., Gally, J.A. and Edelman, G.M. (1969). Deoxyribonuclease IV: a new exonuclease from mammalian tissues. *Proc. Natl. Acad. Sci. USA, 62*, 597-603.

Maga, G., Villani, G., Ramadan, K., Shevelev, I., Tanguy Le Gac, N., Blanco, L., Blanca, G., Spadari, S. and Hubscher, U. (2002). Human DNA polymerase lambda functionally and physically interacts with proliferating cell nuclear antigen in normal and translesion DNA synthesis. *J. Biol. Chem., 277*, 48434-48440.

Matsumoto, Y. (2001). Molecular mechanism of PCNA-dependent base excision repair. Progr. *Nucleic Acids Res. Mol. Biol., 68*, 129-138.

Matsumoto, Y. and Kim, K. (1995) Excision of deoxyribose phosphate residues by DNA polymerase β during DNA repair. *Science, 269*, 699-702.

Matsumoto, Y., Kim, K. and Bogenhagen, D.F. (1994). Proliferating cell nuclear antigen-dependent abasic site repair in Xenopus laevis oocytes - an alternative pathway of base excision DNA repair. *Mol. Cell. Biol., 14*, 6187-6197.

Matsumoto, Y., Kim, K., Hurwitz, J., Gary, R., Levin, D.S., Tomkinson, A.E. and Park, M.S. (1999). Reconstitution of proliferating cell nuclear antigen-dependent repair of apurinic/apyrimidinic sites with purified human proteins. *J. Biol. Chem., 274*, 33703-33708.

Matsumoto, Y., Kim, K., Katz, D.S. and Feng, J.-A. (1998). Catalytic center of DNA polymerase β for excision of deoxyribose phosphate groups. *Biochemistry, 37*, 6456-6464.

Minin, A.A., Ivanenkov, V.V. and Meshcheryakov, V.N. (1987). Purification of actin at different stages of loach development. *Biochemistry (Moscow), 52*, 342-347.

Nealon, K., Nicholl, I.D. and Kenny, M.K. (1996). Characterization of the DNA polymerase requirement of human base excision repair. *Nucleic Acids Res., 24*, 3763-3770.

Neifakh, A.A. (1959). The use of the method of radioactive inactivation of nuclei for investigation of their functions in early development of fish. *Zh. Obsh. Biol. (in Russian), 20*, 202-213.

Otterlei, M., Warbrick, E., Nagelhus, T.A., Haug, T., Slupphaug, G., Akbari, M., Aas, P.A., Steinsbekk, K., Bakke, O. and Krokan, H.E. (1999). Post-replicative base excision repair in replication foci. *EMBO J., 18*, 3834-3844.

Pascucci, B., Stucki, M., Jonsson, Z.O., Dogliotti, E. and Hubscher, U. (1999) Long patch base excision repair with purified human proteins. *J. Biol. Chem., 274*, 33696-33702.

Peskin, A.V., Sharova, N.P., Dimitrova, D.D., Stolyarov, S.D. and Filatova L.S. (1997). Effect of oxidative stress on DNA polymerases in the embryogenesis of the loach Misgurnus fossilis. *Dokl. Acad. Nauk (in Russian), 355*, 72-75.

Petermann, E., Keil, C. and Oei, S.L. (2006). Roles of DNA ligase III and XRCC1 in regulation the switch between short patch and long patch BER. *DNA repair (Amst), 5*, 544-555.

Piersen, C.E., Prasad, R., Wilson, S.H. and Lloid, R.S. (1996). Evidence for an imino intermediate in the DNA polymerase beta deoxyribose phosphate excision reaction. *J. Biol. Chem., 271*, 17811-17815.

Prasad, R., Beard, W.A., Chyan, J.Y., Maciejewski, M.W., Mullen, G.P. and Wilson, S.H. (1998). Functional analysis of the amino-terminal 8-kDa domain of DNA polymerase β as revealed by site-directed mutagenesis. DNA binding and 5′-deoxyribose phosphate lyase activities. *J. Biol. Chem., 273*, 11121-11126.

Prasad, R., Beard, W.A. and Wilson, S.H. (1994). Studies of gapped DNA substrate binding by mammalian DNA polymerade beta. Dependence on 5′-phosphate group. *J. Biol. Chem., 269*, 18096-18101.

Prasad, R., Singhal, R.K., Srivastava, D.K., Molina, J.T., Tomkinson, A.E. and Wilson, S.H. (1996). Specific interaction of DNA polymerase β and DNA ligase I in a multiprotein base excision repair complex from bovine testis. *J. Biol. Chem., 271*, 16000-16007.

Satoh, M.S. and Lindahl, T. (1992). Role of poly(ADP-ribose) formation in DNA repair. *Nature, 356*, 356-358.

Sekelsky, J.J., Burtis, K.C. and Hawley, R.S. (1998). Damage control: the pleiotropy of DNA repair Genes in Drosophila melanogaster. *Genetics, 148*, 1587-1598.

Sharova, N.P. (2005). How does a cell repair damaged DNA? *Biochemistry (Moscow), 70*, 341-359.

Sharova, N.P., Dimitrova D.D. and Mikhailov V.S. (1995). Purification and characterization of DNA polymerase ζ from mature oocytes of the bony fish Misgurnus fossilis (loach). *Biochemistry (Moscow), 60*, 1445-1455.

Sharova, N.P., Eliseeva, E.D. and Mikhailov, V.S. (1992). Changes in the activity of two forms of DNA polymerase α during embryogenesis of the teleost fish Misgurnus fossilis. *Biochemistry (Moscow), 57*, 644-649.

Sharova, N.P., Eliseeva, E.D. and Mikhailov, V.S. (1993). Interrelationship between two forms of DNA polymerase α from eggs of the teleost fish Misgurnus fossilis (loach). *Biochemistry (Moscow), 58*, 1453-1462.

Sharova, N.P. and Mikhailov, V.S. (1995). DNA polymerase α in early development of the teleost fish Misgurnus fossilis (loach). *Biochemistry (Moscow), 60*, 1125-1131.

Singhal, R.K., Prasad, R. and Wilson, S.H. (1995). DNA polymerase β conducts the gap-filling step in uracil-initiated base excision repair in a bovine testis nuclear extract. *J. Biol. Chem., 270*, 949-957.

Sobol, R.W., Horton, J.K., Kuhn, R., Gu, H., Singhal, R.K., Prasad, R., Rajewsky, K. and Wilson, S.H. (1996). Requirement of mammalian DNA polymerase β in base excision repair. *Nature, 379*, 183-186.

Sweasy, J.B., Lang, T. and DiMaio, D. (2006). Is base excision repair a tumor suppressor mechanism? *Cell Cycle, 5*, 250-259.

Tom, S., Henricksen, L.A. and Bambara, R.A. (2000). Mechanism whereby proliferating cell nuclear antigen stimulates flap endonuclease I. *J. Biol. Chem., 275*, 10498-10505.

Tom, S., Henricksen, L.A., Park, M.S. and Bambara, R.A. (2001). DNA ligase I and proliferating cell nuclear antigen form a functional complex. *J. Biol. Chem., 276*, 24817-24825.

Tomkinson, A.E., Chen, L., Dong, Z., Leppard, J.B., Levin, D.S., Mackey, Z.B. and Motycka, T.A. (2001). Completion of base excision repair by mammalian DNA ligases. *Progr. Nucleic Acids Res. Mol. Biol.*, 68, 151-164.

Waga, S. and Stillman, B. (1998). The DNA replication fork in eukaryotic cells. *Annu. Rev. Biochem., 67*, 721-751.

Wiebauer, K. and Jiricny, J. (1990). Mismatch-specific thymine DNA glycosylase and DNA polymerase β mediate the correction of G-T mispairs in nuclear extracts from human cells. *Proc. Natl. Acad. Sci. USA, 87*, 5842-5845.

Wilson, D.M. III and Thompson, L.H. (1997) Life without DNA repair. *Proc. Natl. Acad. Sci. USA, 94*, 12754-12757.

Wood, R.D. and Shivji, M.K.K. (1997) Which DNA polymerases are used for DNA-repair in eukaryotes? *Carcinogenesis, 18*, 605-610.

Wu, X., Li, J., Li, X., Hsieh, C.L., Burgers, P.M. and Lieber, M.R. (1996). Processing of branched DNA intermediates by a complex of human FEN-1 and PCNA. *Nucl. Acids Res., 24*, 2036-2043.

In: New Research on DNA Repair
Editor: Breehn R. Landseer, pp. 251-272

ISBN 1-60021-385-5

Chapter IX

ROLES OF PLANT TRANSCRIPTION FACTOR IIH SUBUNIT HOMOLOGS IN NUCLEOTIDE EXCISION REPAIR AND TRANSCRIPTION

***Desma M. Grice*[1]*, Megan J. Osmond*[1]*, Michael Stewart*[1]*, Edward J. Vonarx*[2] *and Bernard A. Kunz*[1]**

[1]School of Life and Environmental Sciences, Deakin University, Geelong, Victoria 3217, Australia;

[2]School of Applied Sciences, Gordon Institute of TAFE, Private Bag 1, Geelong Mail Centre, Victoria 3221, Australia.

ABSTRACT

Eukaryotic general transcription factor (TF) IIH is composed of ten proteins, seven of which are also required for nucleotide excision repair (NER) of UV radiation-induced DNA damage in human cells. Although plant counterparts of these seven human TFIIH subunits have been identified by database screening, only the roles in NER of the *Arabidopsis thaliana* XPD and p44 homologs as well as its two XPB paralogs have been investigated. Here we review studies that addressed the repair and transcription functions of those proteins, as well as new findings concerning Arabidopsis homologs of the human TFIIH components p62 and TTDA. We and others have determined that AtXPB1, AtXPB2, AtXPD and AtGTF2H2 can function in NER of UV photoproducts, but UV treatment does not increase *AtXPB1, AtXPB2, AtXPD, AtGTF2H2, AtTFB1A* or *AtTFB5* transcript levels. Furthermore, AtGTF2H2 also can perform all essential transcription functions of yeast Ssl1 (p44 ortholog) in its absence, and AtXPD also can operate in transcription in yeast. In contrast, the newly isolated *AtTFB1A* cDNA does not complement the UV sensitivity or transcription defect conferred by a conditional yeast *tfb1* mutation, possibly because of the relatively low level of homology between Tfb1 and AtTFB1A. However, we have detected interactions between AtGTF2H2, AtXPD, AtTFB1A and AtTFB5, as well as interactions between these plant proteins and various of their yeast homologs in two-hybrid assays. Collectively, these observations suggest that AtXPB1, AtXPB2, AtXPD, AtGTF2H2, AtTFB1A and AtTFB5 may participate in NER and transcription in plants as part of TFIIH, and underpin the initial development of a structural model for its NER subassembly.

INTRODUCTION

Exposure to solar ultraviolet (UV) radiation is especially problematic for plants because of their obligate phototrophy and sessile lifestyle. The UV wavelengths in sunlight can decrease plant genome stability, growth and productivity (Teramura and Sullivan, 1994; Ries et al., 2000; Xiong and Day, 2001), effects at least partly due to the induction of UV photoproducts that block DNA replication and transcription (Strid et al., 1994; Britt, 1996). One approach plants use to minimize their damage load is to accumulate UV-absorbing secondary metabolites in their epidermal cells (Frohnmeyer and Staiger, 2003). Despite such shielding, UV radiation reaches the DNA in subepidermal levels. Thus, plants further counter the onslaught of DNA damage by monomerizing UV-induced pyrimidine dimmers via light-dependent photoreactivation, and removing UV photoproducts from DNA by nucleotide excision repair (NER) (Britt, 1996, 1999; Frohnmeyer and Staiger, 2003). Although NER is the most well-studied means of eliminating UV photoproducts in eukaryotes, its mechanism has not been elucidated for plants. In this chapter, we examine evidence implicating Arabidopsis (*Arabidopsis thaliana*) homologs of transcription factor (TF) IIH subunits in NER and transcription. We briefly review evidence supporting such roles, describe the isolation of cDNAs encoding Arabidopsis homologs of human p62 and TTDA, identify interactions of these homologs with other potential plant TFIIH subunits, and apply the interaction data to begin developing a model of the plant TFIIH core complex that may function in NER.

TFIIH is Heteromeric and Multifunctional

Among eukaryotes, NER has been most extensively characterized in the budding yeast *Saccharomyces cerevisiae* and human cells (Friedberg et al., 2006). It involves several stages: lesion recognition, incision of the damaged DNA strand on each side of the lesion, release of the lesion-carrying oligonucleotide, repair synthesis and ligation to seal the repair patch. The concerted performance of these steps is biochemically complex with NER pathways for transcriptionally active or inactive DNA in both organisms probably requiring more than 30 proteins, many of which operate in multimeric complexes (Friedberg et al., 2006; Fuss et al., 2006). These complexes assemble sequentially at sites of DNA damage (Volker et al., 2001; Reidl et al., 2003; Tapias et al., 2004), a strategy that may increase the efficiency of repair (Politi et al., 2005).

One of the NER components recruited to damage sites following lesion recognition but prior to DNA incision is TFIIH (Reidl et al., 2003). It is composed of ten subunits (we use the human designations for the TFIIH subunits except where noted) believed to form two groups. Core TFIIH (XPB, p62, p52, p44, p34) is linked to the cyclin activating kinase complex (cdk7, cyclin H, MAT1) by the XPD subunit which also interacts with the tenth TFIIH component, TTDA (human p8) (Chang and Kornberg, 2000; Schultz et al., 2000; Zurita and Merino, 2003; Coin et al., 2006). In this chapter we consider XPB, XPD, p62, p52, p44, p34, and TTDA, the seven that are required for NER and, except for TTDA, are essential for viability because they are also necessary for transcription (Feaver et al., 2000; Takagi et al., 2003; Giglia-Mari et al., 2004, 2006).

TFIIH is the only one of six basal factors required for transcription of protein-encoding genes by RNA polymerase (Pol) II that is known to have enzymatic activity. XPB and XPD are ATP-dependent $3' \rightarrow 5'$ and $5' \rightarrow 3'$ helicases, respectively, whose roles in NER are to catalyze unwinding of the DNA encompassing the lesion. XPB also promotes incision 5′ to the DNA lesion by the ERCC1-XPF endonuclease (Coin et al., 2004). Additionally, Ssl1, the *Saccharomyces cerevisiae* ortholog of p44, has an E3 ubiquitin ligase activity which, although not essential for NER, is associated with damage-induced transcription of DNA repair genes (Takagi et al., 2005). XPB activity is regulated by phosphorylation, TTDA, and interaction with p52 (Jawhiri et al., 2002; Coin et al., 2004, 2006), whereas XPD is regulated by p44, with which it interacts (Coin et al., 1998; Seroz et al., 2000). In addition, two-hybrid assays or co-immunoprecipitation experiments showed interactions among p44, XPD, p62 and XPB, between p62 and p34, and between TTDA and XPD (Iyer et al., 1996; Seroz et al., 2000; Coin et al., 2006).

Analysis of TFIIH or its subunits indicates remarkable, organism-dependent, versatility. TFIIH functions in NER of transcriptionally active or inactive DNA, transcription by RNA Pol II and RNA Pol I, cell cycle control, and regulation of nuclear receptor activity (Bradsher et al., 2002; Hoogstraten et al., 2002; Iben et al., 2002; Chen et al., 2003; Zurita and Merino, 2003; Compe et al., 2005; Russell and Zomerdijk, 2005; Sarkar et al., 2005). It has also been implicated in the transcriptional response to DNA damage and translation initiation (Gulyas and Donahue, 1992; Yoon et al., 1992; Takagi et al., 2005).

Plant Homologs of TFIIH Subunits XPB, XPD and p44 Act in Repair or Transcription

UV-sensitive Arabidopsis and rice (*Oryza sativa*) mutants with reduced UV photoproduct excision, and a UV-resistant *Arabidopsis* mutant with increased removal of UV-induced DNA damage have been isolated, indicating that NER can contribute to the survival of UV-irradiated plants (Britt et al., 1993; Hidema et al., 1997; Fidantsef et al., 2000; Tanaka et al., 2002; Liu et al., 2003). Interestingly, NER may function primarily on nuclear genomes in proliferating plant cells (Cannon et al., 1995; Chen et al., 1996; Kimura et al., 2004; Quaite et al., 1994), where TFIIH would be required for both transcription and repair. Indeed, bioinformatic analyses have identified potential plant homologs of all seven TFIIH subunits that function in NER (The Arabidopsis Genome Initiative, 2000; Costa et al., 2001a; Kunz et al., 2002, 2005). To date, Arabidopsis genes that specify two counterparts of XPB (*AtXPB1*, *AtXPB2*) or single homologs of XPD (*AtXPD*) or p44 (*AtGTF2H2*) have been characterized at the cDNA level (Ribiero et al., 1998; Costa et al., 2001b; Liu *et al.*, 2003; Morgante et al., 2005; Vonarx et al., 2006).

XPB and XPD feature ATPase and helicase motifs that are conserved in AtXPB1, AtXPB2 and AtXPD (Ribiero et al., 1998; Liu *et al.*, 2003), consistent with the plant proteins operating in NER and transcription. Indeed, expression of *AtXPB1* and *AtXPB2* cDNAs in yeast cells partially complemented the UV sensitivity conferred by deletion of *RAD25* (Costa et al., 2001b; Morgante et al., 2005), which encodes the *S. cerevisiae* XPB ortholog, and an *AtXPD* mutation decreased UV resistance and excision of UV photoproducts in Arabidopsis (Liu et al., 2003). The latter phenotypes were complemented by introduction of wildtype *AtXPD* DNA, indicating that AtXPD functions in NER (Liu et al., 2003). That AtXPD also can act in transcription was shown by expressing the *AtXPD* cDNA in yeast cells to complement a thermolabile RNA Pol II transcription defect due to a conditional mutation in the yeast XPD ortholog *RAD3* (Vonarx et al., 2006). Similarly, individual residues or sequence features (e.g., COOH-terminal C4 finger and C4C4 RING zinc-binding motifs) necessary for interaction of p44 with XPD, p62 or p34, stimulation of the XPD helicase, or ubiquitin ligase activity are conserved in AtGTF2H2 (Vonarx et al., 2006). Remarkably, expression of the *AtGTF2H2* cDNA in yeast cells prevented, respectively, the UV sensitivity or inviability associated with mutation or deletion of *SSL1*, which encodes the p44 ortholog (Vonarx et al., 2006), demonstrating that AtGTF2H2 can operate in NER and transcription. Collectively, these observations, plus the interactions of AtXPD with AtGTF2H2 or Ssl1 and AtGTF2H2 with Rad3 (Vonarx et al., 2006), suggest that AtXPB1/2, AtXPD and Atp44 function in Arabidopsis as part of TFIIH.

Arabidopsis has Two Genes that Encode p62 Homologs

Arabidopsis chromosomes 1 and 3 each contain a gene (At1g55750 and At3g61420, respectively) predicted to encode a p62 homolog (The Arabidopsis Genome Initiative, 2000). Although a second gene (At1g55680) on chromosome 3 was also identified as encoding a third p62 homolog, our initial analysis indicated that the predicted protein has less homology than the other two, is smaller by at least 121 amino acids, lacks all characteristic p62 motifs (see below) and contains a WD-40 repeat domain not present in p62 homologs. For these reasons, it was not considered further.

A preliminary alignment of the predicted At1g55750 and At3g61420 cDNA sequences taken from the Arabidopsis genome database revealed that they are 85% identical. However, there were three major regions of sequence discrepancy that prompted us to examine the predicted gene structures (Fig. 1A). This comparison indicated that At1g55750 has 20 exons whereas At3g61420 has only 18, with exons corresponding to 1 and 17 in At1g55750 apparently missing. However, alignment of the affected areas revealed that a sequence in the 5' untranslated region (UTR) of At3g61420 is 74% identical to exon 1 of At1g55750 but lacks the start codon and flanking sequences. Similarly, a segment corresponding to exon 17 of At1g55750 is present in the intron preceding exon 17 in At3g61420. We also noted that intron sequences preceding exon 19 of At1g55750 were detected at the 3' end of exon 16 and the 5' end of exon 17 in At3g61420 (Fig. 1A). Analysis of the exon-intron junctions showed variation between the homologs at splice sites for the exons that differ in At1g55750 and At3g61420, whereas exons that coincide in the homologs have little or no difference between their splice sites. Consequently, alterations in splicing likely account for two of the three large regions of sequence variation between the predicted cDNAs, and the inclusion of intron sequences in exons 16 and 17 of At3g16420. We also determined that the 15 nucleotides preceding the stop codon in At1g55750 are absent in At3g61420. Furthermore, the 3' UTR of At1g55750 contains a segment downstream of the stop codon with 87.5% identity to the region encompassing the stop codon in At3g61420. However, there is no sequence in At3g61420 corresponding to the region between that segment and the stop codon in At1g55750. The latter features suggest that the difference at the 3' end of exon 18 in At3g61420 relative to the 3' end of exon 20 in At1g55750 may be the result of a recombination event that deleted the missing sequence. Owing to the various differences, the predicted At3g61420 protein would lack p62 motif II at its COOH-terminus and the first third of the NH_2-terminal phosphotyrosine binding/pleckstrin homology (PTB/PH) domain essential for NER (Fig. 1B) (see below).

The extent of their predicted cDNA sequence identity, plus the differences in their proteins, suggests At1g55750 and At3g61420 are paralogs. Arabidopsis has undergone several instances of polyploidism, and duplication blocks have been identified, mapped and ordered into different classes based upon their estimated age of duplication (The Arabidopsis Genome Initiative, 2000; Vision et al, 2000; Simillion, 2002; Blanc et al., 2003; Bowers et al., 2003; Blanc and Wolfe, 2004). At1g55750 is located within a duplication block (here designated A for ease of discussion) on chromosome 1, and block A and another block (here designated B) on chromosome 3 are the result of a duplication event (Vision et al., 2000; Blanc et al., 2003). Surprisingly, At3g61420 is not located within block B but is present in a second block (here designated C), unrelated to block A, on chromosome 3. This indicates that if block B originally had a copy of At1g55750 it has been lost, or that At1g55750 was not initially present in block A or block B. The absence of sequence overlap between blocks A and C suggests that a smaller unmapped duplication event within larger duplication blocks, either A and C or B and C, may be responsible for At1g55750 and At3g61420. However, we detected no sequence similarities in genes, or their predicted proteins, flanking At1g55750 and At3g61420. This is not unexpected since selection pressure immediately following a duplication event may result in differential gene retention at each site leading to a lack of congruency surrounding the duplicate genes, or even loss of one of the gene copies, leaving no similarity to be identified (Lockton and Gaut, 2005). It is not possible to discern the sequence of events that gave rise to At1g55750 and At3g61420 since blocks A, B and C are dated to the past 50 million years indicating that they are all due to relatively recent events (Vision et al., 2000). Nonetheless, the alignments of the predicted cDNA or protein sequences, and gene structure comparisons make it seem reasonable to think that At1g55750 served as a template for duplication and that the At3g61420 protein would not function in NER. Thus, further efforts focused on At1g55750.

Primers whose design was based on the At1g55750 sequence were used to amplify from reverse-transcribed leaf mRNA a 1,776 bp Arabidopsis cDNA that encompassed a single open reading frame. Comparison of the cDNA and genomic sequences verified that the corresponding gene, *AtTFB1A*, is comprised of 20 exons and 19 introns. The context of the translation initiation codon (GAUAUGGCU) in *AtTFB1A* differs from the consensus dicot sequence A(A/C)aAUGGCu (Joshi et al., 1997; Lukaszewicz et al., 2000) in having a G and a U at positions -3 and -1, respectively. The significance of these replacements is unknown, but purines were detected at position -3 in 82% of 3643 dicot mRNA sequences analyzed (Joshi et al., 1997). *AtTFB1A* has a second potential start codon 87 nucleotides downstream to the first, but its sequence context (GGGAUGUUG) bears no resemblance to the consensus lacking all conserved nucleotides outside of the start codon, although there is a purine at -3.

AtTFB1A encodes a predicted protein (591 amino acids (aa), 67.5 kDa) larger than human p62 (548 aa, 62 kDa) and smaller than yeast Tfb1 (642 aa, 75 kDa), but with similar levels of identity (p62: 23%, Tfb1: 19%) and overall conservation (p62: 63%, Tfb1: 55%). Six motifs designated I-VI are characteristic of p62 homologs (Castro et al., 2002). Motifs VI and II are situated in the COOH-terminal and NH_2-terminal regions, respectively, whereas motifs I, III, V, and IV are clustered towards the center of the protein (Fig. 1, B and C). Motif I contains the highly conserved FWxxF/YF/Y sequence and along with motif IV is within the region of p62 required for binding to XPD (Jawhiri et al., 2004). Motif II may be required for interaction with p44 homologs as deletion of the 114 aa C-terminal segment encompassing motif II in yeast Tfb1 prevents the interaction with Ssl1 (Matsui et al., 1995). All six motifs are at the corresponding positions in AtTFB1A, and exhibit considerable similarity to their human counterparts (48% identity, 93% conservation on average) including near identity to the FWxxF/YF/Y sequence (Fig.1C). Human p62 has an NH_2-terminal PTB/PH domain that is essential for NER because it interacts with the 3' endonclease XPG (Gervais et al., 2004). However, this region is not required for TFIIH assembly and is dispensable for basal transcription. The PTB/PH aa sequence adopts a β-sandwich fold made up of two orthogonal β-sheets comprised of seven β-strands and two α-helices (Gervais et al., 2004). A well-conserved form of this sequence is present in AtTFB1A, but the predicted At3g61420 protein lacks the 31 residues encompassing the first and second β-strands which are critical for PTB/PH domain formation (Gervais et al., 2004).

AtTFB1A Does Not Restore Transcription or UV Resistance When Expressed in a Conditional Yeast *tfb1* Mutant

The yeast *tfb1-101* allele has a deletion of the COOH-terminal 114 aa which confers thermolabile transcription by RNA Pol I and II and so prevents growth at the non-permissive temperature of 37°C (Matsui et al., 1995; Wang et al., 1995; Iben et al., 2002) (Fig. 2, A and B, rows 1 and 2). To determine if AtTFB1A can function in transcription, we introduced the *AtTFB1A* cDNA into the *tfb1-101* mutant on the yeast expression vector pYES-DEST^{TM}52 (Invitrogen) and assessed its ability to complement the temperature-sensitive growth defect. Insert expression was verified by reverse transcription of isolated RNA, PCR-amplification of the relevant cDNA and cDNA sequencing.

Furthermore, control reactions showed that RNA preparations did not contain plasmid DNA, and insert-specific amplification products were not generated from strains lacking the plasmid. All inserts used in this assay have native stop codons and should be expressed as native proteins. Although expression of yeast *TFB1* permitted growth at 37°C (Fig. 2, A and B, row 3), expression of *AtTFB1A* did not restore growth at the restrictive temperature (Fig. 2, A and B, row 4).

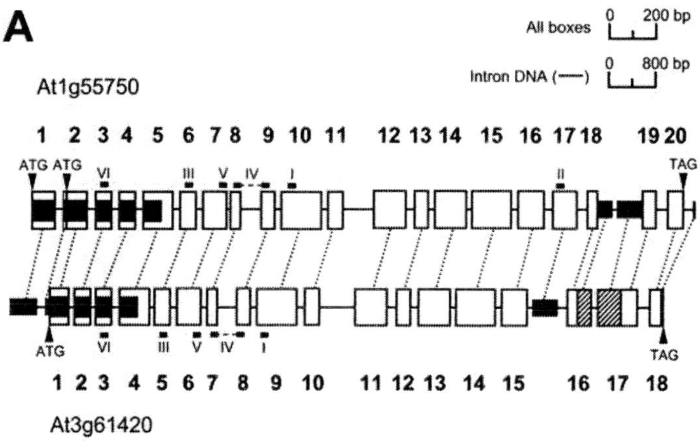

Figure 1A. Arabidopsis p62 homologs and conserved features in AtTFB1A. Structures of At1g55750 and At3g61420 predicted to encode p62 homologs. The first and last exons (white boxes) of At1g55750 are predicted to contain, respectively, a 47 bp 5' UTR and a 256 bp 3' UTR which have been omitted from the diagram. Exons are numbered above or below the structures, and the homologous sequences in the two structures are linked by dotted lines. The positions of the conserved motifs I through VI are indicated by the black bars immediately above At1g55750 or below At3g61420, and the locations of the ATG and TAG codons are pinpointed by the black triangles. The thin black lines indicate intron sequences. Black boxes in one homolog indicate intronic sequence that is exonic sequence in the other homolog. The gray boxes encode residues forming the PTB/PH domain.

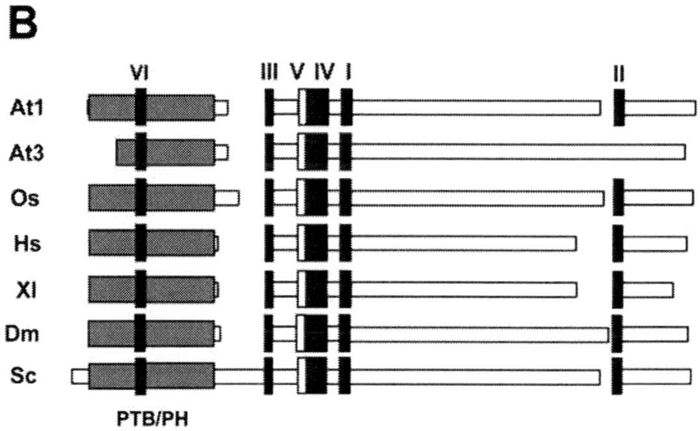

Figure 1B. Arabidopsis p62 homologs and conserved features in AtTFB1A. Schematic alignment of p62 homologs. The large black, gray or white boxes indicate regions of homology whereas the smaller white boxes indicate unique sequences, and gaps are introduced between some boxes for optimal alignment. Labels as in A except At1: At1g55750. Accession numbers: *Oryza sativa* (Os), XM481749; *Homo sapiens* (Hs), P32780; *Xenopus laevis* (Xl), AAH46266.1; *Drosophila melanogaster* (Dm), AAF58253; *Saccharomyces cerevisiae* (Sc), AAA35143.

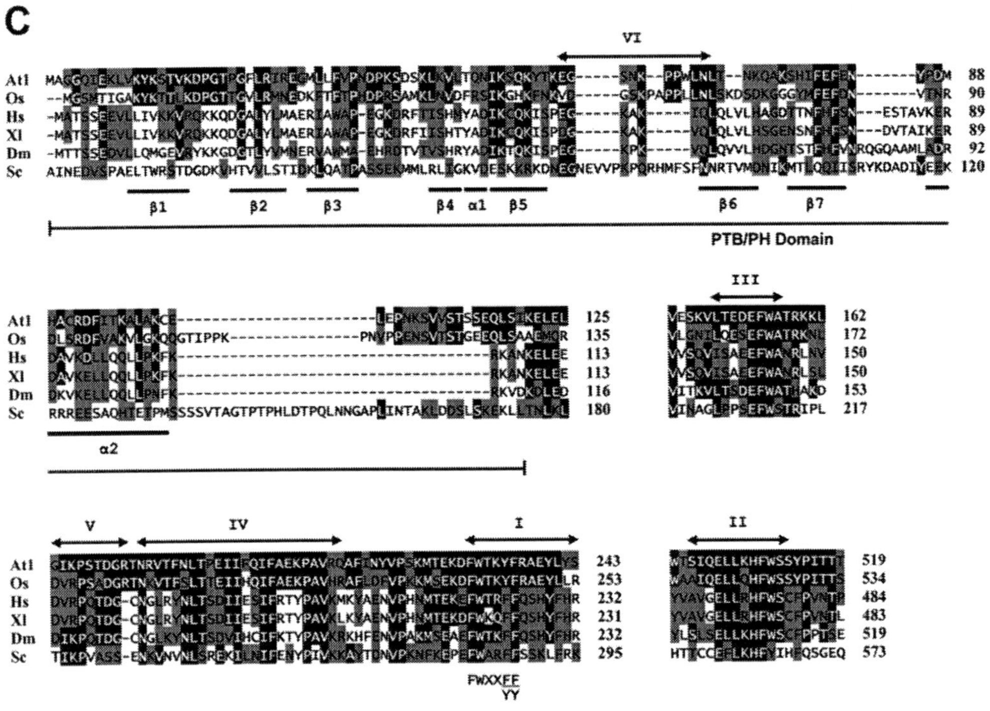

Figure 1C. Arabidopsis p62 homologs and conserved features in AtTFB1A. Alignments of regions containing the PTB/PH domain and the six conserved p62 motifs. Databases were screened via BLAST (Altschul et al., 1997), primers designed or amino acid sequences predicted using GeneRunner 3.05 (www.generunner.com/), and sequences aligned using CLUSTALW (Thompson et al., 1994). The *AtTFB1A* cDNA and the yeast *TFB1* gene were amplified from reverse-transcribed Arabidopsis leaf RNA or from yeast genomic DNA (Vonarx et al., 2002), respectively, using primers AtTFB1A-F1 (5'-CCACCCTACAGTTGGATATGGCT-3') and AtTFB1A-R1 (5'-TTTTTGTCCTCTATACGTA-3') or ScTFB1-F1 (5'-CACCAAGTAAACCATAATGTCA-3') and ScTFB1-R3 (5'-CAACTGGAGAGCCT TTAT-3'), respectively, for topoisomerase I-mediated directional cloning into pENTR/D-TOPO® (Invitrogen). Residues identical to or conserved with those in AtTFB1A are highlighted in black or gray, respectively, and non-conserved residues are highlighted in white. Motifs I-VI are overlined with double-headed arrows. The position of the PTB/PH domain in human p62 is underlined. The seven β-strands and two α-helices involved in the formation of the PTB/PH domain are indicated by thick underlines. The highly conserved FWxxF/YF/Y sequence is indicated below motif I. The accession numbers are as for part A.

The failure of *AtTFB1A* to complement the transcription defect at 37°C could have been due to the Tfb1-101 protein preventing incorporation of AtTFB1A into yeast TFIIH by competing or interacting with it. To test this possibility, we made use of the *tfb1-101* allele being on the single copy, centromeric plasmid pRS313 in a yeast strain (YSB260) deleted for *TFB1* and *HIS3*. This vector carries *HIS3* so that it can be maintained selectively by cell growth in histidine omission medium. In the presence of histidine, however, pRS313 is lost from about 40% of cells following ten generations of growth (Sikorski and Hieter, 1989). If AtTFB1A can substitute for yeast Tfb1 in transcription, viable cells lacking pRS313 should be present following growth in medium containing histidine. Since pRS313 contains the *HIS3* gene, which confers the ability to grow in the absence of histidine, loss of pRS313 would result in histidine auxotrophy. When YSB260 carrying *tfb1-101* on pRS313 and the yeast *TFB1* gene on the expression vector pYES-DEST^{TM}52 was grown in histidine-containing medium, pRS316 was lost from 47% of the cells (Table 1). (Figure 2, c and d, row 6), but they were viable at 30°C and 37°C in the presence of histidine indicating complementation of the *tfb1* deletion solely by *TFB1* on pYES-DEST^{TM}52 (Fig. 2, e and f, row 6). In contrast, only 1 of 1,500 colonies arising from YSB260 cells carrying *AtTFB1-1* on pYES-DEST^{TM}52 appeared to have lost pRS313 (Table 1). However, amplification of the *HIS3* gene and the *tfb1-101* allele from cells in this colony showed that pRS313 had not been lost, and suggested that the inability of the cells to grow on histidine omission medium was most likely due to a mutation in *HIS3*. Clearly, *AtTFB1A* is unable to substitute for *TFB1* in the absence of *tfb1-101* and therefore the cells do not lose pRS313, retain the ability to grow in the absence of histidine at 30°C and are inviable at 37°C (Fig. 2, c and d, row 7).

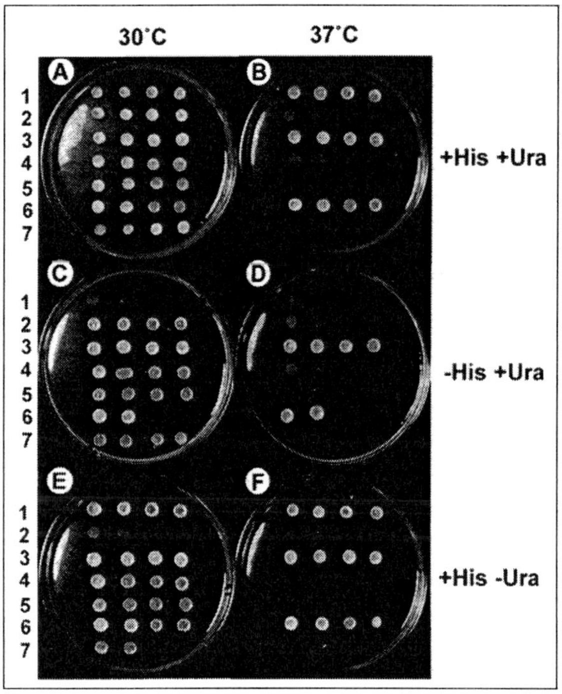

Figure 2. *AtTFB1A* does not complement the transcriptional defects in the yeast *tfb1-101* mutant. The *AtTFB1A* cDNA and yeast *TFB1* gene were transferred from pENTR/D-TOPO® to pYES-DEST^{TM}52 following the manufacturer's instructions, and insert-bearing vectors, or the empty vector, were transformed into strain YSP260 (*MATa, ura3-52, leu2-3,112, his3-Δ200, tfb-101 Δ::LEU2, (pRS313-tfb1-Δ101)*. Strain YSB207 (*MATa, ura3-52, leu2-3,112, his3- Δ 200, tfb1 Δ::LEU2, (pRS316-TFB1)* is isogenic to YSB260 and was used as a control. The strains were grown overnight in, and spotted on to, appropriately supplemented minimal expression medium [Vonarx et al., 2002]. Plates A and B contain histidine (+His) and uracil (+Ura), C and D lack histidine (-His) and contain uracil (+Ura), and E and F contain histidine (+His) and lack uracil (-Ura). The plates were then incubated at 30°C (A, C, E) or 37°C (B, D, F) for three days. In this system, the *TFB1* allele is carried on the pRS316 plasmid, *tfb1-101* is carried on pRS313 and the *AtTFB1A* cDNA or yeast *TFB1* gene is carried on pYES-DEST^{TM}52. These vectors permit strain growth in the absence of uracil, histidine or uracil, respectively. On the plates, the strains are numbered as follows. 1. YSB207: TFB1 (Ura+); 2. YSB260:

tfb1-101 (His+); 3. YSB260: *tfb1-101* (His+) + yeast *TFB1* (URA+); 4. YSB260: *tfb1-101* (His+) + yeast *AtTFB1-1* (URA+); 5. YSB260: *tfb1-101* (His+) + empty pYES-DEST™52 vector (URA+); 6. YSB260: *tfb1-101* (His+) + yeast *TFB1* (URA+); 7. YSB260: *tfb1-101* (His+) + yeast *AtTFB1-1* (URA+). The strains in rows 6 and 7 were grown for >10 generations under conditions non-selective for the vectors prior to spotting on to the plates. If they lost pRS313 (His+) or pYES-DEST™52 (URA+) they would be unable to grow in the absence of histidine or uracil, respectively.

Table 1. Elimination of the pRS313 vector by growth under non-selective conditions.

	Medium					
	Nutrient Medium		Uracil Omission Medium		Histidine Omission Medium	
Insert on Expression Vector	Number of Colonies	Percent Plasmid Retention	Number of Colonies	Percent Plasmid Retention	Number of Colonies	Percent Plasmid Retention
AtTFB1A	1500	100	141	9.4	1	0.07
TFB1	1430	100	5	0.34	676	47.3

Yeast strain YSB260 containing the pRS313 plasmid was grown in 100 ml nutrient medium (non-selective medium) for two days at 30°C with shaking. Cell titers were determined, and appropriately diluted aliquots were plated on nutrient medium plates which were incubated at 30°C for two days. Colonies that emerged were replica-plated onto histidine or uracil omission medium and the plates were incubated for two days at 30°C. The percent plasmid retention was then determined.

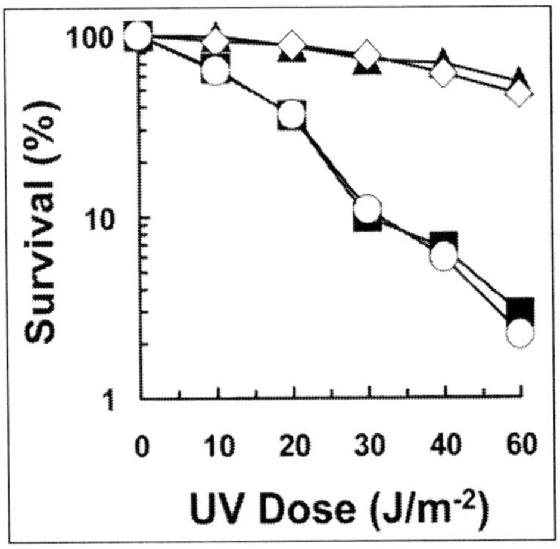

Figure 3. *AtTFB1A* does not restore UV resistance when expressed in the *tfb1-101* mutant. Yeast strains were grown in appropriately supplemented minimal expression medium [Vonarx et al., 2006] at 30°C with shaking to 5 x 10^6 cells/ml as determined using a Coulter counter. Diluted culture aliquots were spread on expression medium, the plates were UV-irradiated [Vonarx et al., 2002] and cell survival was determined after 5 days incubation in the dark at 30°C. Data are the means of at least three independent experiments for each strain tested, and error bars (standard deviations) are smaller than the data points. Wildtype (*TFB1*) strain YSB207 (▲), isogenic *tfb1-101* mutant YSB260 carrying yeast *TFB1* on pYES-DEST™52 (◇), *tfb1-101* mutant carrying *AtTFB1A* on pYES-DEST™52 (■), *tfb1-101* mutant carrying pYES-DEST™52 with no insert (O).

At 30°C the *tfb1-101* allele confers increased UV sensitivity due to a transcription-independent NER defect (Matsui et al., 1995; Wang et al., 1995). To determine if *AtTFB1A* can function in NER, we tested the effect on viability of expressing the *AtTFB1A* cDNA on pYES-DEST™52 in UV-C irradiated *tfb1-101* cells. Transcription of the insert was confirmed as described above. As anticipated, *tfb1-101* sensitized yeast cells to UV (254 nm), and this phenotype was not influenced by the presence of pYES-DEST™52 alone (Fig. 3). Expression of yeast *TFB1* fully restored UVB resistance to the wildtype level (Fig. 3), but *AtTFB1A* was unable to complement the UV sensitivity of the *tfb1-101* mutant.

AtTFB5 Encodes a TTDA Homolog

By screening the Arabidopsis genome database with the human TTDA and yeast Tfb5 (TTDA ortholog) sequences, we identified the At1g12400 locus on chromosome 1 as the sole sequence potentially encoding a TTDA homolog. As the predicted gene structure features two 5' UTRs separated by an intron, primers were designed to amplify a 219 bp cDNA containing only the open reading frame plus stop codon from reverse-transcribed leaf mRNA. Comparison of the cDNA and genomic sequences supported the predicted gene structure of three exons and two introns with the open reading frame beginning within the second exon (Fig. 4A). The context of the translation initiation codon (CUU**AUG**GUU) in *AtTFB5* differs at most flanking positions from the consensus dicot sequence A(A/C)aAUGGCu, except for the G and U at positions +4 and +6, respectively (Joshi et al., 1997; Lukaszewicz et al., 2000). Human TTDA is present in all cellular TFIIH and in a free pool distributed between the nucleus and cytoplasm (Coin et al., 2006; Giglia-Mari et al., 2006). These observations plus the need for transcription in plant meristematic tissue make it seem unlikely that the less than optimal start codon context of *AtTFB5* reflects a requirement for inefficient translation.

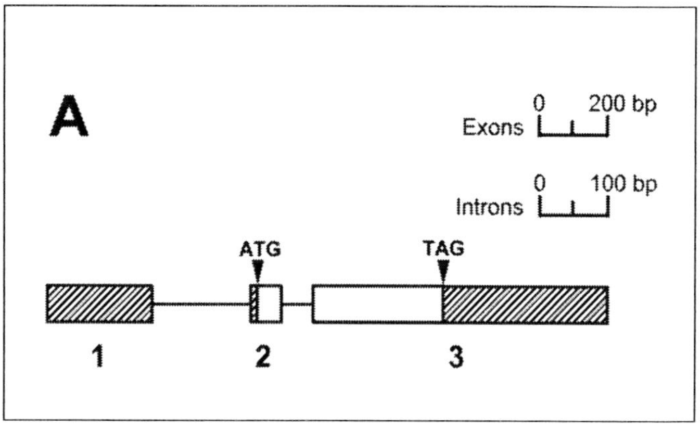

Figure 4A. Arabidopsis TTDA homolog and conserved features in AtTFB5. Structure of At1g12400 predicted to encode a TTDA homolog. The entire first exon, and the first 9 bp of exon two are predicted to be 5' UTR regions (hatched boxes), and the final 234 bp of the third exon is predicted to be a 3' UTR. The positions of the ATG and TAG codons are indicated by black triangles. The thin black lines indicate intron sequences.

B

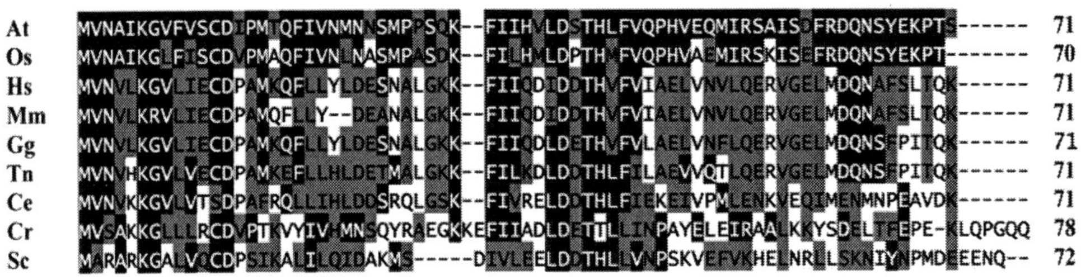

Figure 4B. Arabidopsis TTDA homolog and conserved features in AtTFB5. Alignments of TTDA homologs. The *AtTFB5* cDNA and the yeast *TFB5* gene were amplified from reverse-transcribed Arabidopsis leaf RNA or from yeast genomic DNA (Vonarx et al., 2002), respectively, using primers AtTFB5-F1 (5'-CACCATGGTTAATGCAATCAAAGG-3') and AtTFB5-R1 (5'-ATGTGATCTAAGAAGTGGG-3'), or ScTFB5-F1 (5'-CACCATGGCTAGAGCAAGAAA-3') and ScTFB5-R1 (5'-ATTACTGATTTTCTTCTTC-3'), respectively, for topoisomerase I-mediated directional cloning into pENTR/D-TOPO®. Residues identical to or conserved with those in AtTFB5 are highlighted in black or gray, respectively, and non-conserved residues are highlighted in white. Accession numbers: *Oryza sativa* (Os), XP478800; *Homo sapiens* (Hs), AAH38848; *Mus musculus* (Mm), NP076886; *Gallus gallus* (Gg), CAG31773; *Tetraodon nigroviridis* (Tn), CAG03600; *Caenorhabditis elegans* (Ce), AAF59588; *Chlamydomonas reinhardtii* (Cr), AAP12520; *Saccharomyces cerevisiae* (Sc), NP_076886.

AtTFB5 encodes a predicted protein 71 aa in length with a molecular mass of 8 kDa, which compares very favorably with the human (71 aa, 8 kDa) and yeast (72 aa, 8.2 kDa) proteins. An alignment of TTDA homologs indicated that AtTFB5 exhibits greater homology with the predicted rice protein, (77% identity, 87% conservation) than with vertebrate (34-41% identity, 44-54% conservation) or lower eukaryote (27%-41% identity, 44%-54% conservation) homologs (Fig. 4B). The first ten residues of human TTDA are essential for stimulation of XPB helicase activity and so NER (Coin et al., 2006). This region is 100% conserved in AtTFB5 and exhibits 60% identity with the corresponding human TTDA sequence.

Expression of *AtTFB1A* and *AtTFB5* is not Induced by UV

Within 1 kb 5' to the start ATG, the promoter regions of *AtTFB1A* and *AtTFB5* contain *cis*-elements including ACGT, ACCTA, (ACACGT), MRE [(T/C)AN(A/C)ACCTA(C/A)C], SBF-1 [(G/T(A/T)(A/G)TNGTTAA(A/T)$_3$] or TCA (TCATCTTCTT) implicated in light regulation or response to stresses such as UV (Brosché et al., 2002) Thus, RNA levels were quantitated via PCR to determine whether UV (254 nm) irradiation of two-week old plants had any impact on the expression of *AtTFB1A* or *AtTFB5*. In neither case did UV doses of 750 Jm^{-2} or 1500 Jm^{-2} cause any significant change in mRNA steady state levels by up to 9 h post-irradiation (Table 2). However, these doses were sufficient to alter gene expression, and any changes in expression should have been detectable by the time points chosen, as indicated by a previous study (Molinier et al., 2005) in which a lower 254 nm UV dose of 600 Jm^{-2} modified gene expression in Arabidopsis by 30 min, 1 h or 6 h after irradiation. Previously, transcription of *AtXPB1*, *AtXPB2*, *AtGTF2H2* and *AtXPD* was shown not be UV-inducible (Morgante et al., 2005; Vonarx et al., 2006). Together with the results of this study, these observations indicate that UV does not increase the expression of Arabidopsis homologs of the TFIIH core complex subunits at the transcriptional level.

Table 2. Quantitative PCR analysis of AtTFB1A and AtTFB5 gene expression.

Gene	Dose (Jm^{-2})	Hours Post-Irradiation	Delta C_T	Standard Error
AtTFB1A	0	3	19.5	0.1
	0	9	18.4	0.6
	750	3	20.3	0.3
	750	9	18.2	0.4
	1500	3	18.2	0.3
	1500	9	18.8	0.5
AtTFB5	0	3	15.8	0.5
	0	9	15.0	0.8
	750	3	16.3	0.4
	750	9	15.2	0.9
	1500	3	15.8	0.5
	1500	9	16.0	0.6

Fourteen day-old plants were exposed to 750 or 1500 Jm^{-2} 254 nm UV radiation. 750 Jm^{-2} was the lowest dose at which minimal phenotypic effects, primarily yellowing at leaf edges, became evident by five days post-irradiation. At 1500 Jm^{-2} the effects were more pronounced after five days but the plants survived and grew normally. Control plants were treated exactly the same as irradiated plants except for UV exposure. Leaf tissue was harvested after 3 h or 9 h post-irradiation incubation under yellow light to avoid photoreactivation of UV photoproducts, and frozen in liquid nitrogen. RNA was isolated, assessed for concentration, purity and integrity, and treated with RNase-free DNase I [Vonarx et al., 2006]. *AtTFB1A*, *AtTFB5* and 18S rRNA cDNAs were quantitatively amplified [Vonarx et al., 2006] from reverse-transcribed RNA using primers AtTFB1A-FP (5'-CGGCATCAGGTTTCTCTAC-3') and AtTFB1A-RP (5'-TGTCCTCTATACGTAGCCG-3'), AtTFB5-FP (5'-GCAATCAAAGGAGTGTTCG T-3') and AtTFB5-RP (5'-ATCTGCTCTACATGTGGTTG-3'), or 18S-F (5' AGTATGGTCGCAAGGC TGA 3') and 18S-R (5' TGTCAATCCTTACTATGTCTG 3'), respectively. Amplicon lengths and identities were confirmed by non-quantitative PCR, size-evaluation and sequencing. Appropriate linearity tests were performed to ensure that differences in the threshold cycle values for the cDNAs remained constant despite variation in the amount of starting material. Primer concentrations used gave the lowest threshold cycle values and no amplification in no-template controls. Delta C_T values are the means of three biological replicates, each normalized to the C_T for 18S rRNA in the same reaction. They represent the threshold cycle of detection for each gene of interest. One cycle difference corresponds to a doubling of the original mRNA population.

AtTFB1A and AtTFB5 Interact with TFIIH Subunit Homologs

Analysis of human and yeast TFIIH using electron microscopy, co-immunoprecipitation or yeast two-hybrid assays has demonstrated that human p62 or yeast Tfb1 interact with p44 and XPD or Ssl1 and Rad3, respectively (Bardwell et al., 1994; Feaver et al., 1993, 2000; Iyer et al., 1996), and TTDA interacts with XPD (Coin et al., 2006). In this study, we used the yeast two-hybrid system to test for intracellular interaction of AtTFB1A and AtTFB5 with each other, as well as with AtGTF2H2 and AtXPD and their yeast homologs. The yeast reporter strain Mav203, carrying the two-hybrid vector pDEST™22 with the *AtTFB1A* ORF fused to the GAL4 transcriptional regulator DNA activation domain (AD), or pDEST™32 with the *AtTFB5* ORF fused to the GAL4 DNA binding domain (BD), was transformed with pDEST™32 having the *AtGTF2H2, AtTFB5, AtXPD, RAD3, SSL1* or *TFB5* ORF fused to the GAL4 BD, or pDEST™22 bearing the *AtGTF2H2, AtXPD, RAD3* or *SSL1* ORF fused to the GAL4 AD, respectively. Protein interactions were detected by screening transformants for the ability to grow on histidine omission medium. None of the plant or yeast fusion proteins conferred histidine prototrophy when expressed alone, nor did they interact with the GAL4 AD or BD as appropriate (data not shown). However, AtTFB1-1 interacted with AtGTF2H2 (weakly), AtXPD and AtTFB5, and their yeast counterparts at 30°C (Fig 5) and 37°C (data not shown). The latter observation indicated that the interaction of AtTFB1A with the yeast TFIIH subunits was not destabilized under the conditions used to assay for complementation of the transcription defects in the *tfb1-101* mutant. We also found that AtTFB5 interacted with AtXPD and AtGTF2H2 and their yeast homologs (Table 3). It is important to note that because the two-hybrid interactions occur in cells containing TFIIH components, interactions between the proteins being tested might be mediated by another yeast TFIIH subunit. For example, co-immunoprecipitation experiments suggest that XPD may act as a bridge between p44 and TFB1 since it interacts with both subunits (Iyer et al., 1996). Thus, the two-hybrid results should be interpreted cautiously.

Table 3. Summary of two-hybrid interactions involving AtTFB5.

Arabidopsis or Yeast Protein		
DNA-binding Domain	Activation Domain	Interaction
AtTFB5	AtGTF2H2	+
AtTFB5	Ssl1	+
AtTFB5	AtXPD	+
AtTFB5	RAD3	+
AtTFB5	No insert	-
No insert	AtTFB5	-

Protein-protein interactions were determined qualitatively by scoring growth of cells carrying pairs of two-hybrid vectors on histidine omission medium.

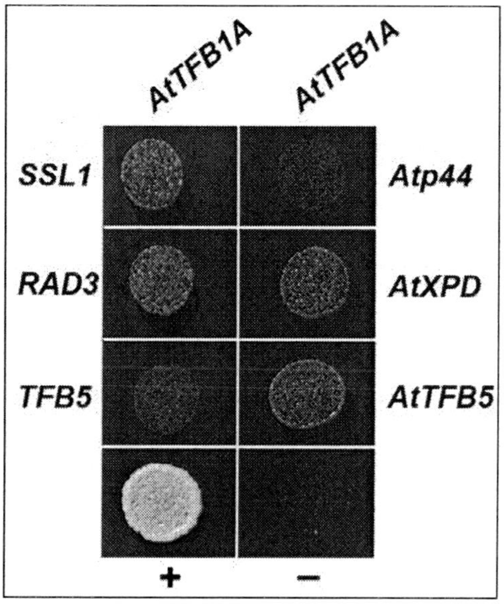

Figure 5. Interaction of AtTFB1A with plant and yeast TFIIH subunit homologs. Intracellular protein-protein interactions were assessed using a yeast two-hybrid assay (Pro-Quest™ Two-Hybrid System, Invitrogen™). Inserts were transferred from pENTR/D-TOPO® to the two-hybrid vectors pDEST™22 or pDEST™32 to fuse them with the *GAL4* activation or DNA-binding domains, respectively. Pairwise combinations of the fusion vectors were transformed into MaV203, and fusion protein interactions screened following the manufacturer's instructions. Fusion protein interaction activates a *HIS3* reporter resulting in growth on histidine omission medium containing 7.5 mM 3-amino-1,2,4 triazole to reduce leaky expression of the *HIS3* reporter. The "+" and "-" signs indicate positive and negative interaction controls provided with the system used.

Towards a Structure for the Plant TFIIH NER Core Complex

Human TFIIH features a ring-like assembly that involves six (XPB, XPD, p62, p55, p44, p34) of the seven subunits required for NER, and a similar complex is formed by some of the corresponding yeast TFIIH components (Chang and Kornberg, 2000; Schultz et al., 2000). The model of the human core TFIIH complex shown in Fig. 6A is consistent with mapping of TFIIH subunits within these structures and demonstrated interactions among p44, XPD, p62 and XPB, or between p62 and p34, XPB and p52, and TTDA and XPD (Iyer et al., 1996; Seroz et al., 2000; Coin et al., 2006). We have determined that AtGTF2H2 interacts with AtXPD (Vonarx et al., 2006), they both interact with AtTFB1 and ATTFB5, and the latter two proteins interact. As mentioned above, however, the possibility of bridging interactions in two-hybrid assays must be kept in mind. Even so, our results are consistent with the reported interactions of the human TFIIH components. On this basis, and given the similar sizes of the plant TFIIH subunit homologs relative to their human counterparts, we have been able to overlay AtGTF2H2, AtTFB1A, AtTFB5 and AtXPD onto the model of the human TFIIH core complex (Fig. 6B). Consequently, we anticipate that interactions similar to those involving XPB, p52 and p34 also may be found for their Arabidopsis homologs, and Arabidopsis TFIIH components required for NER may assemble into a ring-like structure.

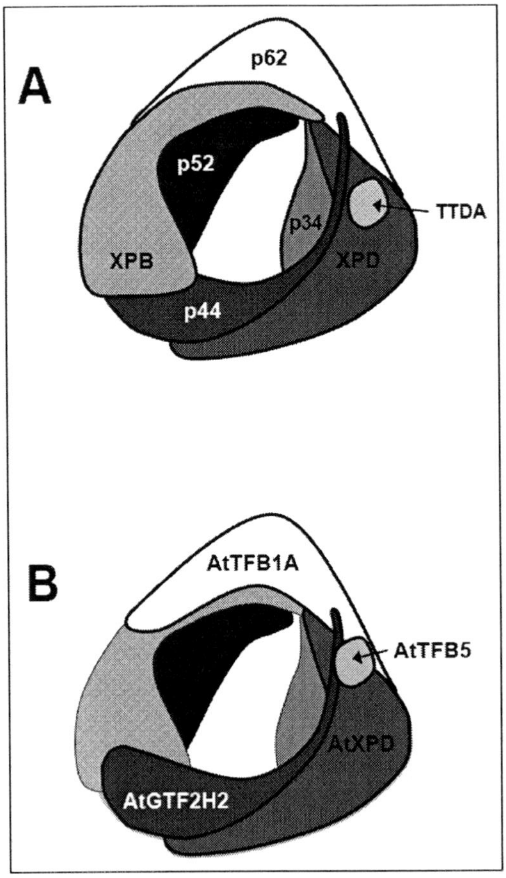

Figure 6. Preliminary model of Arabidopsis TFIIH. A, Positions of human TFIIH components compiled from interaction studies and electron microscopy. B, Positions of AtGTF2H2, AtTFB1A, AtTFB5 and AtXPD based on interaction studies described in this chapter and in Vonarx et al. [2006].

Conclusions

Human p62 and TTDA participate in transcription and NER as part of TFIIH and interact with other TFIIH components including p44 or XPD, respectively. In this chapter, we identified two Arabidopsis loci, At1g55750 and At3g61420, as potentially encoding p44 homologs, and a third locus, At1g12400, predicted to encode the TTDA homolog, AtTFB5. Analysis of the predicted cDNAs, gene structures and protein sequences for the former two loci suggested that At1g55750 is more likely to encode a functional p44 homolog as the At3g61420 protein was missing critical motif sequences. However, expression of the corresponding cDNA, *AtTFB1A*, in yeast cells did not complement the transcription defects or UV sensitivity conferred by the *tfb1-101* mutation, although AtTFB1-1 and AtTFB5 appeared to interact with each other and the putative Arabidopsis TFIIH components AtGTF2H2 and AtXPD, as well as their yeast counterparts Ssl1 and Rad3. The inability of AtTFB1A to function in yeast was not due to the failure of the vector to express the insert or of the transcript to be translated in yeast. Transcription was confirmed via cDNA amplification, and other studies showed the *AtGTF2H2, AtPOLH, AtREV7, AtXPF* and *AtXPD* cDNAs can complement the phenotypes of the corresponding yeast mutants when transcribed from a near identical vector (Vonarx et al., 2002, 2006; H. J. Anderson, personal communication), indicating that the plant transcripts are translated to produce functional proteins. In addition, when *AtTFB1A* was expressed in yeast two-hybrid vectors, positive interaction results were obtained demonstrating that the *AtTFB1A* transcript can be translated in yeast. The inability to eliminate the *tfb1-101* allele in the presence of *AtTFB1A* suggests that competition between AtTFB1A with the Tfb1-101 protein for incorporation into yeast TFIIH, or an inhibitory interaction between AtTFB1A and TB1-101, was not responsible for the incapacity of AtTFB1A to complement the *tfb1-101* phenotype. Furthermore, the two-hybrid data also argue against the latter scenario as AtTFB1A and yeast TFB1 did not interact (data not shown). Although AtTFB1A interacted with yeast TFIIH components at 30°C and 37°C, the interactions might not have been functional. Alternatively, the lack of complementation might have been due to an inability to associate with other untested yeast TFIIH components such as Rad25, whose human homolog, XPB, has been shown to interact with human p62 (Iyer et al., 1995). The relatively low homology of AtTFB1A and yeast Tfb1 tends to support these possibilities. Finally, interaction of AtTFB1A and AtTFB5 with other plant and yeast TFIIH subunits indicates that these proteins may operate as part of TFIIH in plants, and on the basis of these interactions it has been possible to begin developing a model of the plant TFIIH core assembly expected to function in NER. So far the model strongly resembles that for the human TFIIH core complex suggesting that this subassembly will also be found to take on a ring-shaped structure in plants.

REFERENCES

Altschul, S. F., Madden, T. L., Schaffer, A. A., Zhang, J., Zhang, Z., Miller, W. & Lipman, D. J. (1997) Gapped BLAST and PSI-BLAST: a new generation of protein database search programs. *Nucleic Acids Res., 25*, 3389-3402.

Bardwell, L., Bardwell, A. J., Feaver, W. J., Svejstrup, J. Q., Kornberg, R. D. & Friedberg, E. C. (1994) Yeast RAD3 protein binds directly to both SSL2 and SSL1 proteins: Implications for the structure and function of transcription/repair factor b. *Proc. Natl. Acad. Sci. USA, 91*, 3926-3930.

Blanc, G., Hokamp, K. & Wolfe, K. H. (2003) A recent polyploidy superimposed on older large-scale duplications in the *Arabidopsis* genome. *Genome Res., 13*, 137-144.

Blanc, G. & Wolfe, K. H. (2004) Functional divergence of duplicated genes formed by polyploidy during *Arabidopsis* evolution. *Plant Cell, 16*, 1679-1691.

Bowers, J. E., Chapman, B. A., Rong, J. & Paterson, A. H. (2003) Unravelling angiosperm genome evolution by phylogenetic analysis of chromosomal duplication events. *Nature, 422*, 433-438.

Bradsher, J., Auriol, J., Proietti de Santis, L., Iben, S., Vonesch, J. L., Grummt, I. & Egly, J. M. (2002) CSB is a component of RNA Pol I transcription. *Mol. Cell, 10*, 819-829.

Britt, A. B. (1996) DNA damage and repair in plants. *Annu. Rev. Plant Physiol. Mol. Biol., 47*, 75-100.

Britt, A. B. (1999) Molecular genetics of DNA repair in higher plants. *Trends Plant Sci.,* 4, 20-25.

Britt, A. B., Chen, J. J., Wykoff, D. & Mitchell, D. (1993) A UV-sensitive mutant of *Arabidopsis* defective in the repair of pyrimidine-pyrimidinone (6-4) dimers. *Science, 261*, 1571-1574.

Brosché, M., Schuler, M. A., Kalbina, I., Connor, L. & Strid, Å. (2002) Gene regulation by low level UV-B radiation: identification by DNA array analysis. *Photochem. Photobiol. Sci., 1*, 656-664.

Cannon, G. A., Hedrick, L. A. & Heinhorst, S. (1995). Repair mechanisms of UV-induced DNA damage in soybean chloroplasts. *Plant Mol. Biol. 29*, 1267-1277.

Castro, J., Merino, C. & Zurita, M. (2002) Molecular characterization and developmental expression of the TFIIH factor p62 gene from *Drosophila melanogaster*: effects of the UV light sensitivity of a p62 mutant fly. *DNA Repair, 5*, 359-368.

Chang, W.-H. & Kornberg, R. D. (2000) Electron crystal structure of the transcription factor and DNA repair complex, core TFIIH. *Cell, 102*, 609-613.

Chen, J. J., Jiang, C. Z.., & Britt, A. B. (1996) Little or no repair of cyclobutyl pyrimidine dimers is observed in the organelle genomes of the young Arabidopsis seedling. *Plant Physiol., 111*, 19-25.

Chen, J., Larochelle, S., Li, X. & Suter, B. (2003) Xpd/Ercc2 regulates CAK activity and mitotic progression. *Nature, 424*, 228-232.

Coin, F., Auriol, J., Tapias, A., Clivio, P., Vermeulen, W. & Egly, J.-M. (2004) Phosphorylation of XPB helicase regulates TFIIH nucleotide excision repair activity. *EMBO J., 23*, 4835-4846.

Coin, F., De Santis, L. P., Nardo, T., Zlobinskaya, O., Stefanini, M. & Egly, J. M. (2006) P8/TTD-A as a repair-specific TFIIH subunit. *Mol. Cell, 21*, 215-226.

Coin, F., Marinoni, J.-C., Rodolfo, C., Fribourg, S., Pedrini, A. M. & Egly, J.-M. (1998) Mutations in the XPD helicase gene result in XP and TTD phenotypes, preventing interaction between XPD and the p44 subunit of TFIIH. *Nature Genet, 20*, 184-188.

Compe, E., Drané, P., Laurent, C., Diderich, K., Braun, C., Hoeijmakers, J. H. J. & Egly, J.-M. (2005) Dysregulation of the peroxisome proliferators-activated receptor target genes by XPD mutations. *Mol. Cell. Biol, 25*, 6065-6076.

Costa, R. M. A., Lima, W. C., Vogel, C. I. G., Berra, C. M., Luche, D. D., Medina-Silva, R., Galhardo, R. S., Menck, C. F. M. & Olivera, V. R. (2001a) DNA-repair-related genes in sugarcane expressed sequence tags (ESTS). *Genet. Mol. Biol, 24*, 131-140.

Costa, R. M. A., Morgante, P. G., Berra, C. M., Nakabashi, M., Brumeau, D., Bouchez, D., Sweder, K. S., Van Sluys, M.-A. & Menck, C. F. M. (2001b) The participation of *AtXPB1*, the *XPB/RAD25* homologue gene from *Arabidopsis thaliana*, in DNA repair and plant development. *Plant J., 28*, 385-395.

Feaver, W. J., Huang, W., Gileadi, O., Myers, L., Gustafsson, C. M., Kornberg, R. D. & Friedberg, E. C. (2000) Subunit interactions in yeast transcription/repair factors TFIIH, requirement for Tfb3 subunit in nucleotide excision repair. *J. Biol. Chem., 275*, 5941-5946.

Feaver, W. J., Svejstrup, J. Q., Bardwell, L., Bardwell, A. J., Buratowski, S., Gulyas, K. D., Donahue, T. F., Friedberg, E. C. & Kornberg, R. D. (1993) Dual roles of a multiprotein complex from *S. cerevisiae* in transcription and DNA repair. *Cell, 75*, 1379-1387.

Fidantsef, A. L., Mitchell, D. L., & Britt, A. B. (2000). The *Arabidopsis UVH1* gene is a homolog of the yeast repair endonuclease *RAD1*. *Plant Physiol., 124*, 579-586.

Friedberg, E. C., Walker, G. C., Siede, W., Wood, R. D., Schultz, R. A. & Ellenberger, T. DNA Repair and Mutagenesis (2nd edition). Washington, DC: *American Society for Microbiology*; 2006.

Frohnmeyer, H. & Staiger, D. (2003) Ultraviolet-B radiation-mediated responses in plants. Balancing damage and protection. *Plant Physiol., 133*, 1420-1428.

Fuss, J. O. & Cooper, P. K. (2006) *DNA repair: dynamic defenders against cancer and aging.* PLoS Biol, 4(6): e203. DOI:10.1371/journal.pbio.0040203.

Gervais, V., Lamour, V., Gaudin, F., Thierry, J. C. & Kieffer, B. (2001) Assignment of the 1H, 15N, 13C resonances of the N-terminal domain of the human TFIIH p62 subunit. *J. Biomol. NMR, 19*, 281-282.

Giglia-Mari, G., Coin, F., Ranish, J. A., Hoogstraten, D., Theil, A., Wijgers, N., Jaspers, N. G. J., Raams, A., Argentini, M., van der Spek, P. J., Botta, E., Stefanini, M., Egly, J.-M., Aebersold, R., Hoeijmakers, J. H. J. & Vermeulen, W. (2004) A new, tenth subunit of TFIIH is responsible for the DNA repair syndrome trichthiodystrophy group A. *Nature Genet., 36*, 714-719.

Giglia-Mari, G., Miquel, C., Theil, A. F., Mari, P. O., Hoogstraten, D., Ng, J. M. Y., Dinant, C., Heijmakers, J. H. J. & Vermeulen, W. (2006) Dynamic interaction of TTDA with TFIIH is stabilized by nucleotide excision repair in living cells. *PloS Biol. 4(6)*: e156. DOI: 10.1371/journal.pbio.0040156.

Gulyas, K. D. & Donahue, T. F. (1992) *SSL2*, a suppressor of a stem-loop mutation in the *HIS4* leader encodes the yeast homolog of human *ERCC-3*. *Cell, 69*, 1031-1042.

Hidema, J., Kumagai, T., Sutherland, J. C. & Sutherland, B. M. (1997) Ultraviolet B-sensitive rice cultivar deficient in cyclobutyl pyrimidine dimer repair. *Plant. Physiol., 113*, 39-44.

Hoogstraten, D., Nigg, A. L., Heath, H., Mullenders, L. H. F., van Driel, R., Hoeijmakers, J. H. J., Vermeulen, W. & Houtsmuller, A. B. (2002) Rapid switching of TFIIH between RNA polymerase I and II transcription and DNA repair in vivo. *Mol. Cell, 10*, 1163-1174.

Iben, S., Tschochner, H., Bier, M., Hoogstraten, D., Hozak, P., Egly, J.-M. & Grummt, I. (2002) TFIIH plays an essential role in RNA polymerase I transcription. *Cell, 109*, 297-306.

Iyer, N., Reagan, M. S., Wu, K.-J., Canagarajah, B. & Friedberg, E. C. (1996) Interactions involving the human RNA polymerase II transcription/nucleotide excision repair complex TFIIH, the nucleotide excision repair protein XPG, and Cockayne syndrome group B (CSB) protein. *Biochemistry, 35*, 2157-2167.

Jawhiri, A., Boussert, S., Lamour, V., Atkinson, R. A., Kieffer, B., Poch, O., Potier, N., Dorsselaer, A., Moras, D. & Poterszman A (2004) Domain architecture of the p62 subunit from the human transcription/repair factor TFIIH deduced by limited proteolysis and mass spectrometry analysis. *Biochemistry, 43*, 14420-14430.

Jawhiri, A., Lainé, J.-M., Dubaele, S., Lamour, V., Poterszman, A., Coin, F., Moras, D. & Egly, J.-M. (2002) p52 mediates XPB function within the transcription/repair factor TFIIH. *J. Biol. Chem., 277*, 31761-31767.

Joshi, C. P., Zhou, H., Huang, X. & Chiang, V. L. (1997) Context sequences of translation initiation codon in plants. *Plant Mol. Biol., 35*, 993-1001.

Kimura, S., Tahira, Y., Ishibashi, T., Mori, Y., Mori, T., Hashimoto, J. & Sakaguchi, K. (2004) DNA repair in higher plants: photoreactivation is the major DNA repair pathway in non-proliferating cells while excision repair (nucleotide excision repair and base excision repair) is active in proliferating cells. *Nucleic Acids Res., 32*, 2760-2767.

Kunz, B. A., Anderson, H. J., Creighton, J. C., McCarthy, L. J., Straffon, A. F. L., Tabone, E. K. & Vonarx E. J. Plant genes implicated in nucleotide excision repair or translesion synthesis. In: Ruiz-Rubio M, Alejandre-Duran E, Roldan-Arjona T editors. *Recent research developments in DNA repair and mutagenesis*, Kerala, India: Research Signpost; 2002, 111-139.

Kunz, B. A., Anderson, H. J., Osmond, M. J. & Vonarx E. J. (2005) Components of nucleotide excision repair and DNA damage tolerance in *Arabidopsis thaliana. Environ. Mol. Mutagenesis, 45*, 115-127.

Liu, Z., Hong, S. W., Escobar, M., Vierling, E., Mitchell, D. L., Mount, D. W. & Hall, J. D. (2003) Arabidopsis *UVH6*, a homolog of human *XPD* and yeast *RAD3* DNA repair genes, functions in DNA repair and is essential for plant growth. *Plant Physiol., 132*, 1405-1414.

Lockton, S. & Gaut, B. S. (2005) Plant conserved non-coding sequences and paralogue evolution. *Trends Genet., 21*, 60-65.

Lukaszewicz, M., Feuermann, M., Jerouville, B., Stas, A. & Boutry, M. (2000) In vivo evaluation of the context sequence of the translation initiation codon in plants. *Plant Sci., 154*, 89-98.

Matsui, P., Jacqueline, D. & Buratowski, S. (1995) An interaction between the TFB1 and SSL11 subunits of yeast TFIIH correlates with DNA repair activity. *Nucleic Acids Res., 23*, 767-772.

Molinier, J., Oakeley, E. J., Niederhauser, O., Kovalchuk, I. & Hohn, B. (2005) Dynamic response of plant genome to ultraviolet radiation and other genotoxic stresses. *Mutat. Res., 571*, 235-247.

Morgante, P. G., Berra, C. M., Nakabashi, M., Costa, R. M. A., Menck, C. F. M. & Van Sluys, M.-A. (2005) Functional XPB/RAD25 redundancy in Arabidopsis genome: characterization of AtXPB2 and expression analysis. *Gene, 344*, 93-103.

Politi, A., Moné, M. J., Houtsmuller, A. B., Heinrich, R. & van Driel, R. (2005) Mathematical modelling of nucleotide excision repair reveals efficiency of sequential assembly strategies. *Mol. Cell, 19*, 679-690.

Quaite, F. E., Takayanagi, S., Ruffini, J., Sutherland, J. C. & Sutherland, B. M. (1994) DNA damage levels determine cyclobutyl pyrimidine dimer repair mechanisms in alfalfa seedlings. *Plant Cell, 6*, 1635-1641.

Reidl, T., Hanaoka, F. & Egly, J.-M. (2003) The comings and goings of nucleotide excision repair factors on damaged DNA. *EMBO J., 22*, 5293-5303.

Ribiero, D. T., Machado, C. R., Costa, R. M. A., Praekelt, U. M., Van Sluys, M.-A. & Menck, C. F. M. (1998) Cloning of a cDNA from *Arabidopsis thaliana* homologous to the human *XPB* gene. *Gene, 208*, 207-213.

Ries, G., Heller, W., Puchta, H., Sandermann, H., Seidlitz, H. K. & Hohn, B. (2000) Elevated UV-B radiation reduces genome stability in plants. *Nature, 406*, 98-101.

Russell, J. & Zomerdijk, J. C. B. M. (2005) RNA-polymerase-I-directed rDNA transcription, life and works. *Trends Biochem. Sci., 30*, 87-96.

Sarkar, A. H., Tsutakawa, S. E., Kostek, S., Ng, C., Shin, D. S., Peries, M., Campeau, E., Tainer, J. A., Nogales, E. & Cooper, P. K. (2005) Recognition of RNA polymerase II and transcription bubbles by XPG, CSB, and TFIIH: insights for transcription-coupled repair and Cockayne syndrome. *Mol. Cell, 20*, 187-198.

Schultz, P., Fribourg, S., Poterszman, A., Mallouh, V., Moras, D. & Egly, J.-M. (2000) Molecular structure of human TFIIH. *Cell*, 102, 599-607.

Seroz, T., Perez, C., Bergmann, E., Bradsher, J. & Egly, J.-M. (2000) p44/SSl1, the regulatory subunit of the XPD/RAD3 helicase, plays a crucial role in the transcriptional activity of TFIIH. *J. Biol. Chem., 275*, 33260-33266.

Sikorski, R.S. & Hieter, P. (1989) A system of shuttle vectors and yeast host strains designed for efficient manipulation of DNA in *Saccharomyces cerevisiae. Genetics, 122*, 19-27.

Simillion, C., Vandepoele, K., Van Montagu, M. C., Zabeau, M. & Van de Peer, Y. (2002) The hidden duplication past of Arabidopsis thaliana. *Proc. Natl. Acad. Sci. U S A, 99*, 13627-13632.

Strid, Å., Chow, W. S. & Anderson, J. M. (1994) UV-B damage and protection at the molecular level in plants. *Photosyn. Res., 39*, 475-489.

Takagi, Y., Komori, H., Chang, W. H., Hudmon, A., Erdjument-Bromage, H., Tempst, P. & Kornberg, R.D. (2003) Revised subunit structure of yeast transcription factor IIH (TFIIH) and reconciliation with human TFIIH. *J. Biol. Chem., 278*, 43897-43900.

Takagi, Y., Masuda, C. A., Chang W.-H., Komori, H., Wang, D., Hunter, T., Joazeiro, C. A. P. & Kornberg, R. D. (2005) Ubiquitin ligase activity of TFIIH and the transcriptional response to DNA damage. *Mol. Cell., 18*, 237-243.

Tanaka, A., Sakamoto, A, Ishigaki, Y., Nikaido, O., Sun, G., Hase, Y., Shikazono, N., Tano, S. & Watanabe, H. (2002) An ultraviolet-B-resistant mutant with enhanced DNA repair in Arabidopsis. *Plant Physiol.*, 9, 64-71.

Tapias, A., Auriol, J., Forget, D., Enzlin, J. H., Schärer, O. D., Coin, F., Coulombe, B. & Egly, J.-M. (2004) Ordered conformational changes in damaged DNA induced by nucleotide excision repair factors. *J. Biol. Chem.*, 279, 19074-19083.

Teramura, A. H. & Sullivan, J. H. (1994) Effects of UV-B radiation on photosynthesis and growth of terrestrial plants. *Photosynth. Res., 39*, 463-473.

The Arabidopsis Genome Initiative (2000) Analysis of the genome sequence of the flowering plant *Arabidopsis thaliana. Nature, 408*, 796-815.

Thompson, J. D., Higgins, D. G. & Gibson, T. J. (1994) CLUSTAL W: improving the sensitivity of progressive multiple sequence alignment through sequence weighting, position specific gap penalties and weight matrix choice. *Nucleic Acids Res., 22*, 4673-4680.

Vision, T. J., Brown, D. G. & Tanksley, S.D. (2000) The origins of genomic duplications in *Arabidopsis*. *Science, 290*, 2114-2117.

Volker, M., Mone, M. J., Karmakar, P., van Hoffen, A., Schul, W., Vermeulen, W., Hoeijmakers, J. H., van Driel, R., van Zeeland, A. A. & Mullenders, L. H. (2001) Sequential assembly of the nucleotide excision repair factors in vivo. *Mol. Cell, 8*, 213-224.

Vonarx, E. J., Howlett, N. G, Schiestl, R. H. & Kunz, B. A. (2002) Detection of *Arabidopsis thaliana AtRAD1* cDNA variants and assessment of function by expression in a yeast *rad1* mutant. *Gene, 296*, 1-9.

Vonarx, E. J., Tabone, E. K., Osmond, M. J., Anderson, H. J. & Kunz, B. A. (2006) Arabidopsis homologue of human transcription factor IIH/nucleotide excision repair factor p44 can function in transcription and DNA repair and interacts with AtXPD. *Plant J., 46*, 512-521.

Wang, Z., Buratowski, S., Svejstrup, J. Q., Feaver, W. J., Wu, X., Kornberg, R. D., Donahue, T. F & Friedberg, E. C. (1995) The yeast *TFB1* and *SSL1* genes, which encode subunits of transcription factor IIH, are required for nucleotide excision repair and RNA polymerase II transcription. *Mol. Cell. Biol., 15*, 2288-2293.

Xiong, F. S. & Day, T.A. (2001) Effect of solar ultraviolet-B radiation during springtime ozone depletion on photosynthesis and biomass production of Antarctic vascular plants. *Plant Physiol., 125*, 738-751.

Yoon, H., Miller, S. P., Pabich, E. K. & Donahue, T. F. (1992) *SSL1*, a suppressor of a *HIS4* 5'-UTR stem-loop mutation, is essential for translation initiation and affects UV resistance in yeast. *Genes Dev., 6*, 2463-2477.

Zurita, M. & Merino, C. (2003) The transcriptional complexity of the TFIIH complex. *Trends Genet., 19*, 578-584.

In: New Research on DNA Repair
Editor: Breehn R. Landseer, pp. 273-294
ISBN 1-60021-385-5

Chapter X

Repair of Radiation-Induced DNA Double-Strand Breaks in Human Cells: History, Progress and Controversies

Aurélie Joubert[1,2] *and Nicolas Foray*[1,*]
[1]Inserm, U647, ID17, European Synchrotron Radiation Facility, Grenoble, France
[2]Institut de Radioprotection et de Sûreté Nucléaire, Fontenay-aux-Roses, France

Abstract

Among the large spectrum of DNA damage induced by radiation, DNA double-strand breaks (DSBs) are considered, to date, as the key-lesions responsible for the cell killing. However, all along these last 50 years, although it was always intuitive for radiobiologists, such conclusion has resulted from a long and difficult process of development of new techniques and novel concepts. In this review, we endeavored to translate these different steps of thinking, notably by strengthening on the number of technical artifacts, biases due to yeast and rodent data, the variety of phenotypes observed in human cells and the extra-complexity encountered in human tumors. Particularly, the most recent techniques based on immunofluorescence have revolutionized the picture of the molecular events occurring early after irradiation but has also raised the crucial question of the choice of the technique assessing DSB repair and its specificity for particular steps of the repair process. While the homologous recombination repair pathway is predominant in yeasts, its importance in normal and tumor human cells appears to be less and less obvious, raising the problem of the existence of randomized and/or uncontrolled events that may produce much more errors than one could have imagined from the yeasts. DSB repair impairments are now observed in a plethora of various genetic diseases, leading to radiosensitivity, immunodeficiency and sometimes cancer-proneness, but the low frequency and the pleiotropism of such diseases make difficult the development of a *single* predictive assay.

* Correspondence concerning this article should be addressed to Dr. Nicolas Foray, INSERM, U647, ID17, European Synchrotron Radiaiton Facility, Rue Horowitz, 38043, Grenoble, France. Email: foray@esrf.fr

Lastly, the balance, or rather the interplay, between recombination and DNA end-joining repair pathways appears to be strongly dependent upon cell cycle, genotype and oxidative stress that can accompany the DSBs formation. Hence, although DSB repair appears, to date, undoubtedly crucial for cell survival, further investigations are still needed to offer to physicians some tools to secure clinical transfer of anti-cancer approaches by avoiding the risk of deleterious adverse effects of radiation.

Keywords: DNA double-strand breaks, DNA repair, DNA damage, Radiosensitivity

1. Introduction

Among the large spectrum of DNA damage induced by radiation, DNA double-strand breaks (DSBs) are considered, to date, as the key-lesions responsible for lethality. However, for the particular case of human cells, such statement still reminds to be demonstrated more rigorously. In fact, although there is an intuitive consensus among radiobiologists for considering unrepaired and/or misrepaired DSBs as responsible for human radiosensitivity and cancer proneness, nearly all the proposed arguments are indirect and mix analogies and syllogisms. Here are the most representative ones: 1) Radiosensitive yeast and rodent mutants are systematically deficient in DSB repair. Human cells *are believed* to obey the same law; 2) A number of physical and chemical agents like hypoxia, hyperthermia, linear energy transfer influence cell survival. They impact also on formation and repair of DSBs but not significantly on the other radiation-induced DNA damage. Therefore, more than the other DNA damage, DSBs *are likely to be* the molecular bases of cell killing effect induced by radiation; 3) Micronuclei and chromosomal breaks have long been shown to be correlated to the loss of clonogenic capacity. Since micronuclei and chromosomal breaks are the result of unrepaired DSBs, unrepaired DSBs *are supposed* to be correlated to cell survival as well; 4) A number of *in vitro* cell free assays performed with protein extracts provided from radiosensitive mutants of yeast and rodents or from human diseases associated with radiosensitivity show systematically impaired DSBs repair process. This argument can be *reasonably extrapolated* to *in vivo* conditions [1-3]. Hence, although the conclusions of these four points are fully documented, they are not sufficient to definitively establish DSBs as the key-lesions responsible for radiation-induced lethality in human cells. In addition to this *dialectic* insufficiency, the influence of yeast and rodent models, the choice of the numerous techniques assessing DSBs and the specificities of human tumor cells have contributed to make more complex the quest of the molecular bases of human cellular radiosensitivity. All along this review, we endeavored to reflect the most representative episodes of a fascinating history that is still in progress.

2. The Repair Rate of DSBs as the Best Endpoint?

2.1. Rare But Dangerous

In front of the relative fragility of the above arguments, one could logically argue that lethal effect of ionizing radiation is explained by the most frequent radiation-induced DNA damage (*i.e.* base damage) rather than DSBs. The historical counter-example against this hypothesis is the case of cells from ataxia telangiectasia (caused by *ATM* mutations), the human genetic disease associated with the most severe sensitivity to ionizing radiation : *ATM*-mutated cells are DSB repair deficient but do not show any significant photosensitivity like it would be the case if they were base damage repair deficient [4,5].

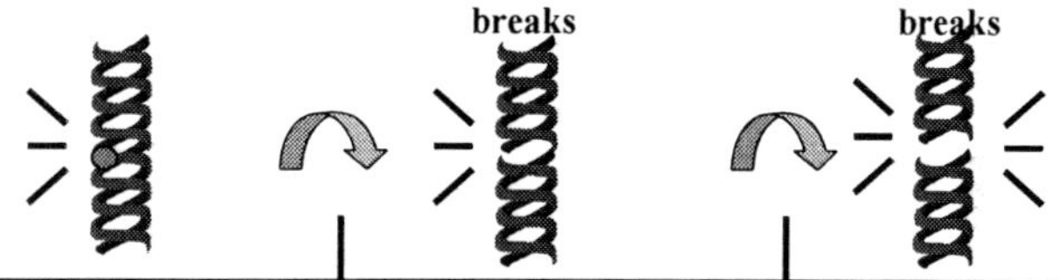

	Base damage	DNA single-strand breaks	DNA double-strand breaks
Energy required	**>1 eV/nm^3**	**>10 eV/nm^3**	**>100 eV/nm^3**
Incidence	**~ 10000 / Gy**	**~ 1000 / Gy**	**~ 40 / Gy**
50% repaired in:	**5-10 min**	**10-20 min**	**> 50 min**
Repaired by:	**Excision-Resynthesis**	**Recombination**	**End-Joining**

Figure 1: The major radiation-induced DNA damage and their physical and biological specificities: the minimal energy deposed as physical cluster required for their formation, their occurrence per X-rays Gy, their average repair half-time and the DNA repair pathways they can trigger.

Ionizing radiation produce a large spectrum of energy microdepositions, that, through excitations and ionisations, lead to the formation of free radicals species that create DNA damage [6]. The formation of DSBs requires more than about 100 eV deposed per nm^3 whereas base damage require about 100 times less energy density (Figure 1). Ionizing radiation produce much more small energy microdepositions than dense ones. DSBs are therefore logically less frequent than base damage [7]. As if the famous principle of conservation of energy was applicable to radiobiology, the more energy required for the DNA damage formation, the longer time required for its repair. In human diploid cells, 50% of DSBs are repaired in about 1 h whereas the great majority of the base damage possess a repair half-time of less than ten minutes [8]. Hence, the less frequent the radiation-induced DNA lesions, the more slowly repairable (Figure 1). Figure 1 summarizes the frequency of each type of radiation-induced DNA damage and their correspondence with energy microdepositions. However, this table would not be complete if the conversion between DNA base damage and DNA strand breaks was not mentioned. In fact, base damage are generally repaired by excision-resynthesis pathways that consist in converting base damage into DNA single-strand breaks (Figure 1). In particular, delayed DSBs occur after exposure to UV (generally some hours post-irradiation) and may also (but indirectly) contribute to the cellular response to UV [9]. Altogether, these results show that immediate or delayed DSBs are important molecular events for responding to radiation. Through all these examples, it is also suggested that the *fate (i.e. the result of repair)*, but not necessarily the *yield* of DNA damage, may condition cell survival after irradiation.

2.2. A More Direct Argument?

Recent literature fortunately provides more direct proofs supporting that unrepaired DSBs are responsible for the lethal effects of ionizing radiation in human cells. By using a microinjection technology, Huang et al. inserted a yeast chromosome fragment into a non-transformed human fibroblast. When injected into cytoplasm, the fragment did not trigger any significant biological effect. Conversely, the injection of the fragment into cell nucleus resulted in cell cycle arrest and clonogenic death if it was not integrated into the genome (*i.e.* repaired) by the fibroblast. These data suggest that unrepaired DNA fragments, even from a foreign DNA, do impact upon cell viability [10]. Such conclusions have been corroborated by data from transfection of linearized plasmids into human cells and strengthen the importance of cell nucleus as a vital target after irradiation [10].

2.3. The Fate or the Yield?

In 1985, by using the neutral elution technique, Ian Radford demonstrated that more DSBs were induced by radiation in radiosensitive and DSB repair deficient rodent mutants than their in radioresistant wild-type counterparts [11]. Extensively used in the 1980's, the neutral elution technique consists in the discrimination of DNA fragments size after eluting them through polycarbonate filters. Radford noticed an intriguing symmetry between DSB induction curves and survival curves for all the rodent cell lines tested. Radford hypothesized that these so-called *mirror curves* reflected a parallel between the yield of radiation-induced DSBs and the clonogenic response to radiation [11]. This hypothesis was inasmuch well accepted in the community that some authors reported also higher yields of induced DSBs in some human radiosensitive tumors [12]. Few months after, the elution technique was demonstrated to be particularly sensitive to the composition of the lysis solution. Notably, the shape of DSB induction curves was strongly influenced by pH and the nature of detergent [1,13]. With neutral elution technique, the radiosensitive *ATM*-mutated cells and radioresistant controls showed the same yield of radiation-induced DSBs [12].

To date, it is quite well established that there is no correlation between the yield of DSBs *induced* and the surviving fraction of irradiated human cells [12]. However, this episode raised two conclusions:

- Techniques assessing DSBs may highlight a particular biochemical step of the complex DNA repair process and their choice is therefore far from trivial.
- The observations that some human tumors show higher yields of induced DNA damage may reflect the impact of cell cycle, chromatin condensation, DNA conformation upon DSB repair process [14].

3. Homologous or Non-Homologous?

3.1. The Relative Importance of Homologous Recombination

In the 1980's, the essential of our knowledge about DSB repair was essentially deduced from yeasts and bacteria. Particularly, the role of the recA bacteria protein appeared crucial in a process called *homologous* recombination (HR), because of the vicinity of homologous strands during mitosis and meiosis [15]. The relative importance of homologous recombination was considerably amplified since such repair pathway was also observed in proliferating mammalian cells (namely in S-G2/M), and especially in human tumors [16]. HR is the predominant DSB repair pathway in prokaryotes and lower eukaryotes. It appears therefore logical that yeasts mutants of Rad51 and Rad52 (both repair proteins essential for HR) elicit the most severe radiosensitivity encountered in yeasts [16]. It was then tempting to extrapolate these conclusions to mammalians by considering HR as a crucial DSB repair pathway as well. However, the great majority of mammalian cells are quiescent (HR was not found to be active in G0/G1) and the most radiosensitive mammalian cells do not hold mutations in RAD51 nor in RAD52, the human *Rad51* nor *Rad52* homologs, respectively. Besides, no genetic disease associated with *RAD51* and *RAD52* mutations has been identified yet). Unlike bacteria and yeasts, another DSB repair pathway, different from HR, was therefore to be considered to explain cellular radiosensitivity of mammalian cells.

3.2. Homologous and Non-Homologous Recombination: an Historical Semantic Error?

The findings with HR were so impressive that any other DSB repair pathway discovered after HR was necessarily expected to be *non-homologous*. Hence, after the discovery of the DNA end-joining process, it appeared logical to add "*non-homologous*" to "*DNA end-joining*". We think personally that this nomenclature is a semantic disaster, not only for teaching radiobiology but also for the common sense: what would mean an *homologous* DNA end-joining pathway? The NHEJ pathway consists in joining the *opposite DNA ends*. Conversely, the terms of homologous and non-homologous refer to *opposite DNA strands*. Hence, still to date, the majority of reports evoke the duality between DSB repair pathways by opposing HR and NHEJ and, overall, by omitting that recombination can be non-homologous, *without being NHEJ* [17-19]. Like in a number of cases in science, nomenclature and use of terms outside from their historical context are sources of misinterpretations...

3.3. And if Recombination was a Random Repair Process?

Recombination is a complex process with successive molecular steps such as DNA damage recognition, nuclease, strand-exchange, re-polymerization activities, etc... and whose hierarchy reminds to be elucidated [16]. To date, except the fact that homologous strands are close in mitosis and meiosis, there is no direct proof for the existence of an *intelligent* repair process that would consist in the *exact* replacement of the lacking sequences. Conversely, can *recombination* (term used in general) be the result of a *random* process based on the insertion of DNA sequences *chosen randomly*? In another words, is recombination a *random* repair process that would become *homologous* in some particular situations like in yeast, bacteria, mitosis and meiosis? Since its discovery, HR has been logically considered as a high-fidelity process. By definition, the absolute fidelity of repair requires: 1) the same amino acid sequence; 2) the same DNA fragment size to be inserted in the break site. This double requirement is inasmuch easily reached as homologous strands are close (*i.e.* in mitosis, meiosis and in small genomes like in bacteria and yeasts). This is also the case in *in vitro* cell-free systems that are based on 100-10000 kb DNA vectors [20]. In mammalians, genome is 100 to 1000 times longer and non-coding sequences are much more frequent than coding ones. Since physical impact of radiation is a stochastic process, DSBs are induced more frequently in non-coding than in coding sequences. A random recombination process would have therefore more chance to manage a break into non-coding sequence and more chance to insert another non-coding sequence into the break site. The practical consequence of such phenomenon would be an increasing amount of the most frequent non-coding sequences in the genome but a weak impact upon viability. The existence of such phenomenon has been fully demonstrated in certain thalassemia in which the frequency of *Alu* sequences, the most frequent ones in human genome, is abnormally high [21]. The relative frequency of Alu sequences is about 10^{-4} in humans, it is 100 to 1000 times higher in small genomes organisms and in kilobases plasmid vectors from *E. Coli*. The fidelity of a random (*non-homologous*) repair process is inasmuch high as genome length is low [22]. Hence, it is reasonable to hypothesize that recombination repair process is a non-random process *per se*. In some limited cases, either when strands are particularly close (like in mitosis and meiosis) or when organisms or assays involve a reduced genome length, a random (*non-homologous*) process may become *homologous artificially*. Lastly, outside from these numerical considerations, a random repair process would naturally lead to accumulation of errors, transformation and cancer: this is exactly what is observed in mammalians whereas there is still no evidence of any cancer in bacteria or yeasts... Hence, all along the next paragraphs, we will deliberately use the term *recombination* as a general term describing a random DNA repair process based on DNA damage recognition and strand exchange, without any assumptions about its homologous or non-homologous properties.

4. A Careful Analysis of DSB Repair Curves

Since the development of the first techniques assessing DSBs, the great majority of authors endeavored to establish a correlation between DSB repair rate and cell survival [2,23]. However, four major problems have limited the impact of such approach: 1) the difficulty in collecting a large number of cell lines in order to thoroughly investigate a wide spectrum of cellular radiosensitivity; 2) the interpretation of the shape of DSB repair curves and how the different DSB repair pathways may modify it; 3) the plethora of techniques assessing DSB repair defects; 4) the weakness of the evidence for a *quantitative* correlation between cellular radiosensitivity and DSB repair. For example, despite similar cellular radiosensitivity, the amount of unrepaired DSB is 2-4 times higher in *LIG4*-mutated than in *ATM*-mutated cells [4,24]. While the first point still reminds to be solved for practical and ethical reasons, let focus on the last three ones.

4.1. First Difficulty: the Extra-Complexity of DSBs

According their localization in chromatin, the size of the breaks and/or the degree of the chromatin compaction, each DSB does not raise the same *steric* problem on DNA. To this *biological* complexity, one must add the diversity provided by physics. Radiation produce a wide spectrum of energy microdepositions differing by their energy density and their frequency. It was shown that dense and complex microdepositions lead to DNA damage that are more slowly repairable than those produced by small energy clusters [7,25]. Hence, like any other radiation-induced DNA damage, DSBs describe a large and continuous spectrum of *reparabilities* [23,25]. In parallel to the concept of *intrinsic radiosensitivity*, one might evoke the notion of *intrinsic reparability*: some DSBs may be repaired in some minutes, some other in some hours. Surprisingly, whereas such extra-complexity of DSBs suggest a large spectrum of molecular and cellular responses, one of the first models explaining the shape of DSB repair curves was based on *two types of DSBs* only, each characterized by a *single and constant* repair half-time [1,26]. This *dual* approach was well accepted inasmuch as eye-fittings to biexponential models were satisfying. Consequently, a number of models were proposed with *fastly/slowly repairable* [1,27], *repairable/non-repairable* [27,28], *repaired/misrepaired* [29,30], *simple/complex* [31] or *cooperative/non-cooperative* damage [32]. It is noteworthy that this concept of duality fitted well with the blunt and cohesive endonuclease restriction sites. Substantial efforts were provided to link these models to formulas describing the occurrence of radiation-induced chromosomal aberrations and cell killing [33-35]. However, no real consensus about the nature of *the two types* of DSBs was reached. Conversely, experimental evidence supported that *fast* and *slow* repair half-times strongly depend upon repair times chosen by authors [25,36,37]. Further biostatistical investigations demonstrated that the shape of DSB repair curves is more consistent with the diversity of energy microdepositions and that *multi-* rather than *bi*-exponential formulas are more relevant to describe it [23].

4.2. Second Difficulty: Gaps or DNA Fragments?

The term *DSBs* grossly means gaps (holes) in DNA. However, it is conceptually and technically difficult to quantify gaps (i.e. *absence* of DNA). By contrast, it may be easier to measure the corresponding DNA fragments. Hence, the first techniques developed to assess DSBs were based on the discrimination of the radiation-induced DNA fragments by their size. This was the case of sucrose gradient sedimentation, neutral elution and pulsed field gel electrophoresis (Figure 2) [1,12]. With such techniques, *DSB induction* consists in an increase of the yield of DNA fragments and the decrease of their average size. Conversely, *DSB repair* results in the decrease of the number of DNA fragments and the increase of their average size. Such principle was thought to provide a great advantage: to be independent of the nature of the DSB repair pathways, since only the *repaired* or *unrepaired* substates are considered [23,38]. By contrast, as far as the quality of DSB repair influences cell survival, such techniques do not permit the evaluation of DSB repair *fidelity*.

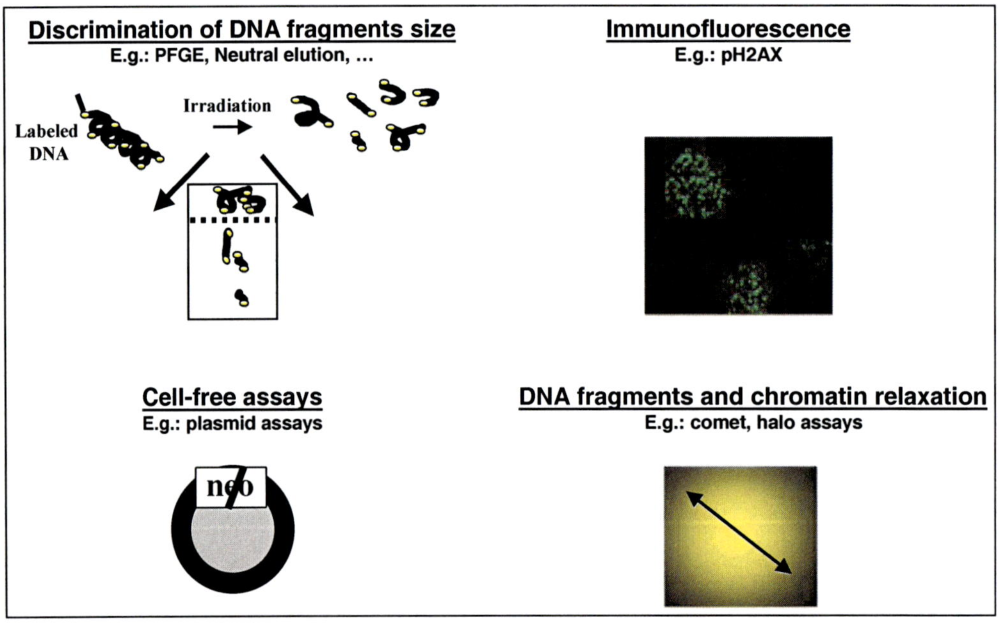

Figure 2: The major techniques assessing DSBs. Techniques based on the discrimination of DNA fragments size by electrophoresis, elution or sedimentation require high radiation doses. Immunofluorescence allows the detection of DSBs inside individual cell nucleus but the chosen fluorescence marker is specific of particular step of the DSB repair process. Cell-free assays allow a feedback analysis of the quality of repair (notably by sequencing) but strongly depend upon the protocol for preparing cell extracts. Comet and halo assays use the principles of pulsed field gel electrophoresis but chromatin relaxation may bias data interpretation.

The migration of long DNA fragments is limited. For example, if one considers that 40 DSBs are produced per Gy per human diploid cell, the average DNA fragment size is about 150 Mb per Gy for a genome of 6000 Mb (the length of human chromosome 1 is estimated to be 122 Mb). DNA fragment of such size cannot migrate in a pulsed field electrophoresis gel (with this technique, the *cut-off* size is about 10-15 Mb) [4,25]. Consequently, radiobiologists used high initial radiation doses to increase the DNA fragments population sufficiently: the initial radiation dose used for the first DSB repair curves were much larger (30-100 Gy) than those used from survival curves (1-5 Gy) [8,12]. Such protocols raised therefore another conceptual problem: can one establish a correlation between DSB repair and survival curves with so large differences between radiation doses range? Further, is it relevant to assess the DSB repair rate of dead cells? We endeavored to reply to such questions in the next paragraph.

4.3. Third Difficulty: to Repair During Death...

Since the yield of unrepaired DSBs decreased with repair time by obeying an exponential-like law, a number of authors hypothesized a progressive *saturation* of the repair protein pools after a too large amount of DNA damage [23,39]. Interestingly, with 40 DSBs produced per Gy per cell, there are always more DNA repair proteins in cell nucleus than radiation-induced DSBs. However, this rapid explanation reminds insufficient since one can always argue about the activity, the availability and the accessibility of these repair proteins to respond to DSBs. However, the concept of *saturation* suggested clearly that the DSB repair rate decreases while initial dose of radiation increases. In fact, in mammalians at least, the DSB repair rate dose not change significantly up to doses larger than 500 Gy [25,40]. Besides, when DSB repair rate is expressed as percentage of unrepaired DSBs, DSB repair rate does not depend upon the technique chosen neither: consequently, DSB repair rate obtained after 30 Gy can be extrapolated to lower doses without gross mistakes [12,23]. This apparent paradox is notably well illustrated by the fact that DSB repair data obtained from pulsed field gel electrophoresis (that requires doses higher than 10 Gy) and from anti-pH2AX immunofluorescence (that requires doses lower than 10 Gy) are similar. [2,40]. As a result, a given cell line can be characterized by its own DSB repair rate expressed as percentage of unrepaired DSB, *independently of the initial dose and the technique chosen.* These findings raise two conclusions:

- DSB repair rate is independent of the initial radiation dose and the theory of *saturation* is not relevant.
- Cells do repair their DNA even for high radiation doses while they are dead clonogenically.

This last point reflects the actual difference existing between a *biochemical process* like DNA repair, whose effects are generally proportional to radiation dose, and the *conventional* definition of the radiobiological *cellular* death as the loss of clonogenic capacity. To measure DSB repair of dead cells is therefore relevant and, anyway, how can we measure DSB repair rate without killing cells?

4.4. Fourth Difficulty: the Diversity of the DSB Repair Rates in Human Cells

Let us focus now on the shape of DSB repair curves of radiosensitive cells. Without denying the pioneers' efforts, very few DSB repair data in mammalians was available before the eighties [12]. In 1984, Jeggo and Kemp established cell lines from Chinese hamster ovary (CHO) treated to various protracted genotoxic treatments (UV, alkylating drugs, etc). Screened for their radiosensitivity, 8 cell lines were called x-rays sensitive (*xrs*) cell lines and were numbered *xrs1* to *xrs8* [41]. The most radiosensitive cell lines of this collection (*xrs5* and *xrs6*), showed about 1% cell survival after 2 Gy (SF2). These two cell lines were studied so intensively that radiobiologists forgot the existence of the 6 others (that were *less radiosensitive)*. While the wild-type controls (CHO-K1) completed their DSB repair in the first 12 h post-irradiation, the DSB repair rate of *xrs5* and *xrs6* was found significantly slower, with an arrest of repair by 4 h post-irradiation corresponding to about 40 % of still unrepaired DSBs [41]. Hence, together with yeasts data, rodent data suggested that cellular radiosensitivity corresponds to an *extreme deficiency* of DSB repair: the relationship between radiosensitivity and DSB repair was logically considered as an *all-or-none* phenomenon [1].

Again, the picture offered by human cells was clearly different. The human model equivalent to radiosensitive *xrs* rodent mutants is ataxia telangiectasia caused by *ATM* mutations. The *ATM*-mutated and *xrs5* cells elicit similar SF2 (about 1 %) [12,41,42]. However, the first assessements of the DSB repair in *ATM*-mutated cells showed *normal* rates [4]. This discrepancy with rodent data (and this failure of the *all-or-none* theory) was so impressive that some authors did not hesitate, either to reconsider DSBs as the key-lesions for lethal effect of ionizing radiation or to evoke species-specific mechanisms [43]. However, cells from ataxia telangiectasia patients do elicit a number of chromosomal aberrations and are actually hyper-radiosensitive: where was the bias?

Rodent cells repair their radiation-induced DSBs about two times faster than human cells. Hence, for some obvious practical reasons, the first assessments of DSB repair rate in human cells did not exceed 12 h post-irradiation [12]. The extent of the repair time range tested in human cells up to 24 h post-irradiation and the application of a new generation of pulsed field gel electrophoresis technique led to another surprise: the *ATM*-mutated cells repaired *faster* than controls in the first 4 h *and slower* in the 4-24 h repair range post-irradiation (Figure 3). At 24 h, only 10-15% of DSBs were still unrepaired (40% for *xrs5* and *xrs6* cell lines) [4]. Hence, how to explain a simple link between radiosensitivity and DSB repair rate if, for a similar radiosensitivity (SF2 of about 1%), DSB repair rates are not similar? In addition, how to explain the accelerated DSB repair in the first 4 h post-irradiation?

The picture became more and more complex when a new radiosensitive human fibroblast cell line, 180BR, became available. The 180BR cell line derived from a non-ataxic patient who succumbed to his anti-leukemia treatment. While the 180BR cells showed a similar radiosensitivity as the *ATM*-mutated and a DSB repair rate that was similar to that of *xrs5* and *xrs6* (Figure 3) [44]. It was therefore hypothesized that the radiosensitivity of *ATM*-mutated cells is explained by other impairments than DSB repair rate [43].

To clarify a very complex situation, the only solution was to gather all the radiosensitive human cell lines available, independently of any genetic disease, and to measure their DSB repair rate. By gathering the largest collection of human non-transformed fibroblast cell lines, representing 8 human syndromes and covering the largest spectrum of human radiosensitivity investigated, four major families of DSB repair curves were identified (Figure 3):

- A DSB repair rate that decreases continuously with repair time. DSB repair is nearly completed after 24 h post-irradiation with 50% repaired DSBs after about 1 h. This is generally the case of cells from apparently healthy patients showing normal radioresistance.
- A very slow DSB repair rate already observed from 20 min post-irradiation, that remains constant in the 4-24 h post-irradiation range and leading to 30-40% of unrepaired DSB at 24 h post-irradiation. This is notably the case of *LIG4*- and *DNA-PK*-mutated cells. The shape of the resulting repair curves is very similar to that of rodent *xrs* mutants.
- A DSB repair rate that is abnormally faster than controls in the 2-8 h post-irradiation range, that decreases thereafter by reaching 0-5% of unrepaired DSBs at 24 h post-irradiation. This is notably the case of cells showing high cancer proneness, weak but significant sensitivity to ionizing radiation and high sensitivity to alkylating agents like cells from Fanconi anemia, Bloom's syndrome and some cases of Xeroderma Pigmentosum.
- A DSB repair rate that is abnormally faster than controls in the 2-8 h post-irradiation range, like the previous case, but that decreases thereafter up to 10-20% of unrepaired DSB at 24 h post-irradiation, as if it describes the combination of the last two features. This is notably the case of cells from ataxia telangiectasia.

5. SF2 IS CORRELATED TO UNREPAIRED DSBS

By plotting the survival at 2 Gy (SF2) against the corresponding percentage of unrepaired DSBs at 24 h, a significant correlation appeared: SF2 was found to be inversely proportional to unrepaired DSBs, suggesting a large continuum in the molecular and cellular responses to radiation but also the existence of some practical limits (reflected by the asymptotes) for radiosensitivity, radioresistance, DSB repair proficiency and DSB repair deficiency (Figure 4). Such biostatistical law obviously explains better why cells with similar radiosensitivity may elicit strongly different DSB repair rate like *ATM*- and *LIG4*-mutated cells. A number of interesting conclusions can be deduced from such findings.

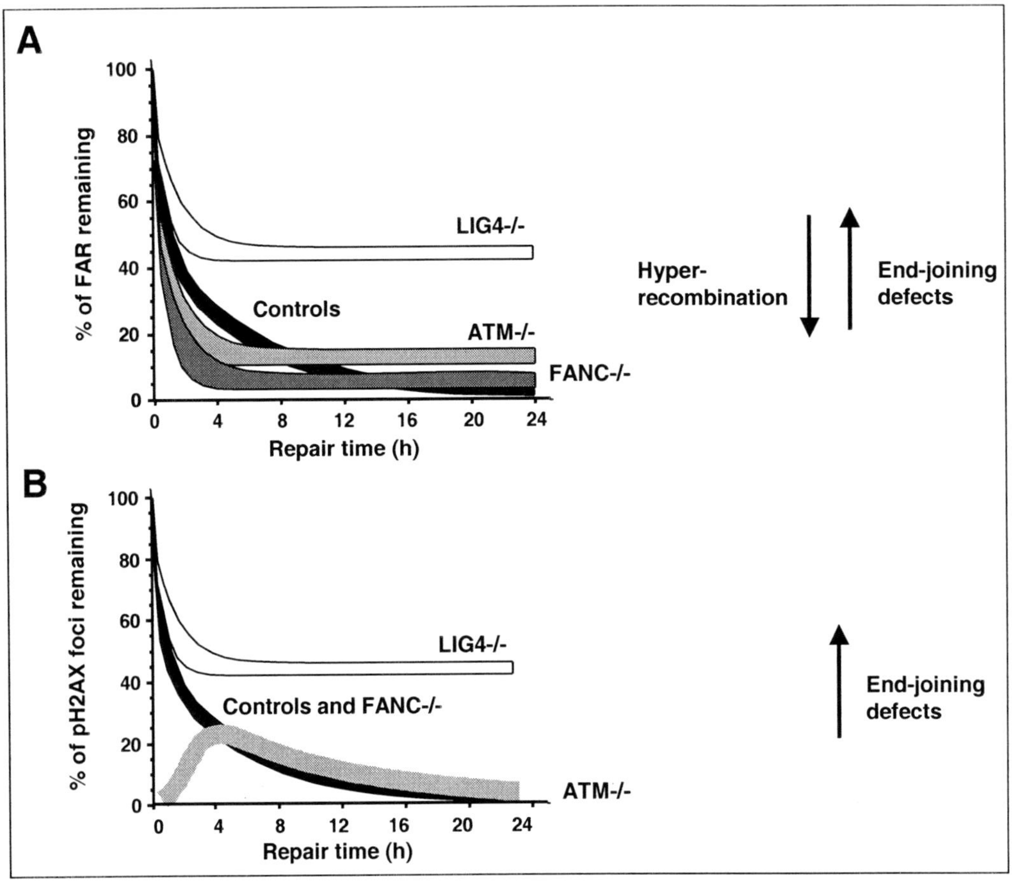

Figure 3: Comparisons between DSB repair data obtained with pulsed field gel electrophoresis (expressed as fraction of activity released (FAR)) (panel A) and with anti-pH2AX immunofluorescence (expressed as nuclear foci scored inside cell nuclei) (panel B). Four cases were tested: radioresistant control, *LIG4-*, *ATM-*, *FANC*-mutated cells representing the radiosensitivity groups I, III and II, respectively. Arrows indicate the impact of hyperrecombination and end-joining defects on the shape of repair curves.

5.1. Two Distinct Groups of Radiosensitivity

By analyzing the clonogenic response to radiation of the cells belonging to the collection described above, three distinct groups of clonogenic response were observed :

- The radioresistant controls formed one group (group I) of SF2dp (dp is for *delayed plating*) ranging from 45.2 to 65.6%.
- The *ATM*-mutated and the *LIG4*-mutated cell lines formed a group (group III) of extremely radiosensitive cells with SF2dp ranging from 3.7 to 7.4%.
- All the remaining cell lines showed intermediate radiosensitivity and formed another group (group II) significantly distinct from the two others (Figure 4).

Interestingly, SF2 of human fibroblasts were found always in the 1-72 %, range, suggesting that some DSBs are *systematically repaired* whatever the genotype (maybe due to chemical DNA end-rejoining). Furthermore, some other DSBs are *systematically unrepaired* as well (maybe due to *in vitro* cell culture conditions that not ensure absolute viability for all the irradiated cells). However, further investigations are required to better understand such findings. Finally, it is noteworthy that a similar radiosensitivity range is observed with human tumors and rodent cell lines but radioresistance upper limit is a bit higher than human fibroblasts (the highest SF2dp is about 80%) and the two distinct groups of radiosensitivity are not significantly different likely because cell cycle distribution add intermediate SF2 values (Figure 4).

5.2. A Universal Correlation for Mammalians?

Clonogenic survival appeared to be inversely proportional to the yield of unrepaired DSBs. It is also the case with chromosome breaks assessed by the premature condensation technique. These correlations remain significant, even when controls and *group III* cell lines are excluded. Our findings indicate that the yield of unrepaired DNA damage leading to one lethal event after irradiation is 1.2 ± 0.6 chromosomal breaks and 3.5 ± 1.5 DSBs per cell, in agreement with the concept that one chromosome fragment causing death originates from one (terminal deletion) or two intrachromosomal DSBs (interstitial deletion or ring). In 1987, Cornforth and Bedford concluded that one unrepaired CB per cell is sufficient for cell killing but their result was obtained from one cell line only. Here, our findings show that this correlation is relevant for human fibroblasts, *whatever their genotype* [45]. What about rodent and human tumor cell lines?

The *xrs5* and *xrs6* cell lines and their radioresistant counterparts obey this correlation. Interestingly, we carefully reviewed the other rodent cell lines screened for their radiosensitivity, notably the other *xrs* family members that was forgotten by radiobiologists. Our conclusions are that like human fibroblasts, rodent cell lines described a continuous spectrum of radiation response, as well, not only for clonogenic response but also for DSB repair. Furthermore, like group II- and *ATM*-mutated cells, some rodent cell lines exhibit a marked radiosensitivity and a weak but significant DSB repair defect. It is noteworthy that such review on DSB repair data in rodents definitively eliminates the possibility of an *all-or-none* phenomenon between intrinsic radiosensitivity and DSB repair defect.

Even if human tumors obey grossly the same correlation, cell cycle and genomic instability add an extra-complexity to the shape of DSB repair curves and make more difficult a precise estimation of the amount of unrepaired DSBs. Interestingly, by using pulsed field gel electrophoresis or any other technique assessing DSBs by discriminating radiation-induced fragments according their size, DSB repair deficiency never exceeds 50% of unrepaired DSBs at 24 h post-irradiation [12,14]. This last remark suggests that about 50% of the radiation-induced DSBs are repaired *anyway* (maybe by chemical re-association process), explaining maybe the minimal fraction of cell surviving evoked above (at least 1% after 2 Gy). Altogether, these findings suggest that such inverse correlation between SF2 and the percentage of unrepaired DSBs presented here is relevant for mammalians in general (Figure 4) [2].

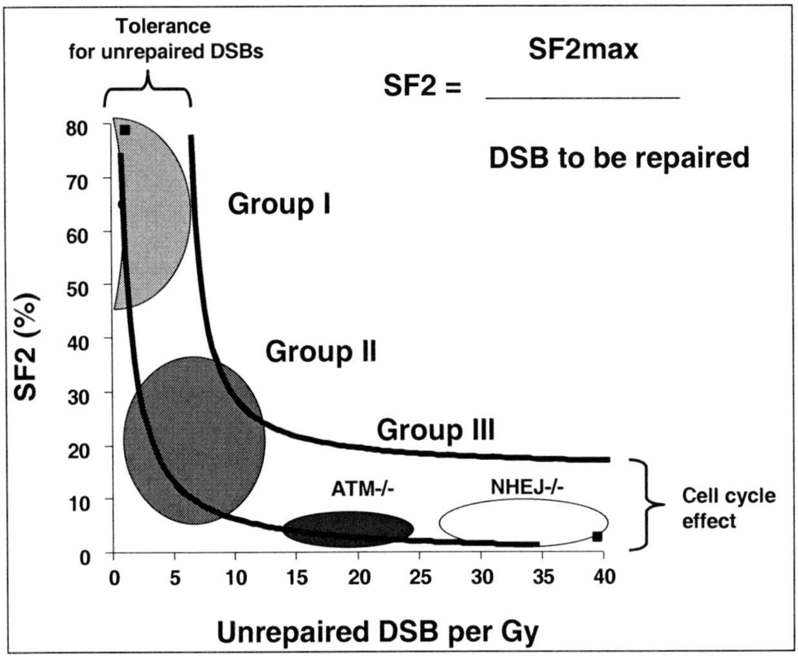

Figure 4: Relationship between survival fraction at 2 Gy (SF2) and unrepaired DSB per Gy calculated from pulsed field gel electrophoresis data. Confident zones indicate the group of radiosensitivity (group I : SF2 ranging from 45.2 to 65.6%; group II: SF2 ranging from 45.2 to 7.4%; group III: SF2 ranging from 3.7 to 7.4%. Group III contains ATM-mutated cells but also any cells in which NHEJ repair proteins like LIG4 or DNA-PK are mutated). Black squares represent the most radioresistant and the most radiosensitive rodent cell lines (CHO-K1 and CHO-xrs5, respectively). SF2 was found to be inversely proportional to unrepaired DSBs. Solid lines result from a general data fitting and represent the lower and the upper limits of the correlation. Lower limit solid line essentially reflects data from untransformed human fibroblast cell lines in plateau phase of growth. Upper limit reflects rather data from growing human tumor cell lines. It is noteworthy that It is noteworthy that, to the notable exception of group II confident zone that migrates up to group I one, the use of anti-*pH2AX* immunofluorescence do not change the shape of this correlation (data not shown).

5.3. Impact of the Choice of the Technique?

The above correlation has been obtained with pulsed field gel electrophoresis. As mentioned above, the DSB repair rate is not dependent upon the technique as far as DSB repair data are expressed as *percentage* of unrepaired DSBs [12]. It has been recently demonstrated that radiation-induced DSBs can be determined from the number of nuclear foci formed by the phosphorylation of the variant histone H2AX (pH2AX) and are easily quantifiable by using immunofluorescence (Figure 2) [40]. Anti-*pH2AX* immunofluorescence is now extensively used for assessing DSB repair data. Besides, anti-*pH2AX* immunofluorescence is becoming the favorite technique of radiobiologists, inasmuch DSB repair curves assessed with this technique were believed to be similar to those obtained with pulsed field gel electrophoresis. However, again, the success of anti-*pH2AX* immunofluorescence still raises the question of the choice of the technique and its impact upon the shape of DSB repair shape. We applied anti-*pH2AX* immunofluorescence to all the fibroblast cell lines of the collection. Surprisingly, while radioresistant control and *LIG4*-mutated cells showed the same DSB repair rate as that obtained with pulsed field gel electrophoresis, pH2AX repair data from group II and *ATM*-mutated cells were clearly different (Figure 3): Group II cells elicit a *normal* DSB repair rate with nearly no pH2AX foci remaining at 24 h post-irradiation and with *no accelerated repair* rate between 2 and 8 h post-irradiation. Furthermore, *ATM*-mutated cells do not show pH2AX immediately after irradiation. The yield of pH2AX foci increases progressively up to a maximum at 4 h and decreased thereafter to about 10% of pH2AX foci remaining.

There are therefore three major consequences for the prediction of radiosensitivity with the anti-*pH2AX* immunofluorescence technique :

- the DSB repair rate of group II cells appears to be normal with the anti-*pH2AX* immunofluorescence.
- the accelerated DSB repair rate observed between 2 and 8 h post-irradiation in group II and *ATM*-mutated cells is not predicted by anti-*pH2AX* immunofluorescence.
- although SF2 remains inversely proportional to the pH2AX foci remaining at 24 h, such correlation cannot predict the weak but significant radiosensitivity of group II cells, rendering group II and group I undistinguishable.

Once again, technical artifacts lead to misinterpretations of the DSB repair curves of human cells. How to overcome such problems?

6. DSB Repair Curves: the Solution is in the Shape!

6.1. To Repair or not to Repair?

Let's go back to the pulsed field gel electrophoresis data: a high percentage of DSB remaining at 24 h post-irradiation like that observed in 180BR cells means that some unrejoined DNA fragments still exist after irradiation. It is therefore reasonable to imagine that such unrepaired DNA fragments propagate all along the cell cycle, leading to mitotic death, notably by producing micronuclei. Besides, used here intuitively, the term *unrejoined* fragment describes the natural consequence of the *defect* of a DSB repair pathway that would consist in a rejoining of the DNA ends. Here, *DSB repair defect* means *absence of repair*. If such DSB repair pathway exists, its defect would likely lead to the formation of unrejoined DNA fragments, reflected by the decrease of the DSB repair rate, to cellular death and therefore to radiosensitivity (Figure 3).

Conversely, the accelerated DSB repair rate observed between 2 and 8 h with pulsed field gel electrophoresis does not reflect an *absence of repair* but rather its *deregulation*. Hence, the DSB repair pathway cannot be based on DNA-ends rejoining since its deregulation would necessarily consist in some unrejoined DNA fragments. Furthermore, if this *deregulation* would be an *overactivity* of the DNA rejoining process, it would necessarily result in a radioresistance: group II and *ATM*-mutated, in which the accelerated DSB repair has been observed, are however weakly and highly radiosensitive, respectively (Figures 3 and 4). Consequently, another DSB repair pathway should function concomitantly with the DNA ends-rejoining process and participate also (but in a lesser extent) to the cellular response to radiation. This deregulation would consist in an accelerated DSB repair rate and would be reflected with pulsed field gel electrophoresis by the gross decrease of DSB repair curves. Interestingly, unlike syndromes associated with DNA ends-rejoining defects, cancer proneness is a major common clinical feature of group II syndromes and ataxia telangiectasia. Therefore, it is tempting to consider that such accelerated DSB repair would result in the generation of errors, responsible for cellular transformation and cancer rather than a significant impact upon radiosensitivity.

6.2. Two Different DSB Repair Pathways but two Types of Impairments!

The DSB repair pathway mentioned above that consists in DNA-ends rejoining obviously corresponds to the so-called *non-homologous end-joining* (NHEJ). Its defect leads to unrejoined DNA fragments and severe radiosensitivity. This situation naturally corresponds to the cells mutated in LIG4, DNA-PK and Ku80 proteins (*i.e.* human 180BR fibroblasts, human MO59J glioma and rodent xrs5 and xrs6 mutants, respectively) that are all major NHEJ actors. In addition, since NHEJ participates also to the immunoglobulins formation through V(D)J recombination, human syndromes with NHEJ defect are also associated with immnodeficiency. Lastly, since phosphorylation of H2AX histone has been show to be an essential molecular event of NHEJ, it is logical that both pulsed field gel electrophoresis and anti-*pH2AX* immunofluorescence techniques provide the same DSB repair rate and therefore the same quantitative information.

Is the second DSB repair pathway mentioned above *homologous recombination* (HR)? HR is strongly dependent upon RAD51 and RAD52 proteins and these proteins are not active in quiescent fibroblasts (Figure 5). On another hand, it is important to remind here that no syndrome associated with defect in HR has been identified yet, likely because HR is required for mitosis and meiosis and that any lack of HR would necessarily prevent cell viability. Lastly, a defect of HR would not correspond to an accelerated DSB repair rate. Hence, another DSB repair pathway, different from NHEJ and HR, with another kind of impairment than a simple lack of activity, is therefore to be considered to explain and relate the radiosensitivity of group II cells.

Far from a defect in HR, a number of authors have reported abnormally high recombination rates both *in vitro* and *in vivo* in some human cells. As a high-fidelity repair process, high rates of HR would lead to a larger radioresistance and absence of cancer proneness. By contrast, high rates of a *random recombination* repair pathways can lead to errors accumulations, may participate to cell survival and contribute to cancer proneness. As mentioned above, recombination is a complex succession of DNA damage sensoring, nuclease activity, strand exchange and resolving, polymerization and/or ligation. It is easy to imagine a deregulation at any steps of such process, notably during the DNA strand cleavage.

Hence, *hyperrecombination* may describe high rates of random recombination process, active at least in G0/G1, independently of RAD51 and RAD52 proteins and whose biological consequence is the generation of errors and cancer proneness. But with which proteins?

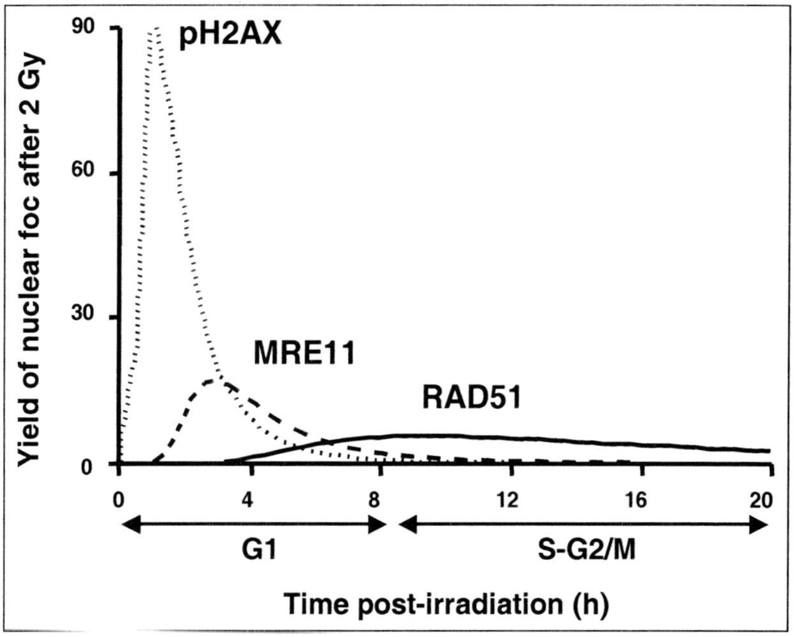

Figure 5: Yields of nuclear foci in radioresistant non-transformed human fibroblasts exposed to X-rays (2 Gy) observed after anti-*pH2AX*, *MRE11* and *RAD51* immunofluorescence. Arrows indicate that RAD51 foci are observed only in S-G2/M cells.

6.3. MRE11-Dependent Repair Pathway as an Alternative to NHEJ?

The MRE11 protein is very conserved through species. In mammalians, it was shown to be associated with the NBS1 and RAD50 proteins by forming the MNR complex. Interestingly, the MRE11 protein was demonstrated to possess endo- and exo-nuclease activity, generally required for a *recombination-like* repair process. In 2001, Mirzoeva and Petrini pointed out the existence of different patterns of nuclear foci formed by the MRE11 protein [46]. Our preliminary experiments in radioresistant group I fibroblasts revealed that MRE11 foci appeared after 2 Gy from 2 to 8 h (Figure 5). The striking temporal coincidence between the occurrence of MRE11 foci and the accelerated DSB repair rate observed with pulsed field gel electrophoresis prompted us to examine MRE11 foci after exposure to radiation in representative cell lines of the collection. The group II and *ATM*-mutated cells did not show MRE11 foci. By contrast, the *LIG4*-mutated cell lines showed significantly *more* MRE11 foci than controls. Interestingly, abnormally high number of MRE11 foci was also observed in the NHEJ-deficient M059J cells.

The disappearance of pH2AX foci is generally considered to be a consequence of DSB repair [40]. Conversely, a hyperrecombination phenotype was systematically observed when MRE11 foci are *absent* or *impaired* (*ATM*-mutated and group II cells) [2]. Interestingly, hyperrecombination and spontaneous breaks with large cell nuclei are common features of group II cells. These features might be by an uncontrolled nuclease activity of MRE11. However, that aberrant MRE11 cannot mediate only a simple degradation of DNA ends since it would lead to more lethal events in group II cells and lead to a *severe* rather than a *weak* radiosensitivity. Hence, our findings suggest that *absent* MRE11 foci after irradiation may be a signature of a recombination-like process that acts in G0/G1 in parallel to NHEJ and participates in cell survival. Its lack of control would be responsible of genomic instability and cancer proneness but with a weak impact upon radiosensitivity. Since ATM is required also for MRE11 foci formation, ATM would act upstream the MRE11-dependent repair pathway. Further investigations are however needed to consolidate such assumptions.

Interestingly, an *excess* of MRE11 foci was systematically observed in NHEJ-deficient cells in agreement with preliminary observations in *LIG4*-mutated cells, in rodent mutants and after treatments to cis-platin and iodides, that are known to be NHEJ inhibitors [2]. These findings are therefore consistent with a model in which ATM mediates a *compensatory/rescue switch* between NHEJ and MRE11-mediated repair pathway. When the NHEJ pathway is impaired, an *ATM*-dependent compensatory switch would permit some DSBs to be repaired by MRE11-dependent pathway, leading to an excess of MRE11 foci. Since NHEJ is the major DSB repair pathway in mammalians, such impairment would lead to an excess of unrepaired DNA fragments and an extreme radiosensitivity, as predicted by pulsed field gel electrophoresis or anti-*pH2AX* immunofluorescence. Conversely, the impairment of the MRE11-mediated pathway would lead to absent MRE11 foci, causing hyperrecombination and weak radiosensitivity. This defect is not observed by anti-pH2AX immunofluorescence. Lastly, in *ATM*-mutated cells, few DSBs are repaired by an impaired NHEJ and the rescue to the MRE11-dependent pathway cannot be triggered, leading to both severe radiosensitivity and cancer-proned hyperrecombination.

7. CONCLUSIONS

The careful review of literature dealing with DSB repair in human cells highlights the numerous traps in which radiobiologists can fall: 1) the influence of concepts based on data obtained with other species (*e.g. recombination*); 2) the use of terms taken outside their historical concepts (*e.g. homologous*); 3) the choice of techniques that may amplify one molecular step and hide all the others (*e.g.* pulsed field gel electrophoresis *vs.* anti-pH2AX immunofluorescence). All these difficulties permit to draw an impressive picture of the progress that remains to be pursued, more than one century after the discovery of X-rays, to better understand the biological effects of ionizing radiation on human cells and their clinical consequences. Further than the beauty of demonstrations and the pleasure to solve enigma, let us not forget that such question should provide elements to reply to two major questions: how to develop efficient anti-cancer treatments and how to estimate the biological risk after an exposure to radiation of individuals, whatever medical or casual?

ACKNOWLEDGEMENTS

We would like to thank Drs. E.P Malaise, V. Favaudon, J. Balosso and B. Kysela for fruitful discussions. A.J. was supported by the Institut de Radioprotection et de Sûreté Nucléaire (IRSN, France). This work was also supported by the Association Pour la Recherche sur l'Ataxie-Telangiectasie (APRAT), Association pour la Recherche contre le Cancer (ARC), the Electricité de France (Comité de Radioprotection), the ETOILE Project (Région Rhône-Alpes and Lyon University).

REFERENCES

[1] Iliakis, G. The role of DNA double strand breaks in ionizing radiation-induced killing of eukaryotic cells. *Bioessays*, 1991 13, 641-648.

[2] Joubert, A; Gamo, K; Bencokova, Z; Gastaldo, J; Rénier, W; Chavaudra, N; Favaudon, V; Arlett, C; Foray, N. DNA double-strand break repair in syndromes associated with acute radiation response: a balance between DNA-PK- and MRE11-dependent pathways, *in press*.

[3] Steel, G. *Basic Clinical Radiobiology*, 3rd edition, p. 262: Arnold Publishers, 1993.

[4] Foray, N; Priestley, A; Alsbeih, G; Badie, C; Capulas, EP; Arlett, CF; Malaise, EP. Hypersensitivity of ataxia telangiectasia fibroblasts to ionizing radiation is associated with a repair deficiency of DNA double-strand breaks. *Int J Radiat Biol*, 1997 72, 271-283.

[5] Lobrich, M; Jeggo, PA. The two edges of the ATM sword: co-operation between repair and checkpoint functions. *Radiother Oncol*, 2005 76, 112-118.

[6] Ward, JF. Radiation mutagenesis: the initial DNA lesions responsible. *Radiat Res*, 1995 142, 362-368.

[7] Goodhead, DT. The initial physical damage produced by ionizing radiations. *Int J Radiat Biol*, 1989 56, 623-634.

[8] Frankenberg-Schwager, M. Review of repair kinetics for DNA damage induced in eukaryotic cells in vitro by ionizing radiation. *Radiother Oncol*, 1989 14, 307-320.

[9] Garinis, GA; Mitchell, JR; Moorhouse, MJ; Hanada, K; de Waard, H; Vandeputte, D; Jans, J; Brand, K; Smid, M; van der Spek, PJ; Hoeijmakers, JH; Kanaar, R; van der Horst, GT. Transcriptome analysis reveals cyclobutane pyrimidine dimers as a major source of UV-induced DNA breaks. *Embo J*, 2005 24, 3952-3962.

[10] Huang, LC; Clarkin, KC; Wahl, GM. Sensitivity and selectivity of the DNA damage sensor responsible for activating p53-dependent G1 arrest. *Proc Natl Acad Sci U S A*, 1996 93, 4827-4832.

[11] Radford, IR. The level of induced DNA double-strand breakage correlates with cell killing after X-irradiation. *Int J Radiat Biol Relat Stud Phys Chem Med*, 1985 48, 45-54.

[12] Foray, N; Arlett, CF; Malaise, EP. Radiation-induced DNA double-strand breaks and the radiosensitivity of human cells: a closer look. *Biochimie*, 1997 79, 567-575.

[13] Radford, IR. Lysis solution composition and non-linear dose-response to ionizing radiation in the non-denaturing DNA filter elution technique. *Int J Radiat Biol,* 1990 57, 479-483.

[14] Chavaudra, N; Bourhis, J; Foray, N. Quantified relationship between cellular radiosensitivity, DNA repair defects and chromatin relaxation: a study of 19 human tumour cell lines from different origin. *Radiother Oncol*, 2004 73, 373-382.

[15] Kawabata, M; Kawabata, T; Nishibori, M. Role of recA/RAD51 family proteins in mammals. *Acta Med Okayama*, 2005 59, 1-9.

[16] Dudas, A; Chovanec, M. DNA double-strand break repair by homologous recombination. *Mutat Res*, 2004 566, 131-167.

[17] Rothkamm, K; Kruger, I; Thompson, LH; Lobrich, M. Pathways of DNA double-strand break repair during the mammalian cell cycle. *Mol Cell Biol*, 2003 23, 5706-5715.

[18] Udayakumar, D; Bladen, CL; Hudson, FZ; Dynan, WS. Distinct pathways of nonhomologous end joining that are differentially regulated by DNA-dependent protein kinase-mediated phosphorylation. *J Biol Chem*, 2003 278, 41631-41635.

[19] Allen, C; Halbrook, J; Nickoloff, JA. Interactive competition between homologous recombination and non-homologous end joining. *Mol Cancer Res*, 2003 1, 913-920.

[20] Baumann, P; West, SC. DNA end-joining catalyzed by human cell-free extracts. *Proc Natl Acad Sci U S A*, 1998 95, 14066-14070.

[21] Nicholls, RD; Fischel-Ghodsian, N; Higgs, DR. Recombination at the human alpha-globin gene cluster: sequence features and topological constraints. *Cell*, 1987 49, 369-378.

[22] Thyagarajan, B; McCormick-Graham, M; Romero, DP; Campbell, C. Characterization of homologous DNA recombination activity in normal and immortal mammalian cells. *Nucleic Acids Res*, 1996 24, 4084-4091.

[23] Foray, N; Charvet, AM; Duchemin, D; Favaudon, V; Lavalette, D. The repair rate of radiation-induced DNA damage: a stochastic interpretation based on the gamma function. *J Theor Biol*, 2005 236, 448-458.

[24] Badie, C; Iliakis, G; Foray, N; Alsbeih, G; Cedervall, B; Chavaudra, N; Pantelias, G; Arlett, C; Malaise, EP. Induction and rejoining of DNA double-strand breaks and interphase chromosome breaks after exposure to X rays in one normal and two hypersensitive human fibroblast cell lines. *Radiat Res*, 1995 144, 26-35.

[25] Foray, N; Monroco, C; Marples, B; Hendry, JH; Fertil, B; Goodhead, DT; Arlett, CF; Malaise, EP. Repair of radiation-induced DNA double-strand breaks in human fibroblasts is consistent with a continuous spectrum of repair probability. *Int J Radiat Biol*, 1998 74, 551-560.

[26] Radivoyevitch, T; Hoel, DG; Hahnfeldt, PJ; Rydberg, B; Sachs, RK. Recent data obtained by pulsed field gel electrophoresis suggest two types of double-strand breaks. *Radiat Res*, 1998 149, 52-58.

[27] Curtis, SB. Lethal and potentially lethal lesions induced by radiation--a unified repair model. *Radiat Res*, 1986 106, 252-270.

[28] Sontag, W. A discrete cell survival model including repair after high dose-rate of ionizing radiation. *Int J Radiat Biol*, 1997 71, 129-144.

[29] Tobias, CA. The repair-misrepair model in radiobiology: comparison to other models. *Radiat Res Suppl*, 1985 8, S77-95.

[30] Cucinotta, FA; Nikjoo, H; O'Neill, P; Goodhead, DT. Kinetics of DSB rejoining and formation of simple chromosome exchange aberrations. *Int J Radiat Biol*, 2000 76, 1463-1474.

[31] Stewart, RD. Two-lesion kinetic model of double-strand break rejoining and cell killing. *Radiat Res*, 2001 156, 365-378.

[32] Ostashevsky, JY. A model relating cell survival to DNA fragment loss and unrepaired double-strand breaks. *Radiat Res*, 1989 118, 437-466.

[33] Curtis, SB. Mechanistic models. *Basic Life Sci*, 1991 58, 367-382; discussion 382-366.

[34] Kendal, WS. Technical report. The application of probability-generating functions to linear-quadratic radiation survival curves. *Int J Radiat Biol*, 2000 76, 581-587.

[35] Guerrero, M; Stewart, RD; Wang, JZ; Li, XA. Equivalence of the linear-quadratic and two-lesion kinetic models. *Phys Med Biol*, 2002 47, 3197-3209.

[36] Kysela, BP; Michael, BD; Arrand, JE. Relative contributions of levels of initial DNA damage and repair of double strand breaks to the ionizing radiation-sensitive phenotype of the Chinese hamster cell mutant, XR-V15B. Part I. X-rays. *Int J Radiat Biol*, 1993 63, 609-616.

[37] Foray, N; Badie, C; Alsbeih, G; Fertil, B; Malaise, EP. A new model describing the curves for repair of both DNA double-strand breaks and chromosome damage. *Radiat Res*, 1996 146, 53-60.

[38] Foray, N; Fertil, B; Alsbeih, MG; Badie, C; Chavaudra, N; Iliakis, G; Malaise, EP. Dose-rate effect on radiation-induced DNA double-strand breaks in the human fibroblast HF19 cell line. *Int J Radiat Biol*, 1996 69, 241-249.

[39] Goodhead, DT. Saturable repair models of radiation action in mammalian cells. *Radiat Res Suppl*, 1985 8, S58-67.

[40] Rothkamm, K; Lobrich, M. Evidence for a lack of DNA double-strand break repair in human cells exposed to very low x-ray doses. *Proc Natl Acad Sci U S A*, 2003 100, 5057-5062.

[41] Jeggo, PA; Kemp, LM. X-ray-sensitive mutants of Chinese hamster ovary cell line. Isolation and cross-sensitivity to other DNA-damaging agents. *Mutat Res*, 1983 112, 313-327.

[42] Taylor, AM; Harnden, DG; Arlett, CF; Harcourt, SA; Lehmann, AR; Stevens, S; Bridges, BA. Ataxia telangiectasia: a human mutation with abnormal radiation sensitivity. *Nature*, 1975 258, 427-429.

[43] Painter, RB; Young, BR. Radiosensitivity in ataxia-telangiectasia: a new explanation. *Proc Natl Acad Sci U S A*, 1980 77, 7315-7317.

[44] Badie, C; Iliakis, G; Foray, N; Alsbeih, G; Pantellias, GE; Okayasu, R; Cheong, N; Russell, NS; Begg, AC; Arlett, CF; et al. Defective repair of DNA double-strand breaks and chromosome damage in fibroblasts from a radiosensitive leukemia patient. *Cancer Res*, 1995 55, 1232-1234.

[45] Cornforth, MN; Bedford, JS. A quantitative comparison of potentially lethal damage repair and the rejoining of interphase chromosome breaks in low passage normal human fibroblasts. *Radiat Res*, 1987 111, 385-405.

[46] Mirzoeva, OK; Petrini, JH. DNA damage-dependent nuclear dynamics of the Mre11 complex. *Mol Cell Biol*, 2001 21, 281-288.

In: New Research on DNA Repair
Editor: Breehn R. Landseer, pp. 295-324
ISBN 1-60021-385-5

Chapter XI

DNA Double-Strand Breaks Repair Pathways in the Framework of Chromatin

Tom C. Karagiannis[1,2,*] *and Assam El-Osta*[3]

[1]Molecular Radiation Biology, Trescowthick Research Laboratories, Peter MacCallum Cancer Centre, East Melbourne, Victoria, Australia

[2]Department of Pathology, The University of Melbourne, Parkville, Victoria, Australia

[3]Epigenetics in Human Health and Disease, Baker Medical Research Institute, The Alfred Medical Research and Education Precinct, Prahran, Victoria, Australia

Abstract

DNA double-strand breaks are defined are two simultaneous nicks in opposite strands of the DNA helix, which are in sufficiently close proximity to one another that base-pairing and chromatin structure are unable to maintain the broken DNA ends juxtaposed. These lesions arise in the course of normal and pathological endogenous DNA metabolism or can result from exogenous agents such as ionizing radiation and certain chemotherapeutic drugs. They represent the most severe type of DNA damage with respect to preservation of genomic integrity and there is growing evidence suggesting that in higher eukaryotes aberrant repair of double-strand breaks is a key event in carcinogenesis. Therefore, cells have evolved complex mechanisms that include cell cycle arrest, activation and increased expression of various genes including those associated with DNA repair and in certain cases induction of the apoptotic pathway, to respond to double-strand breaks. Given the intense research effort in this field in the past few years our knowledge of the molecular details relating to the detection, signalling and repair of double-strand breaks has greatly increased. Here we discuss recent progress in our understanding of molecular responses to DNA double-strand breaks with a particular emphasis on homologous recombination and non-homologous end-joining, the major

* Correspondence concerning this article should be addressed to Dr. Tom Karagiannis, Peter MacCallum Cancer Centre, Research, Level 2, Locked Bag No. 1, A'Beckett Street, East Melbourne, 8006, VIC, Australia. E-mail: tom.karagiannis@petermac.org.

double-strand break repair pathways. In addition, the role of changes in chromatin architecture in response to double-strand breaks is currently emerging and already numerous links between changes in chromatin – resulting from both the contribution of histone tail modifications and chromatin remodelling complexes – and repair pathways have been made. Therefore, the importance of epigenetic modifications in the double-strand break repair pathways will also be discussed.

Keywords: double-strand breaks, signal transduction, homologous recombination, non-homologous end-joining, chromatin remodelling, histone modifications.

INTRODUCTION

Among the possible DNA lesions, DNA double-strand breaks (DSBs) are considered to be the most lethal and represent the greatest challenge to the maintenance of genomic integrity. Therefore, cells have evolved very complex mechanisms to recognize, signal and potentially repair these lesions. The pathways by which cells respond to DSBs have been the subject of previous reviews, however, in this chapter the molecular details are significantly expanded, revised and updated, particularly with respect to the mechanisms involved in sensing DSBs and the integration of epigenetic factors in the cellular response pathways to DSBs [1,2].

Double-strand breaks are the result of two simultaneous nicks in opposite strands of the DNA helix. A requirement for DSB formation is that the two nicks are in sufficiently close proximity to one another that base-pairing and chromatin structure are unable to maintain the broken DNA ends juxtaposed. It is generally accepted that for a DSB the distance between the breaks on opposite DNA strands needs to be less than 12 base pairs.

It is known that DSBs can be generated following exposure of cells to exogenous agents such as ionizing radiation, radiomimetic drugs and topoisomerase inhibitors [3-6]. In addition, lesions referred to as 'pathological' DSBs can arise from endogenous processes. For example, approximately 1% of inhaled oxygen is converted into oxidative free radicals (reactive oxygen species (ROS)) such as hydrogen peroxide and superoxide anions that have the potential to cause DSBs [7-9]. Indeed, the total induced DNA alterations, including base changes, single-strand breaks (SSBs) and DSBs, from endogenous sources, predominantly ROS, easily outweigh those induced from background γ-irradiation of approximately 1 mGy per year; the ratio of DSBs resulting from endogenous ROS to those from background irradiation is estimated to be about 10^3 [8,9]. Furthermore, pathological DSBs may occur when the replication machinery meets nicks or other types of lesions in the template DNA (stalled or collapsed replication fork) [5,10] and as discussed further below, DSBs may result from incorrect resolution of lesions induced by normal DNA-metabolic proteins such as RAG1 and RAG2 [5,11-13].

Although they represent a major danger to the stability of the genome, DSBs are created deliberately by specific nucleases for defined biological functions. A well characterized example of this is V(D)J recombination in which immunoglobulin heavy chain and light chain variable region genes and T-cell-receptor genes are assembled from germline variable (V), diversity (D), and joining (J) sites in B- and T-lymphocytes, respectively [14-16]. These recombination pathways are initiated by site-specific DSBs that are generated at target loci by the lymphoid-specific RAG1 and RAG2 nuclease complex [14-16]. The sequential two-step process involves firstly nicking one strand to form a SSB between the recombination signal sequences and the coding segment, by hydrolyzing the phosphodiester bond [17]. This forms a new 3′ OH which is used as a nucleophile to attack the second phosphodiester bond in the subsequent reaction. Thus, the second nick produces a hairpin coding end and a blunt, 5′-phosphorylated signal end [17]. Furthermore, physiological DSBs are produced during meiosis to initiate strand exchange between homologous chromosomes and are also generated during Ig class-switch gene rearrangement pathways (i.e. switching from one immunoglobulin isotype to another, for example, IgM to IgG) [5,6,18]. Importantly, these highly regulated DSBs that are introduced in the course of 'normal' programmed DNA metabolism are repaired by many of the same DNA-break processing proteins, as those that repair pathological DSBs. In addition, it has been shown that DSBs may also arise when telomeres become critically shortened leading to activation of DNA double-strand break response pathways and ultimately to apoptosis and/or senescence [19,20].

Cells are generally able to adapt to low levels of irreparable DNA damage however, it is accepted that in certain cases only a single DNA DSB is potentially, cytotoxic if it inactivates an essential gene or can trigger programmed cell-death (apoptosis) [5,21]. DSBs are considered to be particularly dangerous lesions because they are more difficult to repair than other types of DNA damage. For example, single-strand nicks are more easily repaired by any of the excision-repair (including base and nucleotide excision repair) or mismatch repair pathways that involve using the undamaged anti-parallel strand as a template [22,23]. In contrast, DSBs result in loss of integrity of both DNA strands and the biochemical configurations of the broken DNA ends are typically incompatible and rejoining by a simple ligation step is not usually possible to repair the lesion [6].

There is comprehensive experimental evidence supporting a causal relationship between DSBs and genomic instability. Erroneous rejoining of DNA DSBs can induce deletion or insertion mutations and chromosomal translocations that are potentially carcinogenic [24-28]. Indeed, chromosomal translocations involving the Ig and T-cell receptor loci are evident in many cancers of lymphoid origin, suggesting that they are the result of defective resolution of DSBs related to V(D)J recombination [11,14-16]. For example, in the B-cell malignancy, Burkitt's lymphoma, the *c-MYC* gene is often juxtaposed to the Ig heavy-chain genes as a result of incorrect chromosomal recombination (thought to be due to aberrant recombinase activity) [11,14-16]. More generally, the loss or amplification of chromosomal regions that is a feature of many cancer cells may be associated with inappropriate DSB repair. It has been suggested that erroneous DSB repair can lead to carcinogenesis in cases where the deleted genes are associated with tumor suppressor loci or amplified genes encode an oncogene [5,28]. Moreover and as discussed later in this chapter, defects in cellular responses to DSBs including mutations in factors involved in signalling and repair, may increase cancer predisposition or could frequently be the initiating events of carcinogenesis [5,28,29].

CELLULAR RESPONSES TO DOUBLE-STRAND BREAKS

The main cellular responses to DSBs include cell-cycle regulation, transcriptional and post-transcriptional changes of certain genes (including activation of expression of those associated with repair), increases in cellular levels of deoxyribonucleotides and in certain cases induction of apoptosis [5,28]. The overall cellular response DSBs has been considered as a classical signal-transduction cascade [5,28,30]. In other words, DSBs (signals) are detected by sensor proteins that activate protein kinase (transduction) cascades, which result in amplification and diversification of the signal through a series of downstream effector molecules (response). Given that cells have evolved intricate protein pathways, some of which are interlinked with other DNA metabolic processes, to respond to DSBs, it is important to keep in mind the signal-transduction analogy as we discuss individual components of DSB response pathways in detail, for an overall perspective.

RECOGNITION OF DOUBLE-STRAND BREAKS

Although there have been major advances in our knowledge regarding the molecular details of DSB signalling and repair, there are still many gaps in our knowledge with respect to the molecular pathways involved in the detection of these lesions. Nevertheless, in the context of sensing DSBs it is pertinent to consider the checkpoint proteins Rad9, Rad1, Hus1 (which form a heterotrimeric complex commonly known as 9-1-1) and Rad17 which are named after the analogous proteins in *Schizosaccharomyces pombe* [31,32]. The corresponding proteins in *Saccharomyces cerevisiae* are Rad17, Ddc1, Mec3 and Rad24, respectively [33,34]). An important finding in *S. pombe* was that the phosphorylation of Chk1 and Cds1 by Rad3 – a homolog of the key signalling kinases, ataxia telangiectasia mutated (ATM) and ataxia telangiectasia-related (also known as Rad3-related, ATR) in mammalian cells – requires Rad1, Rad9, Rad17, Rad26 and Hus1 [35]. From this observation it was first postulated that this group of proteins might function as DNA damage sensors [35]. It has been shown that the 9-1-1 complex displays sequence and structural homology with proliferating cell nuclear antigen (PCNA) [32,36]. PCNA is a sliding clamp protein that forms a homotrimeric toroidal structure around DNA and tethers DNA polymerase δ during replication [37,38]. These homology analyses together with the findings that Rad1, Rad9 and Hus1 interact with one another in cell lysate and yeast two-hybrid experiments, has prompted the suggestion that 9-1-1 is clamp-like complex that is involved in the detection of damaged DNA [31,39-41]. The finding that 9-1-1 forms nuclear foci in response to DNA damage is consistent with this clamp-protein idea [42].

Clamp-like structures require loading around the DNA by clamp proteins. It is well known, for example, that the clamp loader for PCNA is the heteropentameric structure, replication factor C (RFC) [37,38]. Classical RFC complexes recognize primer-template junctions and load the PCNA clamp around the DNA in an energy-dependent process [37,38]. It follows that the 9-1-1 complex also requires a clamp loader to facilitate its function. Several lines of investigation suggest that the checkpoint protein Rad17 is involved in loading the 9-1-1 clamp around the DNA. Studies indicate that Rad17 exhibits structural homology with RFC and that the *S. cerevisiae* homolog of Rad17 interacts with four of the five RFC subunits [43-47]. Further analyses demonstrate that Rad17 interacts directly with Rad1 and provide evidence that Rad17 interacts with the 9-1-1 clamp in a similar way to the interaction between RFC and PCNA [31].

On the basis of the above observations, a suggested paradigm for the detection of DSBs is that the clamp loader consisting of Rad17 is constitutively bound to chromatin in mammalian cells and recognizes structural modifications induced by DNA damage [31,48]. Following DNA damage the Rad17 clamp loader recruits and fixes the 9-1-1 clamp-proteins around the DNA converting them into extraction-resistant nuclear foci [31,48].

Another candidate for a sensor of DNA damage during replication arrest is the ATR-ATR interacting protein (ATRIP) complex [49]. Findings in *S. pombe* indicate that the ATRIP homolog, Rad26, which is tightly bound to Rad3, is phosphorylated by Rad3 following DNA damage even in the absence of the Rad17 and 9-1-1 complex [50]. Similarly, Ddc2 (also known as Lcd1 and Pie1) the *S. cerevisiae* analog of Rad26, can be phosphorylated without invoking the RFC- and PCNA-like checkpoint proteins following DNA damage [51-53]. These findings suggest that the Rad3-Rad26 complex and analogous *S. cerevisiae* and human complexes could respond to a certain subset of DNA lesions without the need for other early DNA damage sensor elements.

Overall, the data suggest that the Rad17/9-1-1 and ATR-ATRIP complexes represent two independent sensory pathways and that they may function in concert to fully activate checkpoint pathways in response to DNA damage in mammalian cells [54]. It is accepted that the Rad17/9-1-1 and ATR-ATRIP sensory complexes are important for detection of replication associated DSBs and of other aberrant DNA structures (such as gapped DNA) at arrested replication forks [32,48]. The ATR-ATRIP complex preferentially to ultraviolet (UV)-induced DNA damage [55].

Recent evidence indicates that the MRN complex, which is known to be essential for the cellular response to DSBs and is involved in the initial processing of the lesions, acts as a sensor for ATM. The MRN complex has a well-characterized association with the ATM-dependent checkpoint pathway [56-58]. The highly conserved MRN complex consists of the Mre11, Rad50 and Nbs1 (nibrin) proteins [56]. The MRN complex possesses both DNA nuclease and DNA binding capabilities and these properties allow its involvement in the initial processing of DSBs [56]. The Mre11 protein has 3′,5′-exonuclease activity which is thought to create single-stranded DNA by catalyzing the resection of free DNA ends and also possesses endonuclease activity that cleaves DNA hairpins [59]. The Rad50 ATPase is a member of the structural maintenance of chromosomes family of proteins. The current paradigm suggests that Rad50 homodimers associate with two Mre11 molecules to form tetrameric MR complexes which bind to DNA via the coiled-coil arms of Rad50 dimers [59]. It is thought that the Nbs1 protein is involved in localizing (C terminus of Nbs1 binds directly with Mre11 [60]) and assembling the complex at DSB ends probably via direct interaction with the core histone variant H2AX [61,62]. Furthermore, *in vitro* studies have indicated that Nbs1 stimulates the enzymatic activities of Mer11/Rad50 [63,64]. Hypomorphic mutations in the *NBS1* and *MRE11* genes lead to the Nijmegen breakage syndrome (NBS) and ataxia-telangiectasia like disease (ATLD), respectively, which share many features with the neurodegenerative condition ataxia telangiectasia (A-T) which results from a deficiency in

ATM [65,66]. It is well-known that Nbs1 is a substrate for ATM phosphorylation in response to DNA damage (ATM phosphorylates Nbs1 on Serine 343 and other residues), and that this event is required for activation of the S-phase checkpoint [67-70]. Similarly, MRE11 is also phosphorylated by ATM in response to DSBs [71,72]. On the basis of these observations the MRN complex has traditionally been assigned a role downstream of ATM in the DSB response pathway. Paradoxically, relatively recent evidence indicates that the MRN complex is required for the activation of the ATM protein [73-77]. Specifically, it has been shown that Nbs1 binds to ATM (the C-terminal 20 amino acids - FXF/Y domain - of Nbs1 have been shown to bind to the N-terminal HEAT repeats of ATM [74,78,79]) and that this interaction is required for activation and recruitment of ATM to sites of DNA damage [73]. Overall, the MRN complex has now been assigned roles both upstream and downstream of ATM and is regarded as a sensor of DSBs.

Double-Strand Break Signalling Cascades: ATM, ATR and the Downstream Substrate, p53, are Central Components of the Double-Strand Break Response

As alluded to earlier in this chapter, ATM (370 kDa, 3056 amino acids) is the primary signal transducer in the DSB response pathway [80,81]. The ATM kinase is a member of the phosphoinositide 3-kinase (PI(3)K) superfamily – sequence similarity to the p110 lipid kinase subunit – but phosphorylates proteins rather lipids [82,83]. The 350 amino acid C-terminal kinase domain is the only segment of the protein with an assigned function. Like other PI(3)K proteins that have analogous kinase activities, ATM phosphorylates target proteins on serine or threonine residues followed by glutamine [84]. Following activation ATM phosphorylates numerous proteins which are involved in cell-cycle regulation, DNA repair and apoptosis [5,28,81,85]. The yeast orthologues of ATM (Tel1p and Mec1p in *S. cerevisiae* and Tel1 and Rad3 in *S. pombe*) have analogous functions and are also involved in genome surveillance and telomere metabolism [28,81,86,87].

Mutations in the *ATM* gene lead to the autosomal recessive genetic disorder ataxia telangiectasia (A-T) [81,88]. A-T is characterized by early onset cerebral ataxia and telangiectasia which arises due to the dilatation of blood vessels [85,89]. It is thought that these conditions arise from the reduced ability to deal with oxidative stress arising from endogenously generated free radicals. Furthermore, immunodeficiency, chromosomal instability and cancer predisposition are key features of this disease [81,90-92]. Leukemias and lymphomas are common in younger A-T patients whereas, older patients also develop non-lymphoid cancers such as breast and stomach carcinomas [81,93]. Parenthetically, *Atm-/-* is lethal in mice within 3 months due to the development of thymic lymphomas [94,95]. It was first reported in the late 1960s that patients with A-T are hypersensitive to ionizing radiation and conventional doses of radiotherapy are life threatening [81,96,97]. It is thought that the A-T phenotype is at least in part due to aberrant recognition and processing of DSBs. Cell culture studies have indicated that ATM deficient cells are particularly susceptible to the effects of ionizing radiation and to radiomimetic drugs [28,98-100]. It has been shown that A-T cells are defective in regulating ionizing radiation induced G_1-S, S and G_2-M cell-cycle checkpoints [28,99,101].

In vivo, ATM molecules exist as inactive dimers or higher-order multimers that dissociate into the active monomers in response to DNA damage [102]. As discussed above, activation it thought to be mediated by the MRN complex [73-77]. However, alternative evidence indicates that ATM may be activated by intermolecular autophosphorylation of serine 1981 by a mechanism that invokes a DSB-induced change in chromatin structure [102]. Evidence for this mechanism will be discussed in more detail later in this chapter.

Nevertheless, upon activation ATM numerous substrates which are involved in ionizing radiation-induced cell-cycle arrest. For example, activated ATM phosphorylates Mdm2 and Chk2 in the G1 checkpoint [103-105]; Nbs1, Brca1, FancD2 and SMC1 in the ionizing radiation-induced S-phase arrest [68,69,106-110]; and Brca1 and Rad17 in the G2/M checkpoint [111,112]. Additional functionally important ATM substrates include, 53BP1 [113,114], RPA [115-117], TopBP1 [118], LKB1 [101] and MDC1/NFBD1 [119-122], all of which been extensively reviewed in the context of cellular responses to DSBs. A relatively recent addition to the list of ATM substrates is BID – a BCL-2 family member [123,124]. It is well known that the BID protein is involved in regulating apoptosis following the activation of cell-surface death receptors [125]. The recent data raise the possibility that BID participates in the repair of DNA as well as in the regulation of cell cycle progression and apoptosis, following ATM-dependent phosphorylation [124]. Other important ATM phosphorylation substrates are p53 [126,127] and H2AX [128,129] and the significance of these proteins in the DNA damage response pathway will be discussed later in this chapter.

It has been found that ATM substrates are phosphorylated in A-T cells, albeit with delayed kinetics [5,130]. Phosphorylation of ATM targets in A-T cells is largely mediated by another PI(3)K family member, ATR. The ATM and ATR kinases are functionally related and phosphorylate many common substrates which are involved in the DNA damage response [131,132]. Most evidence suggests that ATM and ATR function independently but in mainly overlapping pathways of DNA damage signalling, however, it is becoming apparent that they may function in a co-ordinated manner in certain cases [133,134]. Although they are related ATM and ATR do differ significantly and are generally considered to respond to distinct stimuli [135]. The ATM kinase is primarily activated by DSBs induced by ionizing radiation and radiomimetic drugs [135]. The ATR kinase, instead, responds to DNA damage induced by replicative stress and to other forms of DNA damage, predominantly to UV-induced aberrations [135-137]. Furthermore, as discussed above, ATM and ATR are recruited to DSBs by related mechanisms however, the proteins required are different [78]. The ATM kinase is recruited to DSBs by the MRN complex [73,76] whereas, ATR is recruited to RPA-coated single-stranded DNA by ATRIP [49,138].

In contrast to *ATM*, *ATR* is an essential gene for cell proliferation. Homozygous inactivation of *Atr* results in inviable mouse or chicken B lymphocytes and the $Atr^{-/-}$ genotype in mice causes early embryonic lethality [139,140]. This phenotype is due to extensive chromosomal fragmentation, apoptosis and loss of clonogenicity [139,140]. Furthermore, cells are highly sensitive even to partial ATR deficiency and this is highlighted by the phenotype of Seckel cells that express very low levels of ATR [141]. For example, dominant-negative expression of ATR in mammalian cells results in sensitivity to a variety of DNA damaging agents including topoisomerase I and II inhibitors, UV and ionizing radiation and to the DNA replication inhibitor, hydroxyurea [142-144]. These findings are consistent with a central role for ATR in DNA damage signalling and in recognition and repair of stalled DNA replication complexes. Homologs of ATR in *S. cerevisiae* (MEC1p) and *S. pombe* (RAD3) are also key components in DNA damage surveillance and signalling pathways [145,146].

A key downstream substrate of ATM and ATR is the p53 protein [147]. The p53 protein was discovered in 1979 [148-151] and since that time more than 35 000 related papers have been published. Interestingly, even though the p53 molecular pathways have undergone such intense scrutiny there are still certain ambiguities. Nevertheless, once it was established that the *TP53* gene is the most frequently altered gene in human cancers [152,153], it was recognized that p53 is a major component of the DNA damage response pathway [154,155]. Overall, the role of p53 is to preserve the integrity of the genome through an extensive and complicated molecular stress-signalling network which has been the subject of extensive recent review [156-160]. Broadly, p53 which is normally present in cells at low levels, is rapidly accumulated, stabilized and activated through a series of post-translational modifications in response to DNA damage [158]. Once activated, p53 may act either a sequence-specific transcription factor or may function by transcription-independent mechanisms, to modulate the expression of genes involved in cell cycle regulation and cell growth and survival [156,161]. Specifically, p53 may regulate the cell cycle (primarily through the G_1/S checkpoint but also modulates the G_2/M checkpoint [162]), senescence and apoptosis [156]. In addition, several lines of investigation indicate a direct role for p53 in DNA repair and recombination [159].

Double-Strand Break Repair Pathways

Double strand breaks in mammalian cells are repaired by one of two distinct and complementary pathways – homologous recombination (HR) and non-homologous end-joining (NHEJ) These pathways are essential for maintenance of genomic integrity and organismal survival [163,164]. Indeed, inactivation of either pathway leads to embryonic lethality [163]. HR relies on extensive sequence homology between the recombining ends and essentially involves copying the missing information from an undamaged homologous chromosome [165,166]. Thus, it is typically error-free and occurs without the loss of genetic information. In contrast to HR, NHEJ does not require an undamaged DNA molecule and does not rely on extensive sequence homology [6]. Instead, NHEJ involves processing of the broken DNA terminii to make them compatible, followed by a ligation step. In most cases, this pathway results in the loss of a few nucleotides at each broken end [6]. Hence, DSB repair by NHEJ is typically error-prone and is not a perfect process with respect to preserving genomic information. Similarly, a variation of HR known as single-strand annealing (SSA) leads to loss of genetic material [167]. SSA occurs when there are direct repeat sequences on each side of the damaged DNA ends. Following specific resection of both 5'-ends, the complementary sequences are annealed leading to the loss of a repeat sequence and the intervening DNA [5,167].

Overall, the two major DSB repair pathways function in a coordinated but nonoverlapping manner to maintain genomic integrity and suppress genotoxic stress [168-170]. Much research has focussed at understanding how cells 'choose' which particular repair pathway to use however, the decision-making process is still not yet understood. Nevertheless, it is known that the HR pathway is the one predominantly used by simple eukaryotes such as the yeasts *S. cerevisiae* and *S. pombe*, to repair DSBs [163,171]. NHEJ is the more common DSB pathway in higher eukaryotes, and predominates in most stages of the cell-cycle, particularly in the G_1-phase [5,28,172,173]. Furthermore, the NHEJ pathway is essential for repair of DSBs that occur during V(D)J recombination and Ig class-switch recombination in lymphocytes [6]. However, HR is also important for DSB repair in multicellular eukaryotes, particularly during the late S- and G_2-phases of the cell-cycle because of the availability of the sister chromatid [163,174]. Also, HR has a key role in repairing DSBs that occur at stalled replication forks [164]. Furthermore, HR is the predominant pathway in early embryonic development presumably because the main cellular process in this stage is proliferation [163]. Also, HR would be particularly beneficial because the consequences of error-prone repair on genomic fidelity would be too great on the rapidly dividing cells at this stage [163].

Non-Homologous End-Joining Repair Pathway

The core proteins involved in the NHEJ pathway are Ku70, Ku80 (Ku 80 is also called Ku86), the DNA dependent protein catalytic subunit (DNA-PKcs), XRCC4 and DNA ligase IV [6]. It is proposed that the Ku70/Ku86 heterodimer is the first protein to bind to broken DNA terminii at the site of DSBs [175-177]. Ku is abundant in cells and binds to DNA ends with high affinity in a non-sequence dependent manner [175-177]. Binding of Ku to DNA recruits another PI(3)K family member, DNA-PKcs [178-181]. It is noteworthy that in the absence of Ku, DNA-PKcs binds to DNA terminii, albeit with about 100-fold lower affinity [180]. The binding of DNA-PKcs to DNA is a requirement for activation of its serine/threonine kinase activity [181,182]. Some studies suggest that phosphorylation of downstream target substrates by DNA-PKcs is an important component of DNA damage signalling pathway however, the precise contribution of DNA-PKcs is unclear [180,183]. Interestingly, autophosphorylation of DNA-PKcs results in dissociation of the DNA-Ku-DNA-PKcs complex [183]. Following assembly of the DNA-Ku-DNA-PKcs complex and DNA end processing, the DNA terminii are ligated by the XRCC4-DNA-ligase-IV complex [184-186]. The finding that the XRCC4-DNA-ligase-IV complex does not form a stable complex with DNA in the absence of the Ku-complex, has prompted the suggestion that Ku may recruit and load the ligase complex at the sites of DSBs [187-189].

The relatively more recent identification of additional proteins involved in the NHEJ repair pathway has allowed unique connections between mythology and science to be made. The first of these 'mythological' proteins to be identified was Artemis which is involved in DNA end processing in NHEJ [190-193]. The role of the Hellenic Goddess Artemis in ancient mythology - as a guardian of women in childbirth (she was also guardian of wild animals and goddess of hunting) - has been related to the key role of the protein in the development of immune responses and in the DNA damage response pathway [191]. It has been determined that Artemis has specific nuclease activity and that the protein is phosphorylated by and forms a complex with DNA-PKcs [190]. In short, the Artemis-DNA-PKcs complex appears to be exclusively responsible for cleaving DNA hairpins in V(D)J recombination and is a critical component in resecting DNA terminii (removal of 5' and trimming of long 3' overhangs) in NHEJ [190]. The other 'mythological' protein that has been identified very recently is known as Cernunnos-XLF – Cernunnos is the Latin name given to the Celtic god of fertility, animals, wealth and the underworld [194,195]. The protein has been found to be defective in a rare inherited human syndrome characterized by severe immunodeficiency, developmental delay and hypersensitivity to agents that cause DSBs [194,195]. The precise function of Cernunnos-XLF (whose orthologue in other eukaryotes has not yet been identified) in the NHEJ repair pathway is not fully understood however, it has been suggested that it interacts with XRCC4-DNA ligase IV [194,195].

Double-Strand Break Repair by Homologous Recombination

Sequentially, the HR repair process involves the generation of a single-stranded region of DNA, strand invasion, formation of a Holliday junction, DNA synthesis, branch migration and resolution [164]. Initially the repair of DSBs by HR was characterized in bacteria and yeast. Genetic analysis of *S. cerevisiae* indicated that proteins coded by the RAD52 epistasis group of genes – Rad50, Rad51, Rad52, Rad54, Rad55, Rad 57, Rad59, Mre11 and Xrs2 – are important in HR [5,28,196]. In *S. cerevisiae*, an early event in HR is nucleolytic resection of the broken ends in the 5' to 3' direction [197]. This process involves the Rad50-Mre11-Xrs2 complex. The resulting 3' single-stranded DNA overhangs are then bound by Rad51 [197]. This process is facilitated by replication protein A (RPA), Rad54 and Rad52 which is known to bind to DNA termini [198,199]. Parenthetically, it has been suggested that in mammals Rad52 competes with Ku for binding to broken DNA ends [198,199]. This has lead to the proposal that competition between Rad52 and Ku for binding DNA terminii may be the critical event in determining whether NHEJ or HR is used to repair the DSB [198,199]. Nevertheless, following binding of Rad51 to the 3' ends, the RPA coated Rad51 nucleoprotein filament mediates a search for homologous sequences on an undamaged DNA duplex [197]. Once a homologous region is located, Rad51 initiates strand-exchange in which one of the damaged DNA strands invades the homologous duplex forming a D-loop [165]. This process is facilitated by Rad55 and Rad57 in yeast [200]. The 3' terminii then prime new DNA synthesis resulting in the extension of the damaged DNA by polymerases that copy information from the homologous undamaged partner [197]. In meiotic cells, the DNA ends are ligated by DNA ligase I and the Holliday junctions are resolved by nucleolytic cleavage [197]. This process can result in either crossover or non-crossover gene conversion products depending on the mode by which the Holliday junctions are resolved [165,197]. It has been suggested that mitotic recombination may not involve resolution of Holliday junctions rather it may coupled with replication [172].

Studies have identified that mammalian homologs exist for all of the known gene products that are in involved in HR in *S. cerevisiae* indicating that the basic HR pathway is conserved in higher eukaryotes [201-204]. However, it is accepted that the molecular mechanisms are more complex in higher organisms. For example, a number of RAD51 paralogs such as RAD51B, RAD51C and RAD51D and other proteins with some homology to the catalytic domain of RAD51, such as XRCC2 and XRCC3, have been identified in human cells [205]. It has been proposed that these facilitate the function of RAD51 in a similar manner to Rad55 and Rad57 in *S. cerevisiae* [197]. Furthermore, mammalian cells posses additional proteins that are involved in HR that lack direct homologs in yeast. Two important proteins in this regard appear to be the breast cancer susceptibility proteins, BRAC1 and BRCA2 [206].

It is clear that BRCA1 and BRCA2 are essential for efficient HR in mammalian cells. BRCA1 interacts with a wide range of cellular proteins and has multiple functions in DNA metabolism [206]. Therefore, dissecting its role in HR has proved to be a difficult process. To date, the finding that following ATM phosphorylation, BRCA1 and c-ABL phosphorylate RAD51 on several sites is probably the strongest evidence in support of a role for BRCA1 in HR [110,207,208]. Overall, it is believed that BRCA1 has a regulatory role in HR. In contrast a direct role for BRCA2 in HR has been established. BRAC2 interacts with RAD51 (RAD51 binds to a series of BRC motifs on BRCA2) and regulates the binding of RAD51 to DNA as well as the ability of the protein to form nucleoprotein filaments on DNA [209,210]. Notably, the findings that BRCA2 interacts with histone acetylases and that BRCA1 is associated with the SWI/SNF chromosome-remodelling complex and interacts with histone acetylases, suggest that a role of these proteins in HR may involve modulation of chromatin structure [211,212].

Chromatin Structure, Modification and Remodelling

So far in this chapter we have considered the molecular pathways involved in DSB detection, signalling and repair however, for a more realistic perspective we need to invoke an additional level of complexity and to discuss DSB responses in the context of chromatin. Until recently, the functional interplay between DSB signalling and repair pathways and chromatin were poorly understood, however, a multitude of reports have now made a link between epigenetic events and molecular responses to DSBS. Indeed, has been proposed that a specific histone code may exist at DNA breaks which guides the DNA damage response [213,214].

Before we discuss what is known about the link between epigenetic events and DSB response pathways, it is prudent to briefly consider chromatin organization and the dynamic properties of histones. In human cells, about two metres of diploid DNA compacted into a volume of approximately 1000 μm^3 within the nucleus [215,216]. The DNA is compacted and packaged in a highly organized manner with the highly basic core histones (H2A, H2B, H3 and H4), the linker histone (H1) and other accessory proteins into the nucleoprotein structure known as chromatin [215,216]. The basic repeating unit of chromatin, the nucleosome, consists of 146 base pairs of DNA wrapped around octamers of the core histones [217,218]. Briefly, two histone H3 and H4 heterodimers associate to form a tetrameric $(H3\text{-}H4)_2$ molecule [217,218]. The DNA wraps around this molecule and histone H2A and H2B heterodimers associate on either side of the tetramer forming a nucleosome [217,218]. Strings of thousands of nucleosomes are organized onto a continuous DNA helix (typically separated by 10-60 base pairs of linker DNA) [219]. Nucleosomes are further compacted into 30 nm chromatin fibers via linker (H1) histones and looped domains. These are assembled into so-called higher order chromatin structures that are not well defined [219].

The compaction of the genome in chromatin has important implications in DNA-metabolism events including transcription, replication, recombination and DNA repair, which require access to chromosomal DNA [220-222]. Nucleosomes have been traditionally known simply as stable DNA packaging units however, it is now evident that they are dynamic structures that can be altered by at least three different molecular processes [223]. Firstly, the composition of nucleosomes may be modified by the incorporation of histone variants which are thought to have specialized functions [215]. For example, the histone H2A variants H2AZ and H2AX are constitutively expressed and have been shown to have specific functions predominantly related to transcriptional regulation and DNA damage repair, respectively [215]. Similarly, there are numerous sequence variants of histone H1 which differ in abundance in different cell types and in different phases of the cell cycle, differentiation and development [215]. Additionally, the organization of chromatin can be altered by the replacement, repositioning or in certain cases the ejection of nucleosomes by chromatin remodelling complexes [223]. In short, there are at least five families of chromatin remodellers in eukaryotes (the SWI/SNF, ISWI, NURD/Mi-2, INO80 and SWR1), all of which contain ATPase domains and in general, function with other chromatin modifying components to regulate access to DNA [223]. Finally, chromatin structure can be modified by co-ordinated post-translational modifications on histone residues such acetylation of lysines, serine phosphorylation, methylation of lysines and arginines, polyribosylation and ubiquitylation [224,225].

Integration of Epigenetic Events and Responses to Double-Strand Breaks

An early event in response to DSBs is the phosphorylation of the histone variant H2AX [128], which represents about 2-5% of total H2A in mammalian cells [215,226]. The serine-139 residue at the C-terminal domain of H2AX is rapidly phosphorylated (γH2AX) following DSB induction [128,227]. The serine-139 residue is part of a serine-glutamine-glutamate (SQE) consensus motif for phosphorylation by PI(3)Ks including ATM, ATR and DNA-PKcs [128]. Findings indicate that γH2AX extends over a megabase range of chromatin from the site of DSBs and is required for concentration of relevant proteins into irradiation-induced foci [228-231]. Notably, the findings from a recent study lead to the suggestion that γH2AX does not represent the initial signal for recruitment of DSB repair proteins to nuclear foci but is required for concentration of repair proteins in the vicinity of DSBs, possibly via interactions between the phosphorylated histone tail and specific domains of repair signalling factors [228]. Furthermore, it has been shown that γH2AX is important for the recruitment of histone modifiers and chromatin remodellers to the sites of DSBs [213]. Importantly, γH2AX foci have proven to be sensitive cytological markers of DSBs and have provided a new quantitative experimental target for the investigation of DSB induction and repair [228,231,232]. For example, γH2AX have been observed by immunofluorescence and chromatin immunoprecipitation assays (ChIP) in mammalian cells following induction of DSBs with therapeutically relevant doses of ionizing radiation [228,231-233].

Furthermore, histone acetylation has been implicated in DSB repair [234]. The acetylation status of histones is dependent upon the opposing actions of histone acetyltransferase (HAT) and histone deacetylase (HDAC) enzymes [235,236]. These enzymes alter chromatin structure by acetylating and deacetylating, respectively, the NH_2-terminal tails of lysine residues in core histones [237]. Acetylation of lysines by HATs results in neutralization of the positive charge of histones resulting in weakening of histone-DNA interactions [235-237]. Therefore, the net effect of highly acetylated histones is a more open (also referred to as the transcriptionally active) chromatin conformation. On the other hand, removal of acetyl groups from lysine residues by HDACs results in a more compacted (transcriptionally inactive or repressed) chromatin conformation [235-237]. With respect to DSB repair, genetic studies have indicated that acetylation of histone H3 by the type B histone acetyltransferase HAT1p is necessary for DSB repair by HR (but only has a modest effect on NHEJ) [238]. Also, it has been shown that acetylation of histone H4 by the HAT, ESA1, is required for DSB repair via NHEJ and another pathway known as transcription coupled repair [239]. It was shown that a yeast strain with mutated H4 acetylation sites was defective in DSB repair. Furthermore, the importance of ESA1 mediated acetylation was established with the use of esa1 mutants [239]. A subset of mutants carrying the esa-1851 allele were sensitive to the DNA-damaging agents camptothecin and methyl methane sulphonate, due to diminished acetylation of H4 [239]. In addition, a recent study has indicated that histone H4 acetylation by the HAT cofactor Trrap and Tip60 HAT modulates accumulation of repair proteins at the site of DSBs [240]. Intriguingly, the implication from these findings is that cells may use the same basic mechanism to regulate different cellular processes, such as transcription and repair [240].

Changes in chromatin conformation that are mediated by chromatin remodelling complexes have also been implicated in DSB response pathways [241]. The effect of chromatin remodelling complex-mediated changes in DNA topology, which is of importance in DNA processing events including DSB repair, is that it enhances the accessibility of proteins to the DNA within chromatin [223,241]. Specifically, INO80 [242,243], RSC (member of the SWI/SNF family) [244], and SWI/SNF [244] have been shown to be involved in DSB repair pathways and mutants of SWR1 have been shown to be sensitive to the effects of DNA damaging agents [245]. Notably, Rad54 which is essential in HR, is a member of the SWI2/SNF2 family of ATPases [246-248]. Recent studies have indicated that Rad54 has similar functions to the *bona fide* ATP-dependent chromatin remodelling enzymes and modulates chromatin conformation by sliding nucleosomes along DNA [246-248]. Interestingly, Rad54-mediated chromatin remodelling is facilitated by the Rad51 nucleoprotein filament [247]. These findings are consistent with a role for Rad54 in eliminating chromatin compaction-related constraints to DNA strand exchange during the Rad51-mediated homology search step, of HR [246-248]. It is also noteworthy that the findings that BRCA1 is associated with SWI/SNF remodelling complexes and interacts with histone acetylases, suggest a function of the protein in HR may involve modulation of chromatin structure [211,212].

Furthermore, links between changes in epigenetic conformation and the initial detection of DSBs have also been made. The findings from a landmark study provided evidence that in response to genotoxic stress, inactive homodimeric or multimeric ATM molecules are modified by autophosphorylation of the serine-1891 residue resulting in activation of the ATM protein kinase domain [102]. Furthermore, the rapid and widespread dissociation of inactive ATM complexes into the catalytically active monomers (a radiation dose of 0.5 Gy, which induces about 18 DSBs, resulted in the phosphorylation of approximately 50% of ATM molecules within 5 minutes), lead to the conclusion that changes in chromatin structure and/or histone modifications are detected by ATM [102]. As discussed earlier in this chapter, the recent findings implicating the MRN complex as a DSB sensor and as an essential mediator of ATM recruitment to the lesions [76], call some aspects of the ATM autophosphorylation pathway into question [59].

Recently, it has been shown that exposure of the methylated lysine 79 of histone H3 following DSB induction and changes in chromatin conformation is important for the initiation of checkpoint responses [249]. In this study, the structure of the tandem tudor domain of 53BP1 – a highly conserved protein which possesses the properties of a DSB sensor – was solved [249]. The findings indicated that 53BP1 binds to histone H3 methylated on lysine 79 [249]. Interestingly, lysine 79 of histone H3 is constitutively methylated [250,251] and irradiation does not lead to increased methylation levels [249]. Therefore, it was proposed that the induction of DSBs increases the accessibility of the methylated residue to 53BP1 probably through changes in chromatin conformation [249]. Importantly, careful analysis based on the structural features of chromatin [218,252] indicated that due to nucleosome stacking the lysine 79 residue of histone H3 would normally be inaccessible to 53BP1 [249].

The fact that epigenetic events are critical components of the cellular response to DNA damage is obvious (or at least not unexpected) given that detection, signalling and repair of DSBs, like other DNA metabolism events, has to occur in the context of genomic compaction. As discussed above in this chapter, to date, chromatin modifications such as phosphorylation of the histone variant H2AX, acetylation of histones H3 and H4, specific methylation of H3, and chromatin remodelling complex-mediated changes in the epigenome have important roles in cellular responses to DSBs. In general, it is proposed that histone acetylation and the actions of ATP-dependent chromatin remodelling complexes result in a change in chromatin conformation, allowing repair factors to access the damaged DNA within the chromatin. Changes in chromatin conformation, for example exposing the methylated lysine 79 of histone H3, may be significant for initial detection of the lesions and phosphorylation of H2AX is important for concentration of repair factors in DSB induced foci.

CONCLUSION

There continues to be rapid progress in our understanding of the molecular details of the cellular DSB response. In particular, significant recent advances have been made with respect to our knowledge related to the initial detection of DSBs. The Rad17/9-1-1 clamp complex, the ATR-ATRIP complex and the paradoxical nature of the MRN complex, are beginning to shed some light on how different types of DNA damage and aberrant DNA structures may lead to varying cellular response pathways. However, there are still gaps in our knowledge. An interesting issue that requires clarification is cellular fate (DNA repair and survival or death) following DNA damage. It will be important to understand the molecular processes that determine the fate of a particular cell following DNA damage. An important question is whether there is a critical threshold level of DNA damage (obviously at high numbers of DSBs, the damage is irreparable and cell-death is the consequence) or whether the specific type of DNA aberration is also important in determining whether a 'decision' for survival or death is made. Given its central role in virtually all aspects of the cellular DSB response, p53 (gene and protein) is a tempting candidate as a master-regulator, however, the important 'decisions' may be made earlier in the DSB signalling pathway, upstream of p53. Furthermore, it will be interesting to determine precisely how a particular DSB repair pathway is chosen and to understand the actual level of integration of the NHEJ and HR pathways. Another important area that has recently come into light is the integration of epigenetic events with other components of the DSB response pathway. This has added a significant level of complexity to an already intricate system. Numerous chromatin remodelling and histone modification processes have been implicated in the detection, signalling and repair of DSBs. The most useful histone modification, at least from a practical point of view, has proven to be the identification of phosphorylated γH2AX as a sensitive marker of DSBs. This marker has allowed rapid quantitative investigation of DSB induction and repair at relevant doses of ionizing radiation. Overall, it should be noted that although a potential histone code at the sites of DSBs has been suggested, we are still at early stages with respect to characterizing the relationship between epigenetic and DSB signalling and repair pathways. It is expected that the DSB cellular response pathways, particularly in the context of chromatin, will continue to be the focus of intense investigation.

ACKNOWLEDGEMENTS

The support of the Australian Institute of Nuclear Science and Engineering is acknowledged. TCK was the recipient of AINSE awards. Molecular Radiation Biology and Epigenetics in Human Health and Disease Laboratories are supported by the National Health and Medical Research Council of Australia (350359 and 268905).

REFERENCES

[1] Karagiannis TC, El-Osta A. Double-strand breaks: signalling pathways and repair mechanisms. *Cell Mol Life Sci* 2004;61(17):2137-47.

[2] Karagiannis TC, El-Osta A. Epigenetic changes activate widespread signals in response to double-strand breaks. *Cancer Biol Ther* 2004;3(7):617-23.

[3] Morgan WF, Corcoran J, Hartmann A, Kaplan MI, Limoli CL, Ponnaiya B. DNA double-strand breaks, chromosomal rearrangements, and genomic instability. *Mutat Res* 1998;404(1-2):125-8.

[4] Jeggo PA. DNA breakage and repair. *Adv Genet* 1998;38:185-218.

[5] Khanna KK, Jackson SP. DNA double-strand breaks: signalling, repair and the cancer connection. *Nat Genet* 2001;27(3):247-54.

[6] Lieber MR, Ma Y, Pannicke U, Schwarz K. Mechanism and regulation of human non-homologous DNA end-joining. *Nat Rev Mol Cell Biol* 2003;4(9):712-20.

[7] Boveris A, Sies H, Martino EE, Docampo R, Turrens JF, Stoppani AO. Deficient metabolic utilization of hydrogen peroxide in Trypanosoma cruzi. *Biochem J* 1980;188(3):643-8.

[8] Feinendegen LE. Reactive oxygen species in cell responses to toxic agents. *Hum Exp Toxicol* 2002;21(2):85-90.

[9] Pollycove M, Feinendegen LE. Radiation-induced versus endogenous DNA damage: possible effect of inducible protective responses in mitigating endogenous damage. *Hum Exp Toxicol* 2003;22(6):290-306; discussion 307, 315-7, 319-23.

[10] Schar P. Spontaneous DNA damage, genome instability, and cancer--when DNA replication escapes control. *Cell* 2001;104(3):329-32.

[11] Lewis SM, Agard E, Suh S, Czyzyk L. Cryptic signals and the fidelity of V(D)J joining. *Mol Cell Biol* 1997;17(6):3125-36.

[12] Marculescu R, Le T, Simon P, Jaeger U, Nadel B. V(D)J-mediated translocations in lymphoid neoplasms: a functional assessment of genomic instability by cryptic sites. *J Exp Med* 2002;195(1):85-98.

[13] Raghavan SC, Kirsch IR, Lieber MR. Analysis of the V(D)J recombination efficiency at lymphoid chromosomal translocation breakpoints. *J Biol Chem* 2001;276(31):29126-33.

[14] Korsmeyer SJ. Chromosomal translocations in lymphoid malignancies reveal novel proto-oncogenes. *Annu Rev Immunol* 1992;10:785-807.

[15] Rabbitts TH. Chromosomal translocations in human cancer. *Nature* 1994;372(6502).143-9.

[16] Vanasse GJ, Concannon P, Willerford DM. Regulated genomic instability and neoplasia in the lymphoid lineage. *Blood* 1999;94(12):3997-4010.

[17] McBlane JF, van Gent DC, Ramsden DA, Romeo C, Cuomo CA, Gellert M, et al. Cleavage at a V(D)J recombination signal requires only RAG1 and RAG2 proteins and occurs in two steps. *Cell* 1995;83(3):387-95.

[18] Petersen S, Casellas R, Reina-San-Martin B, Chen HT, Difilippantonio MJ, Wilson PC, et al. AID is required to initiate Nbs1/gamma-H2AX focus formation and mutations at sites of class switching. *Nature* 2001;414(6864):660-5.

[19] d'Adda di Fagagna F, Teo SH, Jackson SP. Functional links between telomeres and proteins of the DNA-damage response. *Genes Dev* 2004;18(15):1781-99.

[20] d'Adda di Fagagna F, Reaper PM, Clay-Farrace L, Fiegler H, Carr P, Von Zglinicki T, et al. A DNA damage checkpoint response in telomere-initiated senescence. *Nature* 2003;426(6963):194-8.

[21] Lee SE, Moore JK, Holmes A, Umezu K, Kolodner RD, Haber JE. Saccharomyces Ku70, mre11/rad50 and RPA proteins regulate adaptation to G2/M arrest after DNA damage. *Cell* 1998;94(3):399-409.

[22] Tuteja N, Tuteja R. Unraveling DNA repair in human: molecular mechanisms and consequences of repair defect. *Crit Rev Biochem Mol Biol* 2001;36(3):261-90.

[23] Hanawalt PC. Subpathways of nucleotide excision repair and their regulation. *Oncogene* 2002;21(58):8949-56.

[24] Ferguson DO, Alt FW. DNA double strand break repair and chromosomal translocation: lessons from animal models. *Oncogene* 2001;20(40):5572-9.

[25] Richardson C, Jasin M. Frequent chromosomal translocations induced by DNA double-strand breaks. *Nature* 2000;405(6787):697-700.

[26] Vamvakas S, Vock EH, Lutz WK. On the role of DNA double-strand breaks in toxicity and carcinogenesis. *Crit Rev Toxicol* 1997;27(2):155-74.

[27] Lengauer C, Kinzler KW, Vogelstein B. Genetic instabilities in human cancers. *Nature* 1998;396(6712):643-9.

[28] Jackson SP. Sensing and repairing DNA double-strand breaks. *Carcinogenesis* 2002;23(5):687-96.

[29] van Gent DC, Hoeijmakers JH, Kanaar R. Chromosomal stability and the DNA double-stranded break connection. *Nat Rev Genet* 2001;2(3):196-206.

[30] Fei P, El-Deiry WS. P53 and radiation responses. *Oncogene* 2003;22(37):5774-83.

[31] Rauen M, Burtelow MA, Dufault VM, Karnitz LM. The human checkpoint protein hRad17 interacts with the PCNA-like proteins hRad1, hHus1, and hRad9. *J Biol Chem* 2000;275(38):29767-71.

[32] Venclovas C, Thelen MP. Structure-based predictions of Rad1, Rad9, Hus1 and Rad17 participation in sliding clamp and clamp-loading complexes. *Nucleic Acids Res* 2000;28(13):2481-93.

[33] Kondo T, Matsumoto K, Sugimoto K. Role of a complex containing Rad17, Mec3, and Ddc1 in the yeast DNA damage checkpoint pathway. *Mol Cell Biol* 1999;19(2):1136-43.

[34] Paciotti V, Lucchini G, Plevani P, Longhese MP. Mec1p is essential for phosphorylation of the yeast DNA damage checkpoint protein Ddc1p, which physically interacts with Mec3p. *Embo J* 1998;17(14):4199-209.

[35] O'Connell MJ, Walworth NC, Carr AM. The G2-phase DNA-damage checkpoint. *Trends Cell Biol* 2000;10(7):296-303.

[36] Thelen MP, Venclovas C, Fidelis K. A sliding clamp model for the Rad1 family of cell cycle checkpoint proteins. *Cell* 1999;96(6):769-70.

[37] Mossi R, Hubscher U. Clamping down on clamps and clamp loaders--the eukaryotic replication factor C. *Eur J Biochem* 1998;254(2):209-16.

[38] Stillman B. Smart machines at the DNA replication fork. *Cell* 1994;78(5):725-8.

[39] Hang H, Lieberman HB. Physical interactions among human checkpoint control proteins HUS1p, RAD1p, and RAD9p, and implications for the regulation of cell cycle progression. *Genomics* 2000;65(1):24-33.

[40] St Onge RP, Udell CM, Casselman R, Davey S. The human G2 checkpoint control protein hRAD9 is a nuclear phosphoprotein that forms complexes with hRAD1 and hHUS1. *Mol Biol Cell* 1999;10(6):1985-95.

[41] Volkmer E, Karnitz LM. Human homologs of Schizosaccharomyces pombe rad1, hus1, and rad9 form a DNA damage-responsive protein complex. *J Biol Chem* 1999;274(2):567-70.

[42] Burtelow MA, Kaufmann SH, Karnitz LM. Retention of the human Rad9 checkpoint complex in extraction-resistant nuclear complexes after DNA damage. *J Biol Chem* 2000;275(34):26343-8.

[43] Bluyssen HA, Naus NC, van Os RI, Jaspers I, Hoeijmakers JH, de Klein A. Human and mouse homologs of the Schizosaccharomyces pombe rad17+ cell cycle checkpoint control gene. *Genomics* 1999;55(2):219-28.

[44] Griffiths DJ, Barbet NC, McCready S, Lehmann AR, Carr AM. Fission yeast rad17: a homologue of budding yeast RAD24 that shares regions of sequence similarity with DNA polymerase accessory proteins. *Embo J* 1995;14(23):5812-23.

[45] Li L, Peterson CA, Kanter-Smoler G, Wei YF, Ramagli LS, Sunnerhagen P, et al. hRAD17, a structural homolog of the Schizosaccharomyces pombe RAD17 cell cycle checkpoint gene, stimulates p53 accumulation. *Oncogene* 1999;18(9):1689-99.

[46] Bao S, Chang MS, Auclair D, Sun Y, Wang Y, Wong WK, et al. HRad17, a human homologue of the Schizosaccharomyces pombe checkpoint gene rad17, is overexpressed in colon carcinoma. *Cancer Res* 1999;59(9):2023-8.

[47] Parker AE, Van de Weyer I, Laus MC, Verhasselt P, Luyten WH. Identification of a human homologue of the Schizosaccharomyces pombe rad17+ checkpoint gene. *J Biol Chem* 1998;273(29):18340-6.

[48] Toueille M, El-Andaloussi N, Frouin I, Freire R, Funk D, Shevelev I, et al. The human Rad9/Rad1/Hus1 damage sensor clamp interacts with DNA polymerase beta and increases its DNA substrate utilisation efficiency: implications for DNA repair. *Nucleic Acids Res* 2004;32(11):3316-24.

[49] Cortez D, Guntuku S, Qin J, Elledge SJ. ATR and ATRIP: partners in checkpoint signaling. *Science* 2001;294(5547):1713-6.

[50] Edwards RJ, Bentley NJ, Carr AM. A Rad3-Rad26 complex responds to DNA damage independently of other checkpoint proteins. *Nat Cell Biol* 1999;1(7):393-8.

[51] Paciotti V, Clerici M, Lucchini G, Longhese MP. The checkpoint protein Ddc2, functionally related to S. pombe Rad26, interacts with Mec1 and is regulated by Mec1-dependent phosphorylation in budding yeast. *Genes Dev* 2000;14(16):2046-59.

[52] Rouse J, Jackson SP. LCD1: an essential gene involved in checkpoint control and regulation of the MEC1 signalling pathway in Saccharomyces cerevisiae. *Embo J* 2000;19(21):5801-12.

[53] Wakayama T, Kondo T, Ando S, Matsumoto K, Sugimoto K. Pie1, a protein interacting with Mec1, controls cell growth and checkpoint responses in Saccharomyces cerevisiae. *Mol Cell Biol* 2001;21(3):755-64.

[54] Zou L, Cortez D, Elledge SJ. Regulation of ATR substrate selection by Rad17-dependent loading of Rad9 complexes onto chromatin. *Genes Dev* 2002;16(2):198-208.

[55] Unsal-Kacmaz K, Makhov AM, Griffith JD, Sancar A. Preferential binding of ATR protein to UV-damaged DNA. *Proc Natl Acad Sci U S* A 2002;99(10):6673-8.

[56] D'Amours D, Jackson SP. The Mre11 complex: at the crossroads of dna repair and checkpoint signalling. *Nat Rev Mol Cell Biol* 2002;3(5):317-27.

[57] Petrini JH. The Mre11 complex and ATM: collaborating to navigate S phase. *Curr Opin Cell Biol* 2000;12(3):293-6.

[58] Lavin MF. The Mre11 complex and ATM: a two-way functional interaction in recognising and signaling DNA double strand breaks. *DNA Repair (Amst)* 2004;3(11):1515-20.

[59] Abraham RT, Tibbetts RS. Cell biology. Guiding ATM to broken DNA. *Science* 2005;308(5721):510-1.

[60] Desai-Mehta A, Cerosaletti KM, Concannon P. Distinct functional domains of nibrin mediate Mre11 binding, focus formation, and nuclear localization. *Mol Cell Biol* 2001;21(6):2184-91.

[61] Kobayashi J, Tauchi H, Sakamoto S, Nakamura A, Morishima K, Matsuura S, et al. NBS1 localizes to gamma-H2AX foci through interaction with the FHA/BRCT domain. *Curr Biol* 2002;12(21):1846-51.

[62] Tauchi H, Matsuura S, Kobayashi J, Sakamoto S, Komatsu K. Nijmegen breakage syndrome gene, NBS1, and molecular links to factors for genome stability. *Oncogene* 2002;21(58):8967-80.

[63] Lee JH, Ghirlando R, Bhaskara V, Hoffmeyer MR, Gu J, Paull TT. Regulation of Mre11/Rad50 by Nbs1: effects on nucleotide-dependent DNA binding and association with ataxia-telangiectasia-like disorder mutant complexes. *J Biol Chem* 2003;278(46):45171-81.

[64] Paull TT, Gellert M. Nbs1 potentiates ATP-driven DNA unwinding and endonuclease cleavage by the Mre11/Rad50 complex. *Genes Dev* 1999;13(10):1276-88.

[65] Stewart GS, Maser RS, Stankovic T, Bressan DA, Kaplan MI, Jaspers NG, et al. The DNA double-strand break repair gene hMRE11 is mutated in individuals with an ataxia-telangiectasia-like disorder. *Cell* 1999;99(6):577-87.

[66] Varon R, Vissinga C, Platzer M, Cerosaletti KM, Chrzanowska KH, Saar K, et al. Nibrin, a novel DNA double-strand break repair protein, is mutated in Nijmegen breakage syndrome. *Cell* 1998;93(3):467-76.

[67] Gatei M, Young D, Cerosaletti KM, Desai-Mehta A, Spring K, Kozlov S, et al. ATM-dependent phosphorylation of nibrin in response to radiation exposure. *Nat Genet* 2000;25(1):115-9.

[68] Lim DS, Kim ST, Xu B, Maser RS, Lin J, Petrini JH, et al. ATM phosphorylates p95/nbs1 in an S-phase checkpoint pathway. *Nature* 2000;404(6778):613 7.

[69] Wu X, Ranganathan V, Weisman DS, Heine WF, Ciccone DN, O'Neill TB, et al. ATM phosphorylation of Nijmegen breakage syndrome protein is required in a DNA damage response. *Nature* 2000;405(6785):477-82.

[70] Zhao S, Weng YC, Yuan SS, Lin YT, Hsu HC, Lin SC, et al. Functional link between ataxia-telangiectasia and Nijmegen breakage syndrome gene products. *Nature* 2000;405(6785):473-7.

[71] Yuan SS, Chang HL, Hou MF, Chan TF, Kao YH, Wu YC, et al. Neocarzinostatin induces Mre11 phosphorylation and focus formation through an ATM- and NBS1-dependent mechanism. *Toxicology* 2002;177(2-3):123-30.

[72] Yuan SS, Su JH, Hou MF, Yang FW, Zhao S, Lee EY. Arsenic-induced Mre11 phosphorylation is cell cycle-dependent and defective in NBS cells. *DNA Repair (Amst)* 2002;1(2):137-42.

[73] Cerosaletti K, Wright J, Concannon P. Active role for nibrin in the kinetics of atm activation. *Mol Cell Biol* 2006;26(5):1691-9.

[74] You Z, Chahwan C, Bailis J, Hunter T, Russell P. ATM activation and its recruitment to damaged DNA require binding to the C terminus of Nbs1. *Mol Cell Biol* 2005;25(13):5363-79.

[75] Uziel T, Lerenthal Y, Moyal L, Andegeko Y, Mittelman L, Shiloh Y. Requirement of the MRN complex for ATM activation by DNA damage. *Embo J* 2003;22(20):5612-21.

[76] Lee JH, Paull TT. ATM activation by DNA double-strand breaks through the Mre11-Rad50-Nbs1 complex. *Science* 2005;308(5721):551-4.

[77] Lee JH, Paull TT. Direct activation of the ATM protein kinase by the Mre11/Rad50/Nbs1 complex. *Science* 2004;304(5667):93-6.

[78] Falck J, Coates J, Jackson SP. Conserved modes of recruitment of ATM, ATR and DNA-PKcs to sites of DNA damage. *Nature* 2005;434(7033):605-11.

[79] Nakada D, Matsumoto K, Sugimoto K. ATM-related Tel1 associates with double-strand breaks through an Xrs2-dependent mechanism. *Genes Dev* 2003;17(16):1957-62.

[80] Savitsky K, Bar-Shira A, Gilad S, Rotman G, Ziv Y, Vanagaite L, et al. A single ataxia telangiectasia gene with a product similar to PI-3 kinase. *Science* 1995;268(5218):1749-53.

[81] Khanna KK, Lavin MF, Jackson SP, Mulhern TD. ATM, a central controller of cellular responses to DNA damage. *Cell Death Differ* 2001;8(11):1052-65.

[82] Zakian VA. ATM-related genes: what do they tell us about functions of the human gene? *Cell* 1995;82(5):685-7.

[83] Lavin MF, Khanna KK, Beamish H, Spring K, Watters D, Shiloh Y. Relationship of the ataxia-telangiectasia protein ATM to phosphoinositide 3-kinase. *Trends Biochem Sci* 1995;20(10):382-3.

[84] Kim ST, Lim DS, Canman CE, Kastan MB. Substrate specificities and identification of putative substrates of ATM kinase family members. *J Biol Chem* 1999;274(53):37538-43.

[85] Rotman G, Shiloh Y. ATM: a mediator of multiple responses to genotoxic stress. *Oncogene* 1999;18(45):6135-44.

[86] Hoekstra MF. Responses to DNA damage and regulation of cell cycle checkpoints by the ATM protein kinase family. *Curr Opin Genet Dev* 1997;7(2):170-5.

[87] Bosotti R, Isacchi A, Sonnhammer EL. FAT: a novel domain in PIK-related kinases. *Trends Biochem Sci* 2000;25(5):225-7.

[88] Lavin MF, Shiloh Y. The genetic defect in ataxia-telangiectasia. *Annu Rev Immunol* 1997;15:177-202.

[89] Bentzen SM, Overgaard M. Relationship between early and late normal-tissue injury after postmastectomy radiotherapy. *Radiother Oncol* 1991;20(3):159-65.

[90] Stobo JD, Tomasi TB. Editorial: Aging and the regulation of immune reactivity. *J Chronic Dis* 1975;28(9):437-40.

[91] Hsieh CL, Arlett CF, Lieber MR. V(D)J recombination in ataxia telangiectasia, Bloom's syndrome, and a DNA ligase I-associated immunodeficiency disorder. *J Biol Chem* 1993;268(27):20105-9.

[92] Peterson RD, Kelly WD, Good RA. Ataxia-Telangiectasia. Its Association With A Defective Thymus, Immunological-Deficiency Disease, And Malignancy. *Lancet* 1964;13:1189-93.

[93] Hecht F, Hecht BK. Cancer in ataxia-telangiectasia patients. *Cancer Genet Cytogenet* 1990;46(1):9-19.

[94] Barlow C, Hirotsune S, Paylor R, Liyanage M, Eckhaus M, Collins F, et al. Atm-deficient mice: a paradigm of ataxia telangiectasia. *Cell* 1996;86(1):159-71.

[95] Xu Y, Ashley T, Brainerd EE, Bronson RT, Meyn MS, Baltimore D. Targeted disruption of ATM leads to growth retardation, chromosomal fragmentation during meiosis, immune defects, and thymic lymphoma. *Genes Dev* 1996;10(19):2411-22.

[96] Gotoff SP, Amirmokri E, Liebner EJ. Ataxia telangiectasia. Neoplasia, untoward response to x-irradiation, and tuberous sclerosis. *Am J Dis Child* 1967;114(6):617-25.

[97] Morgan JL, Holcomb TM, Morrissey RW. Radiation reaction in ataxia telangiectasia. *Am J Dis Child* 1968;116(5):557-8.

[98] Painter RB, Young BR. Radiosensitivity in ataxia-telangiectasia: a new explanation. *Proc Natl Acad Sci U S A* 1980;77(12):7315-7.

[99] Kapp LN, Painter RB. Stable radioresistance in ataxia-telangiectasia cells containing DNA from normal human cells. *Int J Radiat Biol* 1989;56(5):667-75.

[100] Painter RB. Radioresistant DNA synthesis: an intrinsic feature of ataxia telangiectasia. *Mutat Res* 1981;84(1):183-90.

[101] Sapkota GP, Deak M, Kieloch A, Morrice N, Goodarzi AA, Smythe C, et al. Ionizing radiation induces ataxia telangiectasia mutated kinase (ATM)-mediated phosphorylation of LKB1/STK11 at Thr-366. *Biochem J* 2002;368(Pt 2):507-16.

[102] Bakkenist CJ, Kastan MB. DNA damage activates ATM through intermolecular autophosphorylation and dimer dissociation. *Nature* 2003;421(6922):499-506.

[103] Maya R, Balass M, Kim ST, Shkedy D, Leal JF, Shifman O, et al. ATM-dependent phosphorylation of Mdm2 on serine 395: role in p53 activation by DNA damage. *Genes Dev* 2001;15(9):1067-77.

[104] Matsuoka S, Rotman G, Ogawa A, Shiloh Y, Tamai K, Elledge SJ. Ataxia telangiectasia-mutated phosphorylates Chk2 in vivo and in vitro. *Proc Natl Acad Sci USA* 2000;97(19):10389-94.

[105] Chehab NH, Malikzay A, Appel M, Halazonetis TD. Chk2/hCds1 functions as a DNA damage checkpoint in G(1) by stabilizing p53. *Genes Dev* 2000;14(3):278-88.

[106] Kim ST, Xu B, Kastan MB. Involvement of the cohesin protein, Smc1, in Atm-dependent and independent responses to DNA damage. *Genes Dev* 2002;16(5):560-70.

[107] Taniguchi T, Garcia-Higuera I, Xu B, Andreassen PR, Gregory RC, Kim ST, et al. Convergence of the fanconi anemia and ataxia telangiectasia signaling pathways. *Cell* 2002;109(4):459-72.

[108] Zhou BB, Chaturvedi P, Spring K, Scott SP, Johanson RA, Mishra R, et al. Caffeine abolishes the mammalian G(2)/M DNA damage checkpoint by inhibiting ataxia-telangiectasia-mutated kinase activity. *J Biol Chem* 2000;275(14):10342-8.

[109] Yazdi PT, Wang Y, Zhao S, Patel N, Lee EY, Qin J. SMC1 is a downstream effector in the ATM/NBS1 branch of the human S-phase checkpoint. *Genes Dev* 2002;16(5):571-82.

[110] Xu B, O'Donnell AH, Kim ST, Kastan MB. Phosphorylation of serine 1387 in Brca1 is specifically required for the Atm-mediated S-phase checkpoint after ionizing irradiation. *Cancer Res* 2002;62(16):4588-91.

[111] Xu B, Kim S, Kastan MB. Involvement of Brca1 in S-phase and G(2)-phase checkpoints after ionizing irradiation. *Mol Cell Biol* 2001;21(10):3445-50.

[112] Bao S, Tibbetts RS, Brumbaugh KM, Fang Y, Richardson DA, Ali A, et al. ATR/ATM-mediated phosphorylation of human Rad17 is required for genotoxic stress responses. *Nature* 2001;411(6840):969-74.

[113] Rappold I, Iwabuchi K, Date T, Chen J. Tumor suppressor p53 binding protein 1 (53BP1) is involved in DNA damage-signaling pathways. *J Cell Biol* 2001;153(3):613-20.

[114] Anderson L, Henderson C, Adachi Y. Phosphorylation and rapid relocalization of 53BP1 to nuclear foci upon DNA damage. *Mol Cell Biol* 2001;21(5):1719-29.

[115] Gately DP, Hittle JC, Chan GK, Yen TJ. Characterization of ATM expression, localization, and associated DNA-dependent protein kinase activity. *Mol Biol Cell* 1998;9(9):2361-74.

[116] Chan DW, Son SC, Block W, Ye R, Khanna KK, Wold MS, et al. Purification and characterization of ATM from human placenta. A manganese-dependent, wortmannin-sensitive serine/threonine protein kinase. *J Biol Chem* 2000;275(11):7803-10.

[117] Wang H, Guan J, Wang H, Perrault AR, Wang Y, Iliakis G. Replication protein A2 phosphorylation after DNA damage by the coordinated action of ataxia telangiectasia-mutated and DNA-dependent protein kinase. *Cancer Res* 2001;61(23):8554-63.

[118] Yamane K, Wu X, Chen J. A DNA damage-regulated BRCT-containing protein, TopBP1, is required for cell survival. *Mol Cell Biol* 2002;22(2):555-66.

[119] Stewart GS, Wang B, Bignell CR, Taylor AM, Elledge SJ. MDC1 is a mediator of the mammalian DNA damage checkpoint. *Nature* 2003;421(6926):961-6.

[120] Shang YL, Bodero AJ, Chen PL. NFBD1, a novel nuclear protein with signature motifs of FHA and BRCT, and an internal 41-amino acid repeat sequence, is an early participant in DNA damage response. *J Biol Chem* 2003;278(8):6323-9.

[121] Lou Z, Minter-Dykhouse K, Wu X, Chen J. MDC1 is coupled to activated CHK2 in mammalian DNA damage response pathways. *Nature* 2003;421(6926):957-61.

[122] Goldberg M, Stucki M, Falck J, D'Amours D, Rahman D, Pappin D, et al. MDC1 is required for the intra-S-phase DNA damage checkpoint. *Nature* 2003;421(6926):952-6.

[123] Kamer I, Sarig R, Zaltsman Y, Niv H, Oberkovitz G, Regev L, et al. Proapoptotic BID is an ATM effector in the DNA-damage response. *Cell* 2005;122(4):593-603.

[124] Zinkel SS, Hurov KE, Ong C, Abtahi FM, Gross A, Korsmeyer SJ. A role for proapoptotic BID in the DNA-damage response. *Cell* 2005;122(4):579-91.
[125] Zha J, Weiler S, Oh KJ, Wei MC, Korsmeyer SJ. Posttranslational N-myristoylation of BID as a molecular switch for targeting mitochondria and apoptosis. *Science* 2000;290(5497):1761-5.
[126] Banin S, Moyal L, Shieh S, Taya Y, Anderson CW, Chessa L, et al. Enhanced phosphorylation of p53 by ATM in response to DNA damage. *Science* 1998;281(5383):1674-7.
[127] Canman CE, Lim DS, Cimprich KA, Taya Y, Tamai K, Sakaguchi K, et al. Activation of the ATM kinase by ionizing radiation and phosphorylation of p53. *Science* 1998;281(5383):1677-9.
[128] Rogakou EP, Pilch DR, Orr AH, Ivanova VS, Bonner WM. DNA double-stranded breaks induce histone H2AX phosphorylation on serine 139. *J Biol Chem* 1998;273(10):5858-68.
[129] Burma S, Chen BP, Murphy M, Kurimasa A, Chen DJ. ATM phosphorylates histone H2AX in response to DNA double-strand breaks. *J Biol Chem* 2001;276(45):42462-7.
[130] Kastan MB, Lim DS, Kim ST, Xu B, Canman C. Multiple signaling pathways involving ATM. *Cold Spring Harb Symp Quant Biol* 2000;65:521-6.
[131] Bakkenist CJ, Kastan MB. Initiating cellular stress responses. *Cell* 2004;118(1):9-17.
[132] Shiloh Y. ATM and related protein kinases: safeguarding genome integrity. *Nat Rev Cancer* 2003;3(3):155-68.
[133] Jazayeri A, Falck J, Lukas C, Bartek J, Smith GC, Lukas J, et al. ATM- and cell cycle-dependent regulation of ATR in response to DNA double-strand breaks. *Nat Cell Biol* 2006;8(1):37-45.
[134] Cuadrado M, Martinez-Pastor B, Murga M, Toledo LI, Gutierrez-Martinez P, Lopez E, et al. ATM regulates ATR chromatin loading in response to DNA double-strand breaks. *J Exp Med* 2006;203(2):297-303.
[135] Abraham RT. Cell cycle checkpoint signaling through the ATM and ATR kinases. *Genes Dev* 2001;15(17):2177-96.
[136] McBean AM, Thoms ML, Albrecht P, Cuthie JC, Bernier R. Serologic response to oral polio vaccine and enhanced-potency inactivated polio vaccines. *Am J Epidemiol* 1988;128(3):615-28.
[137] Rouse J, Jackson SP. Interfaces between the detection, signaling, and repair of DNA damage. *Science* 2002;297(5581):547-51.
[138] Zou L, Elledge SJ. Sensing DNA damage through ATRIP recognition of RPA-ssDNA complexes. *Science* 2003;300(5625):1542-8.
[139] Brown EJ, Baltimore D. ATR disruption leads to chromosomal fragmentation and early embryonic lethality. *Genes Dev* 2000;14(4):397-402.
[140] de Klein A, Muijtjens M, van Os R, Verhoeven Y, Smit B, Carr AM, et al. Targeted disruption of the cell-cycle checkpoint gene ATR leads to early embryonic lethality in mice. *Curr Biol* 2000;10(8):479-82.
[141] O'Driscoll M, Ruiz-Perez VL, Woods CG, Jeggo PA, Goodship JA. A splicing mutation affecting expression of ataxia-telangiectasia and Rad3-related protein (ATR) results in Seckel syndrome. *Nat Genet* 2003;33(4):497-501.

[142] Wright JA, Keegan KS, Herendeen DR, Bentley NJ, Carr AM, Hoekstra MF, et al. Protein kinase mutants of human ATR increase sensitivity to UV and ionizing radiation and abrogate cell cycle checkpoint control. *Proc Natl Acad Sci U S A* 1998;95(13):7445-50.

[143] Cliby WA, Roberts CJ, Cimprich KA, Stringer CM, Lamb JR, Schreiber SL, et al. Overexpression of a kinase-inactive ATR protein causes sensitivity to DNA-damaging agents and defects in cell cycle checkpoints. *Embo J* 1998;17(1):159-69.

[144] Cliby WA, Lewis KA, Lilly KK, Kaufmann SH. S phase and G2 arrests induced by topoisomerase I poisons are dependent on ATR kinase function. *J Biol Chem* 2002;277(2):1599-606.

[145] Cha RS, Kleckner N. ATR homolog Mec1 promotes fork progression, thus averting breaks in replication slow zones. *Science* 2002;297(5581):602-6.

[146] Humphrey T. DNA damage and cell cycle control in Schizosaccharomyces pombe. *Mutat Res* 2000;451(1-2):211-26.

[147] Coutts AS, La Thangue N. The p53 response during DNA damage: impact of transcriptional cofactors. *Biochem Soc Symp* 2006(73):181-9.

[148] Lane DP, Crawford LV. T antigen is bound to a host protein in SV40-transformed cells. *Nature* 1979;278(5701):261-3.

[149] Linzer DI, Levine AJ. Characterization of a 54K dalton cellular SV40 tumor antigen present in SV40-transformed cells and uninfected embryonal carcinoma cells. *Cell* 1979;17(1):43-52.

[150] Kress M, May E, Cassingena R, May P. Simian virus 40-transformed cells express new species of proteins precipitable by anti-simian virus 40 tumor serum. *J Virol* 1979;31(2):472-83.

[151] DeLeo AB, Jay G, Appella E, Dubois GC, Law LW, Old LJ. Detection of a transformation-related antigen in chemically induced sarcomas and other transformed cells of the mouse. *Proc Natl Acad Sci U S* A 1979;76(5):2420-4.

[152] Caron de Fromentel C, Soussi T. TP53 tumor suppressor gene: a model for investigating human mutagenesis. *Genes Chromosomes Ca*ncer 1992;4(1):1-15.

[153] Hollstein M, Sidransky D, Vogelstein B, Harris CC. p53 mutations in human cancers. *Science* 1991;253(5015):49-53.

[154] Lane DP. Cancer. p53, guardian of the genome. *Nature* 1992;358(6381):15-6.

[155] Levine AJ. p53, the cellular gatekeeper for growth and division. *Cell* 1997;88(3):323-31.

[156] Lavin MF, Gueven N. The complexity of p53 stabilization and activation. *Cell Death Differ* 2006;13(6):941-50.

[157] Levine AJ, Hu W, Feng Z. The P53 pathway: what questions remain to be explored? *Cell Death Differ* 2006;13(6):1027-36.

[158] Braithwaite AW, Prives CL. p53: more research and more questions. *Cell Death Differ* 2006;13(6):877-80.

[159] Gatz SA, Wiesmuller L. p53 in recombination and repair. *Cell Death Differ* 2006;13(6):1003-16.

[160] Wiman KG. Strategies for therapeutic targeting of the p53 pathway in cancer. *Cell Death Differ* 2006;13(6):921-6.

[161] Steele RJ, Thompson AM, Hall PA, Lane DP. The p53 tumour suppressor gene. *Br J Surg* 1998;85(11):1460-7.

[162] Stark GR, Taylor WR. Analyzing the G2/M checkpoint. *Methods Mol Biol* 2004;280:51-82.

[163] Cahill D, Connor B, Carney JP. Mechanisms of eukaryotic DNA double strand break repair. *Front Biosci* 2006;11:1958-76.

[164] O'Driscoll M, Jeggo PA. The role of double-strand break repair - insights from human genetics. *Nat Rev Genet* 2006;7(1):45-54.

[165] van den Bosch M, Lohman PH, Pastink A. DNA double-strand break repair by homologous recombination. *Biol Chem* 2002;383(6):873-92.

[166] Cromie GA, Connelly JC, Leach DR. Recombination at double-strand breaks and DNA ends: conserved mechanisms from phage to humans. *Mol Cell* 2001;8(6):1163-74.

[167] Weinstock DM, Richardson CA, Elliott B, Jasin M. Modeling oncogenic translocations: Distinct roles for double-strand break repair pathways in translocation formation in mammalian cells. *DNA Repair (Amst)* 2006.

[168] Couedel C, Mills KD, Barchi M, Shen L, Olshen A, Johnson RD, et al. Collaboration of homologous recombination and nonhomologous end-joining factors for the survival and integrity of mice and cells. *Genes Dev* 2004;18(11):1293-304.

[169] Mills KD, Ferguson DO, Essers J, Eckersdorff M, Kanaar R, Alt FW. Rad54 and DNA Ligase IV cooperate to maintain mammalian chromatid stability. *Genes Dev* 2004;18(11):1283-92.

[170] Richardson C, Jasin M. Coupled homologous and nonhomologous repair of a double-strand break preserves genomic integrity in mammalian cells. *Mol Cell Biol* 2000;20(23):9068-75.

[171] Haber JE. In vivo biochemistry: physical monitoring of recombination induced by site-specific endonucleases. *Bioessays* 1995;17(7):609-20.

[172] Johnson RD, Jasin M. Sister chromatid gene conversion is a prominent double-strand break repair pathway in mammalian cells. *Embo J* 2000;19(13):3398-407.

[173] Liang F, Han M, Romanienko PJ, Jasin M. Homology-directed repair is a major double-strand break repair pathway in mammalian cells. *Proc Natl Acad Sci U S A* 1998;95(9):5172-7.

[174] Takata M, Sasaki MS, Sonoda E, Morrison C, Hashimoto M, Utsumi H, et al. Homologous recombination and non-homologous end-joining pathways of DNA double-strand break repair have overlapping roles in the maintenance of chromosomal integrity in vertebrate cells. *Embo J* 1998;17(18):5497-508.

[175] Falzon M, Fewell JW, Kuff EL. EBP-80, a transcription factor closely resembling the human autoantigen Ku, recognizes single- to double-strand transitions in DNA. *J Biol Chem* 1993;268(14):10546-52.

[176] Mimori T, Hardin JA. Mechanism of interaction between Ku protein and DNA. *J Biol Chem* 1986;261(22):10375-9.

[177] de Vries E, van Driel W, Bergsma WG, Arnberg AC, van der Vliet PC. HeLa nuclear protein recognizing DNA termini and translocating on DNA forming a regular DNA-multimeric protein complex. *J Mol Biol* 1989;208(1):65-78.

[178] Leuther KK, Hammarsten O, Kornberg RD, Chu G. Structure of DNA-dependent protein kinase: implications for its regulation by DNA. *Embo J* 1999;18(5):1114-23.

[179] Chiu CY, Cary RB, Chen DJ, Peterson SR, Stewart PL. Cryo-EM imaging of the catalytic subunit of the DNA-dependent protein kinase. *J Mol Biol* 1998;284(4):1075-81.

[180] Yaneva M, Kowalewski T, Lieber MR. Interaction of DNA-dependent protein kinase with DNA and with Ku: biochemical and atomic-force microscopy studies. *Embo J* 1997;16(16):5098-112.

[181] Hammarsten O, Chu G. DNA-dependent protein kinase: DNA binding and activation in the absence of Ku. *Proc Natl Acad Sci U S A* 1998;95(2):525-30.

[182] West RB, Yaneva M, Lieber MR. Productive and nonproductive complexes of Ku and DNA-dependent protein kinase at DNA termini. *Mol Cell Biol* 1998;18(10):5908-20.

[183] Merkle D, Douglas P, Moorhead GB, Leonenko Z, Yu Y, Cramb D, et al. The DNA-dependent protein kinase interacts with DNA to form a protein-DNA complex that is disrupted by phosphorylation. *Biochemistry* 2002;41(42):12706-14.

[184] Wilson TE, Grawunder U, Lieber MR. Yeast DNA ligase IV mediates non-homologous DNA end joining. *Nature* 1997;388(6641):495-8.

[185] Schar P, Herrmann G, Daly G, Lindahl T. A newly identified DNA ligase of Saccharomyces cerevisiae involved in RAD52-independent repair of DNA double-strand breaks. *Genes Dev* 1997;11(15):1912-24.

[186] Teo SH, Jackson SP. Identification of Saccharomyces cerevisiae DNA ligase IV: involvement in DNA double-strand break repair. *Embo J* 1997;16(15):4788-95.

[187] Li B, Comai L. Functional interaction between Ku and the werner syndrome protein in DNA end processing. *J Biol Chem* 2000;275(37):28349-52.

[188] Chen L, Trujillo K, Sung P, Tomkinson AE. Interactions of the DNA ligase IV-XRCC4 complex with DNA ends and the DNA-dependent protein kinase. *J Biol Chem* 2000;275(34):26196-205.

[189] Nick McElhinny SA, Snowden CM, McCarville J, Ramsden DA. Ku recruits the XRCC4-ligase IV complex to DNA ends. *Mol Cell Biol* 2000;20(9):2996-3003.

[190] Ma Y, Pannicke U, Schwarz K, Lieber MR. Hairpin opening and overhang processing by an Artemis/DNA-dependent protein kinase complex in nonhomologous end joining and V(D)J recombination. *Cell* 2002;108(6):781-94.

[191] Jeggo P, O'Neill P. The Greek Goddess, Artemis, reveals the secrets of her cleavage. *DNA Repair (Amst)* 2002;1(9):771-7.

[192] Rooney S, Alt FW, Lombard D, Whitlow S, Eckersdorff M, Fleming J, et al. Defective DNA repair and increased genomic instability in Artemis deficient murine cells. *J Exp Med* 2003;197(5):553-65.

[193] Rooney S, Sekiguchi J, Zhu C, Cheng HL, Manis J, Whitlow S, et al. Leaky Scid phenotype associated with defective V(D)J coding end processing in Artemis-deficient mice. *Mol Cell* 2002;10(6):1379-90.

[194] Ahnesorg P, Smith P, Jackson SP. XLF interacts with the XRCC4-DNA ligase IV complex to promote DNA nonhomologous end-joining. *Cell* 2006;124(2):301-13.

[195] Buck D, Malivert L, de Chasseval R, Barraud A, Fondaneche MC, Sanal O, et al. Cernunnos, a novel nonhomologous end-joining factor, is mutated in human immunodeficiency with microcephaly. *Cell* 2006;124(2):287-99.

[196] Jackson SP. Detecting, signalling and repairing DNA double-strand breaks. *Biochem Soc Trans* 2001;29(Pt 6):655-61.

[197] Valerie K, Povirk LF. Regulation and mechanisms of mammalian double-strand break repair. *Oncogene* 2003;22(37):5792-812.

[198] Haber JE. Partners and pathwaysrepairing a double-strand break. *Trends Genet* 2000;16(6):259-64.

[199] Haber JE. DNA repair. Gatekeepers of recombination. *Nature* 1999;398(6729):665, 667.

[200] Sung P. Yeast Rad55 and Rad57 proteins form a heterodimer that functions with replication protein A to promote DNA strand exchange by Rad51 recombinase. *Genes Dev* 1997;11(9):1111-21.

[201] Sonoda E, Takata M, Yamashita YM, Morrison C, Takeda S. Homologous DNA recombination in vertebrate cells. *Proc Natl Acad Sci U S A* 2001;98(15):8388-94.

[202] Wood RD, Mitchell M, Sgouros J, Lindahl T. Human DNA repair genes. *Science* 2001;291(5507):1284-9.

[203] Thompson LH, Schild D. Homologous recombinational repair of DNA ensures mammalian chromosome stability. *Mutat Res* 2001;477(1-2):131-53.

[204] Thompson LH, Schild D. Recombinational DNA repair and human disease. *Mutat Res* 2002;509(1-2):49-78.

[205] Liu N, Lamerdin JE, Tebbs RS, Schild D, Tucker JD, Shen MR, et al. XRCC2 and XRCC3, new human Rad51-family members, promote chromosome stability and protect against DNA cross-links and other damages. *Mol Cell* 1998;1(6):783-93.

[206] Powell SN, Kachnic LA. Roles of BRCA1 and BRCA2 in homologous recombination, DNA replication fidelity and the cellular response to ionizing radiation. *Oncogene* 2003;22(37):5784-91.

[207] Chen G, Yuan SS, Liu W, Xu Y, Trujillo K, Song B, et al. Radiation-induced assembly of Rad51 and Rad52 recombination complex requires ATM and c-Abl. *J Biol Chem* 1999;274(18):12748-52.

[208] Cortez D, Wang Y, Qin J, Elledge SJ. Requirement of ATM-dependent phosphorylation of brca1 in the DNA damage response to double-strand breaks. *Science* 1999;286(5442):1162-6.

[209] Jasin M. Homologous repair of DNA damage and tumorigenesis: the BRCA connection. *Oncogene* 2002;21(58):8981-93.

[210] Liu Y, West SC. Distinct functions of BRCA1 and BRCA2 in double-strand break repair. *Breast Cancer Res* 2002;4(1):9-13.

[211] Scully R, Anderson SF, Chao DM, Wei W, Ye L, Young RA, et al. BRCA1 is a component of the RNA polymerase II holoenzyme. *Proc Natl Acad Sci U S A* 1997;94(11):5605-10.

[212] Bochar DA, Wang L, Beniya H, Kinev A, Xue Y, Lane WS, et al. BRCA1 is associated with a human SWI/SNF-related complex: linking chromatin remodeling to breast cancer. *Cell* 2000;102(2):257-65.

[213] van Attikum H, Gasser SM. The histone code at DNA breaks: a guide to repair? *Nat Rev Mol Cell Biol* 2005;6(10):757-65.

[214] Fernandez-Capetillo O, Nussenzweig A. Linking histone deacetylation with the repair of DNA breaks. *Proc Natl Acad Sci U S A* 2004;101(6):1427-8.

[215] Kamakaka RT, Biggins S. Histone variants: deviants? Genes Dev 2005;19(3):295-310.

[216] Butler PJ. The folding of chromatin. *CRC Crit Rev Biochem* 1983;15(1):57-91.

[217] Arents G, Moudrianakis EN. Topography of the histone octamer surface: repeating structural motifs utilized in the docking of nucleosomal DNA. *Proc Natl Acad Sci U S A* 1993;90(22):10489-93.

[218] Luger K, Mader AW, Richmond RK, Sargent DF, Richmond TJ. Crystal structure of the nucleosome core particle at 2.8 A resolution. *Nature* 1997;389(6648):251-60.

[219] Horn PJ, Peterson CL. Molecular biology. Chromatin higher order folding--wrapping up transcription. *Science* 2002;297(5588):1824-7.

[220] Paques F, Haber JE. Multiple pathways of recombination induced by double-strand breaks in Saccharomyces cerevisiae. *Microbiol Mol Biol Rev* 1999;63(2):349-404.

[221] Fyodorov DV, Kadonaga JT. The many faces of chromatin remodeling: SWItching beyond transcription. *Cell* 2001;106(5):523-5.

[222] Langst G, Becker PB. Nucleosome mobilization and positioning by ISWI-containing chromatin-remodeling factors. *J Cell Sci* 2001;114(Pt 14):2561-8.

[223] Saha A, Wittmeyer J, Cairns BR. Chromatin remodelling: the industrial revolution of DNA around histones. *Nat Rev Mol Cell Biol* 2006;7(6):437-47.

[224] Iizuka M, Smith MM. Functional consequences of histone modifications. *Curr Opin Genet Dev* 2003;13(2):154-60.

[225] Spotswood HT, Turner BM. An increasingly complex code. *J Clin Invest* 2002;110(5):577-82.

[226] Vidanes GM, Bonilla CY, Toczyski DP. Complicated tails: histone modifications and the DNA damage response. *Cell* 2005;121(7):973-6.

[227] Rogakou EP, Boon C, Redon C, Bonner WM. Megabase chromatin domains involved in DNA double-strand breaks in vivo. *J Cell Biol* 1999;146(5):905-16.

[228] Celeste A, Fernandez-Capetillo O, Kruhlak MJ, Pilch DR, Staudt DW, Lee A, et al. Histone H2AX phosphorylation is dispensable for the initial recognition of DNA breaks. *Nat Cell Biol* 2003;5(7):675-9.

[229] Paull TT, Rogakou EP, Yamazaki V, Kirchgessner CU, Gellert M, Bonner WM. A critical role for histone H2AX in recruitment of repair factors to nuclear foci after DNA damage. *Curr Biol* 2000;10(15):886-95.

[230] Bassing CH, Chua KF, Sekiguchi J, Suh H, Whitlow SR, Fleming JC, et al. Increased ionizing radiation sensitivity and genomic instability in the absence of histone H2AX. *Proc Natl Acad Sci U S A* 2002;99(12):8173-8.

[231] Celeste A, Petersen S, Romanienko PJ, Fernandez-Capetillo O, Chen HT, Sedelnikova OA, et al. Genomic instability in mice lacking histone H2AX. *Science* 2002;296(5569):922-7.

[232] Pilch DR, Sedelnikova OA, Redon C, Celeste A, Nussenzweig A, Bonner WM. Characteristics of gamma-H2AX foci at DNA double-strand breaks sites. *Biochem Cell Biol* 2003;81(3):123-9.

[233] Karagiannis TC, Harikrishnan KN, El-Osta A. The histone deacetylase inhibitor, Trichostatin A, enhances radiation sensitivity and accumulation of gammaH2A.X. *Cancer Biol Ther* 2005;4(7):787-93.

[234] Hasan S, Hottiger MO. Histone acetyl transferases: a role in DNA repair and DNA replication. *J Mol Med* 2002;80(8):463-74.

[235] Gregory PD, Wagner K, Horz W. Histone acetylation and chromatin remodeling. *Exp Cell Res* 2001;265(2):195-202.

[236] Struhl K. Histone acetylation and transcriptional regulatory mechanisms. *Genes Dev* 1998;12(5):599-606.

[237] Grunstein M. Histone acetylation in chromatin structure and transcription. *Nature* 1997;389(6649):349-52.

[238] Qin S, Parthun MR. Histone H3 and the histone acetyltransferase Hat1p contribute to DNA double-strand break repair. *Mol Cell Biol* 2002;22(23):8353-65.

[239] Bird AW, Yu DY, Pray-Grant MG, Qiu Q, Harmon KE, Megee PC, et al. Acetylation of histone H4 by Esa1 is required for DNA double-strand break repair. *Nature* 2002;419(6905):411-5.

[240] Murr R, Loizou JI, Yang YG, Cuenin C, Li H, Wang ZQ, et al. Histone acetylation by Trrap-Tip60 modulates loading of repair proteins and repair of DNA double-strand breaks. *Nat Cell Biol* 2006;8(1):91-9.

[241] Varga-Weisz P. ATP-dependent chromatin remodeling factors: nucleosome shufflers with many missions. *Oncogene* 2001;20(24):3076-85.

[242] Morrison AJ, Highland J, Krogan NJ, Arbel-Eden A, Greenblatt JF, Haber JE, et al. INO80 and gamma-H2AX interaction links ATP-dependent chromatin remodeling to DNA damage repair. *Cell* 2004;119(6):767-75.

[243] van Attikum H, Fritsch O, Hohn B, Gasser SM. Recruitment of the INO80 complex by H2A phosphorylation links ATP-dependent chromatin remodeling with DNA double-strand break repair. *Cell* 2004;119(6):777-88.

[244] Chai B, Huang J, Cairns BR, Laurent BC. Distinct roles for the RSC and Swi/Snf ATP-dependent chromatin remodelers in DNA double-strand break repair. *Genes Dev* 2005;19(14):1656-61.

[245] Mizuguchi G, Shen X, Landry J, Wu WH, Sen S, Wu C. ATP-driven exchange of histone H2AZ variant catalyzed by SWR1 chromatin remodeling complex. *Science* 2004;303(5656):343-8.

[246] Jaskelioff M, Van Komen S, Krebs JE, Sung P, Peterson CL. Rad54p is a chromatin remodeling enzyme required for heteroduplex DNA joint formation with chromatin. *J Biol Chem* 2003;278(11):9212-8.

[247] Alexeev A, Mazin A, Kowalczykowski SC. Rad54 protein possesses chromatin-remodeling activity stimulated by the Rad51-ssDNA nucleoprotein filament. *Nat Struct Biol* 2003;10(3):182-6.

[248] Alexiadis V, Kadonaga JT. Strand pairing by Rad54 and Rad51 is enhanced by chromatin. *Genes Dev* 2002;16(21):2767-71.

[249] Huyen Y, Zgheib O, Ditullio RA, Jr., Gorgoulis VG, Zacharatos P, Petty TJ, et al. Methylated lysine 79 of histone H3 targets 53BP1 to DNA double-strand breaks. *Nature* 2004;432(7015):406-11.

[250] Feng Q, Wang H, Ng HH, Erdjument-Bromage H, Tempst P, Struhl K, et al. Methylation of H3-lysine 79 is mediated by a new family of HMTases without a SET domain. *Curr Biol* 2002;12(12):1052-8.

[251] Lacoste N, Utley RT, Hunter JM, Poirier GG, Cote J. Disruptor of telomeric silencing-1 is a chromatin-specific histone H3 methyltransferase. *J Biol Chem* 2002;277(34): 30421-4.

[252] Mozziconacci J, Victor JM. Nucleosome gaping supports a functional structure for the 30nm chromatin fiber. *J Struct Biol* 2003;143(1):72-6.

In: New Research on DNA Repair
Editor: Breehn R. Landseer, pp. 325-373
ISBN 1-60021-385-5

Chapter XII

ROLE OF P53 IN THE REPAIR OF IONIZING RADIATION-INDUCED DNA DAMAGE

David Murray*[*] *and Razmik Mirzayans
Department of Oncology, University of Alberta, Cross Cancer Institute, Edmonton, Alberta T6G 1Z2, Canada.

ABSTRACT

The p53 tumor-suppressor protein plays a central role in coordinating the cellular response to DNA damage and in maintaining genomic stability. A variety of genotoxic stresses, including DNA double-strand breaks (DSBs) induced by ionizing radiation, transiently stabilize the wild-type p53 protein, cause it to accumulate in the nucleus, and activate it as a transcription factor. Depending on the extent of DNA damage and the genetic background of the cells, p53 may activate transient cell-cycle checkpoints whose role in maintaining genomic stability is traditionally embedded in the concept that this provides time for DNA-repair enzymes to remove damage from the genome before the cellular decision to either resume proliferation or undergo permanent growth arrest or apoptosis. However, there is a considerable emerging literature to suggest that p53 also directly participates in various DNA-repair pathways, including the base excision repair pathway that, among other functions, removes simple base lesions caused by ionizing radiation, as well as the RAD51-dependent homologous-recombination repair pathway and the non-homologous end-joining pathway, two distinct but complimentary mechanisms for DSB repair. In this article, we review recent experimental data supporting a role for p53 in these various DNA-repair processes, with a focus on the repair of DNA damage caused by exposure of human cells to different doses of ionizing radiation, including doses relevant to cancer therapeutic applications. We will also consider the role of p53 in DNA repair in the context of our current knowledge of low-dose responses to ionizing radiation. Finally, we consider some of the implication of

* Correspondence concerning this article should be addressed to David Murray, Tel.: 780-432-8430; Fax: 780-432-8428; E-mail: davem@cancerboard.ab.ca.

these findings to the relative responses of normal and cancerous cells to ionizing radiation.

Keywords: p53, ionizing radiation, base excision repair, non-homologous end joining, homologous recombination.

1. Activation of the DNA Damage-Response Pathway in Mammalian Cells

The response of mammalian cells to low doses of ionizing radiation (IR) involves highly coordinated, genetically-regulated, signal transduction processes, the purpose of which is two-fold: (a) self preservation, i.e., protection of the genome, suppressing mutation and promoting the survival of individual cells; and (b) self protection, i.e., eliminating irreversibly-damaged cells from the tissue (e.g., Wang and Cho 2004). These responses involve specialized DNA damage-recognition factors or "sensors" that trigger a cascade of signalling events that impact on the expression, levels or activity of various downstream genes and proteins (Jackson 2002). As shown in Figure 1, the DNA damage-response pathways are comprised of four general types of protein: sensors, mediators, transducers, and effectors. Sensor proteins bind to DNA lesions and activate mediators (sometimes called "adaptors") and transducers, such as the phosphatidylinositol 3-kinase-related kinase (PIKK) group of protein kinases, which include ATM (mutated in ataxia telangiectasia), ATR (ATM and RAD3 related), and the catalytic sub-unit of the DNA-dependent protein kinase (DNA-PK_{cs}). The PIKKs, which phosphorylate serine or threonine residues followed by a glutamine (Kastan and Lim 2000), have also been referred to as "master switches" in these pathways (Wang and Cho 2004). Transducer serine/threonine kinases transmit and possibly amplify the signal from the activated PIKKs to the downstream effector proteins, such as the wild-type (WT) p53 tumor-suppressor protein. These effectors execute the various functional outcomes of the damage-response network, such as the activation of cell-cycle checkpoints (resulting in transient growth arrest), DNA repair, apoptosis, and accelerated/premature senescence (i.e., permanent withdrawal from the cell cycle).

As an example, IR-induced DNA double-strand breaks (DSBs) recruit a number of proteins/protein complexes to the damaged region of the genome (the so-called "repair foci") immediately post-irradiation. One such early-arriving protein complex is the meiotic recombination 11 (MRE11)-RAD50-Nijmegen breakage syndrome 1 (NBS1) or "MRN" complex. Although some reports suggest MRN to be a mediator, it does exhibit many properties expected of a damage sensor (Petrini and Stracker 2003, Kurz and Lees-Miller 2004) in addition to having multiple roles in DNA repair and signal transduction. The 9-1-1 (RAD9-HUS1-RAD1) complex, whose structure resembles that of proliferating cell nuclear antigen (PCNA), has been suggested to be an important sensor for some types of DNA damage; notably, it is critical for the phosphorylation and activation of the CHK1 (checkpoint protein-1) kinase by ATR in response to genotoxic agents such as ultraviolet light (UV) that produce bulky/helix-distorting DNA lesions, but whether it acts as a 'sensor' *per se* is uncertain (Parrilla-Castellar *et al.* 2004). In yeast, 9-1-1 does appear to be important for the rejoining of IR-induced DSBs; its role in mammalian DSB repair remains to be clarified, although RAD9-deficient murine embryonic stem cells are radiosensitive (Roos-Mattjus *et al.* 2003). Furthermore, HUS1 was reported to translocate from the cytosol to the nucleus following irradiation of human skin fibroblasts, where it associates with PCNA and RAD9 (Komatsu *et al.* 2000). Other early arrivers at the site of DSB lesions include the mediators/adaptors BRCA1 and BRCA2 (breast cancer 1 and 2), 53BP1 (p53 binding protein 1), MDC1 (mediator of DNA damage checkpoint 1) and SMC1 (structural maintenance of chromatin 1). Mediator/adaptor or "scaffolding" proteins are important for the assembly of signalling complexes (Wang and Cho 2004). Two specific protein-protein interaction domains, namely breast cancer C-terminal (BRCT) and fork-head associated (FHA) domains, occur in a number of these mediator/adaptor proteins (Bork *et al.* 1997, Durocher and Jackson 2002).

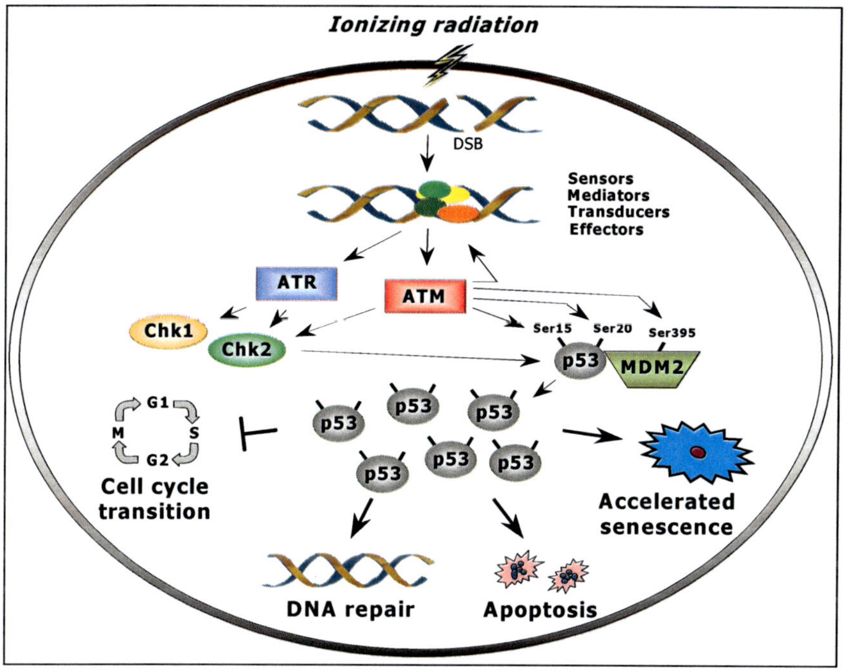

Figure 1. Response of human cells to low doses of ionizing radiation. The DNA damage-response pathways comprise four general types of protein: sensors, mediators (adaptors), transducers, and effectors. Sensor proteins trigger a cascade of signalling events that impact on the expression, levels or activity of numerous downstream genes and proteins. Sensors bind to DNA lesions and activate transducers such as the phosphatidylinositol 3-kinase-related kinase (PIKK) group of protein kinases, including ATM (mutated in ataxia telangiectasia) and ATR (ATM and RAD3 related). The PIKKs are sometimes referred to as "master switches" in the DNA damage-response pathway. Transducer serine/threonine kinases transmit and possibly amplify the signal from the activated PIKKs to the downstream effector proteins, such as the wild-type (WT) p53 tumor-suppressor protein, which undergoes stabilization and extensive post-translational modification (e.g., phosphorylation at serine 15, mediated by the ATM and CHK2 checkpoint kinases), resulting in its activation as a transcriptional transactivator. The effector proteins execute the various functional outcomes of the damage-response network, such as the activation of cell-cycle checkpoints, DNA repair, apoptosis, and accelerated senescence. The MDM2 protein is an E3 ligase that regulates WT p53 levels by various mechanisms (see text, section 3).

These events in turn result in activation of the ATM protein, which is the primary PIKK responder to DSBs (Shiloh 2003a). Two events – autophosphorylation and recruitment – probably combine to regulate the kinase activity of ATM. IR-induced changes in chromatin structure associated with a DSB may serve as the initiating signal for ATM autophosphorylation, which results in activation of the kinase domain and recruitment of ATM to the site of the DSB (Bakkenist and Kastan 2003) where it in turn phosphorylates several of the sensor and mediator/adaptor proteins, including 53BP1, MDC1 and SMC1 (Kurz and Lees-Miller 2004). Targeting of activated ATM to the site of a DSB may be mediated by the MRN complex, possibly in concert with MDC1 (Lee and Paull 2004, Kurz and Lees-Miller 2004). 53BP1 may also be important in the IR-induced activation of ATM (Kurz and Lees-Miller 2004). Other ATM targets include DNA repair-associated proteins such as NBS1, BRCA1, BLM (the helicase that is defective in Bloom syndrome), RAD9, and the protein tyrosine kinase c-ABL, the latter resulting in the phosphorylation of RAD51, which is a critical factor in the rejoining of IR-induced DSBs (see section 10), as well as histone 2AX (H2AX) (for reviews, see Christmann *et al.* 2003, Kurz and Lees-Miller 2004, Niida and Nakanishi 2006). Phosphorylation of H2AX by ATM precedes DSB repair and may serve to attract other mediator/adaptor proteins and repair complexes to the DSB repair focus.

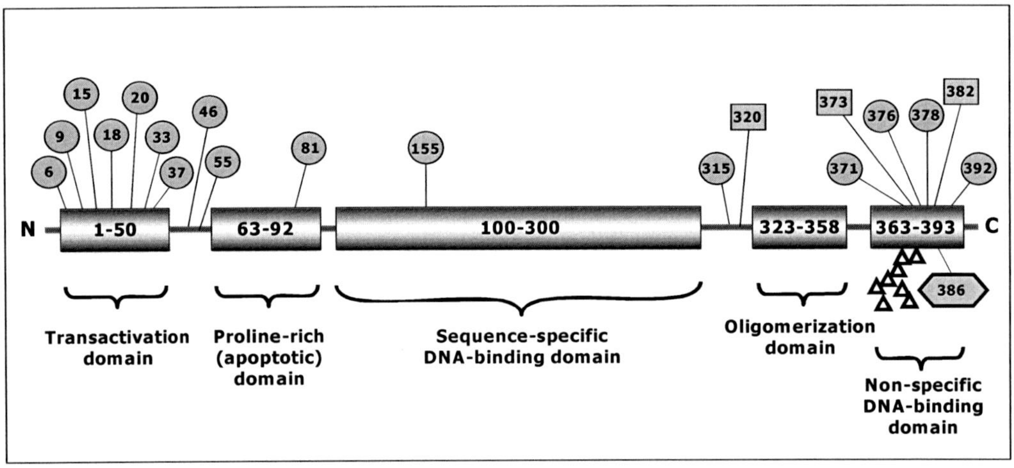

Figure 2. Structural domains and common post-translational modifications of p53. Circles represent serine/threonine phosphorylation sites; squares represent acetylation sites; the hexagon represents the SUMOylation site.

Activated ATM also phosphorylates several key downstream damage-response effectors such as p53 (Kastan *et al.* 1991) either directly or indirectly through another effector kinase, checkpoint protein-2 (CHK2), whose phosphorylation following IR exposure is ATM-dependent (Abraham 2001, Shiloh 2003a, Bartek and Lukas 2003). The activated ATM and CHK2 checkpoint kinases, in concert with other enzymes, confer a complex pattern of residue-specific post-translational covalent modifications on p53, including phosphorylation, acetylation and SUMOylation (e.g., Lambert *et al.* 1998, Appella and Anderson 2001) (Figure 2); this leads to stabilization and increased p53 protein levels and to qualitative changes that result in activation of WT p53 which, as discussed below, participates in various responses to DNA damage. Whereas the ATM-CHK2 axis is the major phosphorylation response pathway following IR exposure, other checkpoint kinases, notably ATR and CHK1, respond to other types of DNA injury in a partially redundant fashion.

2. p53 as a Downstream Effector in the DNA Damage-Response Pathway

p53, once activated by these various post-translational modifications, plays a critical role in coordinating cellular decisions following DNA damage by mediating survival responses such as checkpoint and DNA repair activation as well as death responses such as apoptosis and accelerated senescence (Sengupta and Harris 2005). p53 therefore exerts a dual effect on "repair outcome" by indirectly providing time for repair through checkpoint activation and in some cases, as per the subject of this review, by modulating the repair capability of the cell. Checkpoint activation in the G_1, S and G_2 phases in response to IR transiently delays the progression of damaged cells through the cell cycle, thereby facilitating the repair of DNA damage. The major role of p53 in activating/maintaining the G_1 checkpoint is through its transcriptional transactivation of the *p21*^{waf1} gene that encodes the multifunctional p21^{waf1} protein that is best known as a cyclin-dependent kinase inhibitor but which also acts as a regulator of transcription (Roninson 2003). The G_2 checkpoint has both p53-dependent and -independent components, with transcriptional activation by p53 apparently being necessary to maintain the initially p53-independent G_2/M arrest (Taylor and Stark 2001). The G_2 checkpoint is believed to be particularly relevant to the biological effects of IR exposure.

The impact of p53 on downstream events in the DNA damage-response pathway involves two major types of activity: [i] those mediated directly through p53 interactions with other proteins following its stabilization, which are sometimes referred to as "immediate-early" responses, and [ii] those mediated indirectly through p53's role as a regulator of transcription, i.e., which require new gene expression, and which are often referred to as "early" responses. In general, the former events mediate the transient/rapid responses to IR, whereas events requiring transcription mediate the slower "maintenance" responses. With respect to early/maintenance effects, p53 acts as a transcriptional activator through its N-terminal transactivation domain (Figure 2). Binding to DNA occurs through a homo-tetrameric p53 complex which binds to conserved p53-binding sequences in the promoters of its target genes (Vogelstein *et al.* 2000). p53 transactivates or transrepresses a large number (more than 100) of target genes (Polyak *et al.* 1997, Zhao *et al.* 2000, Vogelstein *et al.* 2000, Mirza *et al.* 2003). Among the genes transactivated by p53 are those encoding the p21^{waf1} protein (see above), *GADD45* (Fornace *et al.* 1989), *SMC1*, pro-apoptotic genes such as *CD95/FAS/APO1*, *BAX*, *NOXA* and *PUMA*, and genes encoding members of the 14-3-3 family of proteins, such as 14-3-3σ, which plays a role in maintaining the G_2/M arrest. Induction of these genes, e.g., *GADD45* (Kastan *et al.* 1992), following IR exposure is defective in cells lacking WT p53. GADD45 protein complexes with PCNA, which can influence some types of DNA repair (e.g., Ford 2005). Other downstream targets of p53 transcriptional regulation involved in DNA repair have also been identified, as discussed below. Genes that are transrepressed by p53 include anti-apoptotic genes such as *BCL2*.

The transcription-independent roles of WT p53 in mediating its downstream effects *in vivo* are less clearly understood. In the case of DNA repair, such events are of two major types: those based on p53-repair protein interactions, and those based on an interaction between p53 and DNA damage itself. Examples of physical interactions between p53 protein and DNA repair-related proteins in response to DNA damage include the WRN helicase which is defective in Werner syndrome cells (Blander *et al.* 1999, Yannone *et al.* 2001), RAD51 (Sturzbecher *et al.* 1996, Buchhop *et al.* 1997), and replication protein A (RPA) (Dutta *et al.* 1993, Miller *et al.* 1997), and are discussed further below. A more direct involvement of p53 in DNA repair is suggested by the known 3′→5′ exonuclease activity of p53, which may play a role in some repair processes, e.g., by providing a proofreading function for DNA polymerase β (POLβ) (Mummenbrauer *et al.* 1996, Huang 1998), and by the possible function of p53 in damage recognition. For example, the regulatory C-terminal domain of p53 can bind sequence-independently to various types of DNA damage or unusual DNA structures such as DNA ends, strand breaks, 3-stranded DNA structures, extra base bulges/deletion-insertion mismatches and Holliday junctions (e.g., Steinmeyer and Deppert 1988, Oberosler *et al.* 1993, Jayaraman and Prives 1995, Bakalkin *et al.* 1995, Reed *et al.* 1995, Lee *et al.* 1995, Lee *et al.* 1997, Yokote *et al.* 1998, Dudenhoffer *et al.* 1998, Albrechtsen *et al.* 1999, Liu and Kulesz-Martin 2001). This activity is distinct from the above-mentioned sequence-specific binding. It is uncertain, however, if the affinity of p53 for some types of damaged versus undamaged DNA is sufficiently specific to be relevant *in vivo*.

Clearly, a number of signals have to be integrated for an effective DNA-damage response, and this function is predominantly the realm of p53. Axiomatically, p53 plays a central role in maintaining genomic stability (Loeb 2001), either by facilitating damage repair or by preventing the proliferation of cells with un-repairable damage by triggering accelerated senescence or apoptosis. It is these diverse critical functions that have earned p53 the title "guardian of the genome" (Lane 1992). It has been widely noted that a downside of investing so many critical homeostatic functions in a single protein is that, should p53 become defective, the cell could be seriously compromised. Indeed, p53 is mutated or inactivated in most human cancers (Levine 1997), which abrogates downstream effector responses and can result in genetic instability.

3. Integration of DNA Repair with Checkpoint Activation

Among the features expected of an integrated DNA damage-response pathway is the existence of feedback mechanisms at the mediator/transducer level that would amplify the initial signal when damage exceeds some threshold level, such that the cell will not over-respond to low levels of injury; this aspect will be discussed further in section 13. Mechanisms are also required to "reset" the machinery and return it to normality following successful repair. Several models have been forwarded to explain how the reversal of cell-cycle checkpoints is temporally coordinated with the repair of DNA lesions. A detailed discussion of these models is beyond the scope of the present chapter; however, the excellent review of Wang and Cho (Wang and Cho 2004) is recommended for this purpose. A common feature of these models is the assumption that feedback signals from the damage-sensing complexes, which we presume to involve p53, will direct either the maintenance of the checkpoint if repair is ongoing or will signal the lifting of checkpoint when the lesion load has been reduced below some "threshold" level, so that the cell can re-enter the cell cycle, resulting in clonogenic survival. Alternatively, if DNA damage persists or is deemed irreparable, either apoptosis or accelerated senescence genes will be induced by p53 (depending on the genetic background), leading to cell death. Unlike the reversible immediate-early and early responses discussed above, these "late" death responses are mostly irreversible. The transition from reversible to irreversible states allows for the temporal regulation of the DNA-damage response (Wang and Cho 2004).

Two extreme models illustrate some of the considerations involved. The first model, the *Integrative Surveillance* model, invokes a hypothetical "brain-like" entity, the "regulatory hub", that continuously monitors the genome for damage and for the status of repair (Wang and Cho 2004). The relative rates of lesion accumulation and removal dictate the response. When the lesion load is increasing (i.e., accumulation exceeds repair), the hub activates the DNA-repair and cell-cycle checkpoint machinery. When lesions are decreasing, (i.e., repair exceeds accumulation), the hub maintains checkpoints and inhibits cell-death pathways. When lesions persist for an extended time, the hub activates cell-death pathways. A second model, the *Autonomous Pathways* model, assumes that independent pathways with built-in feedback mechanisms act autonomously to invoke various responses (Wang and Cho 2004). DNA-damage signals activate DNA repair, checkpoints, or cell death through distinct pathways. The DNA-repair status affects the checkpoint or death pathways solely by eliminating lesions and thus the triggering signals. The checkpoint pathway incorporates a negative feedback loop that proceeds through a series of time-dependent steps, eventually leading to lifting of the checkpoints when the damage signal is eliminated. Similarly, cell death is regulated by feedback loops that require repetitive activation to invoke death pathways. These repeating loops may continue until the damage signal is eliminated. When the feedback mechanism fails, the cell dies.

The molecular basis for some of these negative feedback loops is beginning to be deciphered. For example, resetting cellular p53 levels following IR exposure or other genotoxic stresses is mediated by a well-defined loop. Several cellular proteins, notably MDM2 (an E3 ligase), can cause the degradation of WT p53 through ubiquitination and proteasomal degradation, thereby maintaining its baseline levels and activity at a low level. MDM2 also regulates p53 activity by facilitating its export from the nucleus (Liang and Clarke 2001), and it directly interferes with the ability of p53 to act as a transcriptional regulator (Oren 2003, Michael and Oren 2003, Perry 2004). Following IR exposure, the activated ATM kinase phosphorylates both p53 and MDM2, disrupting their interaction and allowing p53 to accumulate and to gradually activate the transcription of the *MDM2* gene (Shiloh 2003b). The resulting delayed induction of MDM2 protein is believed to counteract the cell-cycle arrest invoked by the initial up-regulation of p53 in response to the IR exposure, ensuring that p53 levels and activity return to basal levels in cells that are destined to survive (Perry 2004). Other modulators of p53 activity include PARC, a parkin-like ubiquitin ligase, which serves as a cytoplasmic anchor for p53 and may catalyze its degradation outside the nucleus (Nikolaev *et al.* 2003), and PIRH2, a p53-induced ubiquitin-protein ligase that promotes p53 degradation (Leng *et al.* 2003).

Recently, the p53-MDM2 feedback loop was examined at the single cell level using cell lines derived from the MCF7 human breast cancer cell line (Lahav *et al.* 2004). IR exposure induced a rapid initial wave of p53 up-regulation, followed ~5-6 h later by its down-regulation coinciding with the induction of MDM2. The peak p53 wave amplitude and duration were independent of dose; however, the number of p53 waves generated by the cell was dose-dependent. Whereas a single wave was apparent following a 2-Gy exposure, a 10-Gy exposure evoked two or more waves of p53 (Lahav *et al.* 2004). The authors suggest that the p53-MDM2 feedback loop may generate a "digital clock" that releases multiple pulses of p53 until the DNA damage is repaired or the cell dies (Lahav *et al.* 2004). One possibility is that repeated rounds of p53 activation overcomes the auto-inhibitory loop, thereby allowing p53 to accumulate to sufficient levels to invoke cell death (Wang and Cho 2004).

4. Mammalian DNA-Repair Pathways

Mammalian cells have evolved elaborate DNA-repair mechanisms to remove potentially harmful lesions from their genome, using either error-free or error-prone pathways. These include the base excision repair (BER), RAD51-dependent homologous recombination (HR) repair and non-homologous end joining (NHEJ) pathways. As will be discussed below, p53 can influence these pathways through various mechanisms. As was noted above, an involvement of p53 in DNA repair may reflect a direct role of the p53 protein through its interaction with DNA lesions or repair proteins, or it may be indirectly mediated through transcriptional regulation.

5. The BER Pathway

The BER pathway effects the removal of many types of IR-induced DNA lesions, including base damage, base loss and single-strand breaks (SSBs) (Hoeijmakers 2001, Christmann *et al.* 2003). At least for some types of IR-induced base damage the BER response appears to be coupled to transcription, i.e., damage is preferentially repaired in the transcribed strand of active genes (Leadon and Cooper 1993, Avrutskaya and Leadon 2000, Le Page *et al.* 2000, Lindahl 2001). BER is a highly coordinated multi-protein "hand-off" process initiated by the action of a damage-specific DNA glycosylase that removes the damaged base by cleaving the N-glycosidic linkage, generating an apurinic/apyrimidinic (AP) or "abasic" site. The major glycosylases include 8-oxoguanine glycosylase 1 (OGG1), which removes 8-oxoguanine (8-oxoG, one of the most prevalent forms of base damage after IR) from DNA (Nishimura 2001), as well as the human homologs of *E. coli* endonuclease VIII (NEIL1, 2 and 3), endonuclease III homolog (hNTH1), uracil-DNA glycosylase (UNG) and N-methylpurine DNA glycosylase (MPG or AAG). There is significant degeneracy among the glycosylases in the lesions they process; for example, a second glycosylase,

OGG2, can also remove 8-oxopurines (Boiteux and Le Page 2001). Some glycosylases (referred to as "type II", and including OGG1) have an associated AP-lyase activity that cleaves the ribose-phosphate backbone 3' to the damaged site, leaving either a modified sugar or a phosphate residue at the 3' terminus. Either an AP endonuclease or AP lyase activity cleaves 5' or 3' to the AP site generated by the glycosylase activity, respectively. The major AP endonuclease in human cells is HAP1, also known as APE1 or REF1.

Because POLβ can only act on clean 3'-hydroxyl termini and because DNA ligases require clean 3'-hydroxyl/5'-phosphate termini, any "dirty" end groups must be removed prior to DNA polymerization and ligation. A number of enzymatic activities contribute to this clean-up process. For example, hydrolysis of the phosphodiester bond 5' to an AP site generates a 5' terminal deoxyribosephosphate residue that can be removed by exonucleases or a deoxyribophosphodiesterase. POLβ itself has a deoxyribophosphodiesterase activity. Another enzyme, polynucleotide kinase-phosphatase (PNKP), has 3' DNA phosphatase and 5' DNA kinase activities, while HAP1 has diesterase and phosphatase activities.

The repair of AP sites involves one of two BER sub-pathways, short- and long-patch (Dianov *et al.* 2003). In short-patch BER the replacement of a single nucleotide is carried out by POLβ using the undamaged strand as the template, and DNA ligase 3 (LIG3) seals the resulting gap. LIG3 interacts with several other BER factors, notably POLβ, XRCC1, and poly(ADP-ribose) polymerase 1 (PARP-1). In long-patch BER, POLβ is displaced from the lesion and up to 15 nucleotides are removed and re-synthesized in a process that requires PCNA, replication factor C (RFC), POLδ or POLε, and flap endonuclease 1 (FEN1), with LIG1 sealing the resulting gap. LIG1 is known to interact with both PCNA and POLβ.

Another important BER protein is XRCC1, a putative scaffold protein that has no intrinsic catalytic activity but rather that, via a multi-protein complex with POLβ, PARP-1, and LIG3, is involved in the detection and coordinated processing of strand breaks (Masson *et al.* 1998, Tomkinson *et al.* 2001). XRCC1 also interacts with PNKP, stimulating the ability of the latter to clean up dirty break termini (Whitehouse *et al.* 2001). In addition to positively influencing BER through its interaction with factors such as LIG3, XRCC1 and POLβ (above), PARP-1, a chromatin-associated enzyme, may also be important for remodelling damaged chromatin (Thraves *et al.* 1985).

6. BER AND P53

Several roles for the WT-p53 protein in regulating BER have now been identified (for reviews, see Sengupta and Harris 2005, Friedberg *et al.* 2006). As noted above, the known ability of p53 to interact with several atypical forms of DNA (see Jayaraman and Prives 1995 and references therein) has led to the suggestion that p53 might directly receive signals at sites of DNA damage, including those undergoing BER (Zhou *et al.* 2001). A number of reports in the literature also relate to direct interactions between p53 and BER proteins that might stimulate or modify the activity of the latter. There were several early indications that the p53 protein might interact directly with BER factors. For example, the HAP1/REF1 BER/redox enzyme was reported to be a potent activator of the DNA-binding activity of several transcription factors, including p53, both *in vitro* and *in vivo*, involving both redox-dependent and -independent mechanisms (Jayaraman *et al.* 1997, Parks *et al.* 1997, Gaiddon *et al.* 1999). HAP1 was shown to interact directly and specifically with p53 *in vivo* in the H24-14 cell line that was derived from the H1299 human non-small cell lung cancer cell line, and which contains an inducible p53 vector (Gaiddon *et al.* 1999). Elevated HAP1 levels were associated with increased transcriptional activation by p53 of several endogenous/exogenous p53-responsive promoters. A HAP1 antisense oligonucleotide transiently introduced into H24-14 cells greatly reduced the p53 responsiveness of the endogenous *p21* gene as well as of an exogenous *p21* promoter-driven reporter construct (Gaiddon *et al.* 1999).

Offer and colleagues (Offer *et al.* 1999) reported that BER activity in extracts from the M1/2 (p53-null) myeloid cell line transfected with temperature-sensitive (*ts*) p53 mutants correlated with the levels of WT p53. These authors used an *in vitro* repair assay in which plasmids containing AP sites were incubated with cell extracts and the incorporation of radiolabeled dGTP was quantified. One mutant, p53*val*135 (codon 135 Ala to Val), exhibited WT-p53 activity at the permissive temperature (32°C). Whereas extracts from cells over-expressing WT p53 displayed an ~2 fold increase in BER activity, extracts of cells that were depleted of p53 using a monoclonal antibody showed a reduced ability to activate BER. Also, clones expressing differing levels of p53*val*135 protein showed roughly corresponding levels of BER activity. In short, at the permissive temperature both mouse and human *ts* mutants regained WT-p53 sequence-specific DNA binding activity for p53-responsive elements and exhibited the ability to stimulate BER activity (Offer *et al.* 1999).

In a subsequent study these authors (Offer *et al.* 2001a) examined whether the BER-stimulating effects of p53 *in vivo* were dependent on its transcriptional-transactivation function using M1/2 cells expressing the human p53-22-23 mutant that is transcriptionally defective. A similar level of BER stimulation was seen in the *in vitro* BER assay for both the *ts* and transcriptionally defective p53 forms, suggesting that this activity is in fact independent of transcription. This view was further substantiated using a reporter-gene assay involving a luciferase-containing probe that had been modified by the introduction of differing numbers of AP sites. Expression of both WT p53 and the p53-22-23 mutant stimulated BER. Thus, using two independent assays it was established that p53 facilitates BER of AP sites both *in vitro* and *in vivo* and that this effect was not dependent on transcriptional transactivation by p53. However, an intact C-terminal domain and central domain were essential for this activity.

Interestingly, the ability of p53 to facilitate BER in extracts derived from the p53-null M1/2 cell line and its *ts* mutant p53*val*135 counterpart under permissive (WT) conditions was highly dependent on the dose of IR. Whereas lower doses of IR up to ~0.5 Gy resulted in an enhancement of BER activity in the *ts* cells under permissive conditions, higher doses caused a rapid reduction in BER activity that was associated with the induction of a p53-dependent apoptotic response (Offer *et al.* 2002). In the $p53^-$ cell line there was no such transition, with BER activity increasing progressively with dose. We will return to this discussion in section 13.

Using the same *in vitro* assay for BER of AP sites by extracts from various types of mammalian lymphoid cells that were synchronized by centrifugal elutriation, the regulation of BER activity by p53 *in vivo* was found to be cell cycle phase-specific (Offer *et al.* 2001b).

In non-irradiated cells, BER activity showed two distinct maxima in the G_0/G_1 and G_2/M phases, whereas S-phase cells showed low levels of BER activity. Exposing the cells to 4 Gy of γ-rays enhanced the G_0/G_1-associated BER activity but decreased the activity of G_2/M cells. This observation was interpreted to suggest that non-irradiated cells use both the G_1 and G_2/M checkpoints to repair base lesions, whereas irradiated cells repair predominantly in G_1. The alterations in BER activity observed post-irradiation were found to be regulated by WT p53, with this effect being eliminated for tumor-derived mutant forms of the protein (Offer *et al.* 2001b). An intriguing observation was that, whereas the level of WT-p53 stabilization following a 4 Gy γ-ray exposure was similar in the different cell-cycle phases, both the activation of p53 (dephosphorylation at serine 376) and its sequence-specific DNA binding activity, as well as apoptosis, were preferentially induced as the cells progressed towards G_2/M.

Purified WT-p53 protein has also been reported to stimulate BER activity in a model *in vitro* reconstitution system involving a 55-bp uracil-containing oligonucleotide substrate (Zhou *et al.* 2001). Using mutant p53 proteins it was established that the intact N-terminal and central domains of p53 were *both* necessary for stimulating BER in this model. Several tumor-derived p53 mutants showed attenuated ability to stimulate BER. Stimulation of BER by WT p53 appeared to involve the ability of p53 to enhance the HAP1-facilitated loading of POLβ onto the AP site, i.e., p53 stabilized the interaction between POLβ and AP sites in the presence of HAP1 (Zhou *et el.* 2001). The involvement of the N-terminal region of p53 in BER was shown to be through its interaction with POLβ, whereas the central DNA-binding region appeared to help localize p53 to sites of BER (Zhou *et el.* 2001).

Using the same uracil-containing substrate, these authors (Zhou *et al.* 2001) found that *in vivo* BER activity in extracts from the LNCaP human prostate cancer cell line correlated closely with the level of WT p53. Endogenous p53 levels were modulated by pre-exposing the cells to γ-rays, which elevated both p53 levels and BER activity. As in the earlier study of Offer *et al.* (1999), extracts of γ-irradiated LNCaP cells that were subsequently immunodepleted of p53 showed a marked (~75%) decrease in BER activity, addition of purified p53 to the depleted extracts restoring their BER activity (Zhou *et al.* 2001).

An important general consideration from the study of Zhou *et al.* (2001) is that, for p53 to stimulate BER activity, it must interact at the rate-limiting step of the pathway for the particular assay system being utilized. In fact, Srivastava and colleagues (1998) had reported that removal of the 5'-terminal deoxyribosephosphate residue was the rate-limiting step of reconstituted BER *in vitro*. Furthermore, earlier studies had shown that HAP1 physically interacts with POLβ, facilitating the loading of POLβ onto the AP site and accelerating removal of the 5'-deoxyribosephosphate by POLβ (Bennett *et al.* 1997). Thus, Zhou and colleagues (2001) concluded that it was reasonable to assume that p53 stimulates BER in this system by stabilizing the interaction between POLβ and AP sites. An interesting question in this context is whether p53 merely plays a stimulatory role in BER, or whether it is required for this activity. Comparison of the response of normal, $p53^-$ and $POL\beta^-/p53^-$ mouse embryo fibroblasts to methyl methanesulfonate (MMS) indicated that cytotoxicity related to 5'-deoxyribosephosphate intermediates was *not* dependent on WT-p53 function, suggesting that the role of p53 in BER was as a facilitator, rather than an essential enzymatic participant (Sobol *et al.* 2003). An interesting aspect of this study was the suggestion that some incomplete BER intermediates may be shuttled into the HR-repair pathway (see section 11), implying some level of crosstalk between the BER and the recombination pathways (Sobol *et al.* 2003).

Whether p53 might play a similar role in sensing base lesions such as 8-oxoG following exposure of cells to reactive oxygen species (ROS) was examined using the paired HCT116 human colon carcinoma cell line (p53-WT) and its isogenic derivative (HCT116 p53$^-$) that lacks p53 (Achanta and Huang 2004). Employing an *in vitro* pull-down assay using an 8-oxoG-containing oligonucleotide as "bait", it was found that p53 was pulled down from cell extracts along with OGG1 and HAP1 (which are involved in the BER of 8-oxoG), suggesting a physical interaction of p53 with these proteins. Pre-treating cells with the superoxide-generating oxidant 2-methoxyestradiol (2ME) prior to protein extraction resulted in a significant increase in binding of p53 to the oligonucleotide, which was accompanied by increased HAP1 binding; in contrast, OGG1 binding was independent of p53 status but was increased by 2ME treatment. p53 also facilitated the excision of 8-oxoG from DNA *in vitro* and in whole cells (Figure 3A), indicating a direct role for p53 in sensing and repairing oxidative DNA damage. Similar results were obtained following treatment of cells with the hydroxyl radical-generating oxidant H_2O_2, where again there was a more rapid removal of 8-oxoG from the DNA of p53$^+$ than p53$^-$ cells (Figure 3B).

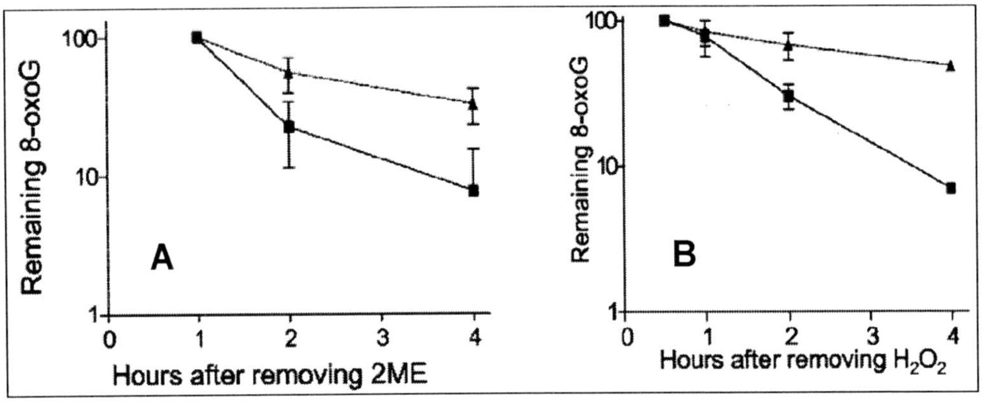

Figure 3. WT p53 facilitates the removal of 8-oxoguanine (8-oxoG) residues from cellular DNA. HCT116 p53$^+$ (■) and p53$^-$ (▲) cells were treated with 3 μM 2ME for 6 h (panel A) or with 500 μM H_2O_2 for 1 h (panel B). The cells were washed with fresh medium then incubated in drug-free medium for the indicated times. The percentage of the remaining 8-oxoG residues in cellular DNA is plotted as a function of post-treatment incubation time. Reproduced from Achanta and Huang (2004), with permission of the authors and publishers.

These authors (Achanta and Huang 2004) also employed an *in vitro* BER assay using a [^{32}P]-labelled 8-oxoG-containing oligonucleotide as a substrate to evaluate whether p53 might enhance the activities of OGG1 and HAP1 in 8-oxoG repair. Whereas p53 did not affect OGG1 or HAP1 activity individually, it significantly enhanced the combined activity of OGG1 and HAP1 in this BER model. These data indicate a possible physical and functional interaction between p53, OGG1, and HAP1. The authors suggest that, following ROS exposure, cellular p53 may localize to the nucleus and interact with and stimulate OGG1 and HAP1 to excise 8-oxoG from DNA.

Using the alkaline comet assay, a marked BER defect was demonstrated in $p53^-$ cells using matched isogenic pairs of WT or $p53^-$ cell lines (the RKO cell system and mouse embryo fibroblasts) treated with MMS (Seo *et al.* 2002b). MMS causes extensive methylation of the N7 position of guanine, as well as other alkylation damages, BER of which is initiated by the MPG glycosylase. As shown in Figure 4, the $p53^-$ cells exhibited an ~3-fold slower repair of MMS-induced lesions. This BER defect did not appear to be related to either cell-cycle arrest, apoptosis or other cell-death pathways. Using a clonogenic survival assay the $p53^-$ cells were found to be more sensitive than p53-WT cells to MMS, consistent with their BER deficit. With respect to the mechanism by which p53 facilitated the BER of methylation damage in this test system, no differences in HAP1 activity or HAP1 protein levels between the $p53^+/p53^-$ cell lines were evident, so p53 did not appear to stimulate BER at the HAP1 incision step (Seo *et al.* 2002b). Similarly, no alterations were seen in FEN-1 or PCNA protein levels. In contrast, POLβ protein levels were greatly reduced in the $p53^-$ cells, implicating POLβ as the p53 target in this process. If correct, then POLβ should be rate limiting for the BER-mediated removal of MMS-induced lesions in this system. That POLβ may indeed be rate-limiting for BER *in vivo* has been suggested previously (Wilson and Kunkel 2000, Sobol *et al.* 2000). As to how p53 might alter POLβ levels, either POLβ could be regulated by p53, or complex formation between p53 and POLβ might stabilize the polymerase, which would then be unstable in p53-deficient cells. Because *POLβ* mRNA levels were similar in $p53^+$ and $p53^-$ cells, the latter explanation was favored by these authors (Seo *et al.* 2002b).

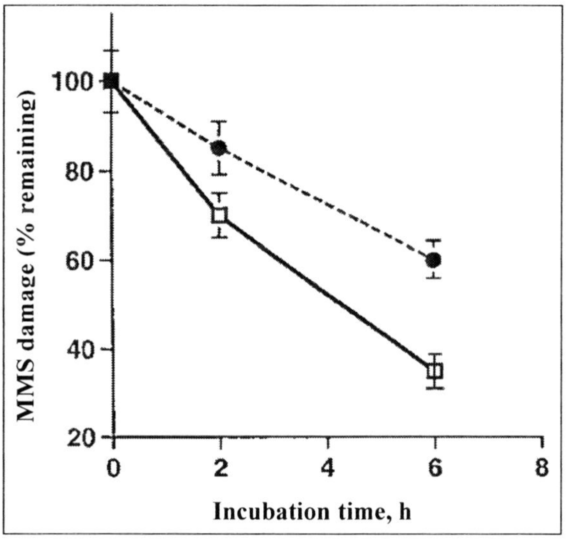

Figure 4. p53-deficient cells show a reduced rate of repair of methyl methanesulfonate (MMS)-induced DNA lesions. p53-WT (□) and p53-mutant ($p53^-$) (●) matched isogenic RKO human colon carcinoma cell lines were treated with 200 μM MMS for 2 h; strand breaks were measured using the alkaline comet assay at different times after termination of the treatment. Reproduced from Seo *et al.* (2002b), with permission of the authors and publishers.

Although few of these reports relate to the effects of IR specifically, the common theme of these various studies using different model systems and DNA-damaging strategies is that the p53 protein does appear to enhance BER of DNA containing uracil, methylated bases or oxidized bases such as 8-oxoG, and it does this at least in part by directly interacting with and stimulating BER factors such as HAP1/OGG1 and POLβ, i.e., through a typical immediate-early response. One discrepancy among these studies relates to whether or not p53-dependent transcriptional transactivation (i.e., early or maintenance responses) plays a major role in stimulating BER. For example, Offer and colleagues suggested that *in vitro*-generated mutant p53 lacking transcriptional activity was as (if not more) effective as the WT protein in stimulating BER of AP sites (Offer *et al.* 2001a). In contrast, Zhou and colleagues (2001) found that transcription-deficient p53 was inefficient at facilitating BER. The reason for this discrepancy is not immediately apparent, although clearly different test systems and assays for BER activity were used.

The transcription-dependent effects of WT p53 may have some role in homeostasis by returning DNA-repair activities to normal once the DNA-damage signal has subsided (see section 3). Thus, Lu and colleagues (Lu *et al.* 2004a, Lu *et al.* 2004b) have shown that WT p53 transcriptionally transactivates the serine/threonine phosphatase PPM1D (also known as WIP1) following exposure to UV or IR. One role of the induction of PPM1D may be to suppress BER by mediating the dephosphorylation of the key BER enzyme UNG2 once stimulated BER activity is no longer required by the cell.

The sequence-specific DNA-binding activity of p53 is greatly dependent on its redox status, being abrogated by protein oxidation (see Jayaraman *et al.* 1997 and references therein). An interesting insight into the relationship between p53 and the BER/redox factor HAP1/REF1 in maintaining this redox balance derives from a study of the cellular effects of selenomethionine (SeMet), which is the main dietary source of selenium (Seo *et al.* 2002a). SeMet is incorporated specifically into cellular selenoproteins such as thioredoxin reductase and glutathione peroxidase, and it regulates p53 through a HAP1-dependent redox mechanism. This study (Seo *et al.* 2002a) used H1299 human lung tumor cells (p53-null) in which WT p53 was transiently expressed. Incubation with SeMet promoted the HAP1-mediated reduction of p53 cysteine residues independently of DNA damage, resulting in p53 activation and enhancement of its transactivation potential (Seo *et al.* 2002a). A dominant-negative HAP1 blocked p53 reduction, as did a small-interfering RNA inhibitor targeted to WT HAP1. Also, thioredoxin reductase activity was enhanced 3-10 fold by SeMet. WT murine fibroblasts treated with SeMet were protected against the cytotoxic effects of UV, 4-nitroquinoline 1-oxide and carboplatin, whereas p53-null mouse embryo fibroblasts were not protected. At the DNA-repair level, SeMet stimulated the nucleotide excision repair (NER) pathway, which operates on bulky DNA lesions induced by UV and some cancer therapeutic agents (e.g., cisplatin), in p53-WT but not $p53^{-}$ cells. The activity of other DNA-repair pathways, such as BER, or the cellular response to agents such as IR or ROS that invoke the BER pathway, were not measured in this study.

7. DSB-Repair Pathways

IR-induced DSBs are repaired by two major pathways: non-homologous end joining (NHEJ) and RAD51-dependent HR-repair. In NHEJ, the two ends of a DSB are ligated without the requirement of sequence homology, such that loss of nucleotides can occur, resulting in "illegitimate" (i.e., error-prone or mutagenic) repair. NHEJ might thus be regarded as a "band-aid" approach to DSB rejoining that can be tolerated to a limited extent because much of the mammalian genome does not code for proteins, although NHEJ can also result in the error-free/precise religation of DSB ends if they are compatible (van Heemst *et al.* 2004, Guirouilh-Barbat *et al.* 2004). In contrast, HR-repair is a more precise high-fidelity process in which the identical undamaged sister chromatid is used as a template to repair the damaged homologous sequence. For this reason, HR-repair is most effective in late-S-phase cells, which consequently are highly radioresistant (Rothkamm *et al.* 2003, Hinz *et al.* 2005), and it is essentially absent in G_1 cells where DSB-mediated HR is inefficient (Richardson *et al.* 1998, Takata *et al.* 1998, Johnson and Jasin 2000). Whereas HR-repair is considered to be the primary method for repairing IR-induced DSBs in lower eukaryotes such as yeast, NHEJ is the major pathway used by mammalian cells, especially in G_1/G_0 phase (Jeggo 1998, Haber 2000, Cromie *et al.* 2001, Jackson 2002). Whether NHEJ and HR-repair are competing or sequential processes is unclear, and understanding how pathway choice decisions are regulated is an important area of research. For example, other than the availability of the sister chromatid (i.e., the phase of the cell cycle), the cellular "decision" between HR and NHEJ could reflect a direct competition of the enzyme complexes for the same type of DSB; alternatively, NHEJ and HR may not compete for the repair of a DSB but rather may act sequentially (e.g., Pierce *et al.* 2001, Delacote *et al.* 2002). Certainly there are common early elements in these two pathways where such decisions might be coordinated; for example, recognition and signalling of DSBs by the MRN complex appears to be an early step in both pathways (Kuschel *et al.* 2002, Valerie and Povirk 2003). The consensus is that the coordinated activity of NHEJ and HR-repair is probably required for repairing an IR-induced DSB with minimal error, with HR-repair possibly being required for the accurate repair of a subgroup of DSBs that can not be handled by NHEJ alone (Valerie and Povirk 2003). As will be discussed below, a number of investigators (e.g., Tang *et al.* 1999, Gebow *et al.* 2000) have suggested that p53 is an important determinant of the relative contributions of these pathways to DSB rejoining.

8. Non-Homologous End Joining

NHEJ proteins catalyze the direct rejoining of broken DNA ends in a process that requires no or as little as one base pair of homology and does not require strand exchanges (Liang *et al.* 1998). The process is driven by the DNA-PK holoenzyme complex, which consists of a regulatory KU sub-unit (which is itself a heterodimer of KU70 and KU80) and a catalytic sub-unit, DNA-PK_{cs}, which is a member of the PIKK family of protein kinases. NHEJ is initiated by the binding of the KU complex to the DSB. Potential roles for KU in DSB rejoining include providing end protection, as well as acting as a scaffolding or alignment factor for holding the broken DNA ends together. The KU complex can translocate from the DNA end in an ATP-independent manner, so several KU molecules can be bound per DSB. KU then recruits DNA-PK_{cs} to the DSB, generating the DNA-PK holoenzyme complex. This event activates the serine/threonine kinase function of DNA-PK_{cs}, which can then phosphorylate its target proteins such as p53, especially when the substrates are bound to the same DNA molecule as DNA-PK. Other *in vitro* substrates for DNA-PK_{cs} include WRN, JUN, FOS, MYC, the C-terminal domain of RNA polymerase II, HSP90, RPA, PARP-1, and KU. DNA-PK_{cs} is also subject to inhibitory autophosphorylation (Chan and Lees-Miller 1996). As with any kinase, whether reported *in vitro* activities occur significantly *in vivo* needs to be rigorously substantiated. There are many reported interactions among these NHEJ proteins. E.g., KU interacts with LIG4 and DNA-PK_{cs} forms contacts with XRCC4 (Hsu *et al.* 2002, Calsou *et al.* 2003). KU70 also interacts with the MRE11 protein, an interaction that may serve to recruit the MRN complex, which is involved in micro-homology searching (Paull and Gellert 2000), into the NHEJ pathway (Goedecke *et al.* 1999).

Other proteins involved in NHEJ include XRCC4 and LIG4, which associate strongly with each other and mediate tail-removal and gap-filling/ligation functions (Jeggo 1998). The association of XRCC4, which is also a target of the DNA-PK_{cs} kinase *in vitro*, with LIG4 increases the ability of the latter protein to bind to and ligate broken DNA ends (Critchlow *et al.* 1997, Lee *et al.* 2004). As with BER, a prerequisite for ligation is the processing of "dirty" ends to generate "clean" 5'-phosphate and 3'-hydroxyl termini. This step involves various proteins such as PNKP (which interacts with XRCC4 when phosphorylated by casein kinase II (Koch *et al.* 2004)), FEN1, the MRN complex (which has exonuclease, endonuclease and helicase activities and removes excess DNA at 3' flaps), and Artemis, another NHEJ protein that is phosphorylated and potentially activated by ATM. Artemis exists in a complex with DNA-PK_{cs} and exhibits single-strand-specific 5'→3' exonuclease activity, but on binding to and being phosphorylated by DNA-PK_{cs} it acquires endonucleolytic activity on 5' and 3' ends as well as on hairpins (Jeggo and Lobrich 2005). The role of the MRN complex in DSB repair *in vivo* is complicated, and extensive discussion is beyond the scope of this article; the magnum opus of Friedberg and colleagues (Friedberg *et al.* 2006) should be consulted for details. Note that, during the processing of DSB termini, the opportunity arises for base pairing within flanking regions (e.g., microhomologies of 1-4 base pairs) which may influence the outcome of the DSB fusion process (Roth and Wilson 1986). PARP-1 may also play a role in NHEJ, considering that DNA-PK_{cs} activity is enhanced by ribosylation and that PARP1 itself is subject to phosphorylation by DNA-PK_{cs} (Ruscetti *et al.* 1998). The crosstalk and mutual activation of PARP-1 and DNA-PK_{cs} may be involved in DSB recognition.

9. NHEJ AND P53

p53 appears to engage in crosstalk with both of the major DSB-repair pathways, NHEJ and HR-repair. As with BER, this crosstalk may involve three different levels of p53 activity that operate on different timescales: (i) rapid effects mediated by the direct interaction of the p53 protein with damaged or intermediate forms of DNA or (ii) direct interaction between p53 and DSB-repair proteins; and (iii) delayed effects mediated by the transcriptional transactivation of repair-related genes by activated p53. In addition, the NHEJ protein DNA-PK_{cs} can itself phosphorylate and activate p53, which can transactivate downstream target genes and invoke late responses such as apoptosis (Shieh *et al.* 1997, Wang *et al.* 2000), although the role of DNA-PK_{cs} in the activation of apoptotic pathways is contentious (Jimenez *et al.* 1999, Jhappan *et al.* 2000). A number of *in vitro* and *in vivo* model systems for NHEJ activity have been utilized to study these interactions, many of the former involving a restriction enzyme-induced DSB to parody features of those DSBs induced by IR. The common feature of these assays is that they report on some type of DNA end-joining event that does not involve extensive regions homology at the DNA ends and/or the availability of an intact homologous sequence or repeat sequence.

Several studies using *in vitro* end-joining assays or reporter-gene reactivation or integration assays have suggested that WT p53 *inhibits* or *suppresses* NHEJ. In the first such study, an *EcoR*I-linearized pUC19 DNA substrate was used to compare *in-vitro* end-joining by cellular extracts from TK6 (p53-WT), WI-L2-NS (p53-mutant) and TK6-E6-5e (p53-null) cells, which were derived from the same donor (Bill *et al.* 1997). There was ~2-fold higher plasmid end-joining activity in extracts from p53-deficient WI-L2-NS and TK6-E6-5e cells compared with extracts from p53-WT TK6 cells. Using a lacZ reporter gene, no significant difference was seen in repair fidelity (i.e., in misrepair frequency) among the three cell lines, and each cell line yielded similar repair/misrepair products. Thus, it was concluded that WT p53 suppressed end-rejoining activity while not influencing the legitimacy or type of rejoining events (Bill *et al.* 1997). Mallya and Sikpi (1998) compared the ability of non-irradiated or irradiated TK5 (p53-WT) and WTK1 (p53-mutant) human lymphoblastoid cell lines to process a pZ189 plasmid containing an *EcoR*I-induced DSB. They observed a similar baseline level of DSB rejoining in non-irradiated p53-WT and p53-mutant host cells; however, when the cells were exposed to 0.5 Gy of γ-rays immediately before probe transfection, the level of end-joining activity was increased by ~2 fold in p53-mutant host cells but was slightly *de*creased in p53-WT host cells, suggesting that WT p53 suppresses some aspect of DSB repair. Although irradiation of the host cells did affect the fidelity of DSB rejoining, p53 did not appear to influence this process. In a follow-up study, these authors compared the abilities of non-irradiated and irradiated p53-WT TK6 human lymphoblastoid cells and their E6-transformed derivatives TK6-E6-20C (p53-WT) and TK6-E6-5E ($p53^-$) to repair *EcoR*I-linearized pZ189 plasmid and the luciferase-reporter plasmid pGL3 (Mallya and Sikpi 1999). The constitutive end-joining activity of the non-irradiated p53-WT host cells was again 2-3 fold higher than for $p53^-$ cells. As in the earlier study, IR exposure selectively increased the DSB-rejoining capacity of cells lacking functional p53, supporting the suggestion that WT p53 somehow suppresses DSB rejoining following irradiation. Unlike rejoining capacity, changes in DSB-rejoining fidelity were similar in p53-WT and $p53^-$ host cells.

Based on these observations, it was hypothesized that WT p53 might also suppress the integration of plasmid DNA into host-cell genomes (Lee *et al.* 1999). Long-term transfection frequencies (which should reflect plasmid integration) were evaluated in isogenic hamster cell lines (CHO and CHO-p53) that express WT and mutant p53, respectively, both before and after exposure to IR to increase WT-p53 protein levels. End joining was probed using either pCDNA3 or pRep10 plasmid linearized with *Sca*I or *Sac*I, respectively. These plasmids exhibit little sequence homology with the host genome. Low doses of IR (~50-100 cGy) significantly increased the transfection efficiency in cells expressing mutant p53 while only modestly enhancing the transfection of p53-WT cells. The authors concluded that WT p53 may suppress some types of illegitimate recombination event that do not require extensive homology (Lee *et al.* 1999). They speculate that this activity could reflect either a suppression of incorrect end joining by WT p53 and/or a direct participation of WT p53 in verifying fidelity during end joining, as had been suggested by several other groups (e.g., Bakalkin *et al.* 1994, Brain and Jenkins 1994, Lee *et al.* 1995, Mummenbrauer *et al.* 1996, Sturzbecher *et al.* 1996, Bertrand *et al.* 1997, Xia and Liber 1997, Mekeel *et al.* 1997).

A greatly increased *in-vitro* DNA end-joining activity has been observed in cell-free extracts derived from $p53^-$ mouse embryo fibroblasts than from their $p53^+$ counterparts (Okorokov *et al.* 2002). Whereas extracts from $p53^+$ cells did not exhibit significant end-joining activity, extracts from $p53^-$ cells promoted considerable rejoining of a linear DNA substrate with either 5'-cohesive or blunt DNA ends. Adding purified WT-p53 protein to $p53^-$ extracts inhibited their end-joining activity. The Arg175His p53-mutant protein, which is one of the most common p53 mutations seen in human cancers, showed a more pronounced ability to suppress end-joining by $p53^-$ extracts, whereas this effect was blocked by the Ser15Asp phosphorylation-mimicking mutant protein. Binding to WT p53 *in vitro* was seen to protect DNA ends from exonuclease digestion and to inhibit T4 DNA ligase-dependent end-joining. These authors (Okorokov *et al.* 2002) speculated that WT p53 may similarly suppress DNA end-joining activity *in vivo* in non-stressed cells, and suggest that this activity may involve the ability of p53 to bind to DNA ends and thereby compete with NHEJ proteins. An inhibitory effect of WT p53 for donor sequence-free rejoining of plasmid DNA ends (i.e., NHEJ-like activity) was also reported by Akyuz and colleagues (Akyuz *et al.* 2002). Increased levels of cellular DSB repair were also reported in rat embryo fibroblasts when their WT-p53 function was abrogated by expression of dominant-negative mutants (Bristow *et al.* 1998). These investigators studied several fibroblast lines transfected with an activated form of the H-*ras* oncogene either alone or in combination with several missense p53-mutant alleles. Using the neutral comet assay following a continuous low dose-rate IR exposure, mutant p53-expressing clones were found to be more proficient at rejoining DSBs than p53-WT clones. Clones expressing mutant p53 also displayed increased radioresistance and decreased IR-induced G_1 arrest, but showed no alteration in their susceptibility to undergoing IR-induced apoptosis.

In marked contrast to the above reports, several other studies have suggested that WT p53 can *stimulate* NHEJ activity in some test systems. For example, Yang *et al.* (1997) investigated the role of p53 in *in-vivo* DNA end joining using a clonal cell line, *ts*FRO, generated by stably transfecting a *ts*-p53 (Val138) expression vector into a p53-null thyroid carcinoma cell line. DNA end-joining activity was assessed by transfecting linearized luciferase plasmid into *ts*FRO cells. Increased luciferase activity was observed when the cells were grown at the permissive temperature, suggesting that WT p53 enhanced end-joining activity while not activating an apoptotic response. Isogenic mouse embryo fibroblast lines of differing p53 status were also used to study the role of p53 in NHEJ of a defined DSB, specifically of cohesive double-stranded ends with a 5′ 4-base overhang (Tang *et al.* 1999). These investigators used an episomal plasmid-reactivation assay involving a *Hind*III-linearized reporter plasmid (pSV2-LUC) containing the luciferase gene. To induce p53, the cells were exposed to 2 Gy of γ-rays prior to transfection of the plasmid substrate. In cells with WT p53, DSB-rejoining activity increased ~3 fold post-irradiation, whereas $p53^-$ cells showed a slightly higher basal level of repair that was not enhanced by IR exposure. The enhancement of DSB rejoining seen in irradiated p53-WT cells was established to be independent of the transcriptional transactivation function of p53 using the p53*val*135 mutant, which is transcriptionally inactive and which, like the isogenic WT-p53 control, evoked an ~2.5-fold increase in DSB rejoining post-irradiation. Enhancement of DSB rejoining did, however, require an intact C-terminal domain and its associated non-specific DNA binding activity, which is known to be required for promoting the annealing of complimentary single-stranded sequences. Thus, the p53-AS mutant (which has an inactivating deletion in this region) failed to enhance DSB rejoining post-irradiation. Whereas p53 stimulated the rejoining of DSBs with cohesive ends, the rejoining of blunt-ended DSBs was not enhanced. Thus, p53 protein appeared to directly enhance DSB rejoining by facilitating the re-annealing of short complementary strands of single-stranded DNA (Tang *et al.* 1999). It has also been suggested that WT p53 may facilitate the precise ligation of a DSB (Lin *et al.* 2003).

Clearly, there are some major inconsistencies in the p53-NHEJ literature. Possibly the only real consensus is that the WT-p53 protein plays some direct role in NHEJ that is independent of its transcriptional-transactivation function. However, in some studies, loss of WT p53 led to increased levels of end-joining activity (Bill *et al.* 1997, Bristow *et al.* 1998, Mallya and Sikpi 1998, Lee *et al.* 1999, Okorokov *et al.* 2002, Akyuz *et al.* 2002), implying that WT p53 suppresses NHEJ, whereas other studies indicated a stimulatory role for WT p53 in NHEJ (Tang *et al.* 1999, Yang *et al.* 1997). Not surprisingly, some studies have concluded that WT p53 has no effect on DSB end-joining (Kohli and Jorgensen 1999). A number of factors might contribute to such contradictory findings. Firstly, comparison of these different data sets is complicated by the variety of cell types employed therein; how a DSB is processed by a cell would be anticipated to depend on the genetic background. The same caveat applies to the variety of DSB end-joining assays used, which often involve differing degrees of homology. It should be emphasized that the ability of these various assays to report on NHEJ *per se* in the cellular context has rarely been established, for example by using repair-deficient mutant cell lines. Also, several of these studies have evaluated only baseline end-joining activity, whereas others have examined IR-induced activity, or both. Additional confounding factors include the possibility that mutant p53 proteins used in some studies may confer a gain-of-function phenotype, and the possibility that cell-cycle distribution could modify the way in which DSBs are processed (Tang *et al.* 1999). Other authors have suggested that WT p53 might act on DNA-end processing mainly through a fidelity-control function, i.e., by suppressing error-prone, but not error-free, NHEJ (Akyuz *et al.* 2002), possibly involving the recognition of heterologies or exonucleolytic proofreading (Tang *et al.* 1999), functions that may differ in importance among the model systems used. It should also be remembered that, as discussed earlier for BER, p53 will necessarily have to stimulate the rate-limiting step of the NHEJ pathway for the particular assay system or cell type under investigation, which may differ for different model systems.

10. The RAD51 Homology-Directed HR-Repair Pathway

Compared to NHEJ, the details of HR-repair are fairly poorly understood (for reviews, see Thompson and Schild 2001, Valerie and Povirk 2003). In contrast to NHEJ, HR-repair depends on extensive regions of homology and provides a mechanism for the error-free repair of DSBs in the S- and G_2-phases, presumably because it optimally involves HR between sister chromatids. Early events in HR-repair include DNA-end processing and single-strand degradation, which likely involve the MRN complex that displays endonuclease, exonuclease and helicase activities and removes excess DNA at 3′ flaps (Paull and Gellert 1999, Trujillo *et al.* 1998) and is possibly involved in the resection of 5' DNA ends. The role of MRN in DSB repair appears to be both complex and multifunctional, as discussed earlier in the context of NHEJ. A candidate protein for removing 5′ flaps from HR-repair intermediates is FEN1. Important HR-repair proteins with regard to pairing and exchanging homologous sequences include RAD51, a DNA-dependent ATPase that catalyses strand exchange between homologous DNA molecules in the presence of RPA (Baumann *et al.* 1996, Benson *et al.* 1998). RAD51 interacts with a number of other proteins, including p53 (see below), RAD52, RAD54, SRS2, and BRCA2 that mediate, orchestrate, or sometimes counteract the activity of RAD51 in HR-related events. The five RAD51 paralogs, RAD51B, RAD51C, RAD51D, XRCC2, and XRCC3, are also involved in HR-repair and facilitate assembly of the RAD51 nucleoprotein filament. Various interactions among RAD51 and its paralogs have been described, suggesting that these proteins may function as a unit (Friedberg *et al.* 2006). Collectively, these proteins function in homology searching and strand-pairing/exchange. RAD51C may also play a role in the resolution of Holliday structures.

The RAD52 protein interacts with RAD51 to promote RAD51 polymerization and thus RAD51-mediated homologous pairing and strand exchange (Benson *et al.* 1998, McIlwraith *et al.* 2000), and it may act as a mediator of RAD51/RPA activity. RAD52 appears to be subject to complex but ill-defined post-translational regulation. RAD52 is phosphorylated following IR exposure, and this modification is required for efficient nuclear DSB repair-focus formation (Kitao and Yuan 2002, Yuan *et al.* 2003). Complex formation between RAD51 and RAD52 following IR exposure appears to be dependent on the co-operative activity of the ATM and c-ABL kinases (Chen *et al.* 1999). Interaction between RAD51 and RAD52 is negatively regulated by UBL1, an ubiquitin-like protein (Li *et al.* 2000). Another HR-repair factor, RAD54, belongs to the SNF2/SWI2 helicase family of DNA-dependent ATPases. RAD54 may function with RAD51 in chromatin remodeling (Kingston *et al.* 1996, Swagemakers *et al.* 1998). RAD54 can bind to RAD51, possibly stabilizing the protein-DNA complex and stimulating homology searching and strand invasion (Mazin *et al.* 2003). In the final stages of HR-repair, DNA polymerases re-synthesize the deleted DNA sequences using the intact homologous sequence as a template. DNA ligases then join the newly synthesized fragments. The Holliday junctions are resolved by endonucleases called "resolvases".

Other key proteins involved in HR-repair include BRCA1 and BRCA2, both of which play direct or indirect roles in DNA-damage responses. BRCA1, which is phosphorylated by ATM and appears to play a signalling role in response to IR, is required for the IR-induced phosphorylation of some ATM substrates, including p53, c-JUN, NBS1, and CHK2 (Kurz and Lees-Miller 2004). Its exact role in DNA repair remains unclear, although BRCA1 interacts with several repair-associated proteins such as the MRN complex. Furthermore, recent studies suggest that BRCA1, via interactions with CHK2 and RAD51, promotes both HR-repair and error-free NHEJ while suppressing error-prone NHEJ events (Zhang *et al.* 2004, Zhang and Powell 2005, Zhuang *et al.* 2006). BRCA2 is believed to be involved in the early stages of HR-repair, where it may play a chaperone-like function in orchestrating the handoff of reaction intermediates (Powell *et al.* 2002). BRCA2 interacts directly with RAD51 and RPA at the site of a DSB, an interaction that may help to load RAD51 onto the DNA (Marmorstein *et al.* 1998, Valerie and Povirk 2003).

11. HR-Repair and p53

WT p53 appears to play a major role in the regulation of HR in non-stressed cells as well as in both HR and HR-dependent DNA-repair responses following exposure of cells to IR and other genotoxic agents. In general, and as discussed below, inactivation of WT p53 results in enhanced levels of spontaneous and stress-induced HR. Many aspects of this regulation have been expertly reviewed elsewhere (e.g., Sengupta and Harris 2005, Friedberg *et al.* 2006), and these articles and the original references cited therein should be consulted for details. In contrast to some p53-regulated processes, WT p53 appears to modulate both spontaneous and IR-induced HR independently of its function as a transcription factor (Sengupta and Harris 2005). Details of the effects of p53 on HR-mediated DSB-repair processes are superficial, in part because of the paucity of appropriate model systems (Valerie and Povirk 2003). As with NHEJ, model systems for studying HR-repair activity typically involve a restriction enzyme-induced DSB to parody DSBs induced by IR or other agents; unlike the NHEJ assays, however, an intact (but dysfunctional) homologous sequence spanning the region of the DSB is provided either *in cis* or *in trans*. HR-repair is monitored by scoring a phenotypic change that is dependent on restoration of the expression of the full-length sequence following a successful exchange between the two overlapping but non-functional sequences. Several of these models involve an I-*Sce*I endonuclease-based system for generating the target DSB (Akyuz *et al.* 2002, Zhang *et al.* 2004). The recombination probes are either episomal or are integrated into the genome by applying selection pressure for a selectable marker within the probe.

The regulation of HR-repair by WT p53 may relate to the effects of p53 itself on HR-repair substrates or intermediates, or it may involve the interaction of p53 with other HR proteins. Probably the best defined and arguably most important such interaction of p53 is with the RAD51 protein, a binding that appears to inhibit RAD51 activity in HR (Sturzbecher *et al.* 1996, Buchhop *et al.* 1997) but possibly stimulates some aspects of HR-repair, particularly those relating to the fidelity of strand annealing and exchanges. Although interactions between p53 and RAD51 have usually been defined using *in vitro* models, the direct interaction with and suppression or stimulation of RAD51 activity (depending on the circumstances) may be the major mechanism through which p53 regulates both HR and HR-repair *in vivo*. A number of activities for p53, both with and without its associate, RAD51, have been described with respect to the processing of DNA ends or structures that may represent substrates for or intermediates in HR-repair. Through its C-terminal domain, the WT-p53 protein itself can bind to and process a broad range of unusual DNA structures such as free ends, bulges and mismatches (e.g., Steinmeyer and Deppert 1988, Jayaraman and Prives 1995, Bakalkin *et al.* 1995, Reed *et al.* 1995, Lee *et al.* 1995, Yokote *et al.* 1998, Albrechtsen *et al.* 1999, Liu and Kulesz-Martin 2001), a function that appears to be modified by its association with RAD51. p53, in association with RAD51, can also bind directly to Holliday junctions and facilitate their cleavage by endonucleases (Sturzbecher *et al.* 1996, Lee *et al.* 1997). WT p53 can sense base mispairings within HR intermediates (Dudenhoffer *et al.* 1999) and, as noted earlier, it possesses an inherent 3′→5′ exonuclease activity that may play a role in the repair of such structures (Oberosler *et al.* 1993, Mummenbrauer *et al.* 1996), and indeed RAD51 has been reported to enhance this activity during the degradation of mismatch-containing heteroduplexes (Susse *et al.* 2000). The repeated observation that WT p53 itself or the p53-RAD51 complex can interact with and degrade some types of HR intermediates (Lee *et al.* 1997, Dudenhöffer *et al.* 1998, Susse *et al.* 2000, Janz *et al.* 2002) has led to the hypotheses that p53 or p53-RAD51 monitors the fidelity of strand-exchange events (e.g., involving base mismatches) by the HR machinery, and further that some p53 mutations result in loss of this proofreading activity (Dudenhoffer *et al.* 1998, Janus *et al.* 1999, Gebow *et al.* 2000, Sengupta and Harris 2005).

In addition to RAD51, WT p53 interacts physically and/or functionally with a number of other proteins involved in HR-repair, including BRCA2, RPA, and the RecQ subfamily of 3'→5' DNA helicases that includes WRN and BLM (Dutta *et al.* 1993, Miller *et al.* 1997, Marmorstein *et al.* 1998, Alberchtsen *et al.* 1999, Blander *et al.* 1999, Wang *et al.* 2001, Yannone *et al.* 2001). WT p53 also binds to RAD54 through its C-terminal domain (Linke *et al.* 2003). These interactions presumably modify and “fine-tune” HR-dependent responses, but their role in regulating the repair of IR-induced DSBs is largely unknown. The likelihood that p53 engages in crosstalk with the WRN and BLM proteins, which are defective in Werner and Bloom syndromes, respectively, is of interest because these helicases are important for HR-repair and genome stability (Hickson 2003, Valerie and Povirk 2003, Sengupta and Harris 2005). These helicases appear to interact with p53 at various levels, including physically and functionally, as well as at the transcriptional and post-transcriptional levels, one outcome being modulation of their helicase activity on Holliday junctions *in vitro* (Blander *et al.* 1999, Spillare *et al.* 1999, Wang *et al.* 2001, Yang *et al.* 2002, Valerie and Povirk 2003).

The phenotypic implications of abrogation of WT-p53 function on HR-dependent processes is best defined for spontaneous HR events. p53, either by itself or through its ability to inhibit RAD51 and/or interact with other HR factors such as RAD54, is well known to suppress certain aspects of HR and thereby prevent the development of genomic instability, consistent with its billing as the “Guardian of the genome” (Sturzbecher *et al.* 1996, Wiesmuller *et al.* 1996, Bertrand *et al.* 1997, Lee *et al.* 1997, Mekeel *et al.* 1997). Consistent with such a role, many human tumor cell lines with mutated or no p53, or mammalian cell lines in which WT p53 had been inactivated using viral oncogenes or dominant-negative p53 mutants, showed a reduced inhibition of RAD51 and/or dramatically increased rates of spontaneous and IR-induced inter- and intra-chromosomal HR (Wiesmüller *et al.* 1996, Sturzbecher *et al.* 1996, Bertrand *et al.* 1997, Honma *et al.* 1997, Buchhop *et al.*

1997, Mekeel *et al.* 1997, Linke *et al.* 1997, Dudenhöffer *et al.* 1998, Dudenhöffer *et al.* 1999, Saintigny *et al.* 1999, Willers *et al.* 2000, Akyuz *et al.* 2002). Transactivation-deficient p53 mutants generally retain most of the WT inhibitory activity (Willers *et al.* 2000, Akyuz *et al.* 2002, Sengupta *et al.* 2003, Linke *et al.* 2003), suggesting that suppression of HR is independent of p53s role as a transcription factor, although transactivation-dependent responses may become more relevant as the degree of sequence homology decreases (Akyuz *et al.* 2002). In keeping with these observations, over-expression of WT p53 suppressed the HR-repair of chromosomal and extra-chromosomal substrates (Wiesmuller *et al.* 1996, Akyuz *et al.* 2002). This scenario does not appear to apply to every type of HR event, including those between some episomal recombination substrates (Willers *et al.* 2001) and IR-induced intra-chromosomal HR in mice (Aubrecht *et al.* 1999). Also, the ability of WT p53 to suppress HR-repair events was optimal between the sequences of short homology, suggesting that WT p53 is more effective in suppressing error-prone HR-repair events (Akyuz *et al.* 2002) and that p53 may create a threshold for HR between short and long homologies (see also Gebow *et al.* 2000).

These observations collectively indicate an important but complex role for the WT-p53 protein in monitoring and restricting strand exchanges between homologous DNA sequences, both alone and through its interactions with other HR factors such as RAD51, essentially controlling the balance between pro- and anti-recombinogenic events. The consensus is that WT p53 plays an important role in fidelity control during HR by restraining strand exchange between sequences of imperfect homology (Dudenhöffer *et al.* 1998, Susse *et al.* 2000, Akyuz *et al.* 2002).

In considering the potential role of WT p53 in regulating HR-repair in irradiated cells, a complication arises from the fact that such cells will contain numerous IR-induced DSBs (~40 per cell per Gy) that, if not rejoined by one repair pathway or another, can also participate in deleterious HR- or NHEJ-mediated interactions with other regions of the genome. Presumably the elevation of WT p53 levels associated with IR exposure (see section 1) serves to suppress this sequence of events. However, many of the recombination proteins (e.g., RAD51) involved in these negative responses that are suppressed by p53 are also involved in DSB-repair pathways, so the cell is faced with something of a quandary: the non-specific inhibition of these activities by WT p53 could also inhibit DSB repair. This quandary appears to be resolved by WT p53 exerting a general suppressive effect on HR while making the HR-repair process (and possibly NHEJ) less promiscuous.

There are several possible strategies by which the various options for processing a cellular DSB might be coordinated. A commitment to one pathway versus another could be dictated by the nature of the initial protein-binding/damage-recognition event. For example, the decision as to whether a DSB is processed by HR-repair or NHEJ has been suggested to involve a competition between the binding of the RAD52 and KU proteins to the DSB (see Okorokov *et al.* 2002, and references cited therein). Alternatively, HR-repair and NHEJ may be coupled, rather than competing, pathways such that intermediate structures in DSB processing may be actively directed to NHEJ or HR-repair depending on the evolving circumstances (Richardson and Jasin 2000). As noted above, a number of proteins are involved in the early stages of both the NHEJ and HR-repair of DSBs and might therefore be considered as candidates for involvement in the regulation and switching of DSB-repair options following IR exposures. For example, the MRN complex seems to be involved in the early stages of both pathways (Kuschel *et al.* 2002, Valerie and Povirk 2003). The WRN protein, which plays a role in HR-repair and interacts with p53 during the DNA-damage response (Blander *et al.* 1999, Yannone *et al.* 2001), also interacts with the KU heterodimer, resulting in increased WRN exonuclease activity, suggesting a link between WRN and NHEJ (Cooper *et al.* 2000, Li and Comai 2000, Valerie and Povirk 2003). WT p53, because of its indispensable role in signal integration in the DNA damage-response pathway, is likely to be a key regulator of the HR-repair versus NHEJ decision, possibly through its direct involvement in sensing DNA damage (Gebow *et al.* 2000). In keeping with such a coordinating role, p53 has been variously reported to either inhibit or stimulate different aspects of both NHEJ and HR-repair in response to DSBs (see above).

One issue of note is the possibility that other mechanisms might be used to rejoin IR-induced DSBs in some situations, including homology-directed single-strand annealing (SSA), a process that shares components with both NHEJ and HR-repair (Valerie and Povirk 2003). It has been suggested that RAD54 may regulate switching decisions between HR-repair and SSA (Dronkert *et al.* 2000). Other studies suggest that, whereas RAD51 appears to be a limiting factor for HR-repair in some types of mammalian cells, RAD52 may be rate-limiting for SSA events, such that excess RAD52 may inhibit RAD51-dependent strand-invasion HR events (Yanez and Porter 1999). Cells may also use recombination mechanisms such as break-induced replication (BIR) to rejoin a DSB under some circumstances, but little is known about this process or the role of p53 therein. Finally, as was noted in section 6, it has been suggested that some incomplete BER intermediates may be switched into the HR-repair pathway, implying some level of crosstalk between the BER and recombination pathways that may involve p53 (Sobol *et al.* 2003).

12. Phenotype of p53-Deficient Human Cells

As discussed above, there is considerable evidence to suggest that p53 participates in DSB rejoining by interacting with various key proteins in both the HR-repair and NHEJ pathways, the two distinct but complimentary mechanisms for DSB repair (e.g., Sedelnikova *et al.* 2002, Fei and El-Deiry 2003). So what is the phenotypic impact of p53 deficiency on the DSB-repair phenotype of human cells? One widely-used model system for such purposes is cells derived from patients with Li-Fraumeni syndrome (LFS), a hereditary cancer-susceptibility syndrome that commonly involves germline p53 mutations, although some LFS patients with WT p53 have been found to have mutations in the *CHK2* gene (Lee *et al.* 2001). Human LFS fibroblasts lacking p53 are defective in global NER of UV-induced cyclopyrimidine dimers, a phenotype that has been attributed to defective repair of non-transcribed sequences (Ford and Hanawalt 1995, Wang *et al.* 1995, Mirzayans *et al.* 1996). This observation is consistent with the known role of p53 in NER (Ford 2005). But what about DSBs induced by biologically-relevant doses of IR? In our laboratory, we examined this question using LFS fibroblast strains harboring either a germ-line (codon 254; strains 2674T and 2675T) or somatic (codon 234; strain 2800T) mutation in one allele of the p53 gene (Mirzayans *et al.* 2006). Using the neutral comet assay, each of the p53-deficient

(2800T, 2674T and 2675T) LFS strains were found to exhibit significantly slower DSB rejoining than the normal human fibroblast strain GM38 following IR exposure. For example, Figure 5A shows the levels of residual DSBs in these various fibroblast strains measured using the neutral comet assay at 24 h after exposure to 8 Gy of γ-rays. By 24 h, essentially all detectable DSBs were rejoined in normal fibroblast cultures, whereas significant levels of DSBs remained unrejoined in the p53-deficient cultures. A similar picture emerged when these LFS strains were evaluated using the γ-H2AX assay in which DSBs are quantified indirectly based on the extent of phosphorylation of the variant histone H2AX (to γ-H2AX), as quantitated using anti-γ-H2AX antibodies (Olive and Bana'th 2004). At 24 h after irradiation, γ-H2AX levels in 2800T, 2674T and 2675T cells were much greater than in normal GM38 fibroblasts (Figure 5B), consistent with the DSB-repair defect suggested using the neutral comet assay.

Although these findings are consistent with a role for WT p53 in DSB rejoining in human fibroblasts, we should note that p53-deficient LFS cells are genetically unstable (e.g., Boyle *et al.* 1998), and this instability could impact on their repair phenotype secondarily to the p53 mutation. Nonetheless, these data do corroborate an earlier study showing that DSB rejoining following exposure to 20 Gy of γ-rays, as measured using the neutral comet assay, was also slower in p53-deficient murine fibroblasts than in their p53-WT counterparts (Al Rashid *et al.* 2005).

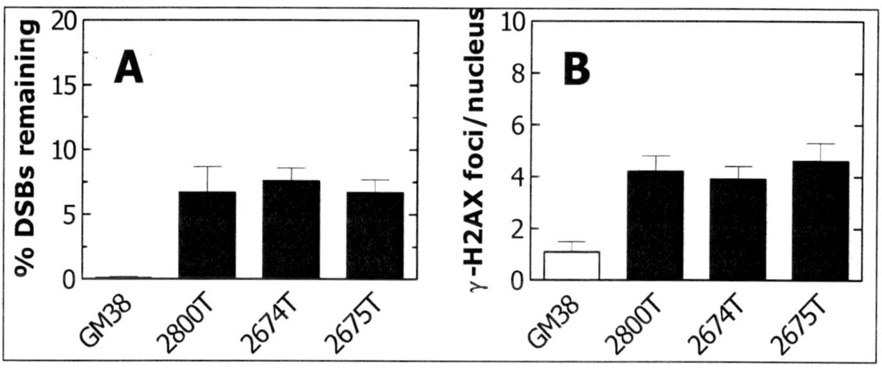

Figure 5. Impact of p53 deficiency on the repair of IR-induced DSBs in the human fibroblast background. DSB rejoining in p53-WT (GM38) and p53-deficient (2800T, 2674T and 2675T LFS) human fibroblasts after exposure to 8 Gy of γ radiation as measured using: [A] the neutral comet assay (bars represent the percentage of the initial lesions remaining at 24 h post-irradiation, means ± SE); or [B] the γ-H2AX focus assay (bars represent the actual level of foci present at 24 h post-irradiation, means ± SE). Adapted from Mirzayans *et al.* (2006).

So what about the phenotypic impact of p53 deficiency in the human tumor-cell background? To address this question, we used the neutral comet and γ-H2AX assays to compare DSB rejoining in the p53-WT human colorectal carcinoma cell line HCT116 and its p53-knockout counterpart, clone 379.2, in which the p53 gene was homozygously disrupted by targeted HR (Bunz *et al.* 1998). The cells were a generous gift of Dr. Bert Vogelstein, Johns Hopkins University, Baltimore, Maryland. The data obtained were very similar to those described above for LFS cells. Whereas the rate of DSB rejoining in HCT116 cultures after exposure to ≤ 8 Gy of γ-rays was similar to that in two normal fibroblast strains, GM38 and

GM10, the p53-deficient 379.2 cells displayed greatly impaired DSB rejoining as evidenced by both assays (Mirzayans *et al.* 2006).

Although it is important not to over-interpret such data, e.g., because of the many known genetic defects in the parental HCT116 cell line, such as its known mismatch repair-deficient status resulting from a hemizygous mutation in h*MLH1* resulting in a truncated/nonfunctional MLH1 protein (Boyer *et al.* 1995), essentially the same result was obtained for the paired $p53^{+}/p53^{-}$ fibroblast and tumor cell lines, strongly suggesting that WT p53 stimulates the overall rate of DSB rejoining in cultured human cells. Whether this effect is universal remains to be established. Certainly some earlier studies of human tumor cell lines with differing p53 status did not support such a role for WT p53 in DSB rejoining, although in these studies DSBs were evaluated using less-sensitive techniques that necessitated the use of very high (e.g., 40 Gy) doses of IR (DiBiase *et al.* 1999, Orre *et al.* 2006). Such inconsistencies concerning the role of p53 in DSB rejoining may therefore reflect differences in the cell types, DSB assays and IR doses used, as well as in the uncertainty as to which pathway is responsible for the observed repair.

13. Dose-Response Relationships for the Activation of DNA-Damage Responses

In terms of developing comprehensive models for the integration of DNA damage-response pathways (Wang and Cho 2004), as well as for comparison among different data sets, it is critical to consider the IR dose levels used in any given study. Of particular interest here are those studies in which a range of low doses was used. Sengupta and Harris (2005) coined the term "cellular rheostat" to describe the role of p53 in this context. They suggest that, after low-dose IR exposures, the WT p53 protein may interact with the DNA-repair machinery and facilitate repair; in contrast, after higher doses (when DNA damage can not be handled by p53 alone), p53 undergoes stabilization and activation as a transcription factor, e.g., for checkpoint genes. The biological objectives of having a graded dose response to IR would be to ensure that: [1] after low doses, the cell will not 'overreact' to low levels of DSBs; [2] after modest doses the cell will have the maximum opportunity to repair and survive; and [3] after high doses the cell will die such that, in multi-cellular systems/tissues, heavily-damaged cells would be eliminated from the population.

Thus, crosstalk between p53 and the DNA-repair machinery, and indeed with any effector response, should intuitively be dependent on the dose of IR used. Certainly the literature on the impact of p53 deficiency on the BER pathway reflects this situation. Thus, as noted in section 6, the ability of p53 to facilitate BER was highly dependent on the dose of IR. Whereas low doses up to ~0.5 Gy (where p53 levels are typically quite low) resulted in enhanced BER activity, higher doses caused a rapid inhibition of BER activity and invoked p53-dependent apoptosis (Offer *et al.* 2002). In $p53^{-}$ cells there was no such transition, with BER activity increasing progressively with dose. In further experiments it was established that the decision of $p53^{+}$ cells to deactivate the BER pathway at higher doses was independent of the decision to execute the apoptotic program. These effects were probably not determined entirely by the cellular levels of p53; in fact, different doses of IR also resulted in different patterns of post-translational modification of p53 that would be expected to impact on its DNA-binding activity and possibly on its ability to crosstalk with the BER pathway. Indeed, p53 appears to selectively bind to different promoters depending on its post-translational modification status (Giaccia and Kastan 1998, Brooks and Gu 2003, Oren 2003). Also of interest in this regard is the observation that the dose-response for the activation of ATM in human cells showed a graded increase over the 10-50 cGy range, above which dose a plateau was reached (Bakkenist and Kastan 2003).

The literature relating to the phenotypic impact of p53 status on DSB rejoining in the biologically-relevant dose range (i.e., below ~10 Gy) is essentially absent because, until recently, cellular DSB assays were too insensitive to be used with such doses. As was noted in section 12, earlier investigations generally employed extremely high (supra-lethal) doses of IR to study DSB rejoining in cells, which could be misleading because dead or dying cells will not necessarily implement appropriate DNA-damage responses. The use of supra-lethal doses might even evoke completely different response pathways, with high doses possibly engaging alternative or redundant ATM-independent pathways that are not relevant at low doses (e.g., Kurz and Lees-Miller 2004). There is indeed some evidence to suggest differences in the repair machinery that processes DSBs in cells exposed to relatively low doses ($\leq$ 10 Gy) versus supra-lethal doses of IR (Nevaldine *et al.* 1997, Mirzayans *et al.* 2004). Such dose-dependent effects may explain some of the discrepancies in the literature relating to the impact of p53 deficiency on DSB rejoining in human cells (see section 12), and which suggest that WT p53 may play a much greater role in DSB rejoining after exposure to physiologically-relevant doses of IR.

There may be important biological consequences of the non-linear nature of these dose-response curves for the activation of different cellular responses to low (single) doses of IR or to low dose rate exposures. Thus, Marples, Joiner and colleagues have reported that many human cell lines exhibit a hypersensitivity to killing by low single doses of IR, followed by a region at doses in the 0.5-Gy range where survival increases before decreasing again above ~1 Gy (Marples *et al.* 1997, Joiner *et al.* 2001). This behavior has become known as "low dose hyper-radiosensitivity-increased radioresistance" (HRS-IRR), and appears to involve inducible cytoprotective mechanisms such as DNA-repair pathways that are triggered above some threshold level of damage.

For cell types that exhibit HRS-IRR, an expectation would be that DSB-repair pathways should be stimulated above the critical dose threshold, i.e., above ~40 cGy. Using the neutral comet assay, we (Murray *et al.* 2006) demonstrated that GM38 human fibroblasts did indeed show more extensive rejoining of DSBs following low doses (below ~40 cGy) than after higher doses (2 Gy) of γ-rays, consistent with this model (Figure 6). Not surprisingly, the HRS-IRR phenotype is highly dependent on cellular p53 status. Thus, a number of cells lines with abrogated p53 (i.e., A549 and T98G cells treated with the p53 inhibitor pifithrin, and the 379.2 HCT116-p53^{-} clone and the 2800T LFS human fibroblast strain described in section 12) were totally lacking the HRS-IRR response, whereas their p53^{+} counterparts all showed a typical HRS-IRR response, indicating a requirement for functional WT p53 therein (Enns *et al.* 2004). One of the tumor lines examined, T98G, in which p53 is mutated at Met-237 resulting in a severe reduction in transcriptional transactivation, showed a full HRS-IRR response, suggesting that p53-dependent transcription may not be essential for this response.

Other researchers have indicated that WT p53 is *not* required for cells to exhibit HRS-IRR, and in fact many human tumor cell lines were found to exhibit a marked HRS-IRR phenotype (Marples *et al.* 2004). The reasons for this discrepancy are not apparent, but clearly require further evaluation, especially if this information is to be applied to tumor radiotherapy. It is also of interest that the HRS-IRR phenomenon exhibits some resemblance to the above-mentioned dose-dependency of BER activity (Offer *et al.* 2002). Furthermore, the shape of the IR dose-response for the serine-1981 autophosphorylation of ATM (Bakkenist and Kastan 2003), the major kinase responsible for p53 activation post-irradiation, is also quite similar to that for HRS-IRR, implicating ATM as a critical regulator of the HRS-IRR response (Marples 2004).

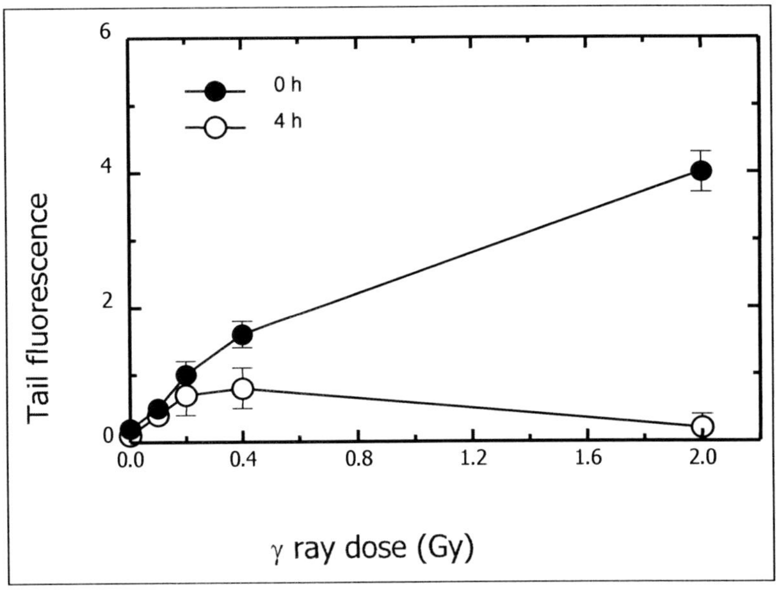

Figure 6. Effect of dose on the extent of repair of IR-induced DSBs in GM38 human normal fibroblasts. DSBs were evaluated by the neutral comet assay in GM38 fibroblasts either immediately after (●) or 4 h after (○) exposure to different doses of γ rays. Taken from Murray *et al.* (2006).

Although space prohibits an extensive discussion of this topic, the response of cells to low doses of IR delivered at low dose rates is also an important area of research. Major questions here include whether such chronic exposures might fail to stimulate the typical DNA damage-response pathways seen after acute doses of IR and how such effects depend on the cells' p53 status (which is critically relevant to cancer therapeutics). Consistent with this thinking, the autophosphorylation of ATM at serine 1981 in human tumor cell lines was much reduced after low dose rate versus acute exposures (Collis *et al.* 2004). The activation of γ-H2AX, an important downstream target of ATM, was also diminished at low dose rate (Collis *et al.* 2004). Similar findings have been reported using human fibroblasts immortalized with the hTERT gene (Ishizaki *et al.* 2004); whereas acute IR exposures produced significant phosphorylation of p53 at serine 15 (which is largely mediated by ATM) and high levels of γ-H2AX foci, these effects were greatly diminished following a protracted low dose rate exposure.

The "adaptive response" is another low-dose phenomenon in which exposure of cells to a low "priming" dose of IR can invoke resistance to subsequent high-dose exposures. Whether DNA repair is enhanced following a priming exposure to a low dose of IR, and whether this enhancement is dependent on p53, are important questions (Murray *et al.* 2006). In some studies, priming treatments were indeed found to enhance DNA repair. For example, priming A549 human lung tumor cells with 25 cGy of γ rays stimulated the repair of thymine glycol when the same cells were exposed to a second "test" dose (2 Gy) of γ rays 3 h later (Le *et al.* 1998). Priming has also been reported to accelerate the repair of cellular DSBs in some studies (e.g., Ikushima *et al.* 1996) but not in others (Murray *et al.* 2006). Thus, like the effects of p53 on DSB rejoining, such effects may depend on the cell type and experimental conditions and/or on the assay being used to detect DNA damage. At least some adaptive responses to IR appear to critically depend on WT p53, which has been suggested to channel IR-induced DSBs into an adaptive "legitimate" repair pathway (Sasaki *et al.* 2002). Also, Michael Weinfeld and his colleagues (personal communication) have observed that the above-mentioned stimulated repair of thymine glycol in p53-WT cells receiving a priming dose of γ rays was not apparent in p53-deficient cells.

14. Processing of Complex Lesions by BER

A unique feature of IR-induced damage to cellular DNA is that, because of the micro-heterogeneous nature of IR tracks, large amounts of energy can be deposited in discrete regions known as spurs, blobs and short tracks; when these events coincide with chromatin, they result in the formation of complex "clustered damages" or multiply damaged sites in which several lesions occur within one or two helical turns of the DNA molecule. These are in fact the "signature" lesions caused by IR (Ward 2000) and they will be relatively difficult for the cell to repair because both strands of the helix are locally damaged. The majority of non-DSB clusters consist of oxidized purines and pyrimidines and AP sites. How are such complex lesions processed by the cellular DNA-repair pathways? If such lesions are recognized first by the BER machinery, this could create a situation in which attempted repair of a complex lesion containing several base damages and/or single-strand breaks by the BER pathway could convert otherwise non-lethal lesions into lethal DSBs or other types of complex lesion. An obvious question is, does anything coordinate this decision, and might p53 play a role in this process? What is known is that the levels of such clustered damage were independent of the cells' p53 status (Bennett *et al.* 2004).

15. Role of Chromatin Structure/Acetylation in DNA Repair, and the Role of p53 Therein

Recent evidence strongly implicates chromatin remodeling and repair-factor accessibility as a key early event in the cellular response to lesions such as DSBs in both yeast and mammalian cells, with post-translational histone modifications, and especially acetylation involving proteins such as the NuA4 histone acetyltransferase (HAT) complex, playing an important role in both the early and late stages (i.e., in post-repair chromatin restoration) of DSB repair (Downs *et al.* 2004, Utley *et al.* 2005). p53 does appear to modulate chromatin accessibility in response to UV (Allison and Milner 2004, Rubbi and Milner 2003) and may play a more general role in chromatin remodeling and in determining the access of repair complexes to DNA damage (Adimoolam and Ford 2003, Ford 2005). Also, binding of p53 to the nuclear matrix increases following genotoxic stress (Jiang *et al.* 2001) and this has been identified as the rate-limiting factor in the repair of higher-order DNA structures (Aranda-Anzaldo *et al.* 1999). Clearly, there are important but complex relationships between histone acetylation and DSB repair, possibly involving p53, that need to be understood in greater detail. p53 itself is also subject to acetylation within its C-terminal regulatory domain (Figure 2) involving the HAT activity of the p53 transcriptional coactivators p300/CBP and PCAF (p300/CBP associated factor) (Sakaguchi *et al.* 1998, North and Hainaut 2000, Appella and Anderson 2000, Grossman 2001). p53 acetylation can be reversed by histone deacetylase complexes such as HDAC1 and SIR2 (Luo *et al.* 2004). IR-induced DNA damage also induces the rapid acetylation of ATM which depends on another HAT, Tip60 (Sun *et al.* 2005).

16. Interplay Between DNA Repair and p53 Status at the Whole-Animal Phenotype Level

An important aspect of WT p53's role as the "Guardian of the genome" (Lane 1992) is its ability to trigger accelerated senescence and apoptosis to prevent the propagation of genetically unstable cells. Indeed, NHEJ-deficient mice generally show an enhanced incidence of cancers if their p53 gene is disrupted (e.g., Nacht *et al.* 1996, Lim *et al.* 2000). Thus, the phenotype of a DNA repair-deficient animal can be critically dependent on p53 status.

17. Nucleotide Excision Repair and p53

The role of p53 in both the global and transcription-coupled pathways of NER are well described. However, the role of NER in the cellular response to IR-induced DNA damage is generally regarded as minor, so this topic will not be discussed further here. For a discussion of such effects, the reader is referred to one of several excellent existing reviews on this subject (Ljungman and Lane 2004, Ford 2005, Sengupta and Harris 2005).

18. Role of p63, p73, and c-ABL

In addition to p53, mammalian cells also contain two p53-related genes, *p63* and *p73*, that encode transcription factors with shared structures and functions (Levrero *et al.* 2000, Melino *et al.* 2002). As with p53, DNA damage promotes the stabilization of the p63 and p73 proteins and activates their transcriptional functions (Gong *et al.* 1999, Flores *et al.* 2002). Some genes are regulated by all three p53 family members whereas others are regulated by p63 or p73 and not by p53 (Melino *et al.* 2002). All three proteins appear to contribute to IR-induced apoptosis (Flores *et al.* 2002). However, our knowledge of their role in DNA repair is minimal at this time. Stabilization and activation of p73 by DNA damage requires the nuclear c-ABL tyrosine kinase, whose activation by IR requires functional ATM (Baskaran *et al.* 1997, Shafman *et al.* 1997, Agami *et al.* 1999, Gong *et al.* 1999, Yuan *et al.* 1999), with a possible contribution from p53-dependent transcriptional activation (Chen *et al.* 2001). Phosphorylation of p73 by c-ABL activates its pro-apoptotic function (Costanzo *et al.* 2002). The CHKl kinase can also phosphorylate and activate p73 (Gonzalez *et al.* 2003).

As with p53, c-ABL is an important integrating factor in the DNA damage-response pathway. As noted above, c-ABL in the nucleus responds to signals from DNA damage and activates apoptosis (Wang 2000). Nuclear c-ABL has also been implicated in the tyrosine phosphorylation of the NHEJ protein DNA-PK_{cs} (Kharbanda *et al.* 1997) and of the HR-repair factors BRCA1 (Foray *et al.* 2002), RAD51 (Yuan *et al.* 1998, Chen *et al.* 1999) and RAD52 (Kitao and Yuan 2002), implying some role in the regulation of DSB repair. However, the impact of these tyrosine-phosphorylation events on phenotypic DNA repair is poorly understood. c-ABL may also be an important regulator of p53 during the cellular response to DSBs. Following its ATM-dependent activation in response to IR exposure (Baskaran *et al.* 1997, Shafman *et al.* 1997), interaction between c-ABL and p53 is enhanced (Yuan *et al.* 1996). c-ABL can contribute to p53 accumulation by preventing its ubiquitination and nuclear export (Sionov *et al.* 2001).

Interestingly, the activation of nuclear c-ABL by DNA damage in fibroblasts is dependent on cell adhesion (Truong *et al.* 2003). Consequently, detached fibroblasts are more resistant to apoptosis caused by anticancer drugs such as cisplatin than are fibroblasts attached to extracellular matrix. Thus, c-ABL appears to represent an integration point for nuclear and extracellular signals in the DNA damage-response pathway (Zhu and Wang 2004). The importance of such effects for IR-mediated events has not been established.

19. Additional Factors

Several additional areas of research are providing potentially very important insight into the integration of DNA repair and checkpoint pathways. For example, aspects of the crosstalk between p53 and DNA-repair pathways may be indirect, involving other cellular proteins such as the E2F family of transcription factors. E2F1 stabilization in response to DNA damage depends on p53 and on phosphorylation by ATM/ATR (Blattner *et al.* 1999, Lin *et al.* 2001). E2F transcriptional targets include DNA-repair genes involved in BER (UNG), NHEJ (DNA-PK_{cs}), and HR-repair (RAD51, RAD54) (Ren *et al.* 2002, Polager *et al.* 2002).

E2F1 is an important contributor to the regulation of p53 activity and it induces the multiple phosphorylation of p53 in a similar manner to DNA damage (Rogoff *et al.* 2002). Another important protein is PARP-1, which has been mentioned several times as a factor in DNA-repair processes, including BER and NHEJ. Several checkpoint and DNA repair-related proteins, such as $p21^{waf1}$, LIG3, XRCC1, DNA-PK_{cs} and KU70, and most importantly from the current perspective, p53, contain a specific poly(ADP-ribose)-binding sequence motif through which PARP-1 could modify their functions (Pleschke *et al.* 2000). Some studies also suggest a role for PARP-1 in regulating the accumulation, activation and DNA binding of p53 following genotoxic stress (Oliver *et al.* 1999). There may therefore be an important interrelationship between p53 and ADP-ribosylation in the integration of the cellular response to IR. Another protein that might influence the crosstalk between p53 and DNA-repair pathways is human DMC1, a RAD51 paralog which forms helical filaments in the presence of ATP and promotes DNA strand transfer in an RPA-dependent reaction (Masson *et al.* 1999, Sehorn *et al.* 2004). Yet another is *KARP-1*, which encodes an isoform of KU80 that is transcribed from an upstream promoter within the same locus. In contrast to *KU80*, *KARP-1* mRNA expression is IR inducible and induction is mediated by p53 and ATM, this observation being consistent with the presence of a p53 binding site within the second intron of the *KARP-1* locus (Myung *et al.* 1998). KARP-1 may play some role in mammalian DSB repair (Myung *et al.* 1998).

20. Conclusion

The WT p53 tumor-suppressor protein plays a broad range of roles in orchestrating the cellular response to DNA damage that have earned it the title of "Guardian of the genome" (Lane 1992). These activities involve direct functions of the p53 protein, acting alone or in combination with its supporting cast of binding partners, or are mediated via the transcriptional transactivation of an array of downstream target genes. In this review we have focused on how the WT-p53 protein crosstalks with the three major DNA-repair pathways involved in processing IR-induced DNA damage, namely BER, NHEJ and HR-repair. Probably the most comprehensible data have been obtained for the BER pathway: this is in large part because of the existence of several powerful *in vitro*/cell-free assays, including reconstitution assays, for studying cell extracts and purified proteins, as well as for following the time-course of specific lesions such as 8-oxoG in cellular DNA under circumstances where a single enzymatic system is known to effect the repair of the lesion under observation. In general, while few studies have examined the repair of IR-induced damage *per se*, the existing literature would suggest that: [1] the WT-p53 protein stimulates the BER of a variety of base lesions; and [2] this stimulation is largely mediated by the direct interaction of the p53 protein with BER factors such as HAP1/OGG1 and POLβ. The role of p53-dependent transcriptional transactivation in stimulating BER is less certain, and will require further evaluation. Studies of the repair of IR-induced lesions such as 8-OxoG and thymine glycol in the genome of mammalian cells with differing p53 status are clearly feasible and would provide major insight into this area.

The potential influence of WT p53 on NHEJ of IR-induced DSBs is less evident. Although WT-p53 protein clearly plays a direct role in NHEJ in simple model systems, several studies indicate that p53 suppresses this pathway, while other studies indicate a stimulatory role. As discussed in section 9, the interpretation of such data is compromised by the range of genetic backgrounds and end-joining assays used. Furthermore, there is no model with which NHEJ activity can be exclusively evaluated in intact cultured cells; simply measuring the removal of total DSBs from the genome (e.g., Figure 5) will represent the combined effect of all pathways that can operate on these lesions. The same caveat applies to assessing the HR-repair of DSBs. Nonetheless, the existing data would suggest that p53 does modulate HR-repair, at least for "model" DSBs, primarily through its interaction with RAD51. The most likely outcome of the p53-RAD51 interaction is an increase in the fidelity with which the HR-repair machinery executes strand-exchange events, in particular by suppressing exchanges between imperfectly-homologous DNA sequences. However, as noted above, such conclusions are based on model restriction enzyme-induced DSBs in artificial probe sequences, and it will be critical to extend such studies to actual IR-induced DSBs within the genome.

From the cancer treatment perspective, a key issue is how defects in DNA damage-response pathway might influence the responses described so far. For example, how might the p53 status of tumor cells impact on their DSB-repair capability? Are the HR-repair and NHEJ processes different in normal versus tumor cells? Certainly a number of differences have emerged between WT and mutant p53 in this regard. Thus, some cancer-derived p53 mutants do not appear to recognize Holliday junctions (Janz *et al.* 2002), and some mutations suppress the ability of the WRN and BLM helicases to bind to and unwind Holliday junctions (Yang *et al.* 2002). Such observations have led to the hypothesis that normal (i.e., p53-WT) cells and tumor cells that lack p53 may use different pathways to repair IR-induced DSBs (Tang *et al.* 1999). In particular, DSBs may be repaired predominantly by NHEJ if WT p53 is available, such as in normal cells, whereas cells lacking WT p53 function (typically tumor cells) may switch to HR as the preferred pathway for DSB repair. A similar role has been suggested for BRCA1 (Zhang *et al.* 2004, Zhuang *et al.* 2006), another protein that is often altered in cancerous versus normal cells. Interestingly, Sah and colleagues (Sah *et al.* 2003) reported that over-expressing WT p53 in a human epidermoid carcinoma cell line, A431, using an adenoviral vector (Ad-p53), resulted in increased radiosensitivity that was associated with a functional impairment of DNA repair in a host-cell reactivation assay and with the decreased expression of NHEJ-related genes such as *KU70*, *DNA-PK* and *XRCC4*. In marked contrast, normal human fibroblasts were not sensitized by Ad-p53 and showed no suppression of their functional DNA-repair capability, although expression of some repair genes was reduced. Thus, the anticipated differential effect of WT p53 on normal and tumor cells may indeed have some clinical ramifications.

Acknowledgments

We thank the Canadian Institutes for Health Research and the Alberta Cancer Board Research Initiative Program for research funding. Thanks from DM to Eleanor S for inspiration.

References

Abraham RT. Cell cycle checkpoint signaling through the ATM and ATR kinases. *Genes Dev 15*, 2177-2196, 2001.

Achanta G, Huang P. Role of p53 in sensing oxidative DNA damage in response to reactive oxygen species-generating agents. *Cancer Res 64*, 6233-6239, 2004.

Adimoolam S, Ford JM. p53 and regulation of DNA damage recognition during nucleotide excision repair. *DNA Repair (Amst.) 2*, 947-954, 2003.

Agami R, Blandino G, Oren M, Shaul Y. Interaction of c-Abl and p73alpha and their collaboration to induce apoptosis. *Nature 399*, 809-813, 1999.

Akyuz N, Boehden GS, Susse S, Rimek A, Preuss U, Scheidtmann KH, Wiesmuller L. DNA substrate dependence of p53-mediated regulation of double-strand break repair. *Mol Cell Biol 22*, 6306-6317, 2002.

Albrechtsen N, Dornreiter I, Grosse F, Kim E, Wiesmüller L, Deppert W. Maintenance of genomic integrity by p53: complementary roles for activated and non-activated p53. *Oncogene 18*, 7706-7717, 1999.

Allison SJ, Milner J. Remodelling chromatin on a global scale: a novel protective function of p53. *Carcinogenesis 25*, 1551-1557, 2004.

Al Rashid ST, Dellaire G, Cuddihy A, Jalali F, Vaid M, Coackley C, Folkard M, Xu Y, Chen BP, Chen DJ, Lilge L, Prise KM, Bazett Jones DP, Bristow RG. Evidence for the direct binding of phosphorylated p53 to sites of DNA breaks in vivo. *Cancer Res 65*, 10810-10821, 2005.

Appella E, Anderson CW. Signaling to p53: breaking the posttranslational modification code. *Pathol Biol 48*, 227-245, 2000.

Appella E, Anderson CW. Post-translational modifications and activation of p53 by genotoxic stresses. *Eur J Biochem 268*, 2764-2772, 2001.

Aranda-Anzaldo A, Orozco-Velasco F, Garcia-Villa E, Gariglio P. p53 is a rate-limiting factor in the repair of higher-order DNA structure. *Biochim Biophys Acta 1446*, 181-192, 1999.

Aubrecht J, Secretan MB, Bishop AJ, Schiestl RH. Involvement of p53 in X-ray induced intrachromosomal recombination in mice. *Carcinogenesis 20*, 2229-2236, 1999.

Avrutskaya AV, Leadon SA. Measurement of oxidative DNA damage and repair in specific DNA sequences. *Methods 22*, 127-134, 2000.

Bakalkin G, Yakovleva T, Selivanova G, Magnusson KP, Szekely L, Kiselva E, Klein G, Telenius L, Wiman KG. p53 binds single-stranded DNA ends and catalyzes DNA renaturation and strand transfer. *Proc Natl Acad Sci USA 91*, 413-417, 1994.

Bakalkin G, Selivanova G, Yakovleva T, Kiseleva E, Kashuba E, Magnusson KP, Szekely L, Klein G, Terenius L, Wiman KG. p53 binds single-stranded DNA ends through the C-terminal domain and internal DNA segments via the middle domain. *Nucleic Acids Res 23*, 362-369, 1995.

Bakkenist CJ, Kastan MB. DNA damage activates ATM through intermolecular autophosphorylation and dimer dissociation. *Nature 421*, 499-506, 2003.

Bartek J, Lukas J. Chk1 and Chk2 kinases in checkpoint control and cancer. *Cancer Cell 3*, 421-429, 2003.

Baskaran R, Wood LD, Whitaker LL, Canman CE, Morgan SE, Xu Y, Barlow C, Baltimore D, Wynshaw-Boris A, Kastan MB, Wang JY. Ataxia telangiectasia mutant protein activates c-Abl tyrosine kinase in response to ionizing radiation. *Nature 387*, 516-519, 1997.

Baumann P, Benson FE, West SC. Human Rad51 protein promotes ATP-dependent homologous pairing and strand transfer reactions in vitro. *Cell 87*, 757-766, 1996.

Bennett PV, Cintron NS, Gros L, Laval J, Sutherland BM. Are endogenous clustered DNA damages induced in human cells? *Free Radic Biol Med 37*, 488-499, 2004.

Bennett RA, Wilson DM 3rd, Wong D, Demple B. Interaction of human apurinic endonuclease and DNA polymerase beta in the base excision repair pathway. *Proc Natl Acad Sci USA 94*, 7166-7169, 1997.

Benson FE, Baumann P, West SC. Synergistic actions of Rad51 and Rad52 in recombination and DNA repair. *Nature 391*, 401-404, 1998.

Bertrand P, Rouillard D, Boulet A, Levalois C, Soussi T, Lopez BS. Increase of spontaneous intrachromosomal homologous recombination in mammalian cells expressing a mutant p53 protein. *Oncogene 14,* 1117-1122, 1997.

Bill CA, Yu Y, Miselis NR, Little JB, Nickoloff JA. A role for p53 in DNA end rejoining by human cell extracts. *Mutat Res 385*, 21-29, 1997.

Blander G, Kipnis J, Leal JF, Yu CE, Schellenberg GD, Oren M. Physical and functional interaction between p53 and the Werner's syndrome protein. *J Biol Chem 274*, 29463-29469, 1999.

Blattner C, Sparks A, Lane D. Transcription factor E2F-1 is upregulated in response to DNA damage in a manner analogous to that of p53. *Mol Cell Biol 19*, 3704-3713, 1999.

Boiteux S, Le Page F. Repair of 8-oxoguanine and Ogg1-incised apurinic sites in a CHO cell line. *Progr Nucleic Acid Res Mol Biol 68*, 95-105, 2001.

Bork P, Hofmann K, Bucher P, Neuwald AF, Altschul SF, Koonin EV. A superfamily of conserved domains in DNA damage-responsive cell cycle checkpoint proteins. *FASEB J 11*, 68-76, 1997.

Boyer JC, Umar A, Risinger JI, Lipford JR, Kane M, Yin S, Barrett JC, Kolodner RD, Kunkel TA. Microsatellite instability, mismatch repair deficiency, and genetic defects in human cancer cell lines. *Cancer Res 55*, 6063-6070, 1995.

Boyle JM, Mitchell EL, Greaves MJ, Roberts SA, Tricker K, Burt E, Varley JM, Birch JM, Scott D. Chromosome instability is a predominant trait of fibroblasts from Li-Fraumeni families. *Br J Cancer 77*, 2181-1292, 1998.

Brain R, Jenkins JR. Human p53 directs DNA strand reassociation and is photolabelled by 8-azido ATP. *Oncogene 9*, 1775-1780, 1994.

Bristow RG, Hu Q, Jang A, Chung S, Peacock J, Benchimol S, Hill R. Radioresistant MTp53-expressing rat embryo cell transformants exhibit increased DNA-dsb rejoining during exposure to ionizing radiation. *Oncogene 16*, 1789-1802 1998.

Brooks CL, Gu W. Ubiquitination, phosphorylation and acetylation: the molecular basis for p53 regulation. *Curr Opin Cell Biol 15*, 164-171, 2003.

Buchhop S, Gibson MK, Wang XW, Wagner P, Sturzbecher HW, Harris CC. Interaction of p53 with the human Rad51 protein. *Nucleic Acids Res 25*, 3868-3874, 1997.

Bunz F, Dutriaux A, Lengauer C, Waldman T, Zhou S, Brown JP, Sedivy JM, Kinzler KW, Vogelstein B. Requirement for p53 and p21 to sustain G2 arrest after DNA damage. *Science 282*, 1497-1501, 1998.

Calsou P, Delteil C, Frit P, Drouet J, Salles B. Coordinated assembly of Ku and p460 subunits of the DNA-dependent protein kinase on DNA ends is necessary for XRCC4-ligase IV recruitment. *J Mol Biol 326*, 93-103, 2003.

Chan DW, Lees-Miller SP. The DNA-dependent protein kinase is inactivated by autophosphorylation of the catalytic subunit. *J Biol Chem 271*, 8936-8941, 1996.

Chen G, Yuan SS, Liu W, Xu Y, Trujillo K, Song B, Cong F, Goff SP, Wu Y, Arlinghaus R, Baltimore D, Gasser PJ, Park MS, Sung P, Lee EY. Radiation-induced assembly of Rad51 and Rad52 recombination complex requires ATM and c-Abl. *J Biol Chem 274*, 12748-12752, 1999.

Chen X, Zheng Y, Zhu J, Jiang J, Wang J. p73 is transcriptionally regulated by DNA damage, p53, and p73. *Oncogene 20*, 769-774, 2001.

Christmann M, Tomicic MT, Roos WP, Kaina B. Mechanisms of human DNA repair: an update. *Toxicology 193*, 3-34, 2003.

Collis SJ, Schwaninger JM, Ntambi AJ, Keller TW, Nelson WG, Dillehay LE, Deweese TL. Evasion of early cellular response mechanisms following low level radiation-induced DNA damage. *J Biol Chem 279*, 49624-49632, 2004.

Cooper MP, Machwe A, Orren DK, Brosh RM, Ramsden D, Bohr VA. Ku complex interacts with and stimulates the Werner protein. *Genes Dev 14*, 907-912, 2000.

Costanzo A, Merlo P, Pediconi N, Fulco M, Sartorelli V, Cole PA, Fontemaggi G, Fanciulli M, Schiltz L, Blandino G, Balsano C, Levrero M. DNA damage-dependent acetylation of p73 dictates the selective activation of apoptotic target genes. *Mol Cell 9*, 175-186, 2002.

Critchlow SE, Bowater RP, Jackson SP. Mammalian DNA double-strand break repair protein XRCC4 interacts with DNA ligase IV. *Curr Biol 7*, 588-598, 1997.

Cromie GA, Connelly JC, Leach DR. Recombination at double-strand breaks and DNA ends: conserved mechanisms from phage to humans. *Mol Cell 8*, 1163-1174, 2001.

Delacote F, Han M, Stamato TD, Jasin M, Lopez BS. An xrcc4 defect or Wortmannin stimulates homologous recombination specifically induced by double-strand breaks in mammalian cells. *Nucleic Acids Res 30*, 3454-3463, 2002.

Dianov GL, Sleeth KM, Dianova II, Allinson SL. Repair of abasic sites in DNA. *Mutat Res 531*, 157-163, 2003.

DiBiase SJ, Guan J, Curran WJ, Iliakis G. Repair of DNA double-strand breaks and radiosensitivity to killing in an isogenic group of p53 mutant cell lines. *Int J Radiat Oncol Biol Phys 45*, 743-751, 1999.

Downs JA, Allard S, Jobin-Robitaille O, Javaheri A, Auger A, Bouchard N, Kron SJ, Jackson SP, Cote J. Binding of chromatin-modifying activities to phosphorylated histone H2A at DNA damage sites. *Mol Cell 16*, 979-990, 2004.

Dronkert ML, Beverloo HB, Johnson RD, Hoeijmakers JH, Jasin M, Kanaar R. Mouse RAD54 affects DNA double-strand break repair and sister chromatid exchange. *Mol Cell Biol 20*, 3147-3156, 2000.

Dudenhoffer C, Rohaly G, Will K, Deppert W, Wiesmuller L. Specific mismatch recognition in heteroduplex intermediates by p53 suggests a role in fidelity control of homologous recombination. *Mol Cell Biol 18*, 5332-5342, 1998.

Dudenhoffer C, Kurth M, Janus F, Deppert W, Wiesmuller L. Dissociation of the recombination control and the sequence-specific transactivation function of p53. *Oncogene 18*, 5773-5784, 1999.

Durocher D, Jackson SP. The FHA domain. *FEBS Lett 513*, 58-66, 2002.

Dutta A, Ruppert JM, Aster JC, Winchester E. Inhibition of DNA replication factor RPA by p53. *Nature 365*, 79-82, 1993.

Enns L, Bogen KT, Wizniak J, Murtha AD, Weinfeld M. Low-dose radiation hypersensitivity is associated with p53-dependent apoptosis. *Mol Cancer Res 2*, 557-566, 2004.

Fei P, El-Deiry WS. P53 and radiation responses. *Oncogene 22*, 5774-5783, 2003.

Flores ER, Tsai KY, Crowley D, Sengupta S, Yang A, McKeon F, Jacks T. p63 and p73 are required for p53-dependent apoptosis in response to DNA damage. *Nature 416*, 560-564, 2002.

Foray N, Marot D, Randrianarison V, Venezia ND, Picard D, Perricaudet M, Favaudon V, Jeggo P. Constitutive association of BRCA1 and c-Abl and its ATM-dependent disruption after irradiation. *Mol Cell Biol 22*, 4020-4032, 2002.

Ford JM. Regulation of DNA damage recognition and nucleotide excision repair: another role for p53. *Mutat Res 577*, 195-202, 2005.

Ford JM, Hanawalt PC. Li-Fraumeni syndrome fibroblasts homozygous for p53 mutations are deficient in global DNA repair but exhibit normal transcription-coupled repair and enhanced UV resistance. *Proc Natl Acad Sci USA 92*, 8876-8880, 1995.

Fornace AJ Jr, Nebert DW, Hollander MC, Luethy JD, Papathanasiou M, Fargnoli J, Holbrook NJ. Mammalian genes coordinately regulated by growth arrest signals and DNA-damaging agents. *Mol Cell Biol 9*, 4196-4203, 1989.

Friedberg EC, Walker GC, Siede W, Wood RD, Schultz RA, Ellenberger T. DNA Repair and Mutagenesis, 2nd edition. ASM Press, Washington DC, 2006.

Gaiddon C, Moorthy NC, Prives C. Ref-1 regulates the transactivation and pro-apoptotic functions of p53 in vivo. *EMBO J 18*, 5609-5621, 1999.

Gebow D, Miselis N, Liber HL. Homologous and non-homologous recombination resulting in deletion: effects of p53 status, microhomology, and repetitive DNA length and orientation. *Mol Cell Biol 20*, 4028-4035, 2000.

Giaccia AJ, Kastan MB. The complexity of p53 modulation: Emerging patterns from divergent signals. *Genes Dev 12*, 2973-2983, 1998.

Goedecke W, Eijpe M, Offenberg HH, van Aalderen M, Heyting C. Mre11 and Ku70 interact in somatic cells, but are differentially expressed in early meiosis. *Nature Genet 23*, 194-198, 1999.

Gong JG, Costanzo A, Yang HQ, Melino G, Kaelin WG Jr, Levrero M, Wang JY. The tyrosine kinase c-Abl regulates p73 in apoptotic response to cisplatin-induced DNA damage. *Nature 399*, 806-809, 1999.

Gonzalez S, Prives C, Cordon-Cardo C. p73alpha regulation by Chk1 in response to DNA damage. *Mol Cell Biol 23*, 8161-8171, 2003.

Grossman SR. p300/CBP/p53 interaction and regulation of the p53 response. *Eur J Biochem 268*, 2773-2778, 2001.

Guirouilh-Barbat J, Huck S, Bertrand P, Pirzio L, Desmaze C, Sabatier L, Lopez BS. Impact of the KU80 pathway on NHEJ-induced genome rearrangements in mammalian cells. *Mol Cell 14,* 611-623, 2004.

Haber JE. Partners and pathways repairing a double-strand break. *Trends Genet 16*, 259-264, 2000.

Hickson ID. RecQ helicases: caretakers of the genome. *Nature Rev Cancer 3*, 169-178, 2003.

Hinz JM, Yamada NA, Salazar EP, Tebbs RS, Thompson LH. Influence of double-strand-break repair pathways on radiosensitivity throughout the cell cycle in CHO cells. *DNA Repair (Amst.) 4*, 782-792, 2005.

Hoeijmakers JHJ. Genome maintenance mechanisms for preventing cancer. *Nature 411*, 366-374, 2001.

Honma M, Zhang LS, Hayashi M, Takeshita K, Nakagawa Y, Tanaka N, Sofuni T. Illegitimate recombination leading to allelic loss and unbalanced translocation in p53-mutated human lymphoblastoid cells. *Mol Cell Biol 17*, 4774-4781, 1997.

Hsu HL, Yannone SM, Chen DJ. Defining interactions between DNA-PK and ligase IV/XRCC4. *DNA Repair (Amst.) 1*, 225-235, 2002.

Huang P. Excision of mismatched nucleotides from DNA: a potential mechanism for enhancing DNA replication fidelity by the wild-type p53 protein. *Oncogene 17*, 261-270, 1998.

Ikushima T, Aritomi H, Morisita J. Radioadaptive response: efficient repair of radiation-induced DNA damage in adapted cells. *Mutat Res 358*, 193-198, 1996.

Ishizaki K, Hayashi Y, Nakamura H, Yasui Y, Komatsu K, Tachibana A. No induction of p53 phosphorylation and few focus formation of phosphorylated H2AX suggest efficient repair of DNA damage during chronic low-dose-rate irradiation in human cells. *J Radiat Res (Tokyo) 45*, 521-525, 2004.

Jackson SP. Sensing and repairing DNA double-strand breaks. *Carcinogenesis 23*, 687-696, 2002.

Janus F, Albrechtsen N, Wiesmuller L, Grosse F, Deppert W. The dual role model for p53 in maintaining genomic integrity. *Cell Mol Life Sci 55*, 12-27, 1999.

Janz C, Susse S, Wiesmuller L. p53 and recombination intermediates: role of tetramerization at DNA junctions in complex formation and exonucleolytic degradation. *Oncogene 21*, 2130-2140, 2002.

Jayaraman J, Prives C. Activation of p53 sequence-specific DNA binding by short single strands of DNA requires the p53 C-terminus. *Cell 81*, 1021-1029, 1995.

Jayaraman L, Murthy KG, Zhu C, Curran T, Xanthoudakis S, Prives C. Identification of redox/repair protein Ref-1 as a potent activator of p53. *Genes Dev 11*, 558-570, 1997.

Jeggo PA. Identification of genes involved in repair of DNA double-strand breaks in mammalian cells. *Radiat Res 150 (5 Suppl.)*, S80-91, 1998.

Jeggo PA, Lobrich M. Artemis links ATM to double strand break rejoining. *Cell Cycle 4*, 359-862, 2005.

Jhappan C, Yusufzai TM, Anderson S, Anver MR, Merlino G. The p53 response to DNA damage in vivo is independent of DNA-dependent protein kinase. *Mol Cell Biol 20*, 4075-4083, 2000.

Jiang M, Axe T, Holgate R, Rubbi CP, Okorokov AL, Mee T, Milner J. p53 binds the nuclear matrix in normal cells: binding involves the proline-rich domain of p53 and increases following genotoxic stress. *Oncogene 20*, 5449-5458, 2001.

Jimenez GS, Bryntesson F, Torres-Arzayus MI, Priestley A, Beeche M, Saito S, Sakaguchi K, Appella E, Jeggo PA, Taccioli GE, Wahl GM, Hubank M. DNA-dependent protein kinase is not required for the p53-dependent response to DNA damage. *Nature 400*, 81-83, 1999.

Johnson RD, Jasin M. Sister chromatid gene conversion is a prominent double-strand break repair pathway in mammalian cells. *EMBO J 19*, 3398-3407, 2000.

Joiner MC, Marples B, Lambin P, Short SC, Turesson I. Low-dose hypersensitivity: current status and possible mechanisms. *Int J Radiat Oncol Biol Phys 49*, 379-389, 2001.

Kastan MB, Lim DS. The many substrates and functions of ATM. *Nat Rev Mol Cell Biol 1*, 179-186, 2000.

Kastan MB, Onyekwere O, Sidransky D, Vogelstein B, Craig RW. Participation of p53 protein in the cellular response to DNA damage. *Cancer Res 51*, 6304-6311, 1991.

Kastan MB, Zhan Q, el-Deiry WS, Carrier F, Jacks T, Walsh WV, Plunkett BS, Vogelstein B, Fornace AJ Jr. A mammalian cell cycle checkpoint pathway utilizing p53 and GADD45 is defective in ataxia-telangiectasia. *Cell 71*, 587-597, 1992.

Kastan MB, Lim DS, Kim ST, Yang D. ATM–a key determinant of multiple cellular responses to irradiation. *Acta Oncol 40*, 686-688, 2001.

Kharbanda S, Pandey P, Jin S, Inoue S, Bharti A, Yuan ZM, Weichselbaum R, Weaver D, Kufe D. Functional interaction between DNA-PK and c-Abl in response to DNA damage. *Nature 386*, 732-735, 1997.

Kingston RE, Bunker CA, Imbalzano AN. Repression and activation by multiprotein complexes that alter chromatin structure. *Genes Dev 10*, 905-920, 1996.

Kitao H, Yuan ZM. Regulation of ionizing radiation-induced Rad52 nuclear foci formation by c-Abl-mediated phosphorylation. *J Biol Chem 277*, 48944-48948, 2002.

Koch CA, Agyei R, Galicia S, Metalnikov P, O'Donnell P, Starostine A, Weinfeld M, Durocher D. Xrcc4 physically links DNA end processing by polynucleotide kinase to DNA ligation by DNA ligase IV. *EMBO J 23*, 3874-3885, 2004.

Kohli M, Jorgensen TJ. Lack of dependence on p53 for DNA double strand break repair of episomal vectors in human lymphoblasts. *Biochem Biophys Res Commun 264*, 702-708, 1999.

Komatsu K, Wharton W, Hang H, Wu C, Singh S, Lieberman HB, Pledger WJ, Wang HG. PCNA interacts with hHus1/hRad9 in response to DNA damage and replication inhibition. *Oncogene 19*, 5291-5297, 2000.

Kurz EU, Lees-Miller SP. DNA damage induced activation of ATM and ATM-dependent signaling pathways. *DNA Repair 3*, 889-900, 2004.

Kuschel B, Auranen A, McBride S, Novik KL, Antoniou A, Lipscombe JM, Day NE, Easton DF, Ponder BAJ, Pharoah PDP, Dunning A. Variants in DNA double-strand break repair genes and breast cancer susceptibility. *Hum Mol Genet 11*, 1399-1407, 2002.

Lahav G, Rosenfeld N, Sigal A, Geva-Zatorsky N, Levine AJ, Elowitz MB, Alon U. Dynamics of the p53-Mdm2 feedback loop in individual cells. *Nature Genet 36*, 147-150, 2004.

Lambert PF, Kashanchi F, Radonovich MF, Shiekhattar R, Brady JN. Phosphorylation of p53 serine 15 increases interaction with CBP. *J Biol Chem 273*, 33048-33053, 1998.

Lane DP. p53, guardian of the genome. *Nature 358*, 15-16, 1992.

Le XC, Xing JZ, Lee J, Leadon SA, Weinfeld M. Inducible repair of thymine glycol detected by an ultrasensitive assay for DNA damage. *Science 280*, 1066-1069, 1998.

Leadon SA, Cooper PK. Preferential repair of ionizing radiation-induced damage in the transcribed strand of an active human gene is defective in Cockayne syndrome. *Proc Natl Acad Sci USA 90*, 10499-10503, 1993.

Lee JH, Paull TT. Direct activation of the ATM protein kinase by the Mre11/Rad50/Nbs1 complex. *Science 304*, 93-96, 2004.

Lee H, Sun D, Larner JM, Wu FS. The tumor suppressor p53 can reduce stable transfection in the presence of irradiation. *J Biomed Sci 6*, 285-292, 1999.

Lee KJ, Jovanovic M, Udayakumar D, Bladen CL, Dynan WS. Identification of DNA-PKcs phosphorylation sites in XRCC4 and effects of mutations at these sites on DNA end joining in a cell-free system. *DNA Repair (Amst.) 3*, 267-276, 2004.

Lee S, Elenbaas B, Levine A, Griffith J. p53 and its 14 kDa C-terminal domain recognize primary DNA damage in the form of insertion/deletion mismatches. *Cell 81*, 1013-1020, 1995.

Lee S, Cavallo L, Griffith J. Human p53 binds Holliday junctions strongly and facilitates their cleavage. *J Biol Chem 272*, 7532-7539, 1997.

Lee SB, Kim SH, Bell DW, Wahrer DC, Schiripo TA, Jorczak MM, Sgroi DC, Garber JE, Li FP, Nichols KE, Varley JM, Godwin AK, Shannon KM, Harlow E, Haber DA. Destabilization of CHK2 by a missense mutation associated with Li-Fraumeni Syndrome. *Cancer Res 61*, 8062-807, 2001.

Leng RP, Lin Y, Ma W, Wu H, Lemmers B, Chung S, Parant JM, Lozano G, Hakem R, Benchimol S. Pirh2, a p53-induced ubiquitin-protein ligase, promotes p53 degradation. *Cell 112*, 779-791, 2003.

Le Page F, Klungland A, Barnes DE, Sarasin A, Boiteux S. Transcription coupled repair of 8-oxoguanine in murine cells: the ogg1 protein is required for repair in nontranscribed sequences but not in transcribed sequences. *Proc Natl Acad Sci USA 97*, 8397-8402, 2000.

Levine AJ. p53, the cellular gatekeeper for growth and division. *Cell 88*, 323-331, 1997.

Levrero M, De Laurenzi V, Costanzo A, Gong J, Wang JY, Melino G. The p53/p63/p73 family of transcription factors: Overlapping and distinct functions. *J Cell Sci 113*, 1661-1670, 2000.

Li B, Comai L. Functional interaction between Ku and the Werner syndrome protein in DNA end processing. *J Biol Chem 275*, 28349-28352, 2000.

Li W, Hesabi B, Babbo A, Pacione C, Liu J, Chen DJ, Nickoloff JA, Shen Z. Regulation of double-strand break-induced mammalian homologous recombination by UBL1, a RAD51-interacting protein. *Nucleic Acids Res 28*, 1145-1153, 2000.

Liang F, Han M, Romanienko PJ, Jasin M. Homology-directed repair is a major double-strand break repair pathway in mammalian cells. *Proc Natl Acad Sci USA 95*, 5172-5177, 1998.

Liang SH, Clarke MF. Regulation of p53 localization. *Eur J Biochem 268*, 2779-2783, 2001.

Lim DS, Vogel H, Willerford DM, Sands AT, Platt KA, Hasty P. Analysis of ku80-mutant mice and cells with deficient levels of p53. *Mol Cell Biol 20*, 3772-3780, 2000.

Lin WC, Lin FT, Nevins JR. Selective induction of E2F1 in response to DNA damage, mediated by ATM-dependent phosphorylation. *Genes Dev 15*, 1833-1844, 2001.

Lin Y, Waldman BC, Waldman AS. Suppression of high-fidelity double-strand break repair in mammalian chromosomes by pifithrin-alpha, a chemical inhibitor of p53. *DNA Repair (Amst.) 2*, 1-11, 2003.

Lindahl T. Keynote: past, present, and future aspects of base excision repair. *Progr Nucleic Acid Res Mol Biol 68*, xvii-xxx, 2001.

Linke SP, Harris MP, Neugebauer SE, Clarkin KC, Shepard HM, Maneval DC, Wahl GM. p53-mediated accumulation of hypophosphorylated pRb after the G1 restriction point fails to halt cell cycle progression, *Oncogene 15*, 337-345, 1997.

Linke SP, Sengupta S, Khabie N, Jeffries BA, Buchhop S, Miska S, Henning W, Pedeux R, Wang XW, Hofseth LJ, Yang Q, Garfield SH, Sturzbecher HW, Harris CC. p53 interacts with hRAD51 and hRAD54, and directly modulates homologous recombination. *Cancer Res 63*, 2596-2605, 2003.

Liu Y, Kulesz-Martin M. p53 protein at the hub of cellular DNA damage response pathways through sequence-specific and non-sequence-specific DNA binding. *Carcinogenesis 22*, 851-860, 2001.

Ljungman M, Lane DP. Transcription - guarding the genome by sensing DNA damage. *Nature Rev Cancer 4*, 727-737, 2004.

Loeb LA. A mutator phenotype in cancer. *Cancer Res 61*, 3230-3239, 2001.

Lu X, Bocangel D, Nannenga B, Yamaguchi H, Appella E, Donehower LA. The p53-induced oncogenic phosphatase PPM1D interacts with uracil DNA glycosylase and suppresses base excision repair. *Mol Cell 15*, 621-634, 2004a.

Lu X, Nguyen TA, Appella E, Donehower LA. Homeostatic regulation of base excision repair by a p53-induced phosphatase: linking stress response pathways with DNA repair proteins. *Cell Cycle 3*, 1363-1366, 2004b.

Luo J, Li M, Tang Y, Laszkowska M, Roeder RG, Gu W. Acetylation of p53 augments its site-specific DNA binding both in vitro and in vivo. *Proc Natl Acad Sci USA 101*, 2259-2264, 2004.

Mallya SM, Sikpi MO. Evidence of the involvement of p53 in gamma-radiation-induced DNA repair in human lymphoblasts. *Int J Radiat Biol 74*, 231-238, 1998.

Mallya SM, Sikpi MO. Requirement for p53 in ionizing-radiation-inhibition of double-strand-break rejoining by human lymphoblasts. *Mutat Res 434*, 119-132, 1999.

Marmorstein LY, Ouchi T, Aaronson SA. The BRCA2 gene product functionally interacts with p53 and RAD51. *Proc Natl Acad Sci USA 95*, 13869-13874, 1998.

Marples B. Is low-dose hyper-radiosensitivity a measure of G2-phase cell radiosensitivity? *Cancer Met Rev 23*, 197-207, 2004.

Marples B, Lambin P, Skov KA, Joiner MC. Low dose hyper-radiosensitivity and increased radioresistance in mammalian cells. *Int J Radiat Biol 71*, 721-735, 1997.

Marples B, Wouters BG, Collis SJ, Chalmers AJ, Joiner MC. Low-dose hyper-radiosensitivity: a consequence of ineffective cell cycle arrest of radiation-damaged G2-phase cells. *Radiat Res 161*, 247-255, 2004.

Masson M, Niedergang C, Schreiber V, Muller S, Menissier-de Murcia J, de Murcia G. XRCC1 is specifically associated with poly(ADP-ribose) polymerase and negatively regulates its activity following DNA damage. *Mol Cell Biol 18*, 3563-3571, 1998.

Masson JY, Davies AA, Hajibagheri N, Van Dyck E, Benson FE, Stasiak AZ, Stasiak A, West SC. The meiosis-specific recombinase hDmc1 forms ring structures and interacts with hRad51. *EMBO J 18*, 6552-6560, 1999.

Mazin AV, Alexeev AA, Kowalczykowski SC. A novel function of Rad54 protein. Stabilization of the Rad51 nucleoprotein filament. *J Biol Chem 278*, 14029-14036, 2003.

McIlwraith MJ, Van Dyck E, Masson JY, Stasiak AZ, Stasiak A, West SC. Reconstitution of the strand invasion step of double-strand break repair using human Rad51 Rad52 and RPA proteins. *J Mol Biol 304*, 151-164, 2000.

Mekeel KL, Tang W, Kachnic LA, Luo CM, DeFrank JS, Powell SN. Inactivation of p53 results in high rates of homologous recombination. *Oncogene 14*, 1847-1857, 1997.

Melino G, De Laurenzi V, Vousden KH. p73: Friend or foe in tumorigenesis. *Nat Rev Cancer 2*, 605-615, 2002.

Michael D, Oren M. The p53-Mdm2 module and the ubiquitin system. *Semin Cancer Biol 13*, 49-58, 2003.

Miller SD, Moses K, Jayaraman L, Prives C. Complex formation between p53 and replication protein A inhibits the sequence-specific DNA binding of p53 and is regulated by single-stranded DNA. *Mol Cell Biol 17*, 2194-2201, 1997.

Mirza A, Wu Q, Wang L, McClanahan T, Bishop WR, Gheyas F, Ding W, Hutchins B, Hockenberry T, Kirschmeier P, Greene JR, Liu S. Global transcriptional program of p53 target genes during the process of apoptosis and cell cycle progression. *Oncogene 22*, 3645-3654, 2003.

Mirzayans R, Enns L, Dietrich K, Barley RD, Paterson MC. Faulty DNA polymerase delta/epsilon-mediated excision repair in response to gamma radiation or ultraviolet light in p53-deficient fibroblast strains from affected members of a cancer-prone family with Li-Fraumeni syndrome. *Carcinogenesis 17*, 691-698, 1996.

Mirzayans R, Pollock S, Scott A, Enns L, Andrais B, Murray D. Relationship between the radiosensitizing effect of wortmannin, DNA double-strand break rejoining, and $p21^{WAF1}$ induction in human normal and tumor-derived cells. *Mol Carcin 39*, 164-172, 2004.

Mirzayans R, Severin D, Murray D. Relationship between DNA double strand break rejoining and cell survival following exposure to ionizing radiation in human fibroblast strains with differing ATM/p53 status: implications for the evaluation of clinical radiosensitivity. *Int J Radiat Oncol Biol Phys,* in press, 2006.

Mummenbrauer T, Janus F, Muller B, Wiesmuller L, Deppert W, Grosse F. p53 protein exhibits 3'-to-5' exonuclease activity. *Cell 85*, 1089-1099, 1996.

Murray D, Wang JYJ, Mirzayans R. DNA repair after low doses of ionizing radiation. *Int J Low Radiat 3*, 255-272, 2006.

Myung K, Braastad C, He DM, Hendrickson EA. KARP-1 is induced by DNA damage in a p53- and ataxia telangiectasia mutated-dependent fashion. *Proc Natl Acad Sci USA 95*, 7664-7669, 1998.

Nacht M, Strasser A, Chan YR, Harris AW, Schlissel M, Bronson RT, Jacks T. Mutations in the p53 and SCID genes cooperate in tumorigenesis. *Genes Dev 10*, 2055-2066, 1996.

Nevaldine B, Longo JA, Hahn PJ. The scid defect results in much slower repair of DNA double-strand breaks but not high levels of residual breaks. *Radiat Res 147*, 535-540, 1997.

Niida H, Nakanishi M. DNA damage checkpoints in mammals. *Mutagenesis 21*, 3-9, 2006.

Nikolaev AY, Li M, Puskas N, Qin J, Gu W. Parc: a cytoplasmic anchor for p53. *Cell 112*, 29-40, 2003.

Nishimura S. Mammalian Oggl/Mmh gene plays a major role in repair of the 8-hydroxyguanine lesion in DNA. *Progr Nucleic Acid Res Mol Biol 68*, 107-123, 2001.

North S, Hainaut P. p53 and cell-cycle control: a finger in every pie. *Pathol Biol 48*, 255-270, 2000.

Oberosler P, Hloch P, Ramsperger U, Stahl H. p53-catalyzed annealing of complementary single-stranded nucleic acids. *EMBO J 12*, 2389-2396, 1993.

Offer H, Wolkowicz R, Matas D, Blumenstein S, Livneh Z, Rotter V. Direct involvement of p53 in the base excision repair pathway of the DNA repair machinery. *FEBS Lett 450*, 197-204, 1999.

Offer H, Milyavsky M, Erez N, Matas D, Zurer I, Harris CC, Rotter V. Structural and functional involvement of p53 in BER in vitro and in vivo. *Oncogene 20*, 581-589, 2001a.

Offer H, Zurer I, Banfalvi G, Reha'k M, Falcovitz A, Milyavsky M, Goldfinger N, Rotter V. p53 modulates base excision repair activity in a cell cycle-specific manner after genotoxic stress. *Cancer Res 61*, 88-96, 2001b.

Offer H, Erez N, Zurer I, Tang X, Milyavsky M, Goldfinger N, Rotter V. The onset of p53-dependent DNA repair or apoptosis is determined by the level of accumulated damaged DNA. *Carcinogenesis 23*, 1025-1032, 2002.

Okorokov AL, Warnock L, Milner J. Effect of wild-type, S15D and R175H p53 proteins on DNA end joining in vitro: potential mechanism of DNA double-strand break repair modulation. *Carcinogenesis 23*, 549-557, 2002.

Olive PL, Bana´th JP. Phosphorylation of histone H2AX as a measure of radiosensitivity. *Int J Radiat Oncol Biol Phys 58*, 331-335, 2004.

Oliver FJ, Menissier-de Murcia J, de Murcia G. Poly(ADP-ribose) polymerase in the cellular response to DNA damage, apoptosis, and disease. *Am J Hum Genet 64*, 1282-1288, 1999.

Oren M. Decision making by p53: Life, death and cancer. *Cell Death Differ 10*, 431-442, 2003.

Orre LM, Stenerlow B, Dhar S, Larsson R, Lewensohn R, Lehtio J. p53 is involved in clearance of ionizing radiation-induced RAD51 foci in a human colon cancer cell line. *Biochem Biophys Res Commun 342*, 1211-1217, 2006.

Parks D, Bolinger R, Mann K. Redox state regulates binding of p53 to sequence-specific DNA, but not to non-specific or mismatched DNA. *Nucleic Acids Res 25*, 1289-1295, 1997.

Parrilla-Castellar ER, Arlander SJ, Karnitz L. Dial 9-1-1 for DNA damage: the Rad9-Hus1-Rad1 (9-1-1) clamp complex. *DNA Repair (Amst.) 3,* 1009-1014, 2004.

Paull TT, Gellert M. Nbs1 potentiates ATP-driven DNA unwinding and endonuclease cleavage by the Mre11/Rad50 complex. *Genes Dev 13*, 1276-1288, 1999.

Paull TT, Gellert M. A mechanistic basis for Mre11-directed DNA joining at microhomologies. *Proc Natl Acad Sci USA 97*, 6409-6414, 2000.

Perry ME. Mdm2 in the response to radiation. *Mol Cancer Res 2*, 9-19, 2004.

Petrini JH, Stracker TH. The cellular response to DNA double-strand breaks: Defining the sensors and mediators. *Trends Cell Biol 13*, 458-462, 2003.

Pierce AJ, Hu P, Han M, Ellis N, Jasin M. Ku DNA end-binding protein modulates homologous repair of double-strand breaks in mammalian cells. *Genes Dev 15*, 3237-3242, 2001.

Pleschke JM, Kleczkowska HE, Strohm M, Althaus FR. Poly(ADP-ribose) binds to specific domains in DNA damage checkpoint proteins. *J Biol Chem 275*, 40974-40980, 2000.

Polager S, Kalma Y, Berkovich E, Ginsberg D. E2Fs up-regulate expression of genes involved in DNA replication, DNA repair and mitosis. *Oncogene 21*, 437-446, 2002.

Polyak K, Xia Y, Zweier JL, Kinzler KW, Vogelstein B. A model for p53-induced apoptosis. *Nature 389*, 300-305, 1997.

Powell SN, Willers H, Xia F. BRCA2 keeps Rad51 in line. High-fidelity homologous recombination prevents breast and ovarian cancer? *Mol Cell 10*, 1262-1263, 2002.

Reed M, Woelker B, Wang P, Wang Y, Anderson ME, Tegtmeyer P. The C-terminal domain of p53 recognizes DNA damaged by ionizing radiation. *Proc Natl Acad Sci USA 92*, 9455-9459, 1995.

Ren B, Cam H, Takahashi Y, Volkert T, Terragni J, Young RA, Dynlacht BD. E2F integrates cell cycle progression with DNA repair, replication, and G(2)/M checkpoints. *Genes Dev 16*, 245-256, 2002.

Richardson C, Jasin M. Coupled homologous and non-homologous repair of a double-strand break preserves genomic integrity in mammalian cells. *Mol Cell Biol 20*, 9068-9075, 2000.

Richardson C, Moynahan ME, Jasin M. Double-strand break repair by interchromosomal recombination: suppression of chromosomal translocations. *Genes Dev 12*, 3831-3842, 1998.

Rogoff HA, Pickering MT, Debatis ME, Jones S, Kowalik TF. E2F1 induces phosphorylation of p53 that is coincident with p53 accumulation and apoptosis. *Mol Cell Biol 22*, 5308-5318, 2002.

Roninson JB. Tumor cell senescence in cancer treatment. *Cancer Res. 63*, 2705-2715, 2003.

Roos-Mattjus P, Hopkins KM, Oestreich AJ, Vroman BT, Johnson KL, Naylor S, Lieberman HB, Karnitz LM. Phosphorylation of human Rad9 is required for genotoxin-activated checkpoint signaling. *J Biol Chem 278*, 24428-24437, 2003.

Roth DB, Wilson JH. Nonhomologous recombination in mammalian cells: role for short sequence homologies in the joining reaction. *Mol Cell Biol 6*, 4295-4304, 1986.

Rothkamm K, Kruger I, Thompson LH, Lobrich M. Pathways of DNA double-strand break repair during the mammalian cell cycle. *Mol Cell Biol 23*, 5706-5715, 2003.

Rubbi CP, Milner J. p53 is a chromatin accessibility factor for nucleotide excision repair of DNA damage. *EMBO J 22*, 975-986, 2003.

Ruscetti T, Lehnert BE, Halbrook J, Le Trong H, Hoekstra MF, Chen DJ, Peterson SR. Stimulation of the DNA-dependent protein kinase by poly(ADP-ribose) polymerase. *J Biol Chem 273*, 14461-14467, 1998.

Sah NK, Munshi A, Nishikawa T, Mukhopadhyay T, Roth JA, Meyn RE. Adenovirus-mediated wild-type p53 radiosensitizes human tumor cells by suppressing DNA repair capacity. *Mol Cancer Ther 2*, 1223-1231, 2003.

Saintigny Y, Rouillard D, Chaput B, Soussi T, Lopez BS. Mutant p53 proteins stimulate spontaneous and radiation-induced intrachromosomal homologous recombination independently of the alteration of the transactivation activity and of the G1 checkpoint. *Oncogene 18*, 3553-3565, 1999.

Sakaguchi K, Herrera JE, Saito S, Miki T, Bustin M, Vassilev A, Anderson CW, Appella E. DNA damage activates p53 through a phosphorylation-acetylation cascade. *Genes Dev 12*, 2831-2841, 1998.

Sasaki MS, Ejima Y, Tachibana A, Yamada T, Ishizaki K, Shimizu T, Nomura T. DNA damage response pathway in radioadaptive response. *Mutat Res 504*, 101-118, 2002.

Sedelnikova OA, Rogakou EP, Panyutin IG, Bonner WM. Quantitative detection of (125) IdU-induced DNA double-strand breaks with gamma-H2AX antibody. *Radiat Res 158*, 486-492, 2002.

Sehorn MG, Sigurdsson S, Bussen W, Unger VM, Sung P. Human meiotic recombinase Dmc1 promotes ATP-dependent homologous DNA strand exchange. *Nature 429*, 433-437, 2004.

Sengupta S, Harris CC. p53: traffic cop at the crossroads of DNA repair and recombination. *Nat Rev Mol Cell Biol 6*, 44-55, 2005.

Sengupta S, Linke SP, Pedeux R, Yang Q, Farnsworth J, Garfield SH, Valerie K, Shay JW, Ellis NA, Wasylyk B, Harris CC. BLM helicase-dependent transport of p53 to sites of stalled DNA replication forks modulates homologous recombination. *EMBO J 22*, 1210-1222, 2003.

Seo YR, Kelley MR, Smith ML. Selenomethionine regulation of p53 by a ref1-dependent redox mechanism. *Proc Natl Acad Sci USA 99*, 14548-14553, 2002a.

Seo YR, Fishel ML, Amundson S, Kelley MR, Smith ML. Implication of p53 in base excision DNA repair: in vivo evidence. *Oncogene 21*, 731-737, 2002b.

Shafman T, Khanna KK, Kedar P, Spring K, Kozlov S, Yen T, Hobson K, Gatei M, Zhang N, Watters D, Egerton M, Shiloh Y, Kharbanda S, Kufe D, Lavin MF. Interaction between ATM protein and c-Abl in response to DNA damage. *Nature 387*, 520-523, 1997.

Shieh SY, Ikeda M, Taya Y, Prives C. DNA damage-induced phosphorylation of p53 alleviates inhibition by MDM2. *Cell 91*, 325-334, 1997.

Shiloh Y. ATM: ready, set, go. *Cell Cycle 2*, 116-117, 2003a.

Shiloh, Y. ATM and related protein kinases: Safeguarding genome integrity. *Nature Rev Cancer 3*, 155-168, 2003b.

Sionov RV, Coen S, Goldberg Z, Berger M, Bercovich B, Ben-Neriah Y, Ciechanover A, Haupt Y. c-Abl regulates p53 levels under normal and stress conditions by preventing its nuclear export and ubiquitination. *Mol Cell Biol 21*, 5869-5878, 2001.

Sobol RW, Prasad R, Evenski A, Baker A, Yang XP, Horton JK, Wilson SH. The lyase activity of the DNA repair protein beta-polymerase protects from DNA-damage-induced cytotoxicity. *Nature 405*, 807-810, 2000.

Sobol RW, Kartalou M, Almeida KH, Joyce DF, Engelward BP, Horton JK, Prasad R, Samson LD, Wilson SH. Base excision repair intermediates induce p53-independent cytotoxic and genotoxic responses. *J Biol Chem 278*, 39951-39959, 2003.

Spillare EA, Robles AI, Wang XW, Shen JC, Yu CE, Schellenberg GD, Harris CC. p53-mediated apoptosis is attenuated in Werner syndrome cells. *Genes Dev 13*, 1355-1360, 1999.

Srivastava DK, Berg BJ, Prasad R, Molina JT, Beard WA, Tomkinson AE, Wilson SH. Mammalian abasic site base excision repair. Identification of the reaction sequence and rate-determining steps. *J Biol Chem 273*, 21203-21209, 1998.

Steinmeyer K, Deppert W. DNA binding properties of murine p53. *Oncogene 3*, 501-507, 1988.

Sturzbecher HW, Donzelmann B, Henning W, Knippschild U, Buchhop S. p53 is linked directly to homologous recombination processes via RAD51/RecA protein interaction. *EMBO J 15*, 1992-2002, 1996.

Sun Y, Jiang X, Chen S, Fernandes N, Price BD. A role for the Tip60 histone acetyltransferase in the acetylation and activation of ATM. *Proc Natl Acad Sci USA 102*, 13182-13187, 2005.

Susse S, Janz C, Janus F, Deppert W, Wiesmuller L. Role of heteroduplex joints in the functional interactions between human Rad51 and wild-type p53. *Oncogene 19*, 4500-4512, 2000.

Swagemakers SM, Essers J, de Wit J, Hoeijmakers JH, Kanaar R. The human RAD54 recombinational DNA repair protein is a double-stranded DNA-dependent ATPase. *J Biol Chem 273*, 28292-28297, 1998.

Takata M, Sasaki MS, Sonoda E, Morrison C, Hashimoto M, Utsumi H, Yamaguchi-Iwai Y, Shinohara A, Takeda S. Homologous recombination and non-homologous end-joining pathways of DNA double-strand break repair have overlapping roles in the maintenance of chromosomal integrity in vertebrate cells. *EMBO J 17*, 5497-5508, 1998.

Tang W, Willers H, Powell SN. p53 directly enhances rejoining of DNA double-strand breaks with cohesive ends in gamma-irradiated mouse fibroblasts. *Cancer Res 59*, 2562-2565, 1999.

Taylor WR, Stark GR. Regulation of the G2/M transition by p53. *Oncogene 20*, 1803-1815, 2001.

Thompson LH, Schild D. Homologous recombinational repair of DNA ensures mammalian chromosome stability. *Mutat Res 477*, 131-153, 2001.

Thraves PJ, Kasid U, Smulson ME. Selective isolation of domains of chromatin proximal to both carcinogen-induced DNA damage and poly-adenosine diphosphate-ribosylation. *Cancer Res 45*, 386-391, 1985.

Tomkinson AE, Chen L, Dong Z, Leppard JB, Levin DS, Mackey ZB, Motycka TA. Completion of base excision repair by mammalian DNA ligases. *Progr Nucleic Acid Res Mol Biol 68*, 151-164, 2001.

Trujillo KM, Yuan SS, Lee EY, Sung P. Nuclease activities in a complex of human recombination and DNA repair factors Rad50, Mre11, and p95. *J Biol Chem 273*, 21447-21450, 1998.

Truong T, Sun G, Doorly M, Wang JY, Schwartz MA. Modulation of DNA damage-induced apoptosis by cell adhesion is independently mediated by p53 and c-Abl. *Proc Natl Acad Sci USA 100*, 10281-10286, 2003.

Utley RT, Lacoste N, Jobin-Robitaille O, Allard S, Cote J. Regulation of NuA4 histone acetyltransferase activity in transcription and DNA repair by phosphorylation of histone H4. *Mol Cell Biol 25*, 8179-8190, 2005.

Valerie K, Povirk LF. Regulation and mechanisms of mammalian double-strand break repair. *Oncogene 22*, 5792-5812, 2003.

van Heemst D, Brugmans L, Verkaik NS, van Gent DC. End-joining of blunt DNA double-strand breaks in mammalian fibroblasts is precise and requires DNAPK and XRCC4. *DNA Repair 3*, 43-50, 2004.

Vogelstein B, Lane D, Levine AJ. Surfing the p53 network. *Nature 408*, 307-310, 2000.

Wang JY. Regulation of cell death by the Abl tyrosine kinase. *Oncogene 19*, 5643-5650, 2000.

Wang JYJ, Cho SK. Coordination of repair, checkpoint, and cell death responses to DNA damage. *Adv Protein Chem 69*, 101-135, 2004.

Wang S, Guo M, Ouyang H, Li X, Cordon-Cardo C, Kurimasa A, Chen DJ, Fuks Z, Ling CC, Li GC. The catalytic subunit of DNA-dependent protein kinase selectively regulates p53-dependent apoptosis but not cell-cycle arrest. *Proc Natl Acad Sci USA 97*, 1584-1588, 2000.

Wang XW, Yeh H, Schaeffer L, Roy R, Moncollin V, Egly JM, Wang Z, Freidberg EC, Evans MK, Taffe BG, Bohr VA, Weeda G, Hoeijmakers JHJ, Forrester K, Harris CC. p53 modulation of TFIIH-associated nucleotide excision repair activity. *Nat Genet 10*, 188-195, 1995.

Wang XW, Tseng A, Ellis NA, Spillare EA, Linke SP, Robles AI, Seker H, YangQ, Hu P, Beresten S, Bemmels NA, Garfield S, Harris CC. Functional interaction of p53 and BLM DNA helicase in apoptosis. *J Biol Chem 276*, 32948-32955, 2001.

Ward JF. Complexity of damage produced by ionizing radiation. *Cold Spring Harb Symp Quant Biol 65*, 377-382, 2000.

Whitehouse CJ, Taylor RM, Thistlethwaite A, Zhang H, Karimi-Busheri F, Lasko DD, Weinfeld M, Caldecott KW. XRCC1 stimulates human polynucleotide kinase activity at damaged DNA termini and accelerates DNA single-strand break repair. *Cell 104*, 107-117, 2001.

Wiesmüller L, Cammenga J, Deppert WW. In vivo assay of p53 function in homologous recombination between simian virus 40 chromosomes. *J Virol 70*, 737-744, 1996.

Willers H, McCarthy EE, Wu B, Wunsch H, Tang W, Taghian DG, Xia F, Powell SN. Dissociation of p53-mediated suppression of homologous recombination from G1/S cell cycle checkpoint control. *Oncogene 19,* 632-639, 2000.

Willers H, McCarthy EE, Hubbe P, Dahm-Daphi J, Powell SN. Homologous recombination in extrachromosomal plasmid substrates is not suppressed by p53. *Carcinogenesis 22*, 1757-1763, 2001.

Wilson SH, Kunkel TA. Passing the baton in base excision repair. *Nat Struct Biol 7*, 176-178, 2000.

Xia F, Liber HL. The tumor suppressor p53 modifies mutational processes in a human lymphoblastoid cell line. *Mutat Res 373*, 87-97, 1997.

Yanez RJ, Porter ACG. Gene targeting is enhanced in human cells overexpressing hRAD51. *Gene Ther 6*, 1282-1290, 1999.

Yang Q, Zhang R, Wang XW, Spillare EA, Linke SP, Subramanian D, Griffith JD, Li JL, Hickson ID, Shen JC, Loeb LA, Mazur SJ, Appella E, Brosh RM Jr, Karmakar P, Bohr VA, Harris CC. The processing of Holliday junctions by BLM and WRN helicases is regulated by p53. *J Biol Chem 277*, 31980-31987, 2002.

Yang T, Namba H, Hara T, Takmura N, Nagayama Y, Fukata S, Ishikawa N, Kuma K, Ito K, Yamashita S. p53 induced by ionizing radiation mediates DNA end-jointing activity, but not apoptosis of thyroid cells. *Oncogene 14*, 1511-1519, 1997.

Yannone SM, Roy S, Chan DW, Murphy MB, Huang S, Campisi J, Chen DJ. Werner syndrome protein is regulated and phosphorylated by DNA-dependent protein kinase. *J Biol Chem 276*, 38242-38248, 2001.

Yokote H, Nishio K, Arioka H, Kurokawa H, Fukuoka K, Fukumoto H, Ishida T, Terada T, Itakura T, Saijo N. The C-terminal domain of p53 catalyzes DNA-renaturation and strand exchange toward annealing between intact ssDNAs and toward eliminating damaged ssDNA from duplex formation through preferential recognition of damaged DNA by a duocarmycin. *Mutat Res 409*, 147-162, 1998.

Yuan SS, Chang HL, Lee EY. Ionizing radiation-induced Rad51 nuclear focus formation is cell cycle-regulated and defective in both ATM(-/-) and c-Abl(-/-) cells. *Mutat Res 525*, 85-92, 2003.

Yuan ZM, Huang Y, Fan MM, Sawyers C, Kharbanda S, Kufe D. Genotoxic drugs induce interaction of the c-Abl tyrosine kinase and the tumor suppressor protein p53. *J Biol Chem 271*, 26457-26460, 1996.

Yuan ZM, Huang Y, Ishiko T, Nakada S, Utsugisawa T, Kharbanda S, Wang R, Sung P, Shinohara A, Weichselbaum R, Kufe D. Regulation of Rad51 function by c-Abl in response to DNA damage. *J Biol Chem 273*, 3799-3802, 1998.

Yuan ZM, Shioya H, Ishiko T, Sun X, Gu J, Huang YY, Lu H, Kharbanda S, Weichselbaum R, Kufe D. p73 is regulated by tyrosine kinase c-Abl in the apoptotic response to DNA damage, *Nature 399*, 814-817, 1999.

Zhang J, Powell SN. The role of the BRCA1 tumor suppressor in DNA double-strand break repair. *Mol Cancer Res 3*, 531-539, 2005.

Zhang J, Willers H, Feng Z, Ghosh JC, Kim S, Weaver DT, Chung JH, Powell SN, Xia F. Chk2 phosphorylation of BRCA1 regulates DNA double-strand break repair. *Mol Cell Biol 24*, 708-718, 2004.

Zhao R, Gish K, Murphy M, Yin Y, Notterman D, Hoffman WH, Tom E, Mack DH, Levine AJ. Analysis of p53-regulated gene expression patterns using oligonucleotide arrays. *Genes Dev 14*, 981-993, 2000.

Zhou J, Ahn J, Wilson SH, Prives C. A role for p53 in base excision repair. *EMBO J 20*, 914-923, 2001.

Zhu J, Wang JY. Death by Abl: a matter of location. *Curr Top Dev Biol 59*, 165-192, 2004.

Zhuang J, Zhang J, Willers H, Wang H, Chung JH, van Gent DC, Hallahan DE, Powell SN, Xia F. Checkpoint kinase 2-mediated phosphorylation of BRCA1 regulates the fidelity of nonhomologous end-joining. *Cancer Res 66*, 1401-1408, 2006.

INDEX

A

D

E

F

G

H

I

J

K

L

R

S

T

U

V